Waste and Waste Management Series

SEWAGE TREATMENT: USES, PROCESSES AND IMPACT

Waste and Waste Management Series

Sewage Treatment: Uses, Processes and Impact
Anna Stephens and Mark Fuller
2009. ISBN: 978-1-60692-959-9

Waste and Waste Management Series

SEWAGE TREATMENT: USES, PROCESSES AND IMPACT

ANNA STEPHENS
AND
MARK FULLER
EDITORS

Nova Science Publishers, Inc.
New York

For permission to use material from this book please contact us:
Telephone 631-231-7269; Fax 631-231-8175
Web Site: http://www.novapublishers.com

LIBRARY OF CONGRESS CATALOGING-IN-PUBLICATION DATA

Stephens, Anna, 1963-
Sewage treatment : uses, processes, and impact / Anna Stephens and Mark Fuller.
p. cm. -- (Waste and waste management)
Includes bibliographical references and index.
ISBN 978-1-60692-959-9 (hardcover : alk. paper)
1. Sewage--Purification. I. Fuller, Mark. II. Title.
TD745.S675 2009
628.3'2--dc22

2009009092

Published by Nova Science Publishers, Inc. ✝ New York

CONTENTS

PREFACE

Sewage is mainly a liquid waste containing some solids, produced by humans. It typically consists of washing water, feces, urine, laundry waste and other material which goes down drains and toilets from households and industry. Sewage sludge is an end product of the wastewater treatment process. This material can be a wonderful source of nutrients for the soil. This book provides current studies and research on the treatment and use of sewage. Using this material as a fertilizer can benefit the environment by turning wastes into valuable resources. These sludges would otherwise have to be disposed of by landfilling, lagooning, incineration, or ocean dumping. On the other hand, heavy metals sometimes found in sewage sludge may present environmental problems as there is filtration of metals towards the aquifers which get contaminated. This book addresses these concerns as well as benefits.

Chapter 1 - The origin of the waters arriving at the Sewage Treatment Plant (STP) is varied (industrial, domestic, stormwater....). Depending on the source, the amount, type and chemical forms of trace metals is going to be different. Cadmium, Cr and Zn used to predominate in the industrial inputs while As, Cu, Hg and Ni are the most dominant in the domestic ones (Jenkins and Russell, 1994; Sörme and Lagerkvist, 2002). The levels of Pb are increased when storm-derived waters enter into the STP.

The degree of purification of the water achieved by each STP depends on the treatment. Normally a primary and a secondary treatment are applied to the water. Only a limited number of STP have the tertiary treatment because it is only economically viable for large industrial units. During the first treatment all the floating solids are eliminated, thus trace elements associated with particles disappear during this step. On the secondary (biological) treatment, metals adsorbed by the bacterial flocs will be removed. Most authors (Stones, 1958; Brown et al., 1973; Oliver and Cosgrove, 1974; Toumi et al., 2003) have addressed the secondary treatment as being more efficient than the first one for most of the metals. Finally, when the tertiary treatment (chemical methods, i.e. co-precipitation) is applied, the efficiency of the removal of metals increases considerably.

In conclusion, highly soluble metals (i.e., Pb and Cd) in the sewage matrix will pass through the works and be discharged in the final effluent at a concentration similar to that in the raw sewage (Lester et al., 1979). Other trace metals (Cu and Mn) will be released with the particles and accumulated in the sludge (Karvelas et al., 2003).

Once the treatment is finished, the problem becomes how or where the sludge and effluents should be disposed. Sludges are used in agriculture to enrich the land with the nutrients present in them. Unfortunatelly, there is a filtration of metals towards the aquifers which get contaminated. Regarding the ¨treated waters¨, these are usually spilled to watercourses like rivers, estuaries or oceans. At this point, the speciation of a metal, rather than its total concentration, is the key to undertanding its toxic effects on the biota (Kelly, 1988; Santos-Echeandía et al., 2008).

Because of the importance of trace metals as environmental contaminants, it is essential to understand fully the fate of the substantial quantities of metals entering sewage treatment works, and the consequences for the environment. It is inevitable that water quality legislation for trace metals will include statements relating to their speciation and that it may be possible to tolerate higher total concentrations of some metals as long as the ¨labile¨ fraction is below a certain level (Florence, 1982).

Chapter 2 - High concentrations of N and P in sewage materials are one of the strongest reasons for their agronomic use. Such residues have been reported to increase yields and supply plant nutrients. However, digested sewage sludges still present highly putrescible organic matter and pathogens and therefore they usually need further sanitation and stabilization. Sanitation processes such as composting, liming, thermal-drying and solar-irradiation are frequently required to make sewage sludge safe enough for its agricultural use. For an appropriate sludge management and due to environmental concerns, it is essential to understand side effects of stabilization processes on sewage sludge's N and P.

This chapter discusses the common ways to stabilize domestic sewage sludge and consequences on N and P contents and availability in the produced biosolids. Stabilization of digested sewage sludge has been reported to affect its N and P fertilizing value. There is not a stabilized sewage material (biosolid) that is totally similar to another or to the fresh sewage sludge. The process that most alters N and P content in digested sludge is composting. Compared to other biosolids, composted biosolid remains as the poorest potential source of N and P for plants. Heat-drying at 250 °C and solar-treatment are the only processes which do not cause significant losses of total-N and total-P in sewage sludge. Nutrient contents and availability have a profound effect on agronomic value of biosolids. Therefore, potential markets should be assessed prior to deciding for a stabilization process of sewage sludge.

Chapter 3 - The accumulation and availability of PTE after application of sewage sludge on agricultural land has become a world wide critical problem. This is due to the risk of environmental problems. The risks of human, crop and/or environmental toxicity posed by these elements are a function of their availability and therefore on their soil speciation. PTE commonly found in sewage-sludge include Cd, Cu, Pb and Zn. The toxicity and mobility of these metals strongly depend on their specific chemical forms and on their binding state in the sludge. Alternatively, soil physical–chemical properties, especially pH, CEC, organic matter and clay content, are also likely to assume great importance in determining PTE behavior. This chapter emphasizes on studies performed on representative soils of the Pampas Region, Argentina, showing the effects of soil application of raw sewage sludge, anaerobically digested sewage sludge and a mixture of sewage sludge and its own incinerated ash. The inorganic fraction became the most prevalent form for sludge-borne PTE, whereas sludge-borne Cd was only found in the residual fraction. The chemical behavior of sludge-borne Pb did not depend on soil physico-chemical characteristics. On the contrary, availability of sludge-borne Cu and Zn depended on soil pH and not in soil texture. The results obtained in this study provide supporting evidence for the protection theory, which hypothesizes that mineral components may compensate for any loss of metal retention capacity caused by mineralization of organic components. We also considered soil application of soluble salts of the four studied metals. The relationship between the soil chemical forms of the four studied metals and plant absorption was also studied. In pot experiments, plant absorption depended on PTE availability. However, field experiments showed very little PTE mobility, availability or accumulation in crops. Under controlled conditions, the use of sewage sludge as a soil amendment did not pose a significant risk of contaminating soil, water or plants. Moreover, the mixture of sewage sludge with its own incinerated ash also offers a potential viable utilization of this organic waste as a soil amendment.

Chapter 4 – The authors environmental sanitation group developed research into the treatment of municipal sewage with the objective of producing effluents for use in agriculture during the period 1999 to 2005. A low-cost pilot-scale treatment system was installed and monitored at Experimental Station for Biological treatment of Sewage (EXTRABES) in Campina Grande City, northeast of Brazil. The system comprised an UASB (Upflow Anaerobic Sludge Blanket Reactor) with a hydraulic flow rate of 20 m^3 day^{-1}, followed by three post-treatment options: a constructed wetland, a gravel bed system and polishing ponds. The effluents were analyzed for physical and chemical parameters as well as for thermotolerant coliform and helminthes eggs. The mean values for nitrogen, phosphorus and potassium in the effluents produced were analyzed and expressed as kgN, kgP_2O_5, kgK_2O per hectare and the effluents used to irrigate cotton and sesame. The results showed that the polishing pond effluent was free of helminth eggs and contained <1000 FCU/100mL of thermotolerant coliforms and was thus suitable for unrestricted irrigation as per WHO guidelines (1989). Due to high pH values, which varied between 8.4 and 9.8, considerable removal of phosphorus and nitrogen occurred, reducing the concentrations of nutrients available for fertirrigation. There was no statistical difference ($p>0.05$) between thermotolerant coliform removal between the planted and unvegetated systems with mean effluent concentrations in both systems of 2500 FCU/100 mL. Thus according to the WHO guidelines these effluents are suitable for fertirrigation of cereals and crops which are not consumed raw. The removal efficiencies of carbonaceous material expressed as COD and VSS in the three systems was considerable (72 to 80 mg/L of COD and 10 mg/L of VSS) giving effluent concentrations between 72 and 80 mg/L for COD and 10 mg/L of VSS, with no significant difference ($p>0.05$) between the treatments. Finally after five years of research the technology was passed to farming families of low-income communities.

Chapter 5 - Activated sludge method is widely used for the biological sewage treatment. Activated sludge contains aerobic microbes including bacteria, fungi, amoeba, and protozoa, which play a critical role for efficient treatment process. Since domestic and industrial wastewaters are known to contain various pollutants, activated sludge is therefore thought to be acclimated to pollutants, and might be an attractive source for isolation of pollutant-degrading microbes. Although a microbial community of activated sludge is recently analyzed using culture-independent methods, enrichment and isolation works still play an important part to study biochemical and physiological characters and examine an application potential of microbes.

The author has isolated several pollutant-degrading microbes including a nonylphenol-degrading bacterium, estradiol-degrading bacterium, and benzophenone-degrading yeast from sewage treatment plants. The author assimilates the pollutants as a sole carbon source.

A bioreactor using the immobilized nonylphenol-degrading bacterium worked properly and removed nonylphenol in wastewater, suggesting that a biotreatment system using a pollutant-degrading microbe can contribute to the preservation of aquatic environments.

The bacterial strains isolated in this study were found sphingomonads. Sphingomonads are well known for various pollutant-degrading activities and were recently reported as ubiquitous microbes in activated sludge. Considering the fact that almost of microbes in the activated sludge have not been isolated yet, an employment of suitable culture conditions can bring a finding of novel pollutant-degrading microbes in an activated sludge.

Chapter 6 - First of all a thermogravimetric (TG) and mass spectrometric (MS) study was carried out of the process of devolatilization of sewage sludge. This involved an analysis of the heating profiles in an inert atmosphere for this residue. The study of mass spectrometry shows major emissions of energy-producing gases although fewer pollutants are emitted. It contents of volatile matter represents a valuable resource, a possible alternative is the conversion into fuels by means of pyrolysis that yields gas, bio-oil and biochar. The composition and heating value of the gas makes it suitable for use as a fuel. The bio-oil may be evaluated as an environmentally-friendly green biofuel. The biochar product is carbon-rich and a potential solid biofuel. Other ways it might be used as a C and N source in soil amendment. The process of thermolysis of sewage sludge in a pilot plant is also described and the results are compared with the others obtained under laboratory conditions (semi-pilot scale). The aim of this study was to get the energetic valorisation of the products; gas and carbonized solid, no liquid fraction was obtained due to the specific characteristics of the plant. The combustion gases were subjected to a cleaning process. Finally an energy production system based on the Rankine cycle is proposed.

Chapter 7 - Most wastewater treatment processes produce a sludge which has to be disposed of. Conventional secondary sewage treatment plants typically generate a primary sludge in the primary sedimentation stage of treatment and a secondary, biological, sludge in final sedimentation after the biological process. The characteristics of the secondary sludge vary with the type of biological process and, often, it is mixed with primary sludge before treatment and disposal. Approximately one half of the costs of operating secondary sewage treatment plants in Europe can be associated with sludge treatment and disposal. Land application of raw or treated sewage sludge can reduce significantly the sludge disposal cost component of sewage treatment as well as providing a large part of the nitrogen and phosphorus requirements of many crops. Very rarely do urban sewerage systems transport only domestic sewage to treatment plants; industrial effluents and storm-water runoff from roads and other paved areas are frequently discharged into sewers. Thus sewage sludge will contain, in addition to organic waste material, traces of many pollutants used in our modern society. Some of these substances can be phytotoxic and some toxic to humans and/or animals so it is necessary to control the concentrations in the soil of potentially toxic elements (PTE) and their rate of application to the soil. The risk to health of chemicals in sewage sludge applied to land has been reviewed by Dean and Suess [1].

Sewage sludge also contains pathogenic bacteria, viruses and protozoa along with other parasitic helminths which can give rise to potential hazards to the health of humans, animals and plants. A WHO (1981) Report on the risk to health of microbes in sewage sludge applied to land identified salmonellae and *Taenia* as giving rise to greatest concern. The numbers of pathogenic and parasitic organisms in sludge can be significantly reduced before application to the land by appropriate sludge treatment and the potential health risk is further reduced by the effects of climate, soil-microorganisms and time after the sludge is applied to the soil. Nevertheless, certain crops, limitations on planting, grazing and harvesting are necessary.

Apart from those components of concern, sewage sludge also contains useful concentrations of nitrogen, phosphorus and organic matter. The availability of the phosphorus content in the year of application is about 50% and is independent of any prior sludge treatment. Nitrogen availability is more dependent on sludge treatment, untreated liquid sludge and dewatered treated sludge releasing nitrogen slowly with the benefits to crops being realised over a relatively long period. Liquid anaerobically-digested sludge has high ammonia-nitrogen content which is readily available to plants and can be of particular benefit to grassland. The organic matter in sludge can improve the water retaining capacity and structure of some soils, especially when applied in the form of dewatered sludge cake.

Chapter 8 - The feasibility to dispose wastewater sewage sludge through thermal degradation methods, was studied in consideration of the recent European environmental directives, which limited the use of this solid waste as fertilizer. In particular, the pyrolysis of sewage sludge was investigated, in order to obtain a gas phase, rich in hydrocarbons, for a fuel gas use. The characterization of this produced gas phase, and the determination of its heat of combustion, are fundamental data for the application of a pyrolysis disposal method. For this reason, thermogravimetric coupled mass spectrometric and thermogravimetric coupled gas chromatographic and mass spectrometric analyses were carried out for the obtainment of an accurate characterization of the sewage sludge pyrolysis process. Tests on a laboratory-scale reactor were furthermore carried out with charges of 0.3 kg of sewage sludge. Analogously to the ashes of an incinerator plant, the pyrolysis solid residue is a waste with high content of hazardous heavy metal ions, and in consequence it must be stabilized. For this aim, vitrification is considered the best method which bears the immobilization of the heavy metal ions into a vitreous matrix. This vitrified material doesn't need landfill collocation, but it can be used as inert additive in ceramics and concretes field. On the basis of these considerations, the disposal of wastewater sewage sludge, through a pyrolysis-vitrification process, was studied, considering this two steps process as an integration of the depuration systems present in the wastewater treatment plant.

In order to apply this disposal treatment for sewage sludge produced by the wastewater treatment plant, the resolution of the mass and energy balances was carried out by applying a thermochemical equilibrium model.

Chapter 9 - During the latest years there has been a growing level of concern related to the adverse effects of chemical compounds on the environment. Pharmaceuticals is one of the categories of pollutants that during the last 20 years have drawn the attention of the scientific community as they are ubiquiteously present in wastewater effluents at trace levels, and several of them may have adverse effects on aquatic organisms and human health. Their transformation pathways during wastewater treatment are the subject of a growing current research, while many gaps still exist in the knowledge of the degree into which they are decomposed during wastewater treatment, as well as their degradation products. Analytical techniques for the determination of pharmaceuticals are being optimized and applied in order to provide answers regarding their behaviour during water treatment and their degradation products. Current research and obtained results on this subject are summarized and discussed in this chapter.

Chapter 10 - The addition of food waste to sewage sludge with low content of heavy metals improved C/N ratio in initial material for bioconversion, increased the potassium content, diluted the content of heavy metals, and reduced the viscosity of sewage sludge. The mixture of sewage sludge and food waste was pre-treated at 100°C for 45 minutes to inactivate the pathogens and parasites, disrupt cells in both food waste and sewage sludge, and partially hydrolyze polymers. Bioconversion of the pre-treated mixture of sewage sludge and vegetable food waste was performed in the rotating bioreactor with air supply and controlled temperature 60°C for 10 days with the addition of starter culture, artificial bulking agent, and pH buffering component. Thermophilic bacterial strain of *Bacillus thermoamylovorans* SW25 with high digestion activity have been selected and used as the starter culture to enhance bioconversion of sewage sludge and food waste. The best organic fertiliser was obtained when the ratio of the thermally pre-treated sewage sludge and food waste was 1:1 by total solids. The reductions of organic matter, 16S rRNA of methanogens, and coenzyme F_{420} fluorescence during 10 days of the bioconversion were 67%, 54%, and 87% of the initial values, respectively. The produced organic fertiliser was a powder with the moisture content of 5%. The content of volatile solids was decreased by 25% of total solids. The plant tests indicated that the product of bioconversion can be graded as good commercial fertiliser. Addition of 1.0% to 1.5% of this fertiliser to the subsoil increased the yield and growth of different plants by 113 to 164%. The product was confirmed as a stable and non-toxic matter for the plants by the phytotoxicity test. Although the intensive aerobic thermophilic process consumes more energy than that of the conventional composting process but it saves time and land area by small footprint bioconversion of the mixture of sewage sludge and food waste into a commercially-compatible fertiliser in 10 days without curing and maturation phases that are required in conventional composting.

Chapter 11 - India faces a turbulent water future. Unless water management practices are changed – and changed soon – India will face a severe water crisis within the next two decades and will have neither the cash to build new infrastructure nor the water needed by its growing economy and rising population. Water is one of the critical inputs for the sustenance of mankind. It is used for both terrestrial and aquatic environments in various activities, balancing the ecological system of global environment. Water is the important natural source, abundant in nature and covers about two-thirds of the earth's surface. However, only 1% of the water resource is available as fresh water (i.e., surface water-rivers, lakes, reams, and ground water) for human consumption and other activities. The major uses of water are for irrigation (30%), thermal power plants (50%), while other uses are domestic (7%) and industrial consumption (~12%) (A. K. De, 2002).The United Nation's report on "Water for People, Water for Life" (the first ever UN system wide evaluation on global water resources-2003) has put India a poor 120th for water quality among 122 nations covered. Only Belgium and Morocco are ranked worse than India. The quality indicator value was based on quality and quantity of fresh water (especially ground water), waste water treatment facilities, legalities like application of pollution regulations, India's quality indicator value stood at -3.1 while for based ranked country Finland it was 1.85.

The UN evaluation also ranked India 133 in a list of 180 countries for its poor water availability (1880m^3 per person per year). Kuwait was ranked the poorest on water availability. Against the National average target of 135 lpcd of water and 180 lpcd per capita in large cities, the per capita availability is low and ranges from 165 lpcd in a few larger town to about 50 lpcd in most smaller towns. The availability of water in urban slums is about 27 lpcd. Urbanisation has given rise to a number of environmental problems such as water supply, wastewater generation and its collection, treatment and disposal in urban areas. In most cases wastewater is let out untreated and it either percolates into the ground and in turn contaminates the groundwater or is discharged into the natural drainage system causing pollution in downstream areas. Sewage and not the industrial pollution accounts for more than 75 per cent of the surface water contamination in India. Due to negligence, groundwater is also increasingly getting contaminated. In India less than 50% of the urban population has access to sewage disposal system. Most of the existing collecting systems discharge directly to the receiving water without treatment. Garbage, domestic and otherwise, is directly dumped into water bodies or roadside, which can often be washed into streams and lakes. The municipalities dispose off their treated or partly treated or untreated wastewater into natural drains joining rivers or lakes or used on land for irrigation or fodder cultivation or into sea or combination of these. Toxic chemicals from sewage water transfer to plants and entire in the food chain and affect public health. Pathogens occurring in the sewage water directly affect the mammals causing severe diseases. About 60 per cent of urban deaths in India are due to lack of safe drinking water facilities. Further deaths due to water borne diseases are second only to malnutrition. It is estimated that around 80% of water consumed by a household is let of to the drains of sewers as wastewater.

There is substantial scope for segregated use of the water for further use for gardening, industrial cooling, street cleaning, vehicular washing, fire fighting, irrigation, yard cleaning, fountains, recreational lakes, etc. Though methods are available to improve the quality of recycled water to potable grade, the lack of social acceptance and prohibitive costs may prevent the adoption of these techniques. The importance of reuse and recycling of treated sewage and industrial effluents has been realized on account of two distinct advantages: reduction of pollution in the receiving water bodies and reduction in the requirement of fresh water for various uses. Reuse of municipal wastewater after necessary treatment to meet industrial water requirement is being practiced in India.

Thus, wastewater can be considered as both a resource and a problem. Wastewater and its nutrient content can be used extensively for irrigation and other ecosystem services. Its reuse can deliver positive benefits to the farming community, society, and municipalities. However, wastewater reuse also exacts negative externality effects on humans and ecological systems, which need to be identified and assessed. Before one can endorse wastewater irrigation as a means of increasing water supply for agriculture, a thorough analysis must be undertaken from an economic perspective as well. In this regard the comprehensive costs and benefits of such wastewater reuse should also be evaluated. Moreover, the economic effects of wastewater irrigation need to be evaluated not only from the social, economic, and ecological standpoint, but also from the sustainable development perspective.

Chapter 12 - China produced about 5 million tons of dry municipal sewage sludge in 2007 and this number is increasing very fast; disposal and utilization of municipal sewage sludge is urgent and promising in China. This paper reviewed origins and sources of municipal sewage sludge, its production amount, disposal technologies, and its utilization fates/prospects. In particular sludge thickening, anaerobic/aerobic digestion, dewatering, drying and incineration technologies were illustrated with country specific principles and parameters; while energy replacement and land applications were elaborated as major utilization fates for municipal sewage sludge with practices done or is under operation in China. Disposal technologies and utilizations under research in China such as ultrasonic treatment, pyrolysis and gasification of municipal sewage sludge, hydrogen production through fermentation of municipal sewage sludge, producing activated carbon with municipal sewage sludge, producing construction and building materials out of municipal sewage sludge were also briefly introduced but not elaborated in this review.

Chapter 13 - Colour and non-biodegradable trace organics present in the treated wastewater from sewage treatment plants (STPs) can affect the options for disposal or recycling. Colour is sometimes a concern because its visual appearance affects public acceptance of using recycled water. In the present study, different Advanced Oxidation Processes (UV/O_3, UV/TiO_2 and $UV/TiO_2/O_3$) in laboratory scale systems were used to study the removal efficiencies of colour and trace organics from the treated effluent. The removal efficiencies of the processes were compared by the analysis of colour, A_{254} (the organics absorption at the wavelength of 254 nm), and total organic carbon (TOC). The effects of ozonation procedure and ozone dose were also investigated. The co-treatment of ozonation and photocatalytic oxidation increased TOC removal rate by nearly 50 % compared to the process in which ozonation was used as a pre-treatment for photocatalysis. This was due to the synergistic effects of photocatalysis and ozonation. The colour removal rate significantly increased when the O_3 dose was increased.

Apart from AOPs, adsorption methods using activated carbons were also investigated for colour removal from recycled water. TiO_2 nanoparticles were loaded on the surface of the granular activated carbons (GAC). This allows the GAC to be regenerated by UV irradiation without the thermal treatment. The adsorption results using a fixed bed column were reported, and the regeneration potential of the activated GAC was also discussed.

Chapter 14 - The photolysis of octylphenol (NP) in pure water and surface water was investigated using a solar simulator and natural sunlight in the absence/presence of dissolved organic matter (DOM), HCO_3^-, NO_3^- and Fe(III) ions. The effects of different parameters such as irradiation source, initial pH, initial concentration of substrate, temperature, and the effect of hydrogen peroxide concentration on photodegradation of octylphenol in aqueous solution have been assessed.

The results indicate that the oxidation rate increases in the presence of H_2O_2, Fe(III) and DNOM. The real solar irradiation proves to be more efficient than solar UV simulated irradiation, but the best OP removal, both as percentage and time, is realised by the medium-pressure mercury lamp (UV-MP). In addition, the disappearance of the estrogenic activity of octylphenol during irradiation using YES test was investigated. Based upon the YES test results, there was a strong decrease of estrogenic activity of octylphenol after 8 h sunlight irradiation in the presence of hydrogen peroxide.

Short Communication - To use sewage sludge, the acid-washing of its incineration residue and the coagulation-sedimentation treatment using the washings were performed in comparison with water supply sludge and red clay incineration residues. The concentration level of total phosphorous (T-P) (221-4,790 mg/l) in the solutions of (1+1) H_2SO_4 extraction from both sludges was higher than those of COD (10.7-127 mg/l) and total nitrogen (T-N) (0.2 mg/l>). As for metals, the concentration level was as follows: Al (874-2,900 mg/l) > Fe (777-1,130 mg/l) > Ca (71.5-614 mg/l) > Mg (21.2-271 mg/l) > Mn (37.7-148 mg/l) > Zn (5.72-139 mg/l) > Cu (3.37-42.7 mg/l) > Ni (2.32-6.64 mg/l) > total chromium (T-Cr) (0.27-2.68 mg/l) > Sb (0.01 mg/l>).

The concentration levels of Cd, Pb and Se were those of effluent standards. In the coagulation-sedimentation treatment at pH 6.0 using the acid-washings, max. 10.0-23.5% of COD, max. 22.9-27.2% of T-N and max. 96.5% of T-P were removed. The remaining rate of T-P using the washings of sewage sludge incineration residue was over 100% because the washings contained much T-P. In the same treatment conditions, the remaining rates of Al, Ca, Cu, Mg, Mn, Ni and Zn were over 100%. However, the concentration levels of these metals in treated water were in conformity with the effluent standards.

In: Sewage Treatment: Uses, Processes and Impact
Editor: Anna Stephens and Mark Fuller
ISBN 978-1-60692-959-9

Chapter 1

THE FATE AND TRANSPORT OF TRACE METALS THROUGH SEWAGE TREATMENT PLANT PROCESSES

Juan Santos-Echeandia*
Marine Research Institute (IIM-CSIC)
Vigo, Spain

ABSTRACT

The origin of the waters arriving at the Sewage Treatment Plant (STP) is varied (industrial, domestic, stormwater....). Depending on the source, the amount, type and chemical forms of trace metals is going to be different. Cadmium, Cr and Zn used to predominate in the industrial inputs while As, Cu, Hg and Ni are the most dominant in the domestic ones (Jenkins and Russell, 1994; Sörme and Lagerkvist, 2002). The levels of Pb are increased when storm-derived waters enter into the STP.

The degree of purification of the water achieved by each STP depends on the treatment. Normally a primary and a secondary treatment are applied to the water. Only a limited number of STP have the tertiary treatment because it is only economically viable for large industrial units. During the first treatment all the floating solids are eliminated, thus trace elements associated with particles disappear during this step. On the secondary (biological) treatment, metals adsorbed by the bacterial flocs will be removed. Most authors (Stones, 1958; Brown et al., 1973; Oliver and Cosgrove, 1974; Toumi et al., 2003) have addressed the secondary treatment as being more efficient than the first one for most of the metals. Finally, when the tertiary treatment (chemical methods, i.e. co-precipitation) is applied, the efficiency of the removal of metals increases considerably.

In conclusion, highly soluble metals (i.e., Pb and Cd) in the sewage matrix will pass through the works and be discharged in the final effluent at a concentration similar to that in the raw sewage (Lester et al., 1979). Other trace metals (Cu and Mn) will be released with the particles and accumulated in the sludge (Karvelas et al., 2003).

Once the treatment is finished, the problem becomes how or where the sludge and effluents should be disposed. Sludges are used in agriculture to enrich the land with the nutrients present in them. Unfortunatelly, there is a filtration of metals towards the aquifers which get contaminated. Regarding the ¨treated waters¨, these are usually spilled

* E-mail: jusae@iim.csic.es

to watercourses like rivers, estuaries or oceans. At this point, the speciation of a metal, rather than its total concentration, is the key to undertanding its toxic effects on the biota (Kelly, 1988; Santos-Echeandía et al., 2008).

Because of the importance of trace metals as environmental contaminants, it is essential to understand fully the fate of the substantial quantities of metals entering sewage treatment works, and the consequences for the environment. It is inevitable that water quality legislation for trace metals will include statements relating to their speciation and that it may be possible to tolerate higher total concentrations of some metals as long as the ¨labile¨ fraction is below a certain level (Florence, 1982).

1. INTRODUCTION

The increasing population of the last decades and the huge quantity of industry developed generates more and more waste materials and products. Wastewater discharged in huge quantities without any treatment represents a source of pollution to the environment. It can cause serious environmental and sanitary problems (Toumi et al., 2003). Because of the importance of trace metals as environmental contaminants, it is essential to understand fully the fate of the substantial quantities of metals entering sewage treatment works, and the consequences for the environment (Lester et al., 1979).

In recognition of the sensitivity of aquatic ecosystems, strict water quality criteria have been set for metals in the United States and many other countries.

Although the traditional role of municipal treatment plants has been to remove organic matter and suspended solids, influent metals are causing increasing concern because:

1) They can inhibit aerobic and anaerobic biological treatment processes, thereby causing possible deteriorations in effluent quality and reduced rates of anaerobic sludge digestion (Kugelman and McCarty, 1965; USPHS, 1965; Poon and Bhayani, 1971; USPHS, 1965; Oleszkiewicz and Sharma, 1989).
2) High metal concentrations in sludges disposed of by land application can injure crops and animals and can enter the human food chain. Nickel appears to be one of the most phytotoxic metals (Berrow and Webber, 1972; Purves, 1979).
3) Metals discharged in the treated sewage can be toxic to aquatic life and can cause natural waters to be unsuitable as potable water sources (Environment Canada, 1979).

The source of metals (domestic activities, industrial activities and stormwaters and diffusive sources) is going to be an important factor in relation to the type of metal and the chemical forms in which they will enter the STP.

Wastewater treatment plants (WTPs) are expected to control the discharge of trace metals to the environment (Karvelas, 2003). Conventional processes in the field of wastewater treatment can be divided into two main phases: (1) generation of suspended solids from colloidal and dissolved solids by physical, chemical and biological means in addition to the already existing suspended solids; (2) separation of suspended solids by chemical and mechanical methods including sedimentation, flotation and filtration (Santarsiero et al., 1998).

According to the different behaviour during sewage treatment processes, trace metals can be classified in two different groups:

- Metals which are highly soluble in the sewage matrix, either as free ions or as a complex, would be expected to pass through the works and be discharged in the final effluent at a similar concentration to that in the raw sewage
- Metals strongly adsorbed upon or complexed by suspended solids. Such a metal will be largely removed in the sewage sludges, either primary or secondary.

The majority of the metals behave in this second manner (Lester et al., 1979) and some of them have an intermediate behaviour.

A second important factor which controls the trace metal content and chemical forms in the final effluent will be the different treatments applied to the waters entering the STP. Before entering the STP a pre-treatment process is applied to the waters. The three main treatments inside the STP are usually denominated primary treatment, secondary treatment and tertiary treatment. The principal parameters controlling metal removal in industrial and municipal wastewater treatment plants include: the type of metal and its form; suspended solids concentrations (Patterson et al., 1983); pH and organic composition of wastewater (Reimers, 1983). Inert solids of primary sludge exhibit different sorption affinities for trace metals than biological sludges (Anon., 1965; Brown and Lester, 1982; Nielsen and Hrudey, 1983; Patterson et al., 1983; Reimers, 1983). Primary treatment removes mainly insoluble (i.e. settleable or adsorbed to settleable particles) forms of trace metals while biological sludges remove the soluble forms, including the organometallic complexes (Sterritt and Lester, 1984).

The first two treatments (primary and secondary) have been the same for many years, but new methods and processes have have begun to be developed in the last years with the intention of reducing the metal content in the two end products in the STP, which are sludge and effluent (Sörme and Lagerkvist, 2002). The method and place of disposal of both end products, which is not regulated enough all over the world, is a crucial aspect in relation to environmental damage and contamination. At certain concentrations, trace metals are toxic to higher organisms, microorganisms and plants. Moreover, the toxicity of trace metals in wastewater is shown to depend on factors like metal species and concentration, pH, sludge concentration, wastewater pollution load (Imai and Gloyna, 1990; Dilek and Yetis, 1992) and solubility of the metal ions (Surittanonta and Sherrod, 1981).

Although continuous improvements are applied to the STP processes and new strict regulations, related to trace metals levels in the effluents from the domestic and industrial sources, have been developed, there is still a need to control the metal content in the final effluents from the STP if we want to preserve the environment.

Thus, the fate and transport of trace elements during the sewage treatment plant processes is going to depend on the sources of these elements, the species in which they are present and the number of different treatments that they will receive. The importance of all these variables and the evolution of the quantity and chemical forms of trace metals during the treatment processes will be explained in the following sections.

2. Variety of Waters Entering STP

The origin of the waters arriving at the Sewage Treatment Plant (STP) is varied: domestic activities (Comber and Gunn, 1996; Sörme and Lagerkvist, 2002), industrial activities (Line et al., 1996) and storm water (Makepeace et al., 1995). Figure 1 shows an example of the wide variety of sources entering a STP in Stockholm. Some STP will only treat one type of source while others will have a mixed source treatment or combined sewers.

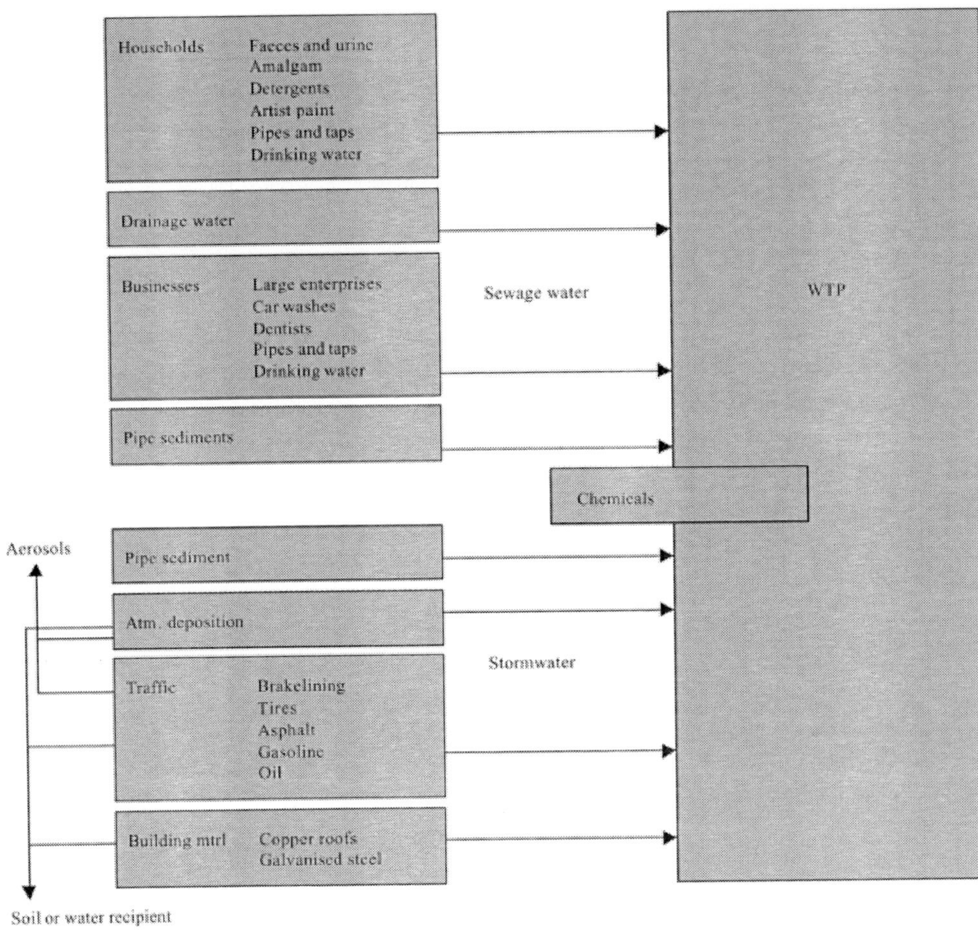

Figure 1. The different sources of trace metals and the possible destination for each source (taken from Sörme and Lagerkvist, 2002 with permission).

Sources are usually monitored when entering the STP. However, some of them arrive without any pre-treatment from the industrial or domestic origin, which can seriously affect the capacity of the treatment inside the STP. In addition to these three sources, we can also find diffuse ones (emissions from different goods in the traffic environment—brake linings, tires, from buildings—roofs, and from households—food) which are difficult to control and monitor (Sörme and Lagerkvist, 2002). Some authors (Makepeace, 1995; Al Bakri et al., 2008) include the store water within the diffuse sources. Furthermore, there is an input of metals (Cr, Ni and Zn) inherent to some tertiary treatment processes (Sörme and Lagerkvist, 2002; Buzier et al., 2006). The high control taken over the metals levels in the known inputs or point sources (domestic and industrial) due to more strict European and world directives makes the input of metals from diffuse sources to STP more important (Comber and Gunn, 1996). Sörme et al. (2002) showed that goods were the main trace metal diffuse source in the case of Stockholm city.

Sometimes, during the high peaks of water disposal from industries and houses (Oliver and Cosgrove, 1974; Golstone et al., 1990; Chipasa 2003) or weather events (Buzier et al., 2006) a high flux of water enters the WTP which is not able to treat the entire volume of water. In these cases, part of this water is disposed without any treatment to the receiving waters. This can have serious consequences for the ambient.

Depending on the source, the amount, type and chemical forms of trace metals is going to be different:

Regarding the *industrial activities and businesses*, we will have different types and fluxes of metals depending on the industrial activities carried out in the area. For example, regarding auto salvage the main inputs are for Zn and Pb. In the metal fabrication industry, Cr (Goldstone et al., 1990a) and Ni used to present high concentrations, while Cu, Pb and Hg are very important in the scrap and recycling industry. Finally, in the wood maintenance industry, As and Cr are two of the metals to take into account (Line et al., 1996). In the past years, Hg used to be an important input in dentist disposal waters but the use of this metal has been restricted or forbidden with the subsequent decrease of Hg in these inputs.

In the case of *domestic sources* (households, drainage water, businesses, pipe sediments) the materials of the pipes that conduct the waters from the houses to the STP is going to be one of the main factors affecting concentrations and type of metals. The occurrence of corrosion in the pipes distribution system may add metals to the finished water (Moore et al., 1975; Boulay and Edwards, 2000). Thus, waters derived from old houses that still have old pipes (made of lead, copper or galvanized iron) will be rich in Cu, Fe, Pb and Zn (Goodman et al., 1984; Alam and Sadiq, 1989; Comber and Gunn, 1996; Sörme and Lagerkvist, 2002). From the last two decades new PVC or concrete pipes are used in the new houses thus, some metal fluxes entering the STP from this source have been reduced. Nevertheless, due to the heat stabilizers contained in the PVC pipes, organotin compounds levels in the waters can increase (Sadiki et al., 1996). In the case of household products, some washing powders and medicated shampoos were shown to have elevated concentrations of copper, zinc and cadmium (Comber and Gunn, 1996) due to the active ingredients as zinc pyrithione included in them. However, more recent ultrapowders, which use zeolites, are low in cadmium. (Stoveland et al., 1980). Faeces are an important input of Cd, Cr, Ni and Zn in the domestic inputs. Levels of these metals are multiplied by a factor of 5–20 with respect to the input water from the general water supply in the city (Comber and Gunn, 1996).

When talking about *storm waters* (pipe sediments, atmospheric deposition, traffic, building materials, runoff), we have to bear in mind that while the first rains after a long period of storm absence will transport high amounts of metals, a prolongated rainy period will have a dilution effect on them. A critical point here will be the building materials of the roofs which are mainly releasing Cu, Pb and Zn (Makepeace et al., 1995; Line at al., 1996; Sörme and Lagerkvist, 2002) with the storm waters. Traffic derived contamination (especially Pb) will find its way to STP as road runoff (Goldstone et al., 1990b).

Many studies have tried to compare the contributions of metals between industrial and administrative or domestic activities (Jenkins and Russell, 1994; Comber and Gunn, 1996). Having a look at some of them, conclusions can be confused. Sometimes the industrial activities seem to be the most important inputs of trace elements to STP (Jenkins and Russell, 1994) while, in other cases, the domestic sources tend to predominate in this contribution (Comber and Gunn,1996).

Nothing far away from the real thing because as we have explained before, many factors affect each type of input. In addition, the industrial activity of each area can be different (Soylak et al., 2001) (Table 1) and thus change the percentages of inputs respect to domestic sources. Or viceversa, the study may be followed in an old house city and percentages of metals due to old pipes will increase respect to industrial inputs. Thus, the main input to each STP is going to depend in a great number of different factors. Chipasa (2003) indicated that the accumulation of different metals in wastewater depends on many local factors such as type of industries in the region, people's way of life and awareness of the impacts done to the environment by careless disposal of wastes.

Once the waters arrive to the STP from various and different unknown sources, it is quite difficult to elucidate if the metal origin is industrial or domestic. Following a meticulous sampling of the waters along the day and during one week it is possible to guess the main input of each metal. When the peak of a metal flux appears during the weekend or night hours, it will presumably be of domestic origin as it is the case of Cu. However, if peaks are predominant during mornings or afternoons in the week will have an industrial origin as it occurs for Cr (Nielsen and Hrudey, 1983; Goldstone, 1990a).

The main lead source will be runoff because the higher concentrations enter the STP just after a store event when the water flux peaks are higher (Lester, 1973; Goldstone et al., 1990b) but this metal can also have an industrial origin when high concentrations of Pb do not match with high water fluxes. A similar behaviour is observed for Cd, Ni and Zn which have a mixed source, in this case from industrial and domestic origin. Next table (Table 2) shows the percentage of each metal associated to each one of the existing sources. Historically, the sources of trace metals to STP have been industrial activities with elements such as Cu, Zn, Ni and Cr. In recent years, industries have to a large degree moved out of the cities and the release of trace metals and other compounds has decreased due to various pre-treatments of the effluent (Figure 2). Whereas the industrial discharge of trace metals to sewers is controllable and is reducing, there remains a significant input from diffuse sources, particularly zinc and copper from domestic sources (Comber and Gunn, 1996).

Table 1. Concentration of some trace metal ions in the waste water simples from various plants in the organizad industrial region of Kayseri (nM). Data are means of three replicates analysis (modified from Soylak et al., 2001)

	Metal-1	Metal-2	Metal-3	Metal-4	Metal-5	Metal-6	Metal-7	Textile-1	Textile-2
Cr	4400							3100	2900
Pb	2600	500	500	1000	1800	100	4500	300	900
Fe	25000	273997	671562	21490	130730	107450	105660	≤3580*	≤3580*
Cu	30217	8341	6767	≤1574*	10387	14164	21403	9600	28643
Cd	≤178*	≤178*	≤178*	≤178*	≤178*	≤178*	≤178*	≤178*	≤178*
Zn	34709	3211	14220	69572	10245	7034	81345	≤306*	≤306*
Ni	38166	6304	4893	1704	13801	18912	30699	1704	1704
Al	42995	2594	39288	1112	2224	29652	182357	≤740*	≤740*

-: Not Determined.

* Less than the detection limit.

Table 2. The contribution in % from different sources and the sum of all sources estimated to Henriskdal, (1999) (Taken from Sörme and Lagerkvist., 2002 with permission)

Source	(%)	Cu (%)	Zn (%)	Pb (%)	Cr (%)	Ni (%)	Cd (%)	Hg
Sewage water	Household	59	30	1	2	16	20	44–47
	Businesses	30	27	38	13	13	32	21
	Drainage water	2	4	1	2	10	3	1
	Pipe sediments	–	–	+	–	–	–	+
Stormwater	Buildings	13–17	24	–	–	–	1	–
	Traffic	5	10–11	9–11	<1	<1	1	–
	Atmos. deposition	<1	2	4	1	1	4	–
	Pipe sediments	–	–	+	–	–	–	(+)
Within the WTP	Chemicals	–	2	–	5	31	–	–
Total from all sources (incl. chemicals)		109–113	99–100	53–55	23	71	60	66–69
Part from sewage water		91	61	40	17	39	55	66–69
Part from stormwater		18–22	36–37	13–15	1	1	5	–

A dash sign (–) indicates that the source is insignificant compared to other sources (in this study). A plus sign (+) indicates that the source probably is significant but the amount has not been estimated.

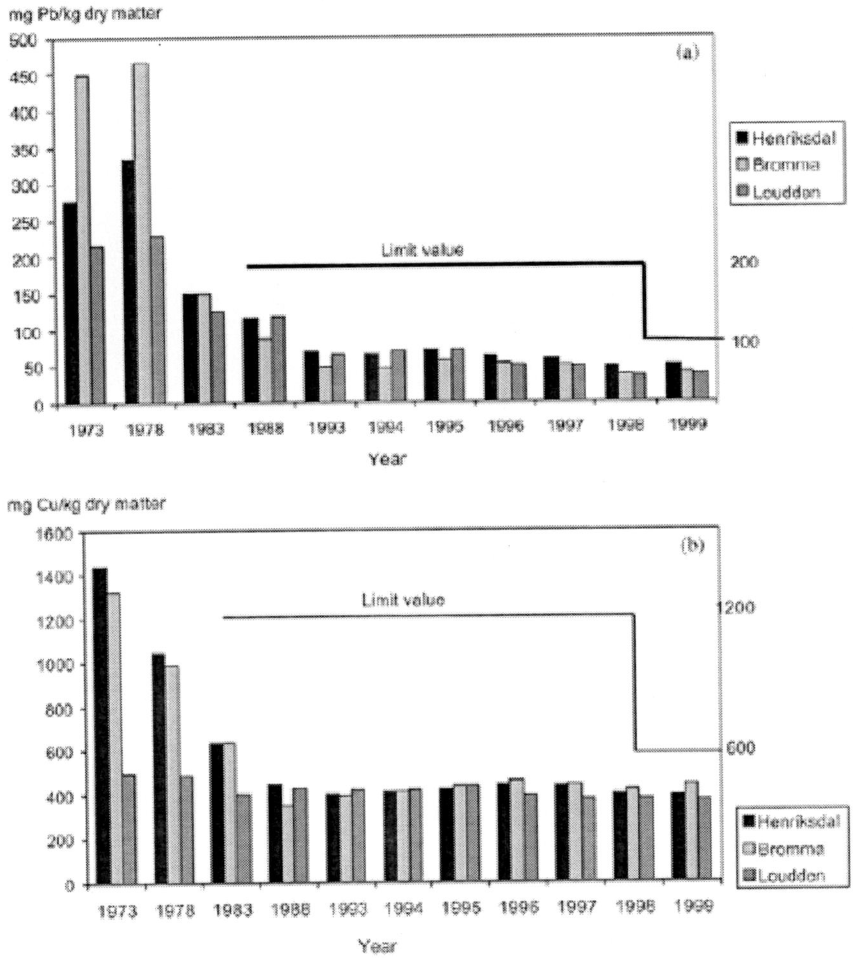

Figure 2. The content of Cu and Pb (mgykg TS) in the sewage sludge of Henriksdal, Bromma and Loudden's WTP, Stockholm, 1973–1999 (Taken from Sörme and Lagerkvist, 2002 with permission).

Table 3 shows the typical ranges of trace elements in the raw waters before being treated inside the STP in different parts of the world. An additional column with the origin of the waters has been included. The range given for each metal is very wide which can be explained by the different origins (industrial, domestic, storm water) of the waters entering the STP. As we will explain during next sections of the chapter when talking about trace metals it is important to have information of the total content of these metals but also about their speciation. The composition of the waters and the distribution of each metal in the different species will be crucial respect to its percentages of removal during different stages of the treatment and also regarding the toxicity of the metal disposed to the environment in the final effluent or sludge. In the next tables we can appreciate the distribution of trace elements between the dissolved and particulate phase (Table 4), different species inside the dissolved phase (Table 5) and particulate phase (Table 6) in the raw sewage.

Table 3. Trace metal concentrations (nM) measured in different sources entering the sewage treatment plant

								Metals (nM)										
Location	**Ag**	**Al**	**As**	**Cd**	**Co**	**Cr**	**Cu**	**Fe**	**Hg**	**Mn**	**Mo**	**Ni**	**Pb**	**Se**	**V**	**Zn**	**Catchment**	**Bibliography**
UK				4-5		6-273	16-913					7-157	92-150			459-4052	Domestic	Comber and Gunn, 1996
USA			53-3150	9-365		346-16638	614-34985					68-5674	188-15555	38-266		2920-154174	Industrial (runoff)	Line et al., 1996
Internacional Review			13-2803	0.4-122153	22-92	19-44239	1-22191	1433-7879656				17-834895	3-125483	6-975	141-167	11-336391	Stormwater	Makepeace, 1995
UK				36-72			504-3447						483-1207				Industrial	Lester et al., 1979
Morocco				0.4-76		8-577	123-976						1-212			1728-12920	Domestic	Toumi et al., 2003
Japan				4.5-34.7		34.6-910	456-5760	1827-78492		89-2093		73.3-532	35-281			662-6835	Domestic	Chino et al., 1991
Greece				20-39		539-1000	692-1794	7038-10154				9712-16868	140-236			5046-9327	Combined sewer	Karvelas et al., 2003
Australia							866-2203						1207-3571				Stormwater	Al Bakri et al., 2008
Canada Review				71-82		38-151952	472-26755					34-76674				413-397400	Combined sewer	Nielsen and Hudrey, 1983
Paris				5-23	8-42	77-269	629-1259	7163-14362				85-256	72-121				Combined sewer	Buzier et al., 2006
USA	16.5-973	20163-64492	19.5-104	1.2-10.14	6.11-26.8	152-182	587-1684			566-5824	86-179	105-203	20.3-42.2		16.3-29.8	905-1758	Combined sewer	Shafer et al., 1998
USA						96-578	47-1023	3581-372492					24-183			122-3073	Stormwater	Tuccillo, 2006
Poland				89-623			787-4407						193-724			4587-12232	Comined sewer	Chipasa, 2003

Table 4. Dissolved (D) and Particulate (P) concentration of trace elements entering the STP. Numbers in bracket represent the percentage of each form

Chemical Form	Cu(nM)	Ni(nM)	Cr(nM)	Pb(nM)	Mn(nM)	Fe(nM)	Cd(nM)	Zn(nM)	Co(nM)	Bibliography
D P	223 (18) 1020 (82)	10364 (79) 2755 (21)	62 (8) 708 (92)	11 (6) 177 (94)	830 (68) 389 (32)	2407 (28) 6189 (72)	4 (12) 26 (88)	504 (7) 6681 (93)		Karvelas et al., 2006
D P	267 (33) 551 (67)	154 (75) 51 (25)	40 (23) 133 (77)	24 (28) 62 (72)		1074 (9) 10744 (91)	4 (80) 1 (20)		14 (48) 15 (52)	Buzier et al., 2006
D P	79 (16) 425 (84)		115 (21) 423 (79)	62 (68) 29 (32)			5 (36) 10 (64)	657 (12) 4847 (88)		Toumi et al., 2003
D P	629 (13) 4249 (87)	426 (63) 255 (37)	385 (57) 289 (43)	386 (13) 2461 (87)			11 (73) 4 (27)	1529 (34) 2905 (66)		Goldstone et al., 1990 (a,b)
D P	3000-4000 (60) 2000-3000 (40)									Kunz an Jardim, 2000

3. Treatment of Waters

The degree of purification of the water achieved by each STP depends on the treatment applied. Normally a primary and a secondary treatment is applied to the water and a limited number of STP have the tertiary treatment because is only economicaly viable for large industrial units.

In the old days, the aim of the STP was to remove the organic matter and suspended solids from wastewater before its disposal. Removal of metals is passive in most treatment systems. In other words, the processes are designed to remove soluble organic matter, particulates, pathogens, and nutrients, and if some removal of metals also happens to occur so much the better (Boulay and Edwards, 2000). Each treatment is devoted to this aim in different way (Figure 3):

- Pre-treatment. Solid materials are removed
- Primary treatment (or mechanical). Removal of readily settleable solids and floatable materials.
- Secondary treatment (or biological). Biodegradable organic matter is reduced from the wastewater.
- Tertiary treatment (or chemical). The addition of organic compounds induces a precipitation of dissolved material.
- Sludge treatment. Gravitational and mechanical thickening of sludge.

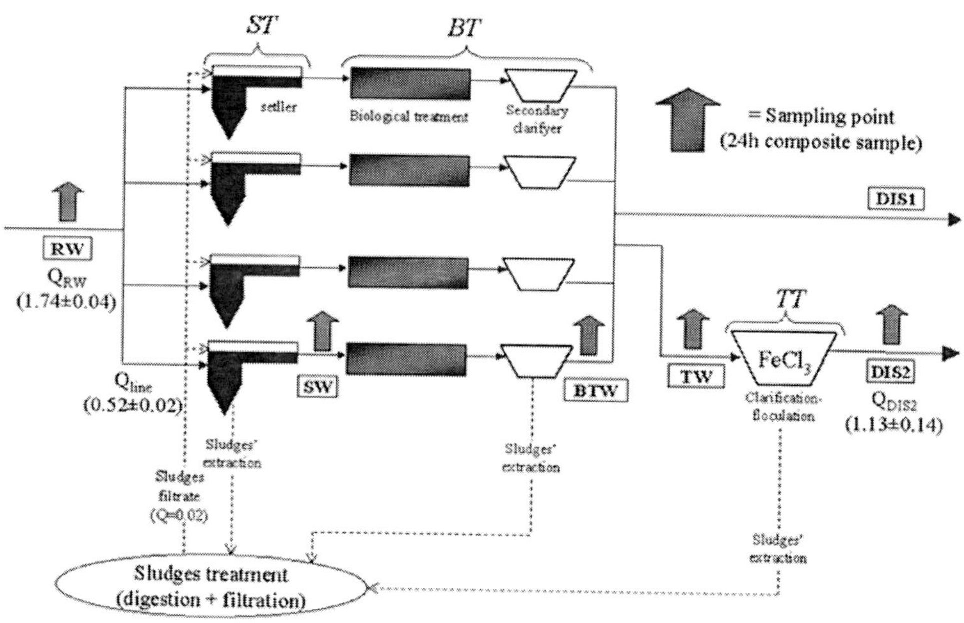

Figure 3. Scheme of the Seine-Awal STP with the Primary treatment (ST), secondary treatment (BT), tertiary treatment (TT) and sludges treatment. Lines indicate wastewater circulation and dashed lines sludge circulation. (Taken from Bouzier et al., 2006 with permission).

In the recent years, the removal of trace elements from wastewater during STP processes is becoming more important. The contamination and its effects (toxicity) when dumped to the environment is of major concern for the actual society than it was some years ago. During the next sections, each treatment will be deeply explained and the effects on trace elements removal of each step will be discussed and quantified.

3.1. Pre-Treatment

Although the removal of trace metals during this stage is scarce, a pre-treatment of all the waters arriving to the STP is necessary to gain a better efficiency of metal removal during the subsequent stages. During this treatment, solid materials like debris, gravel and sand are removed (Chipasa, 2006). In addition, an initial mixing of the waters occurs before passing to the primary treatment. Just after this pre-treatment, waters will enter the STP to be treated.

Metals present in a particulate form concentrate in the smallest fraction ($\leq 63\mu m$) due to their large surface area. As this pre-treatment removes mainly big particles, the metal removal during this pre-treatment will be very low and limited to those present in the particulate forms and associated to big particles. The amount of metals and its speciation is going to be very variable. Metals in the raw waters can be in different forms (trully dissolved, organically and inorganically complexed, colloidal, particulate) and depending on the chemistry of each metal, the predominant species will be different (Tables 4, 5 and 6).

Table 5. Dissolved metal speciation of trace elements entering the STP

Chemical Form	Cu(nM)	Ni(nM)	Bibliography
Inert	236 (88)	60 (39)	Buzier et al.,2006
Labile	31 (12)	94 (61)	
Dissolved	3000	4000	Cantwell et al., 1982
Ligand	53000	10000	Kunkel and Manahan., 1973
Dissolved	5-110	30-80	Sedlak 1997
Ligand1	12-20	20-50	
Ligand2	10	10-30	
Dissolved	3000-4000		Kunz and Jardim., 2000
Ligand	5000-10000		

Table 6. Particulate metal speciation of trace elements entering the STP by applicatipon of the Tessier sequential extraction method (Tessier et al., 1979)

Chemical Form	Cu(nM)	Zn(nM)	Bibliography
Exchangeable	0-8 %	5-18%	Morita and Tsuboi, 2000
Adsorbed	0-18%	0-12%	
Organically bound	5-20%	10-40%	
Carbonate precipitated	20-100%	10-40%	
Sulfide precipitated	40-100%	0-30%	
Residual	0-50%	2-6%	

Metal speciation in raw sewage is very different from that in the natural waters (Stumm and Bilinski, 1973). Most of the soluble metal is complexed by soluble organic ligands and only small quantities of these metals would be in a free ionic form. Copper, nickel and cobalt complexation capacities exceed exceed the typical total concentrations of these metals while complexation capacities for cadmium and lead have been found to be very low (Lewin and Rowell., 1973; Laxen and Harrison, 1981)

3.2. Primary Treatment

The wastewaters enter the STP once they have been pre-treated and mixed. From this moment, the primary treatment is applied to the waters. It mainly consists on the removal of readily settleable solids and floatable materials that have not been removed during the pre-treatment. A secondary function of the process is to reduce the pollution load on the succeeding biological treatment process (Chipasa, 2006). Metal removal during primary sedimentation depends upon the settling of precipitated metals or the association of metals with settleable particulate matter (Goldstone at al., 1990a). Therefore, optimization of metal removal efficiency in the primary clarifier is dependent upon optimization of suspended solids removal. Any factor that affects metal solubilities and settling rates is likely to have a significant influence on metal removal efficiencies (Lester, 1983a).

Metal removal during primary treatment may be attributed to two different processes occurring between the metal and the particles which are:

- Adsorption. Metals are attracted by the suspended solids surface without entering the solids. The removal of Ag, Cd, Co, Cu, Cr, Mn, Ni and Tl were controlled by this mechanism.
- Precipitation. Formation of a chemical precipitate. This was the main removal mechanism for some metals like Bi and Pb

Some metals like Zn and Mo are influenced by both factors (Kempton et al, 1987a; Goldstone et al., 1990b).

As the only way of remove metals during this treatment is by means of its association to particulate material, various studies have investigated the main factors that affect solid removal. The principal determinants of metal removal during primary sedimentation are the efficiency of suspended solids removal and the chemical species of the metal (soluble or insoluble). In addition, these two factors are influenced by several others (Lester, 1983a).

(a) Suspended solids (SS) removal is affected by:

- Design surface loading. It has an adverse effect on suspended solids removal efficiency (Tebbutt and Christoulas, 1975), which is reflected in lower metal removals (Rossin et al., 1983a). However, certain metals are more affected than others, possibly due to their preferential association with different suspended solids fractions (Kempton et al., 1987b). Chen et al., (1974) showed that Cd, Cu, Cr and Zn had a tendency to bind with particles greater than 8 μm in diameter, but Mn, Ni and Pb tended to associate with smaller particles. The unequal association of metals with different particle fraction is a major factor controlling their removal (Kempton et al., 1987b).

- Flow, dry weather or storm. Suspended solids removal efficiencies tended to decrease with an increasing hydraulic loading or flow rates (Kempton et al., 1987a; Rossin et al., 1983a).
- Influent SS concentration. The solids removal efficiencies increased with an increasing influent solids concentration or loading. Several metals (Ag, Cd, Cr, Cu and Tl) appear to be influenced by solids concentration (Kempton et al., 1987b) because higher SS concentrations reduce metal solubilities (Rossin et al., 1983b). This seems to be the most significant factor controlling metal solubility and hence the potential for removal (Kempton et al, 1987a).

(b) Chemical species of metal may depend on:

- Influent metal concentration. An experiment carried out by Kempton et al., 1987a showed that a relatively constant fraction of each metal was in the soluble form irrespective of the total concentration.
- Influent COD. The soluble metal fraction did not appear to be markedly affected by COD, although in the case of copper an increase in COD was accompanied by increasing filterable metal concentration. These data suggest that soluble organic ligands may play a minor role in determining the proportion of metal in the soluble form in raw sewage being the solid phase the dominant controlling factor (Kempton et al., 1987a).
- Hardness, alkalinity and pH of the wastewater.

As a result the removal of trace metals during primary sedimentation varies widely (Table 7).

The fairly wide variations in trace metal removal efficiencies suggest that the operating parameters for primary sedimentation can significantly influence trace metal removal (Kempton et al., 1987a). Studies have shown that, even under similar operating conditions in different STPs, removal efficiencies may vary considerabily (Lester et al., 1979; Stoveland et al., 1979). Even if the studies are carried out in the same treatment plant, the removal percentages can vary between 14-47% (Lester, 1983a) from one day to another.

As it has been pointed out before, the chemical speciation of a metal is going to be a main factor on its removal. The chemical form of the metal in the original discharge and its subsequent speciation in the sewage matrix are important in determining removal (Goldstone et al., 1990a). When the metals enter the primary treatment, they can be present in different chemical forms. Metals can be associated to suspended solids (particulate metals) or dissolved in the sewage matrix (dissolved metals). Between the latter ones, we can find free metal ions, inorganic complexes, organic complexes, bound to high molecular weight organic material, highly dispersed colloids, adsorbed onto colloids and precipitates on organic particles (Stumm and Bilinski, 1973). The matrix of the water present in the STP is going to be a crucial factor in the speciation and behaviour of metals (Lester et el., 1979). Not all the metals behave in the same way. Regarding the distribution of a metal between the dissolved/particulate forms the dissociation constant of a metal is going to be very important for its removal during this primary sedimentation.

Table 7. Precentage of metal removal after the application of the primary treatment in different STP. The last column indicates the average removal for each metal

	% removal Primary treatment												
Cd	60	25		30	43	72	41		43	30	39	33	42
Cr	55	36	28	40		51	55	17		38	68	56	44
Co	50											33	41
Cu	33	70	45	40	61	71	39		61	27	60	40	50
Fe	49									6		32	19
Hg	57	54											56
Mn	33				67					61			47
Pb	66	59	40	50		73	50		67	20		72	56
Ni	15		21	35		23	43	1		22	50	33	27
Zn	54	68	41	35		74	64	17		19	44		46
	(1)	(2)	(3)	(4)	(5)	(6)	(7)	(8)	(9)	(10)	(11)	(12)	(13)

1. Oliver and Cosgrove, 1974
2. Brown et al., 1973
3. Stones, 1977
4. Roberts et al., 1977
5. Chen et al., 1974
6. Lester et al., 1979/Stoveland et al., 1979
7. Goldstone et al., 1990 (a,b)
8. Davis and Jacknow, 1975
9. Lewin and Rowel, 1973
10. Karvelas et al., 2003
11. Nielsen and Hrudey, 1983
12. Buzier et al., 2006
13. Average

This conditional distribution coefficient, known as K_D (L kg^{-1}) is a proportion between the particulate and dissolved concentration of a given metal:

$$K_D = [Mep]/[Med]$$

where, [Mep] = metal concentration in SPM in mol kg^{-1} and [Med] = dissolved metal concentration in mol L^{-1}.

The usual expression of this coefficient is by means of its log K_D, thus, when the log K_D of an element is negative or close to 0, the dissolved form will be the predominant. Positive and higher log K_D values will represent metals with a major affinity for the particulate phase. Table 8 indicates typical Kd values for a number of elements.

Regarding the metal removal during primary treatment, metals with a negative or close to 0 log Kd will be removed in a lower proportion. This can be deduced from a combination of table 7 and 8. Ni in table 7 is the metal removed in a less proportion comparing to the others and the log K_D value for this element is the most negative one.

The removal of trace metals during wastewater treatment is an advantageous consequence of the existence of settleable particles in the influent and conversion of soluble and non-settleable forms during treatment. Thus, the physical removal of settleable forms of trace metals is largely influenced by solids removal (Kempton et al., 1987b).

Table 8. Conditional distribution coefficients for some metals in different STP after the primary treatment

	log K_D				
Cd	0.6-4	3.87-4.5	-0.6	0.3	-0.44
Co			0.03		
Cr	0.3-3.6		0.52	0.57	-0.12
Cu	0.4-3	3.99-4.41	0.31	0.73	0.83
Fe	2.3-5		1		
Mn	-0.86-2.7				
Ni	-0.59-1.84		-0.48		-0.22
Pb	0.9-3.7	4.61-5.35	0.41	-0.33	0.80
Zn	0.9-3.2	3.64-4.42		0.87	0.28
	(1)	(2)	(3)	(4)	(5)

1. Karvelas et al., 2003
2. Cobelo-García et al., 2004
3. Buzier et al., 2006
4. Toumi et al., 2003
5. Goldstone et al., 1990 (a,b)

When talking about the dissolved metal speciation, the chemical forms during this treatment will be less important than during the second treatment because only particulate metals will be removed. The physicochemical forms of trace metals in a complex matrix as raw sewage may be expected to be influenced by soluble and solid-phase ligand concentration. The factors likely to affect the association of previously soluble metals with settleable solids may include the complexation capacity and stability of soluble organic complexes (Rudd et al., 1984), the relative tendencies of metals to form insoluble precipitates directly or to become entrapped by the primary sludge solids (Lester, 1983b) and the forms in which the metals enter the process. The major subfractions present in raw sewage with which trace metals may be associated are surface-bound organic ligands, insoluble salts and inorganic solids, and soluble organic ligands. Dissolved metals which have the capacity for binding to settleable solids will be removed. Soluble species of trace metals are removed with efficiencies consistently less than 1% (Oliver and Cosgrove, 1974). Therefore, any factor that promotes the formation of soluble species, such as the presence of soluble metal-binding agents (Rossin et al., 1983b) will have a direct adverse effect on removal. The oxidation state of a metal is also important at this point. The dominant dissolved species of chromium in raw sewage was the trivalent form (Jan and Young, 1978). This form was largely removed in sedimentation, leaving the hexavalent species in solution (Bailey et al., 1970).

Data about organic complexation of metals in settled sewage are very scarce. Table 9 shows the results reported by Kunkel and Manahan in 1973.

If we compare the data shown in this table 9 with the data reported by this author for a raw sewage and listed in table 5 of this chapter we can observe that both the levels of dissolved copper and complexation capacity have been reduced after primary sedimentation. This behaviour can be explained if the dissolved metal that has been removed was organically bound and by some of the reactions previously explained (organic ligands associated to particles) has been settled down with the suspended particles.

Table 9. Dissolved copper speciation after primary treatment (taken from Kunkel and Manahan, 1973)

Metal	Complexation capacity (nM)	Metal concentration (nM)	Bibliography
Cu	47000	1500	Kunkel and Manahan (1973)

Minimal removal of many soluble metals has been reported after the primary treatment and the proportion of soluble metal to total metal rises during primary sedimentation (Oliver and Cosgrove, 1974). This fact can be explained because metals associated to the particulate phase are mainly removed during this treatment. The next figure shows this phenomenon.

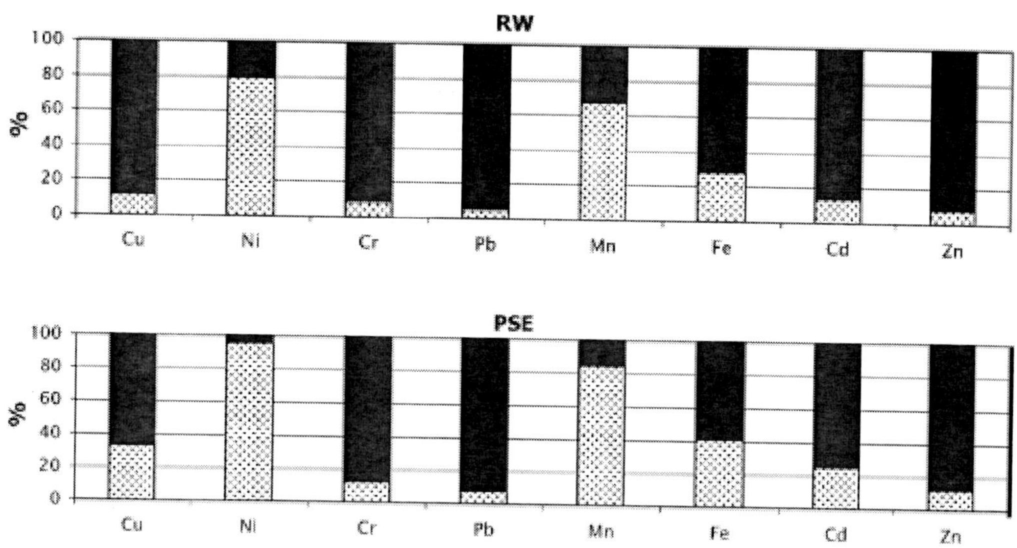

Figure 4. Distribution of trace metals in the dissolved (white) and particulate (black) phase of wastewater (RW: raw wastewater, PSE: primary sedimentation effluent). (Taken from Karvelas et al, 2003 with permission).

In some STPs, trying to optimize the trace metal removal in this primary treatment a reuse of sludges is applied. This procedure (shown in Figure 3 of a previous section) consists in the reutilization of the settled sludges of the primary sedimentation process or other processes (final sludge) to increase metal removal by its association with large particles from the raw sewage. Goldstone et al., 1990a and Buzier et al., 2006 showed that a STP with this type of treatment increased the removal of dissolved metals during this primary stage with respect to other plants in which the settlement of particles was the only procedure applied in order to remove metals (McDermott et al., 1962, 1965; Brown et al., 1973; Chen et al., 1974).

Metal removal during primary sedimentation is important for two reasons:

1. Reduces the metal loading and potential toxic effects of the metals of the biological stage or secondary treatment (Barth et al., 1965; Chipasa, 2006)

2. Contributes to the overall metal removal efficiency of the works carried out in the STP

After this process, the wastewaters will be treated in the consecutive processes (secondary, tertiary,...) while the settled particles that have formed the primary sludge will be separated and treated before its disposal.

3.3. Secondary Treatment

After the primary sedimentation, the wastewaters, free of the most part of the settleable particles enter the secondary treatment. It is a biological treatment process in which biodegradable organic matter that has not been removed during the primary treatment is reduced from the wastewater by means of micro-organism. Bacteria (biomass) are commonly used in this treatment. These bacteria are placed into aerated big tanks provided with oxygen to grow these bacteria which are fed by the dissolved organic matter. After this process, bacteria are separated from the waters by a new settlement process in the secondary clarifier (secondary decantation) (see Figure 3) (Boulay and Edwards, 2000).

We can find a large number of secondary treatments: activated sludge microorganisms (Nielsen and Hrudey., 1983; Karvelas et al., 2003; Chipasa, 2006; Buzier et al., 2006), biological filter plants (Stones, 1977; Goldstone et al., 1990a), high rate algal ponds (Toumi et al., 2003), and new biosorption techniques like fungal biosorption (Kapoor and Viraraghavan, 1995) or membrane bioreactor (Zhang et al., 2005) developed in the last years (Aksu., 2005). The removal efficiencies of two of these methods are compared in Table 10.

Table 10. Removal efficiency for selected metals as a function of the treatment process (taken from Laxen and Harrison, 1981 with permission)

	Removal (%)*		
Metal	Primary sedimentation (Lester *et al.*, 1979)	Activated sludge (Brown & Lester, 1979)	Biological filter (Stones, 1977)
Cd	72 (60–83)†	50 (11–80)‡	
Cu	70 (50–95)	71 (55–93)	59
Pb	73 (66–82)	67 (43–89)	70
Fe		86 (72–97)	62
Mn		17 (6–28)	

* The data from the different references are not strictly comparable.
† Mean and range for 4 values.
‡ Mean and range of averages from 3 to 6 different studies.

Depending on the metal, removal efficiencies are higher with a secondary treatment based on activated sludge process than biological filtration.

Secondary treatment processes can be divided into two types:

- Suspended biomass: Bacteria are suspended in the aquatic medium in contact with the organic pollution by means of agitation (activated sludge, continuous aeration, ...).
- Fixed biomass: Bacteria are fixed in a medium material placed in contact with the water and organic pollution (bacterial layer, bio-discs...).

The basis of these processes is the same (Lester, 1983a). An activated sludge plant may be divided into two essential parts: A biological reactor and a phase separator (Neufeld and Hermann, 1975). In the reactor, or mixed liquor aerator, large populations of micro-organisms grown under aerobic conditions, oxidise the dissolved or suspended organic compounds present in wastewater and convert them into carbon dioxide, water and cellular material. The bacteria and other micro-organisms form aggregates or flocs, a process known as flocculation (Lester, 1983a). Any substance that is adsorbed or absorbed by the bacterial flocs will be removed from the water passing through the activated sludge plant. The mixed liquor then passes into the separator or settling tank. The flocs separate out by sedimentation from the liquid phase and are returned to the reactor.

This secondary treatment was initially developed to remove the dissolved organic matter but a subsequent remove of metals occurs. A combination of flocculation and settling is, therefore, the mechanism by which metals are removed during this process (Goldstone et al., 1990a).

Removal in the activated sludge process is potentially of two types:

- Settlement of insoluble metal
- Adsorption of soluble metal or fine particulate metal by the sludge flocs

Several possible mechanisms for the removal of dissolved and fine particulate metal in activated sludge have been suggested (Brown and Lester, 1979). These are:

1. Physical trapping of precipitated metal in the sludge floc matrix. It is difficult to distinguish between precipitated metals which have settled independently but are removed with the sludge, and precipitated metals that have interacted with the floc matrix and become physically entrapped and settled with the flocs. It seems that both ways of removal are combined and depending on the metal, the preferred form will be one of these two. An experiment developed by Sterritt et al., 1981 showed that Cr(VI), Mn and Ni were mainly removed by adsorption onto the biological solids. Lead and Cr (III) were mainly removed by precipitation
2. Binding of soluble metal by the cell. Different species of bacteria isolated from activated sludge produce extracellular polymers (Wilkinson., 1958; Sutherland, 1972). These can be slime polymers which remain in the dissolved and colloidal phases of the effluent or capsules and microcapsules which remain attached to flocs and settle with the sludges. Several experiments (Morgan, 1961; Friedman and Dugan, 1968; Dugan and Pickrum, 1972) have shown that metals like Fe, Co, Ni, Al, Ca, K, Mg, Mn, Na, Ag, Cu, Cr, Li, Mo, Ti, Hg or Cd are removed by extracellular polymers produced by different species.

3. Accumulation of soluble metal by the cell. This phenomenon is carried out by activated sludge bacteria that are not able to produce capsules or rctracellular material. Metals are accumulated either in the cytoplasm or by adsorption onto the cell wall. Some works have outlined the capacity of removal of Cu and Pb by the cells (Freidman and Dugan, 1968; Peterson, 1971). The metals entering into the cell are transformed as a detoxification mechanism of bacteria. As an example, Hg, Pb and Se have been shown to be methylated by certain micro-organisms (Magos et al., 1964; Jernelov and Martin, 1975; McEntire and Neufeld, 1975; Wong et al., 1975).
4. Volatilization of metal to the atmosphere. Some groups of micro-organisms present in wastewaters are capable of volatilising Hg, Se and Te (Magos et al., 1964; Fleming and Alexander, 1972).

Any factor affecting the flocculation or settling properties of a mixed liquor, for example, loading rate, feed composition (Kiff, 1981), mixing strength, sludge volume index (Magara et al., 1976), will affect the capacity to remove metals (Lester, 1983a).

Many different factors have been shown to affect the removal of metals in the activated sludge process (Lester, 1983a):

- Operating parameters: These affect metal removal by influencing suspended solids removal or the affinity of the metal for the solid phase. Barth et al. (1965) found that the amount of removal of Cu, Zn, Ni and Cr was controlled by the concentration of dissolved oxygen present. Lead was found to be removed more efficiently due to the increased settling time in secondary treatment in comparison with primary treatment (Brown et al., 1973).
- Physical and chemical factors: They play a very important role in determining the affinity of the metal to the biological solids. The pH is one example of these factors. At high pH, considerable precipitation of metal hydroxides occurs, while at low pH there is much interaction between metals and organics (Cheng et al., 1975). The chemical form of the metal (free ion, complexed,..) will be an important factor in the removal during this secondary treatment as we will see below.
- Biological factors: Many workers have demonstrated (Bailey et al., 1970; Ghosh and Zugger, 1973; Cenci and Morozzi, 1977) that for most of the metals only a low concentration of the metal is required to produce a significant reduction in the treatment efficiency (Lester, 1983a). Depending on the biological community that forms the activated sludges, metal toxicity can affect to a different extent. Changes in the communities of organisms may be invoked by metal doses. The presence or absence of a particular species may be indicative of the toxicity of a metal dose. Deflocculation of sludge has been observed at all levels of Zn, Cd and Hg added (0-1000 mg l^{-1}). Severe shockloading was reported to have had ruinous effects in some activated sludge systems (Neufeld and Hermann, 1975; Neufeld, 1976; Oleszkiewicz and Sharma, 1989). It is important to note that the species of the metal present, among others like sludge age, suspended solids concentration and biological community (Guibaud et al., 2003) is an important factor in determining its toxicity.

In the next table (11), removal efficiencies of a number of trace metals after the secondary treatment are shown.

Table 11. Precentage of metal removal after the application of the secondary treatment in different STP. The last column indicates the average removal for each metal

	% removal Secondary treatment									
Cd	50	11		47	63	51	42	55	44	45
Cr	54	78	70	82	33	61	90	50	66	65
Co									39	39
Cu	60	61	55	77	79	43	86	58	72	66
Fe	55			88				21	72	59
Hg	55	17		72						48
Mn	6			28				72		35
Pb	79	43	65	53	73	60	87	31	78	63
Ni	1		45	41	61	9	51	44	31	35
Zn	50	48	60	60	78	57	61	43		57
	(1)	(2)	(3)	(4)	(5)	(6)	(7)	(8)	(9)	(10)

1) Oliver and Cosgrove, 1974
2) Brown et al., 1973
3) Roberts et al., 1977
4) Chen et al., 1974
5) Lester et al., 1979/Stoveland et al., 1979
6) Davis and Jacknow, 1975
7) Goldstone et al., 1990 (a,b)
8) Karvelas et al., 2003
9) Buzier et al., 2006
10) Average

The high variability between pecentages of removal during this secondary treatment (as it occurred for the primary one) are related to the factors that affect metal removal that have been explained before and that may be different among treatment plants. Oliver and Cosgrove (1974) discovered that much of the lead and iron in wastewater was converted into an insoluble form on reacting with municipal sewage; these metals were very efficiently removed. In contrast, Brown et al. (1973) found low cadmium removal efficiencies, and proposed that this was due in part to the high solubility of cadmium ions at pH 7 to 9 (the normal pH of the sewage treatment processes) and o the greater availability of other more concentrated metals which form stable, insoluble complexes with sewage sludges.

Most of the soluble metal in sewage is complexed by soluble organic ligands. The extent to which this may occur may be described by the complexation capacity of the matrix, which is defined as the concentration of metal which may be maintained in solution by soluble ligands (Lester., 1983a). Fulvic and humic acids are natural chelators which may be present in wastewaters (Schnitzer and Khan, 1972; DeWalle and Chian, 1974). Wastewaters contain many other complexes like EDTA that has many industrial uses (Mark et al., 1965) or Nitriloacetic acid (NTA) and sodium tripolyphosphate (STPP) which are used as detergent builders (Stoveland et al., 1979). An example of the speciation capacity of sewage waters after the activated sludge process is shown in table 12.

Table 12. Dissolved copper speciation after secondary treatment (taken from Kunkel and Manahan, 1973)

Metal	Complexation capacity (nM)	Metal concentration (nM)	Bibliography
Cu	14000	750	Kunkel and Manahan (1973)

As it occurred after primary treatment, both the levels of dissolved copper and complexation capacity have been reduced after secondary or activated sludge treatment. The speciation of the metals is going to play an important role during this secondary treatment because those metals where the soluble complexation capacity of the sample exceeds the total metal, significant removal in biological waste water treatment will depend on the dissociation of these complexes to form settleable species (Sterritt and Lester., 1984). Copper and nickel are two examples of this situation but nickel is poorly removed (1-45%) in the activated sludge process (Oliver and Cosgrove, 1974) whereas copper removals average about 50-60% (Lester, 1981; Rossin et al., 1983b) (see Table 11). This can be explained by the more stable binding of copper by the biomass present in the activated sludge process (Cheng et al., 1975). Some studies showed that free Cd represented between 35-41% of the total suggesting that the removal of this metal is largely due to the direct association with the activated sludge biomass. Rossin et al. (1983b) and Alder et al. (1990) indicated that the strong complexing capacity of the domestic or industrial chelators (EDTA, NTA) avoided the removal of metals by maintaining these in solution. At this point, the addition of $FeCl_3$ is used in order to improve trace metal removal complexed by EDTA (Poon and Chu., 1999; Ridge and Sedlak., 2004). Thus, metals are liberated from EDTA complexes because an exchange with Fe(III) occurs, resulting in an uncomplexed metal and $FeEDTA^-$. Another reason why metals are not efficiently removed can be the solubilization of metals (Chen et al., 1974; Stephenson and Lester., 1987; Brown et al., 1973; Buzier et al., 2006).

Apart from its organic complexation, the oxidation state of a metal will also affect its removal efficiency. As an example, Stoveland (1978) demonstrated that removal efficiencies of trivalentchromium were typically between 70 and 90%, while hexavalent chromium removals were typically 20%.

As it has been demonstrated by the previous examples, the speciation is going to be a crucial factor regarding the metal removals. Apart from that, the speciation of a metal is going to have a direct effect on the toxicity for the biomass. This fact can reduce the activity of the biomass present in the activated sludge and thus the efficiency in the removal of metals (Neufeld and Hermann, 1975; Neufeld, 1976; Oleszkiewicz and Sharma, 1989) leading to variation of removal percentages between different plants (Table 11).

It is mandatory to compare the removal efficiencies between the first and the secondary treatment processes. Most authors (Stones, 1958; Brown et al., 1973; Oliver and Cosgrove, 1974; Toumi et al, 2003) have addressed that the secondary treatment is more efficient than the first one for most of the metals (last column of tables 7 and 11). It has been proposed that the greater efficiency of the activated sludge process in metal removal is due to the greater efficiency of suspended solids removal and hence insoluble metal removal in the secondary sedimentation tanks than in primary sedimentation tanks (Brown et al., 1973; Chen et al., 1974).

The next figure (5) shows the partitioning of metals between the dissolved and particulate phases before (PSE) and after (SSE) the secondary treatment. It can be observed that with the exception of some metals like (Cr, Pb or Fe), the proportion of dissolved respect to total metal decreases (Goldstone et al 1990a; Goldstone et al., 1990b; Karvelas et al., 2003). This can be explained because during this secondary process, dissolved species apart from particulate ones are removed. On the contrary, during the primary treatment, metals in the particulate phase where mainly removed.

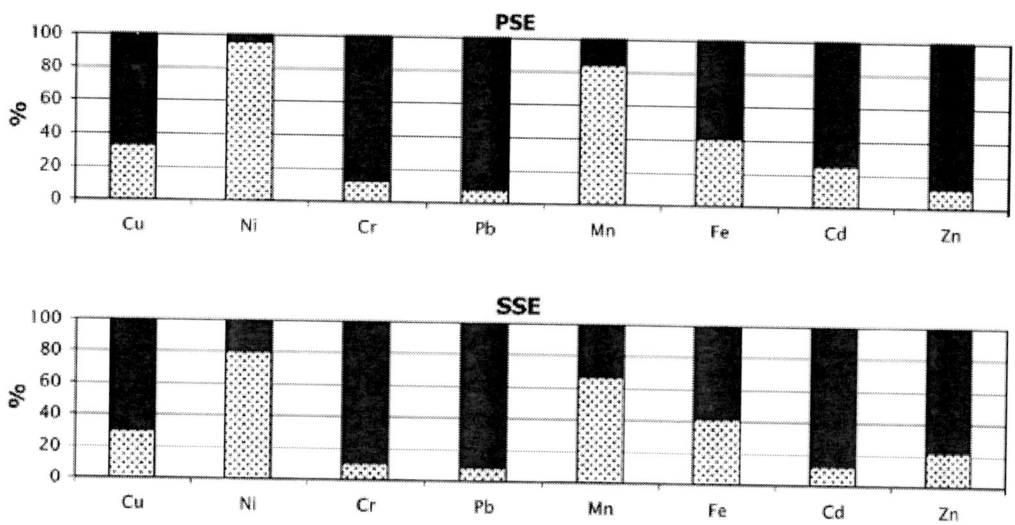

Figure 5. Distribution of trace metals in the dissolved (white) and particulate(black) phase of wastewater (PSE: primary sedimentation effluent, SSE: secondary sedimentation effluent). (Taken from Karvelas et al, 2003 with permission).

3.4. Tertiary Treatment

The secondary treatment was previously considered as a complete process, with its effluent being discharged into the receiving environment after disinfection with chlorine gas (Culp et al., 1978). However, as environmental regulations are getting stringent and introduction of EU Water Framework Directive in 2000 (S. Ntifo, Water Framework Directive, Water UK Policy, 2004. http://www.water.org.uk), secondary sewage effluent was no longer a guarantee for discharge (Henry and Heinke., 1989). Advanced tertiary treatment (physico-chemical process) is therefore, required for further decreasing the residual constituents in secondary sewage effluent (Wu et al., 2005) (Figure 3). Because of the high investment and operational costs involved, tertiary treatment is only applied in big treatment plants and only when the wastewaters are going to be reused in some way (irrigation, some industries…).

This treatment mainly consists in the coagulation and precipitation of pollutant and non-desirable products to make its removal easier. The efficiency on the removal of metals increases considerably.

A variety of technologies have been developed and applied for the treatment of secondary sewage effluents. The commonly used techniques include (Tchobanoglous et al., 2003; Bouzid et al., 2008):

Membrane filtrations with the aid of coagulants. This process is carried out by means of semipermeable membranes (ultrafiltration or reverse osmosis) or by electrodyalisis membranes after getting the pollutants coagulated. The effectiveness of these procedures respect to the removal of metals (like Cd, Co, Cr, Cu, Fe, Ni, Zn etc…) has been tested by many authors (Klischenko et al., 1999; Alonso et al., 2001). An alternative method is the addition of a macroligand that complexes the trace metal in order to increase their molecular weight with a size larger than the pores of the selected membrane that can be retained and which flow out in the retentate; permeate water is then purified from the trace metals (Bodzek et al., 1999; Korus et al., 1999; Bayhan et al., 2001). The macroligands used can be from natural origin like sodium alginate and polysaccharide extracted from brown seaweed (Benbrahim et al., 1998) or cell aggregates (Bayhan et al., 2001) or synthetic macroligands such as carboxyl methyl-cellulose (Petrov et al., 2002; Petrox and Nenov., 2004), polyvinylethylenimin, polyvinyl alcohol, etc.). All this macroligands possess a good complexation ability towards Cu^{2+}, Cd^{2+}, Co^{2+}, Zn^{2+}, Ni^{2+} and Pb^{2+}. The aggregates formed with the macroligands are small and loosely bound; their sedimentation velocities are relatively low. Thus, a process of filtration is also applied to accomplish the objective of removal. A similar thing occurs with the coagulation in which destabilization of colloidal suspensions occurs by neutralizing the forces that keep the suspended particles separated (Rodriguez., 1996; Macarewicz, 2000).

Flocculation in liquid suspensions has been known for decades as a highly useful phenomenon (Bratby., 1980) to remove undesirable materials. In contrast to the process of coagulation, electric charge manipulation is not the dominant mechanism of action in flocculation. In fact, flocculation is possible without significant changes in the particle surface charges. Moreover, flocculation is much more effective than coagulation since the so-called flocs are larger and more strongly bound than the aggregates obtained by coagulation. Thus settlement of the flocs is very quick. Many types of polymers can be used in this process: phosphonomethylated-polyethyleneimine and calcium ions (Cu2+, Co2+, Zn2+, Ni2+, and Pb2+) (Navarro et al., 2003) or polycation-polyanion complex of PEI and its phosphonomethylated derivative (Navarro et al., 2005). The addition of $FeCl_3$ helps in the flocculation process by forming ferric hydroxide complexes of some metals such as Cu and Zn (Poon and Chu., 1999; Ridge and Sedlak., 2004; Buzier et al., 2006).

Chemical oxidation and precipitation. Precipitation of metals as insoluble hydroxides, carbonates or sulfides is widely practiced for treating industrial wastewaters prior to discharge to the environment. Conventional chemical precipitation typically adds a basic salt ($Ca(OH)_2$; Na(OH)) (Meunier et al., 2006), which forms insoluble compounds with dissolved impurities and, with settling providing the primary pollutant removal path. All the trace metals with the exception of chromium form extremely insoluble sulphide salts (Mosey et al., 1971). Thus, the addition of sulphide is also useful to reach the precipitation of some trace metals (Moller at al., 2004). Some metals like Cd, Cr, Cu, Ni, Pb and Zn are efficiently removed by this method (Moller at al., 2004; Maunier et al., 2006). Karthikeyan et al. (1997) showed that the addition of hydrous oxides of iron and aluminium reduced the concentrations of Cu in wastewaters.

Electrocoagulation. An improved alternative of chemical precipitation is electrocoagulation process, in which the liquid is not enriched with anions and salts content does not increase, compared to chemical metal precipitation (Cenkin and Belevstev., 1985). This contributes in producing metallic sludges which are compact using electrocoagulation compared to those generated by chemical precipitation (Persin and Rumeau., 1989) Electrolytic treatment is characterized by simple equipment, brief retention time and easy operation, which would also contribute to reduce the operating cost in large scale application (Rajeshwar and Ibanez., 1997; Wendt and Kreysa., 2001). This method has been evaluated by Golder et al. (2007) for Cr removal and by Heidmann and Calmano (2008) for Zn, Cu, Ni, Ag and Cr obtaining good results.

Activated carbon adsorption. The adsorption is the property that some materials have by which organic molecules extracted from the dissolved phase are fixed in their surface. Activated carbon adsorption has been held to be a particularly competitive and effective process in removing organic and inorganic constituents from the waste effluents (Huang and Vane., 1989; Nakhla et al., 1994; Chu and Hashim., 2000; Lima et al., 2004). However, activated carbon is less economically viable as an adsorbent due to the costly activation and regeneration of the spent carbon and disposal of regenerant wastes (Wu et al., 2005). As a result, over recent years there has been growing interest in using low-cost natural minerals for treating wastewater. Among these minerals is diatomite (Wu et al., 2005), zeolite (Curcovic et al., 1997, Ouki and Kavannagh., 1999), etc.

The use of fly-ash from the incineration is growing in the last year as a new method of adsorption alternative to activated carbon or minerals to remove trace metals like Cu(II) (Shih-Cheng Pan et al., 2003), Cd(II) and Ni(II) (Zhai et al., 2004), Na, K, Ca, Mg (Gascó et al., 2005) and Cu(II), Cd(II), Ni(II) (Bouzid et al., 2008a; Bouzid et al., 2008b). The calcinaed phosphate is also a new adsorbent used for metal removal in watewaters like Pb(II), Cu(II) and Zn(II) (Aklil et al., 2004).

Ion Exchange Resines. It consists in the substitution of one or more ions present in the wastewater by other ions associated to a solid phase fined grained (exchanger), without altering its physical structure. Resines are comonly used and there exist ion and anion exhangers. Because of its high cost, this process is only used when the removal of the pollutant is imponed by its toxicity or if a recovery of a high value product occurs (radioactive isotopes removal, decontamination of waters with mercury, removal of chromates and cyanides, recovery of gold, etc.). The capacity of metal removal (Cd, Cu, Co, Fe, Ni, Pb and Zn) by means of these techniques has been demonstrated in various studies (Bratek et al., 2002; Kolodynska, 2008).

Constructed wetland or wetpark is an artificial marsh or swamp, created for anthropogenic discharge such as wastewater, stormwater runoff or sewage treatment, and as habitat for wildlife, or for land reclamation after mining or other disturbance. Natural wetlands act as biofilters (Upadhyay., 2007), removing sediments and pollutants such as trace metals (Cr, Cd, Cu, Zn, Fe and Ni) from the water, and constructed wetlands can be designed to emulate these features.

The efficiency of the great amount and variety of methods previously described and much other similar that have not been cited due to the high quantity of existing methods are usually tested by means of laboratory experiments. This is useful to evaluate their viability and cost. Depending on this, methods will be applied to real sewage treatment plants but the most part of the sewage treatment plants are provided by a primary and secondary treatment.

Thus, the studies about metal removal efficiencies during tertiary treatment are very scarce. In the following table (13) we can find some examples:

Table 13. Precentage of metal removal after the application of the tertiary treatment in different STP. The last column indicates the average removal for each metal

	% removal Tertiary treatment			
Cd	55		55	55
Cr	91	40	63	65
Co	38			38
Cu	86	84	86	86
Mn	87			87
Pb	82		35	59
Ni	24			24
Zn	39		80	60
	(1)	(2)	(3)	(4)

1. Shafer et al., 1998
2. Buzier et al., 2006
3. Toumi et al., 2003
4. Average

4. Global Vision of the Treatments and Effluent Discharge

Sewage treatment results in the production of a final effluent suitable for discharge to the selected receiving water or reused for irrigation (in most of the cases, a tertiary treatment is mandatory for this last use), and one or more sludges which may require treatment prior to disposal (Lester et al., 1983b). Next Figure represents the percentage of metal released via treated effluent or sludge (primary and activated). Data have been taken from Karvelas et al., 2003 where trace metal fate and transport was studied in a Greek STP. From Figure 6 we can conclude that Fe is mainly released via effluent while Cu and Mn are released via sludge. The rest of the elements (Cr, Pb, Ni, Cd and Zn) are released in the same percentage via both ways. This partition should be related to the partition coefficients of metals (K_d) between the solid phase and the dissolved phase, and the percentages of each metal removed during the wastewater treatment.

These two figures (7 and 8) corroborate that the metals mainly released via sludge (Cu and Mn) present high removal efficiencies and high K_d values while metals released via the effluent (Cd, Ni and Zn) present lower removal efficiencies and K_d. The different treatments applied (primary, secondary or tertiary) to the waste waters reduce trace metal concentrations. Santarsiero et al. (1998) reported that, primary sedimentation affected to a great degree the distribution of Zn, Pb and Cr, while biological processes and secondary sedimentation affected mostly the distribution of Cu, Cd and Ni. As an example we can see the evolution of metals through the different steps in a tertiary treatment plant type sited in the USA (Shafer et al., 1998) (Table 14).

Table 14. Metal removal in various stages of the waste treatment process: Jackson POTW (taken from Shafer et al., 1998 with permission)

	Al	V	Cr	Mn	Fe	Co	Ni	Cu	Zn	As	Mo	Ag	Cd	Pb
Influent	39992	16.2	182	879	34169	6.1	85	1430	1728	73	109	858	2.4	33
First clarifier	39990	14.4	14	1645	16726	10.4	75	233	1789	48	95	97	0.8	23
	(0.1)	(11.5)	(92.6)	(-87.3)	(51.0)	(-72.3)	(12)	(83.7)	(-3.7)	(34.7)	(13.4)	(88.7)	(66.4)	(30.5)
Second clarifier	16530	15	7.4	595	8363	8.9	68	305	1804	36	102	24	1	15
	(58.6)	(-3.7)	(3.4)	(119)	(24.5)	(24.9)	(8.6)	(-5.1)	(-0.9)	(15.9)	(-6.9)	(8.6)	(-8.3)	(24.3)
Final effluent	8487	13.8	5.8	258	7647	6.6	67	280	1926	35	104	8	1.1	12
	(20.1)	(7.1)	(0.8)	(38.5)	(2.1)	(37.8)	(0.5)	(1.8)	(-6.9)	(1.3)	(-1.2)	(1.8)	(-2.4)	(6.8)
Total percent removal	78.8	14.9	96.8	70.7	77.6	(-9.5)	21.1	80.4	-11.5	(51.9)	5.2	99.1	55.7	61.6

Total concentrations in nM. Percent removal of influent metal in parentheses.

Table 15. Trace metal levels in the final effluent of different STP. One of the columns indicates the degree of treatment applied to the wastewaters

								Metals (nM)									
Location	Ag	Al	As	Cd	Co	Cr	Cu	Fe	Hg	Mn	Mo	Ni	Pb	V	Zn	Treatment	Bibliography
U.K.				9-18		385	16-170					170-340	5-135		900-2000	Secondary	Lester et al., 1979
Canada				3		346	190					715			3060	Secondary	Nielsen and Hudrey, 1983
U.K.				5.3		192	1574					392	121		1529	Secondary	Goldstone et al., 1990 (a,b)
USA	0.3-52	333-8500	17-63	0.6-3.3	5-15	6	94-300			78-260	80-140	70-110	3-12	14	275-1900	Tertiary	Shafer et al., 1998
Greece				13		385	520	6800		346		7300	130		4100	Secondary	Karvelas et al., 2003
Morocco				6-7		190-210	63-79						58		1100-1160	Tertiary	Toumi et al., 2003
Paris				2	19	38	126					187	5			Tertiary	Buzier et al., 2006

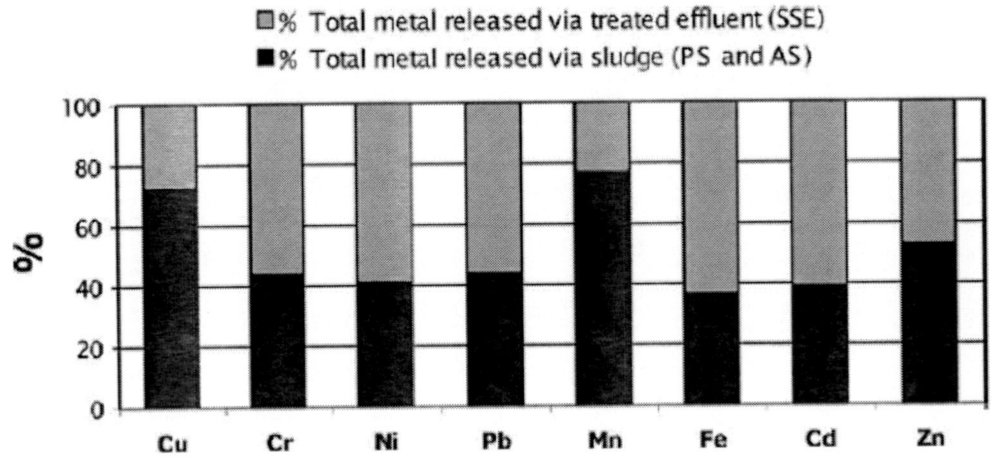

Figure 6. Release of trace metals from the STP via the sludge and the treated effluent streams (taken from Karvelas, 2003 with permission).

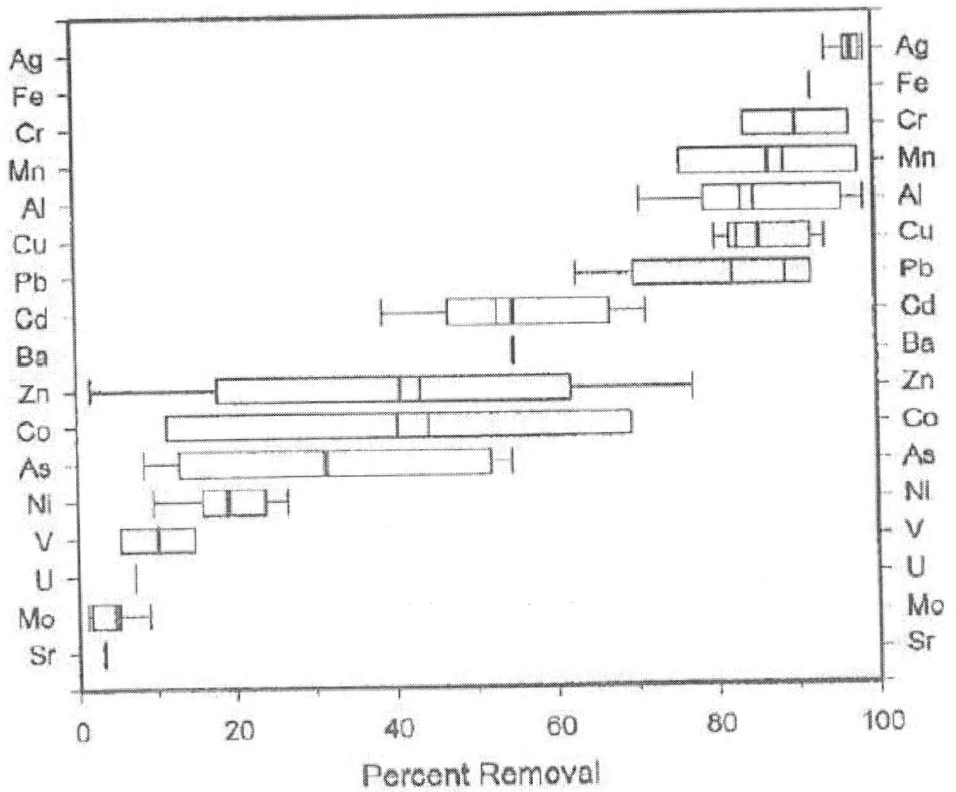

Figure 7. Percent removal of metals in municipal wastewater treatment facilities. Box caps represent observed minimum and maximum values. Heavy line in center of box is mean value (Taken from Shafer et al., 1998 with permission).

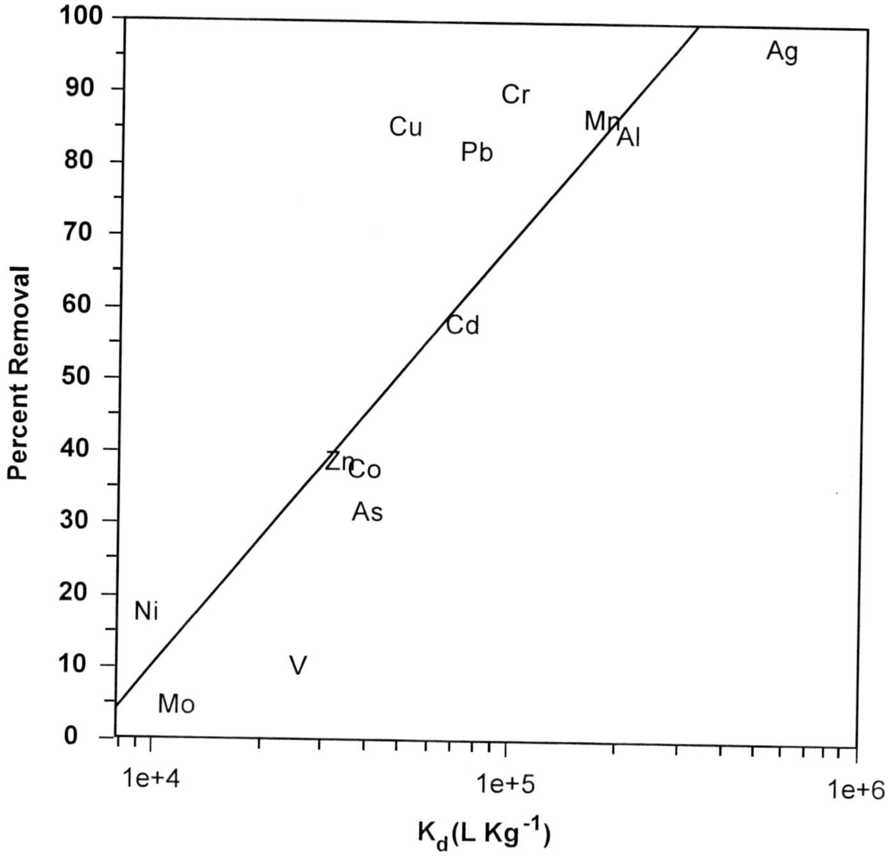

Figure 8. Plot of mean percent removal of metal in wastewater treatment facilities versus mean metal particle partition coefficient in the effluents. Data reflects a 0.4 μm dissolve-particulate phase separation.

From the standpoint of water quality, it is fortunate that the majority of metals entering a STP are concentrated in the sludge. However, significant quantities of trace metals may be discharged to receiving environments (Rossin et al., 1983a). Typical metal concentrations outgoing from different treatment plants are summarized in table (15).

Looking at the ¨treated waters¨, these are usually spilled to watercourses like rivers, estuaries or oceans or reused for irrigation. At this point, the speciation of a metal, rather than its total concentration, is the key to understanding its toxic effects on the biota (Kelly, 1988).

Regarding the percentages of dissolved and particulate metals in the effluent waters, the ratio between dissolved and total metal increases after the treatments (Oliver and Cosgrove., 1974). The following table (16), taken from the data published in Karvelas et al., (2003) agrees with the previous statement.

These results indicate that the sewage treatment processes removes mainly the particulate material as we have explained in previous sections.

With respect to the dissolved phase, free metal ions are relatively scarce in sewage matrices and that soluble and surface bound organic ligands represent the predominant species in many instances. The poor removal efficiencies of some metals in STP may be due to the formation of complexes with soluble ligands which are much more stable than solid phase complexes (Sterritt and Lester., 1984).

Table 16. Trace metal ratios between the dissolved and total concentrations in raw water (RW) and final effluent (FE). Data were taken from Karvelas et al., 2003

	RW (dissolved/total)	FE (dissolved/total)
Cu	0.18	0.3
Cr	0.08	0.1
Ni	0.79	0.8
Pb	0.06	0.08
Mn	0.68	0.65
Fe	0.28	0.4
Cd	0.12	0.1
Zn	0.07	0.22

Cantwell et al., 1982, could not find more than 1% of free Ni in their sewage samples and the same case was for Cu (0.8-3.2%) in studies carried out by Bender et al. 1970 and Patterson 1979. These values are indicating that the sewage water has a large complexation capacity and contains relatively stable complexes.

Few data are available on metals complexation capacities in treated wastewater (Buckley, 1994; van Veen et al., 2002), but the determination of metal speciation in real samples was only scarcely undertaken (Kaplan et al., 1987; Alvarez et al., 1998; Santos-Echeandia et al., 2008). The significance of metal speciation during the treatment process has been addressed by means of laboratory studies: Nielsen et al. (1984) showed that the dissolved Ni was expected to be removed through the reactive free metal species, in activated sludge process, while Stoveland et al. (1979) had shown that this removal is altered by metal complexation. However, direct information on metals speciation during the treatment process is definitely lacking, which does not allow to understand the mechanisms leading to the metal speciation in the treated effluent. A few of the data existing about organic speciation in the effluent of the wastewaters are presented in the next table (17):

Table 17. Dissolved trace metal speciation in the effluent of different STP

Metal	Metal (nM)	CC (nM)	Reference
Cu	1000	1000-9000	Lewin and Rowell., 1973; Laxen and Harrison., 1981b; Patterson., 1979
Co	500	4100	Hanck and Dillard., 1977
Pb	20	20-77	Lewin and Rowell., 1973; Laxen and Harrison., 1981b
Ni	67-160	475-860	Sedlak and Phinney., 1997
Cu	27-150	360-560	Sedlak and Phinney., 1997
Cu	60-310	450-5600	Santos-Echeandia et al., 2008

Many directives have been issued in the last years in order to limit dangerous materials (i.e. trace metals) in the effluent waters disposed into the aquatic environment:

- Department of the environment, The implementation of Directive 76/464/EEC on pollution cause by certain dangerous substances discharged into the aquatic environment of the community, DOE Circular, HMSO, London, 1985.

- S. Ntifo, Water Framework Directive, Water UK Policy, 2004. http://www.water.org.uk.
- Council of European Communities. Directive concerning urban waste water treatment (91/271/ECC). Official Journal L 135/40, 30 May 1991.

But none of them make reference to the different species of metals and only total values are indicated.

5. Sludge Treatment and Disposal

Apart from the treated water, sludge is the other final product originated in the STP after the treatment processes.

As a result of rapid economy growth and development of the countries in the last decades, a large number of new wastewater treatment plants have been developed. The final products of these STP are millions of tonnes of dry sludge. With the adoption of the urban waste water treatment Directive (COUNCIL OF EUROPEAN COMMUNITIES (91/271/EEC)) more sludge has been produced in the last years. For example, the annual production of sludge during 1980 in the U.K. was approximately 1.3 million tonnes (dry solids) (Department of the Environment and National Water Council,1981). In Germany about 1.7 million tonnes were produced annually during the 80s, while in the United Status the 4.5 million tonnes produced in the 80s were doubled during the 90s (Bunch, 1982). Nowadays, more than 50,000 wastewater treatment plants are operating in the European Union, producing more than 8.1 millions tons of dry solids per year (Magarou et al., 1999 de Erikson, 2008). Also, one disposal route (i.e. dumping at sea) ceased in 1998, placing increasing pressures on remaining disposal routes (namely agriculture, landfill and incineration). Therefore, the disposal of the sludge generated from urban wastewater treatment is becoming a crucial environmental problem (Tyagi and Couillard., 1989; Solis et al., 2002; Fuentes et al., 2004) and it must be treated for two main reasons:

- To reduce the volume before its disposal. The transport and disposal of sludges is an expensive process (Lester et al., 1983b).
- To reduce trace metal levels or stabilize them into less bioavailable species. This is mandatory to avoid an environmental damage when sludges are disposed. If sludge is to be used, its pollutants must be minimized in order to avoid risk to health or the environment (Sórme and Lagerkvist, 2002; Directive on the Protection of the environment, and in particular of soil, when sewage sludge is used in agriculture (86/278/EEC). Official Journal L187/6, 4 July 1986).

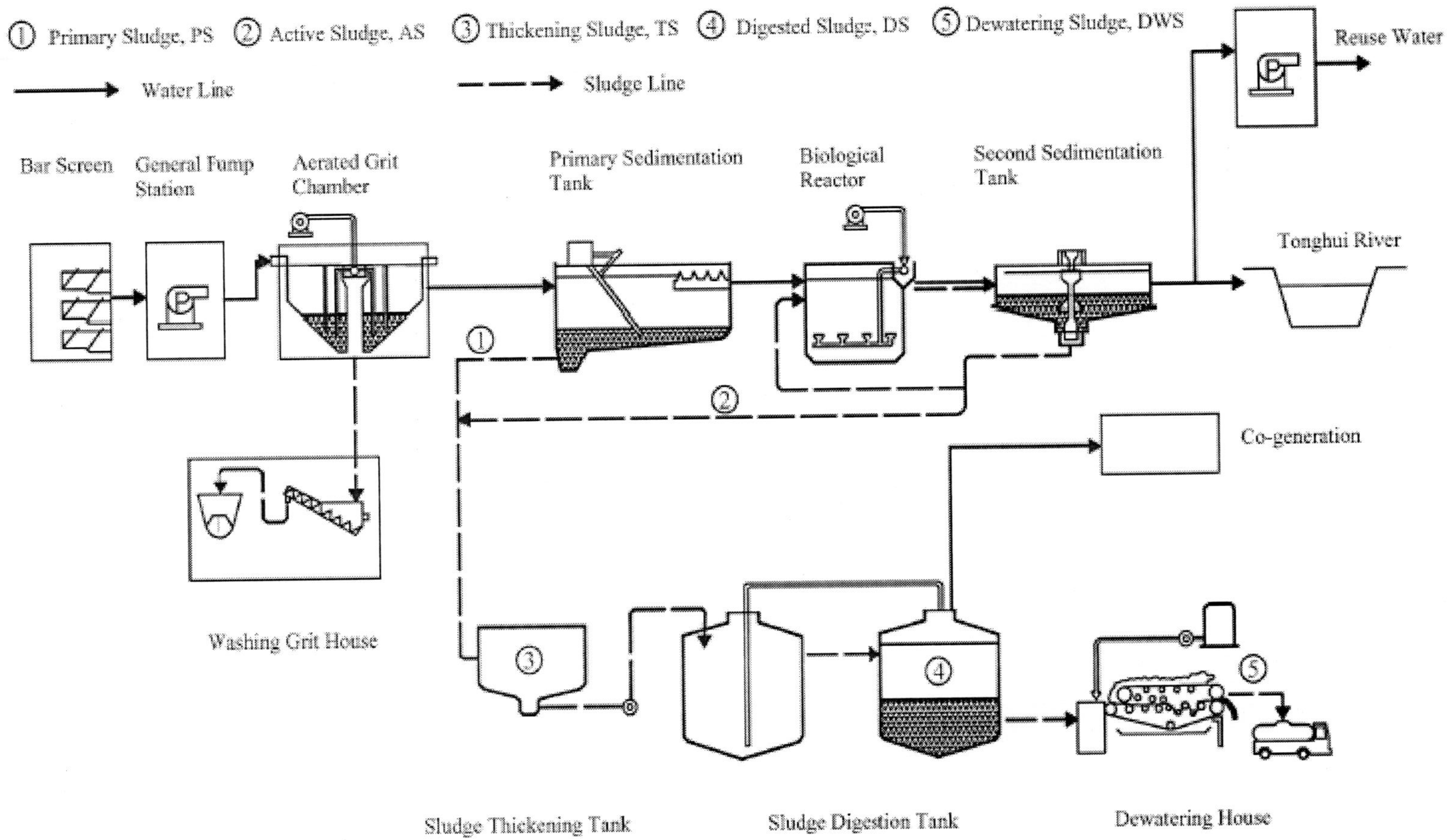

Figure 7. Schematic diagram of Gaobedian wastewater treatment plant (taken from Wang et al., 2006 with permission).

Table 18. Concentrations of trace metals in sewage sludges from various STP (µg/g) (taken from Karvelas et al., 2003 with permission)

Cd	Cr	Cu	Fe	Pb	Ni	Zn	Mn	Reference
4	128	147	46 000	155	34	1878	308	Voutsa et al. (1996)
2.1		370		72	19	1500		Lazzari et al. (2000)
55.7	1228	755	58 328	930	260	12064	3165	Bodzek et al. (1997)
2.78	841	436	22 445	128	622	2049		Scancar et al. (2000)
1.81	60.5			35.9	43.5			Baralikiewicz and Siepak (2001)
	100.1	315.5		54.2	11.9	679.9		Perez et al. (2001)
		1403		412	967			Cornu et al. (2001)
2	552	258	5098	326	41	1739	150	Zorpas et al. (2000)
		934		118	221	1574		Bragato et al. (1998)
1.82	34	287		69		933		Madoni et al. (1999)
		167		75		254		Planquart et al. (1999)
9.2	439	326	16 794	223	36.5	1636	364	Alonso et al. (2002)
3.19	322	239		126.8	317	2451		Abdel-Shafy et al. (1996)

The following processes are applied to the sludges in order to reduce its volume and stabilize the dangerous chemical products (i.e. trace metals) coming from the treated wastewaters (Figure 7):

(a) Thickening by stirring or flotation (gravitational and mechanical)
(b) Digestion, aerobically or anaerobically. The most widely used sludge treatment process is the anaerobic one (Lester et al., 1983b; Manoli and Samara., 1999) which involves microbial decomposition of organic and inorganic matter (Chipasa,2003).
(c) Chemical conditioning with either organic or inorganic materials
(d) Dewatering on drying beds, in filter presses, by vacuum filtration or centrifugation. This step can reduce the original weight of the sludge the dried sludge between (60-80%) (Chipasa., 2003)

Sludges are rich in trace metals as a result of its removal from the wastewater (see Table 18). Thus, the treatment and disposal of trace metal contamined sludges is not without its attendant risks. The stability/immobility of trace metals during sludge treatment and the disposal site are critical in determining the safety of the disposal operation (Lester et al., 1983b).

From the previous table we can state that levels are different between treatment plants for a given metal. This is related to the origin of the waters treated (domestic, industrial or storm water) and the degree of treatment of the waters. Regarding this last factor, trace metal levels in primary sludge (PS) will be different from activated sludge (AS), thickening sludge (TS), digested sludge or dewatering sludge (DS) (see Table 19).

Table 19. Total metal contents in Gaobedian STP system (mg/kg, in dry matter) (taken from Wang et al., 2006 with permission)

Samples	Fe	Mn	Cu	Ni	Zn	Pb	Cr	Mo
PS	16980.0	243.0	58.0	52.6	710.9	35.3	64.7	29.9
AS	12457.6	214.5	67.5	49.9	595.6	28.1	60.0	18.6
TS	16995.7	240.3	55.2	99.2	678.6	33.1	124.7	33.8
DS	18236.5	260.5	58.7	55.3	736.3	41.1	75.0	35.7
DWS	18208.2	257.6	59.6	69.8	746.5	40.2	91.0	32.2
pH<6.5[a]	/[b]	/	800	100	2000	300	600	/
pH ≥ 6.5[a]	/	/	1500	200	3000	1000	1000	/

[a] From "Discharge standard of pollutants for municipal wastewater treatment plant", PR China (GB 18918-2002).
[b] "/" shows no limited values in GB 18918–2002.

Trace metals present in sewage sludge give cause for concern in four areas:

- Human health, when the sludge is applied to agricultural land
- Adverse effects on biological sludge treatment
- Adverse effect on the receiving environment
- Effects on surface and ground water quality when sludge is applied to land

Treatment of sludge is frequently influenced by the final disposal option selected. Sludges can have a number of management options or uses with its own advantages or disadvantages (Tchobanoglous and Burton., 1991):

- Agricultural. It was by far the most used disposal way of sludges in the past due to the potential values of organic matter, N, P, and other plant nutrients present on them that enriched the land (Zufiaurre et al., 1998; Fuentes et al., 2004). However, possible barriers on agricultural application of sewage sludge are worth of further investigations (Zhang and Wu., 2005; Chipasa 2006). One of major restrictive factors for agronomic application of sewage sludges is trace metals retained in them. These metals are non biodegradable and pose eco-toxicity in environment (Lester et al., 1983b). Table 20 shows maximum permissible concentration of metals in sewage sludge for agricultural land application.
- Landfill. If metal levels are high, the sludge is only suitable for reclamation of deteriorated and disturbed lands as well as industrial sludge landfills (Chipasa., 2006). In the next table (21) we can see that the content in metals of the sludges applied for landfill, are higher than the ones use for agriculture
- Incineration in multiple hearth or fluidised bed furnaces. Interest in sludge incineration is increasing because it offers three main advantages: volume reduction of up to 80–90%, destruction of organic micropollutants and pathogens, and energy recovery (Marani et al., 2003).
- Wet air oxidation. Wet air oxidation is an attractive treatment for waste streams which are too dilute to incinerate and too concentrated for biological treatment (Luck., 1999). It can be used for the treatment of the organic compounds in wastewaters and is a destructive wastewater technology based on the oxidation of pollutants at high temperature and high pressure in the liquid phase. Molecular oxygen dissolved in the wastewater reacts with the organic and inorganic pollutants (Genç et al., 2002).

The use of the sludge is going to be determined by the dangerous effects that it presents. For example trace metal concentrations and chemical forms. At present, it is widely accepted that the determination of total trace metal contents in sewage sludge do not provide sufficient information of its potential hazardous effects on environment because the mobility and eco-toxicity of trace metals depend strongly on their specific chemical forms or bindings (Su and Wong., 2004; Fuentes et al., 2004).

Consequently, the studies on trace metal concentrations in sewage sludge and their bioavailability received a great scientific attention in recent decades. With respect to the disposal of sewage sludge to land, the importance of trace metal speciation has already been recognized insofar that it influences bioavailability or potential mobility (Sterritt and Lester, 1980) and provide evidence on the suitability and feasibility of the sewage disposal (Wang et al., 2006).

Table 20. Maximum permissible concentration of metals in sewage sludge for agricultural land application (µg/g dry weight) (taken from Karvelas et al., 2003 with permission)

Country	Cd	Cr	Cu	Pb	Ni	Zn	Reference
Greece/EU	20–40		1000–1750	750–1200	300–400	2500–4000	Council Directive, 86/278/ EEC (1986)
Austria	3	250	500	250	100	1200	Voutsa et al. (1996)
Belgium	5	200	500	1000	100	1500	Voutsa et al. (1996)
Denmark	0.4			120	30		Voutsa et al. (1996)
Germany	10	900	800	900	200	2500	Abdel-Shafy et al. (1996)
France	20	1000	1000	800	200	3000	Planquart et al. (1999)
Luxem-bourg	1.5	100	100	150	50	400	Voutsa et al. (1996)
Holland	1	50	75	100	30	200	Cornu et al. (2001)
Spain	10	400	50	300	120	1100	Voutsa et al. (1996)
Italy	10	600	600	500	200	2500	Lazzari et al. (2000)
Slovenia	5	500	600	500	80	2000	Scancar et al. (2000)
USA	85	3000	4300	840	420	7500	EPA, CFR 503 (1994)

Table 21. Current international legislation concerning levels of trace metals (units are in µg/g dry weight) in sewage sludge and possible final disposition (taken from de Souza Pereira et al., 2005 with permission)

	Metal							
	Ni	Cu	Pb	Zn	Cr	Cd	Hg	Fe
Germany (Agriculture[a])	50	60	100	200	100	1.5	1.0	–
Germany (Landfill[b])	200	1200	1200	3000	1200	20	25	–
EUA (Agriculture)	210	750	150	1400	1500	20	8.0	–
Brazil (Agriculture[c])	25–40	91–190	80–100	470–1512	44–73	1–3	1.5–2.5	–

[a] US EPA/625/R-95/001, and (Müller, 1998, p. 66).

[b] Bundesrepublik Klärschlammverordnung (AbfKläV) of 25.06.1982 and Skrypski-Mäntele (1992, p. 5).

[c] Müller (1998).

The so called BCR sequential extraction method (as described in Ure et al., 1993) is commonly used to characterize the different fractions of metals in sludges (F1: exchangeable/acid soluble fraction; F2: reducible fraction; F3: Oxidizable fraction; F4: residual fraction). When metals are associated to F1, there is a potential risk because they can easily be released to the environment. On the other hand, metals associated to F4, are very stable and use to remain associated to the sludge. In the next figure (8) we can appreciate the different fractions of each metal (Fe, Mn, Cu, Zn, Ni, Cr, Pb and Mo) in the sludge and its evolution during sludge treatment.

From the figure we can see that some elements like Mn, Ni and Zn are associated to F1 while others like Fe, Cu, Cr, Pb and Mo are not present within this fraction. Thus, the first group of elements could represent a risk when sludge is disposed for agricultural use been released to the land and reaching some aquifers. It is important to notice that two of these elements (Ni and Mn) were poorly removed from the dissolved phase during the first treatment of the wastewaters (Table 7) and only when the secondary treatment was applied, they were more efficiently removed from the dissolved phase (Table 11).

Ni and Mn were probably complexed or absorbed to biota during the secondary treatment and will form the first fraction (F1) of the sludges. In addition, the active sludge (AS) had the highest percentage of the mobile and available fraction (F1 and F2) for most of the metals, similar to the results extracted from other studies (Melcer et al., 1992; Alvarez et al., 2002). Active sludge was formed mostly by micro organism floc, and possessed greater surface area than othe sludges.

The large surface area of AS was favoured by metal ions to form loose bindings on sludge floces. These loosely bound metals could be released into water fairly easily, and therefore they were recognized as the easily mobilized metals (Wang et al., 2006).

It is also important to notice that the percentages of Ni, Mn and Zn associated to F1 were decreasing with the treatments (PS, AS, TS, DS, DWS) indicating that the processes are beneficial to preserve the environment from a potential contamination of these metals.

Studies on the chemical fractionation of trace metals in sewage sludge have indicated that the water-soluble, exchangeable and carbonate-bound fractions are major carriers of Ni and Zn, organic matter and sulphides are important carriers of Cr and Cu, iron/manganese oxides and hydroxides are carriers of Pb, Zn and Ni, while a great amount of sludge Fe and Pb is associated with the crystallic phase of sludge particles (Angelidis and Gibbs, 1989; Scancar et al., 2000).

Toxic effects of trace metals may be ameliorated by adding sulphide and raising the pH (Mosey, 1976) and this would also have the effect of reducing metal mobility in the digested sludge produced. In the last years, new methods for reducing trace metal levels in sewage sludge or minimizing the mobility of the most available forms have been developed:

- Composting: Focus on the improvement of the composting process to minimize the mobility of trace metals using various additives is receiving more attention in the last years (Wong et al., 1997; Ho and Qiao, 1998; Fang and Wong, 1999; Zorpas et al., 2000; Chiang et al., 2001). Lime is considered as one of the common amendment materials to co-compost with sewage sludge as it plays a role in trace metal (Cu, Mn, Ni, Pb and Zn) immobilization. The lime amendment was significant in reducing the exchangeable and acid extractable Cu, Mn, Pb and Zn of lime-sludge compost

(Wong and Selvam, 2006). Samaras et al. (2008) used fly ash and lime for immobilizing some metals like Cu, Pb, Zn, Cd, Ni and Cr obtaining good results.

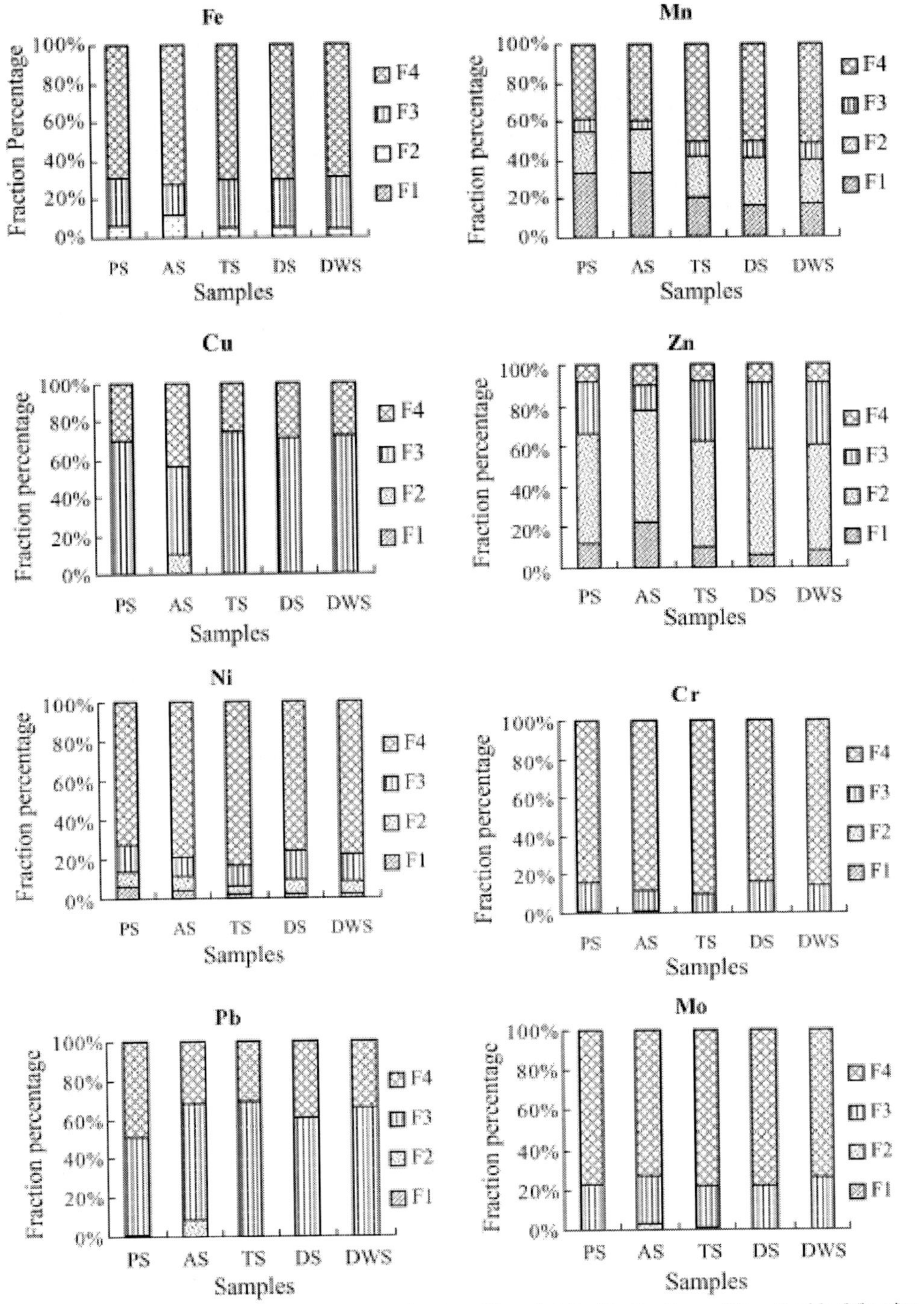

F1: Exchangeable fraction; F2: Reducible fraction; F3: Oxidizable-fraction; F4: Residual fraction.

Figure 8. The percentage distribution of each metal fraction in sludge samples.

- Vermicomposting: Various researchers have recommended that bio-composting of the sewage sludges should be done prior to their application in agricultural fields in order to get a stabilized product and to avoid potential risk of pathogens (Beloso et al., 1993; Gliotti et al., 1997). Use of earthworms for waste management, organic matter stabilization, soil detoxification and vermicompost production has been reported (Bansal and Kapoor, 2000; Kaushick and Garg, 2003). Gupta and Garg (2008) demonstrated that vermicomposting was a good method for stabilization of primary sludge.
- Bioleaching. An effective approach to reduce the toxicity of trace metals in sewage sludge is bioleaching, during which the trace metals can be dissolved under the acidic condition produced by *Thiobacillus* and then be removed through centrifugation or other methods to separate solid and liquid (Chen et al., 2005).
- Cement inmobilisation. Inorganic immobilisation uses cement to solidify certain waste materials (Felix et al., 1997). Katsioti et al. (2008) demonstrated that this method was good for stabilization of sewage sludge, containing trace metals such as Cu, Zn, Pb
- Zeolites. Physicochemical properties of the sewage sludge can be improved by an addition of e.g. diatomaceous earth, artificial zeolites and natural sorbents (clinoptylolite). Trace metals content in the sewage sludge decrease after addition of sorbents, this effect is typical for the immobilization of trace metals on the zeolite surface and pores (Erdem et al., 2004; Sprynskyy et al., 2007). Kosobucki et al. (2008) and Stylianou et al., (2008) showed good resuts for Cu, Pb, Zn, Cd, Ni, Mn and Cr.
- Electrokinetic removal. Electrokinetic technology is considered as one of the most promising methods to remove trace metals from the matter with low hydraulic permeability including contaminated soils and sludges (Acar et al., 1995; Kim et al., 2002). The advantages of electrokinetic technology used for treatment of sewage sludge are the simultaneous removal of different trace metals, high efficiency of removal and short time of the process. Wang et al. (2005) showed high percentages of metal (As, Cr, Ni, Cu, Pb and Zn) removal in sludges.

6. Effects on the Receiving Ambients

The disposal of both final effluent and sludges can have direct effects on the receiving ambients. Waters are normally disposed into a water body (i.e. rivers, lakes or estuaries) and sludges used in agriculture and landfills. The impact that contaminants may have on the receiving water body is dependent on the water use. Two of the major impacts can be on aquatic life in the receiving water body and downstream water users who may use the receiving water body as a raw water source for their drinking water (Makepeace et al., 1995). Large-sized STPs may be significant sources of trace metals to aquatic recipients, therefore it should be taken into account when guidelines for point-source discharges to the aquatic environment are set by local authorities (Karvelas et al., 2003).

Natural water levels of metals and sediment concentration are increased close to the final effluent of a STP (Helz et al., 1975; Bubb and Lester, 1995; Prego et al., 2008) and it is proved that the application of a tertiary treatment decreases these levels or concentrations (Blomqvist et al., 1992).

Moreover, the toxicity of trace metals in wastewater is shown to depend on factors like metal species and concentration, pH, sludge concentration, wastewater pollution load (Imai and Gloyna, 1990; Dilek and Yetis, 1992) and solubility of the metal ions (Surittanonta and Sherrod, 1981).

In the case of the sludges, the risk lies in the metals associated to the sludge that can be released to the ambient and be uptaken by plants or reach the aquifers.

7. Future and Perspectives

Although the regulations for trace metal containing waste disposal have been tightened (European Comission, 2002) it is still necessary to decrease the levels of metals in the final effluents or sludge. Sometimes, it is impossible to improve (mainly due to its cost) the existing sewage treatments to remove metals. In this case, it is important to remove or reduce their quantities in the origin. This involves a limitation of the emission from the main sources (Comber and Gunn, 1996; Aonghusa and Gray, 2002). Comber and Gunn (1996) indicated that is would be easier to control and reduce emissions from the industrial sources than from the domestic ones.

If the STP wants to decrease the metal load on a short-term basis, non-built-in or ´loose´ metals in their control must be tackled. Two different strategies can be taken:

1. Different types of treatment. Improve the existing ones or find new ones. A short-term solution would be investing money in improving the treatment processes already existing, or research and investigate to find new and more efficient techniques of metal removal. Some of these new methods have been described in previous sections.
2. Decrease the metal content in goods. New directives (NSW-EPA, 1998) and studies (Al Bakri et al., 2008) concerning metal levels in the sources could force a reduction of these elements before entering the STP. This would improve the capacity of treatment in the plant. Some authors have investigated the possibility of reducing metals from the source. This is the case of Genc-Fuhrman et al. (2007) who proved the removal of metals in storm water with different sorbents. In the case of domestic sources, a change in some detergent composition (i.e. zeolites), has reduced their metal concentrations. This can only be seen as a long-term solution. Once these sources (industrial and domestic) will be reduced, the problem will be in the diffuse sources that are more difficult to control.

Because of the importance of trace metals as environmental contaminants, it is essential to understand fully the fate of the substantial quantities of metals entering sewage treatment works, the changes in quantity (levels) and quality (speciation) of these metals through sewage treatment works and finally the consequences for the environment. It is inevitable that water quality legislation for trace metals will include statements relating to their speciation and that it may be possible to tolerate higher total concentrations of some metals as long as the "labile" fraction is below a certain level (Florence, 1982). Thus, new directives for limitation of metal levels in the final effluents should follow this address.

References

Acar, Y.B.; Gale, R. J.; Alshawabkeh A.N.; Marks, R.E.; Puppala, S.; Bricka, M. and Parker, R. (1995) Electrokinetic Remediation - Basics And Technology Status. *Journal of Hazardous Materials,* 40 (2): 117-137.

Aklil, A.; Mouflih, M. and Sebti, S. (2004). Removal of Heavy Metal Ions from Water by Using Calcined Phosphate as a New Adsorbent. *Journal of Hazardous Materials*, 112 (3): 183-190.

Aksu, Z. (2005) Application of biosorption for the removal of organic pollutants: a review. *Process Biochemistry*, 40 (2): 997-1026.

Al Bakri, D.; Rahman, R. and Bowling, L. (2008) Sources and management of urban stormwater pollution in rural catchments, Australia. *Journal of Hydrology*, 356 (3-4): 299-311.

Alam, I.A. and Sadiq, M. (1989) Metal contamination of drinking water from corrosion of distribution pipes. *Environmental Pollution*, 57 (2): 167-178.

Alder, A.C.; Siegrist, H.; Gujer, W. and Giger, W. (1990) Behaviour of NTA and EDTA in biological wastewater treatment. *Water Research*, 24 (6):733-742.

Alonso, E.; Santos, A.; Solis, Gj. And Riesco, P. (2001) On The Feasibility Of Urban Wastewater Tertiary Treatment By Membranes: A Comparative Assessment. *Desalination*, 141 (1): 39-51.

Alvarez, E.A.; Mochon, M.C.; Sanchez, J.C.J. and Rodriguez, M.T. (2002) Heavy Metal Extractable Forms in Sludge from Wastewater Treatment Plants. *Chemosphere,* 47 (7): 765-775.

Alvarez, E.A.; Mochon, M.C.; Sanchez, J.C.J. and Rodriguez, M.T. (1998) A Voltammetry Procedure With Medium Exchange For The Speciation Of Heavy Metals In Wastewater Of A Sewage Treatment Plant. *Electroanalysis,* 10 (13): 917-920.

Angelidis, M. and Gibbs, R.J. (1989) Chemistry of metals in anaerobically treated sludges. *Water Research*, 23 (1): 29-33.

Anonymous. (1965) Sewage Processing-Chemical Or Civil. *Chemical And Process Engineering*, 46 (3): 107-and.

Aonghusa, C.N. and Gray, N.F. (2002) Laundry Detergents As A Source Of Heavy Metals In Irish Domestic Wastewater. Journal of Environmental Science and Health *Part A-Toxic/Hazardous Substances and Environmental Engineering*, 37 (1): 1-6.

Bailey, D.A.; Dorrell, J.J. and Robinson, K.S. (1970) The influence of trivalent chromium on the biological treatment of domestic sewage. *Water Pollution Control*, 69: 100-110.

Bansal, S. and Kapoor, K.K. (2000) Vermicomposting Of Crop Residues And Cattle Dung With Eisenia Foetida. *Bioresource Technology*, 73 (2): 95-98.

Barth, E.F.; Ettinger, M.B.; Salotto, B.V. and McDermott, G.N. (1965) Summary report on the effects of heavy metals on the biological treatment process. *Journal of Water Pollution Control Federation*, 37: 86-96.

Bayhan, Y.K.; Keskinler, B.; Çakici, A.; Levent, M. and Akay, G. (2001) Removal of divalent heavy metal mixtures from water by Saccharomyces Cerevisiae using crossflow microfiltration. *Water Research*, 35 (9): 2191-2200.

Beloso, M.C.; Villar, M.C.; Cabaneiro, A.; Carballas, M.; Gonzalezprieto, S.J. and Carballas, T. (1993) Carbon And Nitrogen Mineralization In An Acid Soil Fertilized With Composted Urban Refuses. *Bioresource Technology*, 45 (2): 123-129.

Bender, M.E.; Matson, W.R. and Jordan, R.A. (1970) On Significance Of Metal Complexing Agents In Secondary Sewage Effluents. *Environmental Science and Tech*nology, 4 (6): 520-and.

Berrow, M.L. and Webber, J. (1972) Trace-Elements In Sewage Sludges. *Journal Of The Science Of Food And Agriculture*, 23 (1): 93-and.

Blomqvist, S.; Larsson, U. and Borg, H. (1992) Heavy metal decrease in the sediments of a Baltic bay following tertiary sewage treatment. *Marine Pollution Bulletin*, 24 (5): 258-266.

Bodzek, M.; Korus, I. and Loska, K. (1999) Application of the hybrid complexation-ultrafiltration process for removal of metal ions from galvanic wastewater. *Desalination,* 121 (2): 117-121.

Boulay, N. and Edwards, M. (2000) Copper In The Urban Water Cycle. *Critical Reviews In Environmental Science And Technology*, 30 (3): 297-326.

Bouzid, J.; Elouear, Z.; Ksibi, M.; Feki, A. and Montiel, A. (2008) A Study on Removal Characteristics of Copper From Aqueous Solution by Sewage Sludge and Pomace Ashes. *Journal of Hazardous Materials*, 152 (2): 838-845.

Bratby, J. Coagulation and flocculation, Ch. 8, Upland Press, Croydon, UK (1980).

Bratek, W.; Bratek, K. and Kulazynski, M. (2002) The utilization of waste ion exchange resin in environmental protection. *Fuel Processing Technology*, 77-78: 431-436.

Brown, H.G.; Hensley, C.P.; Mckinney, G.L. And Robinson, J.L. (1973) Efficiency Of Heavy-Metals Removal In Municipal Sewage-Treatment Plants. *Environmental Letters*, 5 (2): 103-114.

Brown, M.J. and Lester, J.N. (1979) Metal removal in activated sludge: the role of bacterial extracellular polymers. *Water Research*, 13(9): 817-837.

Brown, M.J. and Lester, J.N. (1982) Role of bacterial extracellular polymers in metal uptake in pure bacterial culture and activated sludge—I. Effects of metal concentration. *Water Research*, 16(11): 1539-1548.

Bubb, J.M. and Lester, J.N. (1995) The Effect Of Final Sewage Effluent Discharges Upon Thc Behavior And Fate Of Metals In A Lowland River System - A Question Of Dilution. *Environmental Technology*, 16 (5): 401-417.

Buckley, J.A. (1994) The bioavailability of copper in wastewater to Lemna minor with biological and electrochemical measures of complexation. *Water Research*, 28 (12): 2457-2467.

Bunch, R.L. (1982) Sewage Sludge Dilemma of the Eighties. *Studies in Environmental Science,* 19: 69-81.

Buzier, R.; Tusseau-Vuillemin, M.H.; Martin dit Meriadec, C.; Rousselot, O. and Mouchel, J.M. (2006) Trace metal speciation and fluxes within a major French wastewater treatment plant: Impact of the successive treatments stages. *Chemosphere,* 65 (11): 2419-2426.

Cantwell, F.F.; Nielsen, J.S. and Hrudey, S.E. (1982) Free Nickel Ion Concentration In Sewage By An Ion-Exchange Column-Equilibration Method. *Analytical Chemistry*, 54 (9): 1498-1503.

Cenci, G. and Morozzi, G. (1977) Evaluation of the toxic effect of Cd^{2+} and $Cd(CN)_4^{2-}$ ions on the growth of mixed microbial population of activated sludges. *The Science of The Total Environment,* 7 (2): 131-143.

Cenkin, V.E. and Belevtsev, A.N. (1985) Electrochemical Treatment Of Industrial Waste-Water. *Effluent and Water Treatment Journal*, 25 (7): 243-and.

Chao, W.; Xiao-Chen, L.; Pei-Fang, W.; Li-Min, Z. and Hai-Tao, M. (2006) Extractable Fractions of Metals in Sewage Sludges from Five Typical Urban Wastewater Treatment Plants of China. *Pedosphere*, 16 (6): 756-761.

Chen, K.Y.; Young, C.S.; Jan, T.K. and Rohatgi, N. (1974) Trace-Metals In Wastewater Effluents. *Journal Water Pollution Control Federation*, 46 (12): 2663-2675.

Chen, Y.X.; Hua, Y.M.; Zhang, S.H. And Tian, G.M. (2005) Transformation Of Heavy Metal Forms During Sewage Sludge Bioleaching. *Journal Of Hazardous Materials*, 123 (1-3): 196-202.

Cheng, M.H.; Patterson, J.W. and Minear, R.A. (1975) Heavy-Metals Uptake By Activated-Sludge. *Journal Water Pollution Control Federation*, 47 (2): 362-376.

Chiang, K.Y.; Yoi, S.D.; Lin, H.N. and Wang, K.S. (2000) Stabilization Of Heavy Metals In Sewage Sludge Composting Process. *Water Science and Technology*, 44 (10): 95-100.

Chino, M.; Moriyama, K.; Saito, H. and Mori, T. (1991) The Amount Of Heavy-Metals Derived From Domestic Sources In Japan. *Water Air And Soil Pollution*, 57-8: 829-837.

Chipasa, K.G. (2003) Accumulation and fate of selected heavy metals in a biological wastewater treatment system. *Waste Management*, 23(2): 135-143.

Chu, K.H. and Hashim, M.A. (2000) Adsorption Of Copper(Ii) And Edta-Chelated Copper(Ii) Onto Granular Activated Carbons. *Journal Of Chemical Technology And Biotechnology*, 75 (11): 1054-1060.

Cobelo-Garcia, A.; Prego, R. and Labandeira, A. (2004) Land Inputs Of Trace Metals, Major Elements, Particulate Organic Carbon And Suspended Solids To An Industrial Coastal Bay Of The Ne Atlantic. *Water Research*, 38 (7): 1753-1764.

Comber, S.D.W. and Gunn, A.M. (1996) Heavy Metals Entering Sewage-Treatment Works From Domestic Sources. *Journal Of The Chartered Institution Of Water And Environmental Management*, 10 (2): 137-142.

Council of European Communities. Directive concerning urban waste water treatment (91/271/ECC). *Official Journal* L 135/40, 30 May 1991.

Council of European Communities. Directive on the protection of the environment, and in particular of soil, when sewage sludge is used in agriculture (86/278/ECC). *Official Journal L* 187/6, 4 July 1986.

Culp, R,L., Wesner, G.M. and Culp, G.L., Handbook of Advanced Wastewater Treatment, 2nd ed., Van Nostand Reinhold Company, New York, 1978.

Curkovic, L.; Cerjanstefanovic, S. and Filipan, T. (1997) Metal Ion Exchange By Natural And Modified Zeolites. *Water Research*, 31 (6): 1379-1382.

Davis, J.A. and Jacknow, J. (1975) Heavy-Metals In Wastewater In 3 Urban Areas. *Journal Water Pollution Control Federation*, 47 (9): 2292-2297.

Department of the Environment and National Water Council, 1981. Report of the Sub-Committee on the disposal of sewage sludge to land, standing technical Committee report No. 20, National Water Council, London.

Department of the Environment. The implementation of Directive 76/464/EEC on pollution cause by certain dangerous substances discharged into the aquatic environment of the community, DOE Circular, HMSO, London, 1985.

de Souza Pereira, M. and Kuch, B. (2005) Heavy metals, PCDD/F and PCB in sewage sludge samples from two wastewater treatment facilities in Rio de Janeiro State, Brazil *Chemosphere*, 60 (7): 844-853.

Dewalle, F.B. and Chian, E.S.K. (1974) Kinetics Of Formation Of Humic Substances In Activated-Sludge Systems And Their Effect On Flocculation. *Biotechnology And Bioengineering,* 16 (6): 739-755.

Dilek, F.B. and Yetis, U. (1992) Effects Of Heavy-Metals On Activated-Sludge Process. *Water Science And Technology*, 26 (3-4): 801-813.

Dugan, P.R. and Pickrum, H.M. (Removal of mineral ions from water by microbially produced polymers, in Proc. 27th Ind. Waste Conf., Purdue Univ., Eng. Ext. Ser. No. 141, 1972, 1019.

Erdem, E.; Karapinar, N. and Donat, R. (2004) The removal of heavy metal cations by natural zeolites. *Journal of Colloid and Interface Science*, 280 (2): 309-314.

Eriksson, E.; Christensen, N.; Schmidt, J.E. and Ledin, A. (2008) Potential priority pollutants in sewage sludge. *Desalination,* 226 (1-3): 371-388.

European Commission, 2002. DG ENV. E3, Project ENV.E3/ ETU/2000/0058, *Heavy Metals in Waste,* Final Report.

Fang, M. and Wong, J.W.C. (1999) Effects of lime amendment on availability of heavy metals and maturation in sewage sludge composting. *Environmental Pollution*, 106 (1): 83-89.

Felix, F.; Fraaji, A.L.A. and Hendriks, Ch.F. Inorganic immobilisation of waste materials. *Studies in Environmental Science*, 71: 769-780.

Fleming, R.W. and Alexande, M. (1972) Dimethylselenide And Dimethyltelluride Formation By A Strain Of Penicillium *Applied Microbiology*, 24 (3): 424-and.

Florence, T.M. (1982) Development of physico-chemical speciation procedures to investigate the toxicity of copper, lead, cadmium and zinc towards aquatic biota. *Analytica Chimica* Acta, 141 (1): 73-94.

Friedman, B.A. and Dugan, P.R. (1968) Identification Of Zoogloea Species And Relationship To Zoogloeal Matrix And Floc Formation. *Journal Of Bacteriology*, 95 (5): 1903-and.

Fuentes, A.; Lloréns, M.; Sáez, J.; Soler, A.; Aguilar, M.I.; Ortuño, J.F. and Meseguer, V.F. (2004) Simple and sequential extractions of heavy metals from different sewage Sludges. *Chemosphere*, 54 (8): 1039-1047.

Gascó, G.; Méndez, A. and Gascó, J.M. (2005) Preparation of carbon-based adsorbents from sewage sludge pyrolysis to remove metals from water. *Desalination,* 180 (1-3): 245-251.

Genç, N.; Yonsel, S.; Dagasan, L. and Onar, A.N. (2002) Wet oxidation: a pre-treatment procedure for sludge. *Waste Management*, 22 (6): 611-616.

Genc-Fuhrman, H.; Mikkelsen, P.S. and Ledin, A. (2007) Simultaneous removal of As, Cd, Cr, Cu, Ni and Zn from stormwater: Experimental comparison of 11 different sorbents. *Water Research*, 41 (3): 591-602.

Ghosh, M.M. and Zugger, P.D. (1973) Toxic Effects Of Mercury On Activated-Sludge. Process. *Journal Water Pollution Control Federation*, 45 (3): 424-433.

Gliotti, P.L.; Businelli, D. and Machioni, A. (1997) Composition changes of dissolved organic matter in a soil amended with municipal waste compost, *Soil Sciences*, 162: 919–926.

Golder, A.K.; Samanta, A.N. and Ray, S. (2007) Removal of trivalent chromium by electrocoagulation. *Separation and Purification Technology*, 53 (1): 33-41.

Goldstone, M.E.; Kirk, P.W.W. and Lester, J.N. (1990a) The behaviour of heavy metals during wastewater treatment I. Cadmium, chromium and copper. *The Science of The Total Environment,* 95: 233-252.

Goldstone, M.E.; Kirk, P.W.W. and Lester, J.N. (1990b) The behaviour of heavy metals during wastewater treatment II. Lead, nickel and zinc. *The Science of The Total Environment,* 95: 253-270.

Goodman, A.H. (1984) Contamination of water within buildings. *Public Health Engineering*, 12: 23-25.

Guibaud, G.; Tixier, N.; Bouju, A. and Baudu, M. (2003) Relation between extracellular polymers' composition and its ability to complex Cd, Cu and Pb. *Chemosphere*, 52 (10): 1701-1710.

Gupta, R. and Garg, V.K. (2008) Stabilization of Primary Sewage Sludge During Vermicomposting. *Journal of Hazardous Materials*, 153 (3): 1023-1030.

Hanck, K.W. and Dillard, J.W. (1977) Determination Of Complexing Capacity Of Natural-Water By Cobalt(Iii) Complexation. *Analytical Chemistry*, 49 (3): 404-409.

Heidmann, I. and Calmano, W. (2008) Removal of Zn(II), Cu(II), Ni(II), Ag(I) and Cr(VI) present in aqueous solutions by aluminium electrocoagulation. *Journal of Hazardous Materials,* 152 (3): 934-941.

Helz, G.R.; Huggett, R.J. and Hill, J.M. (1975) Behavior of Mn, Fe, Cu, Zn, Cd and Pb discharged from a wastewater treatment plant into an estuarine environment. *Water Research,* 9 (7): 631-636.

Henry, J.G. and Heinke, G.W. *Environmental Science and Engineering*, Prentice-Hall, Englewood Cliffs, NJ, 1989, p.728.

Ho, G. and Qiao, L. (1998) Chromium speciation in municipal solid waste: Effects of clay amendment and composting. *Water Science and Technology*, 38 (2): 17-23.

Huang, C.P. and Vane, L.M. (1989) Enhancing As5+ Removal By A Fe2+-Treated Activated Carbon. Research Journal Of The Water Pollution Control Federation, 61 (9-10): 1596-1603.

Imai, A. and Gloyna, E.F. (1990) Effects of pH and oxidation state of chromium on the behavior of chromium in the activated sludge process. *Water Research*, 24 (9): 1143-1150.

Jan, T.K. and Young, D.R. (1978) Chromium Speciation in Municipal Wastewaters And Seawater. Journal Water Pollution Control Federation, 50 (10): 2327-2336.

Jenkins, D. and Russell, L.L. (1994) Heavy-Metals Contribution Of Household Washing Products To Municipal Waste-Water. *Water Environment Research*, 66 (6): 805-813.

Jernelov, A. and Martin, A.L. (1975) Ecological Implications Of Metal Metabolism By Microorganisms. *Annual Review Of Microbiology*, 29: 61-77.

Kaplan, D.; Abeliovich, A. and Ben-Yaakov, S. (1987) The fate of heavy metals in wastewater stabilization ponds. *Water Research*, 21 (10): 1189-1194.

Kapoor, A. and Viraraghavan, T. (1995) Fungal Biosorption - An Alternative Treatment Option For Heavy Metal Bearing Wastewaters: A Review. *Bioresource Technology*, 53 (3): 195-206.

Karthikeyan, K.G.; Elliott, H.A. and Cannon, F.S. (1997) Adsorption And Coprecipitation Of Copper With The Hydrous Oxides Of Iron And Aluminum. *Environmental Science and Technology*, 31 (10): 2721-2725.

Karvelas, M.; Katsoyiannis, A. and Samara, C. (2003) Occurrence and fate of heavy metals in the wastewater treatment process. *Chemosphere*, 53, (10): 1201-1210.

Katsioti, M.; Katsiotis, N.; Rouni, G.; Bakirtzis, D. and Loizidou, M. (2008) The effect of bentonite/cement mortar for the Stabilization/Solidification of sewage sludge containing heavy metals. *Cement and Concrete Composites*, In Press.

Kaushik, P. and Garg, V.K. (2003) Vermicomposting Of Mixed Solid Textile Mill Sludge And Cow Dung With The Epigeic Earthworm Eisenia Foetida. *Bioresource Technology*, 90 (3): 311-316.

Kelly, M. (1988) *Mining and the Freshwater environment*. Elsevier, Essex, UK.

Kempton, S.; Sterritt, R.M. and Lester, J.N. (1987a) Heavy-Metal Removal In Primary Sedimentation .1. The Influence Of Metal Solubility. *Science Of The Total Environment*, 63: 231-246.

Kempton, S.; Sterritt, R.M. and Lester, J.N. (1987b) Heavy-Metal Removal In Primary Sedimentation .2. The Influence Of Metal Speciation And Particle-Size Distribution. *Science Of The Total Environment*,63: 247-258.

Kiff, R.J. (1978) A study of the factor affecting bioflocculation in the activated sludge process. *Water Pollution Control*, 77: 464-470.

Kiff, R.J. and Fahim, F.A. (1981). Factors Affecting The Distribution Of Coliform Organisms In Activated-Sludge Plant Effluents. *Environmental Technology Letters*, 2 (9): 393-398.

Klischenko, R.; Kornilovich, B.; Chebotaryova, R. and Linkov, V. (1999) Purification of galvanic sewage from metals by electrodialysis. *Desalination*, 126 (1-3): 159-162.

Kołodyńska, D.; Hubicka, H. and Hubicki, Z. (2008) Sorption of heavy metal ions from aqueous solutions in the presence of EDTA on monodisperse anion exchangers. *Desalination*, 227 (1-3): 150-166.

Korus, I.; Bodzek, M. and Loska, K. (1999) Removal of zinc and nickel ions from aqueous solutions by means of the hybrid complexation–ultrafiltration process. *Separation and Purification Technology*, 17 (2): 111-116.

Kosobucki, P.; Kruk, M. and Buszewski, B. (2008) Immobilization of selected heavy metals in sewage sludge by natural zeolites. *Bioresource Technology*, 99 (13): 5972-5976.

Kugelman, I.J. and McCarty, P.L. (1965) Cation toxicity and stimulation in anaerobic waste treatment. *Journal of Water Pollution Control Federation*, 37: 97-116.

Kunkel, R. and Manahan, S.E. (1973) Atomic-Absorption Analysis Of Strong Heavy-Metal Chelating-Agents In Water And Waste-Water. *Analytical Chemistry*, 45 (8): 1465-1468.

Kunz, A. and Jardim, F. (2000) Complexation and adsorption of copper in raw sewage. *Water Research*, 34 (7): 2061-2068.

Laxen, D.P.H. and Harrison, R.M. (1981) The physicochemical speciation of Cd, Pb, Cu, Fe and Mn in the final effluent of a sewage treatment works and its impact on speciation in the receiving river. *Water Research*, 15 (9): 1053-1065.

Lester, J.N. (1981) Removal of heavy metals in conventional wastewater treatment. *Proc. Int. Conf. Heavy metals in the environment*, Amsterdam, pp. 104-113.

Lester, J.N. (1983a) Significance and behaviour of heavy metals in waste water treatment processes I. Sewage treatment and effluent discharge. *The Science of The Total Environment,* 30: 1-44.

Lester, J.N.; Harrison, R.M. and Perry, R. (1979) The balance of heavy metals through a sewage treatment works I. Lead, cadmium and copper. *The Science of The Total Environment,* Volume 12, Issue 1, May 1979, Pages 13-23.

Lester, J.N.; Sterritt, R.M. and Kirk, P.W.W. (1983b) Significance and behaviour of heavy metals in waste water treatment processes II. Sludge treatment and disposal. *The Science of The Total Environment*, 30: 45-83.

Lewin, V.H. and Rowell, M.J. (1973) Trace Metals In Sewage Effluent. *Effluent and Water Treatment Journal*, 13 (5): 273-277.

Lima, I.M.; Marshall, W.E. And Wartelle, I.Y.H. (2004) Hardware-Based Granular Activated Carbon For Metals Remediation. *Journal American Water Works Association*, 96 (7): 95-102.

Line, D.E.; Arnold, J.A.; Jennings, G.D. and Wu, J. (1996) Water Quality Of Stormwater Runoff From Ten Industrial Sites. *Water Resources Bulletin*, 32 (4): 807-816.

Luck, F. (1999) Wet air oxidation: past, present and future. *Catalysis Today*, 53 (1): Pages 81-91.

Macarewicz, E. and Mikulski, T. (2000) Electrochemical properties and the coagulation stability of acrylate polyelectrolytes. *Pol. J. Appl. Chem.*, 44: 205.

Magara, Y.; Nambu, S. and Utosawa, K. (1976) Biochemical and physical properties of an activated sludge on settling characteristics. *Water Research*, 10 (1): 71-77.

Magarou, P. Urban wastewater in Europe: what about sludge, in: *Proceeding of Workshop Problems around Sludge*, Stressa, Italy, 1999.

Magos, L.; Clarkson, T.W. And Tuffery, A.A. (1964) Volatilization Of Mercury By Bacteria. *British Journal Of Industrial Medicine*, 21 (4): 294-and.

Makepeace, D.K.; Smith, D.W.; Stanley, S.J. (1995) Urban Stormwater Quality - Summary Of Contaminant Data. *Critical Reviews In Environmental Science And Technology*, 25 (2): 93-139.

Manoli, E. and Samara, C. (1999) Occurrence And Mass Balance Of Polycyclic Aromatic Hydrocarbons In The Thessaloniki Sewage Treatment Plant. *Journal Of Environmental Quality*, 28 (1): 176-187.

Marani, D.; Braguglia, C.M.; Mininni, G. and Maccioni, F. (2003) Behaviour of Cd, Cr, Mn, Ni, Pb, and Zn in sewage sludge incineration by fluidised bed furnace. *Waste Management*, 23 (2): 117-124.

McDermott, G.N.; Barth, E.F.; Salotto, V. and Ettinger, M.B. (1962) Zinc in relation to activated sludge and anaerobic digesting processes. *Proc. 17th Ind. Waste Conf.*, Purdue Univ., Eng. Ext. Ser. No. 112:461-475.

McDermott, G.N.; Post, M.A.; Jackson, B.N. and Ettinger, M.B. (1965) Nickel in relation to activated sludge and anaerobic digestion processes. *Journal of Water Pollution Control Federation,* 37: 163-177.

McEntire, F.E. and Neufeld, R.D. (1975) Microbial methylation of mercury: A survey. *Water Pollution Control*, 74: 465-470.

Melcer, H.; Monteith, H. and Nutt, S.G. (1992) Activated-Sludge Process Response To Variable Inputs Of Heavy-Metals. *Water Science And Technology*, 25 (11): 387-394.

Meunier, N.; Drogui, P.; Montane, C.; Hausler, R.; Mercier, G. and Blais, J.F. (2006) Comparison Between Electrocoagulation And Chemical Precipitation For Metals Removal From Acidic Soil Leachate. *Journal Of Hazardous Materials*, 137 (1): 581-590.

Moller, A.; Grahn, A. And Welander, U. (2004) Precipitation Of Heavy Metals From Landfill Leachates By Microbially-Produced Sulphide. Environmental Technology, 25 (1): 69-77.

Moore, W.A.; McDermott, G.N.; Post, M.A.; Mandia, J.W. and Ettinger, M.B. (1961) Effects of chromium on the activated sludge process. *Journal of Water Pollution Control Federation*, 33: 54-72.

Morgan, G.B. (1961) The adsorption of radioisotopes in certain micro-organisms. *Q. J. Florida Acad. Sci.*, 24: 94-100.

Morita, H. and Tsuboi, H. (2000) Basic Investigation On The Chemical Forms Of Heavy Metals In A Sewage Treatment Plant. *Water Science and Technology*, 42 (9): 159-165.

Mosey, F.E. (1976) Assesment of the maximum concentration of heavy metals in crude sewage which will not inhibit the anaerobic digestion of the sludge. *Water Pollution Control*, 75: 10-18.

Mosey, F.E.; Swanick, J.D. and Hughes, D.A. (1971) Factors affecting the availability of heavy metals to inhibit anaerobic digestion. *Water Pollution Control*, 70: 668-678.

Nakhla, G.; Abuzaid, N. And Farooq, S. (1994) Activated Carbon Adsorption Of Phenolics In Oxic Systems Effect Of Ph And Temperature-Variations. *Water Environment Research*, 66 (6): 842-850.

Navarro, R.R.; Wada, S. and Tatsumi, K. (2005) Heavy Metal Precipitation By Polycation-Polyanion Complex Of Pei And Its Phosphonomethylated Derivative. *Journal Of Hazardous Materials,* 123 (1-3): 203-209.

Neufeld, R.D. (1976) Heavy Metals-Induced Deflocculation Of Activated-Sludge. *Journal Water Pollution Control Federation*, 48 (8): 1940-1947.

Neufeld, R.D. and Hermann, E.R. (1975) Heavy-Metal Removal By Acclimated Activated-Sludge. *Journal Water Pollution Control Federation*, 47 (2): 310-329.

Nielsen, J.S. and Hrudey, S.E. (1983) Metal loadings and removal at a municipal activated sludge plant. *Water Research*, 17 (9): 1041-1052.

Nielsen, J.S.; Hrudey, S.E. and Cantwell, F.F. (1984) Role Of The Free Metal-Ion Species In Soluble Nickel Removal By Activated-Sludge. *Environmental Science and Technology*, 18 (11): 883-886.

NSW-EPA, 1998. Managing Urban Stormwater*: Council Handbook, Environment Protection Authority*, Chatswood, Sydney.

Ntifo, S. Water Framework Directive, Water UK Policy, 2004. http://www.water.org.uk

Oleszkiewicz, J.A. and Sharma, V.K. (1990) Stimulation and inhibition of anaerobic processes by heavy metals—A review. *Biological Wastes*, 31 (1): 45-67.

Oliver, B.G. and Cosgrove, E.G. (1974) The efficiency of heavy metal removal by a conventional activated sludge treatment plant. *Water Research*, 8 (11): 869-874.

Ouki, S.K. and Kavannagh, M. (1999) Treatment Of Metals-Contaminated Wastewaters By Use Of Natural Zeolites *Water Science And Technology*, 39 (10-11): 115-122.

Pan, S.C.; Lin, C.C. And Tseng, D.H. (2003) Reusing Sewage Sludge Ash As Adsorbent For Copper Removal From Wastewater. *Resources Conservation And Recycling*, 39 (1): 79-90.

Patterson, J.W. (1979) Parameters influencing metals removal in POTW´s. *Proc. 8th National Conf.*, Municipal Sludge Management, pp. 82-90.

Persin, F. and Rumeau, M. Le traitement ′electrochimique des eaux et des effluents, Tribune de l'Eau 42 (1989) 45–56.

Peterson, P.J. (1971) Unusual accumulation of elements by plants and animals. *Sci. Prog.* Oxford, 59: 505-526.

Petrov, S. and Nenov, V. (2004) Removal and recovery of copper from wastewater by a complexation-ultrafiltration process. *Desalination*, Volume 162, 10 March 2004, Pages 201-209.

Petrox, S.; Nenov, V. and Vasilev, S. Divalent heavy metal removal from water by complexationultrafiltration. *Proc. 5th Internat. Conf. on Membranes in Drinking* and *Industrial Water Production*, Mulheim, Ruhr, Germany, 2002, pp. 245-252.

Poon, C.P.C. and Bhayani, K.H. (1971) Metal Toxicity To Sewage Organisms. *Journal Of The Sanitary Engineering Division-Asce*, 97 (Nsa2): 161-and.

Poon, C.S. and Chu, C.W. (1999) The use of ferric chloride and anionic polymer in the chemically assisted primary sedimentation process. *Chemosphere*, 39 (10): 1573-1582.

Prego, R.; Filgueiras, A.V. and Santos-Echeandia, J. (2008) Temporal and spatial changes of total and labile metal concentration in the surface sediments of the Vigo Ria (NW Iberian Peninsula): Influence of anthropogenic sources. *Marine Pollution Bulletin*, 56 (5): 1031-1042.

Purves, D.; Wixson, B.G. And Davies, B.E. (1979) Environmental Geochemistry And Health - Discussion Held 15-16 March 1978 - General Discussion. Philosophical Transactions Of The Royal Society Of London Series B-*Biological Sciences*, 288 (1026): 213-214.

Rajeshwar, K. and Ibanez, J. Environmental Electrochemistry—Fundamentals and Applications in Pollution Abatement, Academic Press, San Diego, CA, 1997, 776 p.

Ridge, A.C. and Sedlak, D.L. (2004) Effect of ferric chloride addition on the removal of Cu and Zn complexes with EDTA during municipal wastewater treatment. *Water Research,* 38 (4): 921-932.

Reimers, R.S., Little, M.D., Englande, A.J., McDonnell, D.B., Bowman, D.D. and Hughes, J.M. 1983 Final Project Report to US-EPA. Investigation of Parasites in Sludges and Disinfection Techniques.

Roberts, P.; Hegi, H.R.; Weber, A. and Krahenbuhl, H.R. (1977) Metals In Municipal Wastewater And Their Elimination In Sewage-Treatment. Progress In *Water Technology,* 8 (6): 301-306.

Rodriguez, F.J., *Principles of Polymer Systems*, 4th edn., Taylor and Francis, Washington DC, 1996.

Rossin, A.C.; Perry, R. and Lester, J.N. (1983a) The removal of nitrilotriacetic acid and its effect on metal removal during biological sewage treatment: *Part 1—adsorption and acclimatisation Environmental Pollution Series A, Ecological and Biological*, 29 (4): 271-302.

Rossin, A.C.; Perry, R. and Lester, J.N. (1983b) The removal of nitrilotriacetic acid during biological sewage treatment and its effect on metal removal: *Part 2–Under shock loading. Environmental Pollution Series A, Ecological and Biologica*l, 30 (1):1-26.

Rudd, T.; Sterritt, R.M. and Lester, J.N. (1984) Formation and conditional stability constants of complexes formed between heavy metals and bacterial extracellular polymers. *Water Research,* 18 (3): 379-384.

Sadiki, A.I.; Williams, D.T.; Carrier, R. And Thomas, B. (1996) Pilot Study On The Contamination Of Drinking Water By Organotin Compounds From Pvc Materials. *Chemosphere*, 32 (12): 2389-2398.

Samaras, P.; Papadimitriou, C.A.; Haritou, I. and Zouboulis, A.I. (2008) Investigation of sewage sludge stabilization potential by the addition of fly ash and lime. *Journal of Hazardous Materials*, 154 (1-3) June 2008, Pages 1052-1059.

Santarsiero, A.; Veschetti, E.; Donati, G. and Ottaviani, M. (1998) Heavy Metal Distribution in Wastewater from a Treatment Plant. *Microchemical Journal*, 59 (2): 219-227.

Santos-Echeandia, J.; Laglera, L.M.; Prego, R. and van den Berg, C.M.G. (2008) Copper speciation in continental inputs to the Vigo Ria: Sewage discharges versus river fluxes. *Marine Pollution Bulletin*, 56 (2): 308-317.

Scancar, J.; Milacic, R.; Strazar, M. and Burica, O. (2000) Total metal concentrations and partitioning of Cd, Cr, Cu, Fe, Ni and Zn in sewage sludge. *The Science of The Total Environment*, 250 (1-3): 9-19.

Schnitzer, M. and Khan, S.U. (1972) *Humic Substances in the Environment*. Marcel Dekker Inc., New York.

Sedlak, D.L.; Phinney, J.T. and Bedsworth, W.W. (1997) Strongly Complexed Cu And Ni In Wastewater Effluents And Surface Runoff. *Environmental Science and Technology*, 31 (10): 3010-3016.

Shafer, M.M.; Overdier, J.T. and Armstong, D.E. (1998) Removal, Partitioning, And Fate Of Silver And Other Metals In Wastewater Treatment Plants And Effluent-Receiving Streams. *Environmental Toxicology And Chemistry*, 17 (4): 630-641.

Solis, G.J.; Alonso, E. and Riesco, P. (2002) Distribution Of Metal Extractable Fractions During Anaerobic Sludge Treatment In Southern Spain Wwtps. *Water Air And Soil Pollution*, 140 (1-4):139-156.

Soon-Oh Kim, Seung-Hyeon Moon, Kyoung-Woong Kim, Seong-Taek Yun. (2002) Pilot scale study on the ex situ electrokinetic removal of heavy metals from municipal wastewater sludges. *Water Research*, 36 (19): 4765-4774.

Sorme, L. and Lagerkvist, R. (2002) Sources of heavy metals in urban wastewater in Stockholm. *The Science of The Total Environment*, 298 (1-3):131-145.

Soylak, M.; Uzun, A. and Elci, L. (2001) Chemical Analysis Of Industrial Waste Water Samples From Some Metal And Textile Plants In The Organized Industrial Region Of Kayseri, Turkey. *Kuwait Journal Of Science and Engineering*, 28 (1): 151-160.

Sprynskyy, M.; Kosobucki, P.; Kowalkowski, T. and Buszewski, B. (2007) Influence of clinoptilolite rock on chemical speciation of selected heavy metals in sewage sludge. *Journal of Hazardous Materials*, 149 (2): 310-316.

Stephenson, T. and Lester, J.N. (1987) Heavy metal behaviour during the activated sludge process II. Insoluble metal removal mechanisms. *The Science of The Total Environment,* 63: 215-230.

Sterritt, R.M. and Lester, J.N. (1980) The Value Of Sewage-Sludge To Agriculture And Effects Of The Agricultural Use Of Sludges Contaminated With Toxic Elements - A Review. *Science Of The Total Environment*, 16 (1): 55-90.

Sterritt, R.M. and Lester, J.N. (1984) Significance And Behavior Of Heavy-Metals In Wastewater-Treatment Processes .3. Speciation In Waste-Waters And Related Complex Matrices. *Science Of The Total Environment*, 34 (1-2): 117-141.

Sterritt, R.M.; Brown, M.J. and Lester, J.N. (1981) Metal Removal By Adsorption And Precipitation In The Activated-Sludge Process. *Environmental Pollution Series A-Ecological And Biological*, 24 (4): 313-323.

Stones, T. (1958). Inst. Sewage Purif., 82.

Stones, T. (1977) Study On Loading And Performance Of Sewage-Treatment Plants. *Effluent and Water Treatment Journal*, 17 (7): 352-353.

Stoveland, S.; Astruc, M.; Lester, J.N. and Perry, R. (1979) The balance of heavy metals through a sewage treatment works II. Chromium, nickel and zinc. *The Science of The Total Environment,* 12 (1): 25-34.

Stoveland, S.; Astruc, M.; Perry, R. and Lester, J.N. (1978) Rapid Flameless Atomic-Absorption Analysis Of Metallic Content Of Sewage Sludges .2. Chromium, Nickel And Zinc. *Science Of The Total Environment*, 9 (3): 263-269.

S. Stoveland, J. N. Lester, R. Perry (1980). The influence of zeolite type A on heavy metal transfer in the activated sludge process. *Environment International*, 3 (1): 49-58.

Stumm, W. and Bilinski, H. (1973) Trace metals in natural waters; difficulties in interpretation arising from our ignorance on their speciation. *Adv. In Water Pollut. Res. Proc. Of 6th Int. Conf.*, 39-49.

Stylianou, M.A.; Inglezakis, V.J.; Moustakas, K.G. and Loizidou, M.D. (2008) Improvement of the quality of sewage sludge compost by adding natural clinoptilolite. *Desalination,* 224 (1-3): 240-249.

Su, D.C. and Wong, J.W.C. (2004) Chemical Speciation And Phytoavailability Of Zn, Cu, Ni And Cd In Soil Amended With Fly Ash-Stabilized Sewage Sludge. *Environment International,* 29 (7): 895-900.

Surittanonta, S. and Sherrad, J.H. (1981) Activated sludge nickel toxicity studies. *Journal of Water Pollution Control Federation*, 53: 1314–1322.

Sutherland, I. W. (1972) Bacterial Exopolysaccharides. *Adv Microb Physiol*, 8: 143-213.

Tchobanoglous, G. and Burton, F.L., *Wastewater Engineering: Treatment, disposal and Reuse*, McGraw-Hill Inc., New York, 1991.

Tchobanoglous, G.; Burton, F.L. and Stensel H.D., *Wastewater Engineering: Treatment and Reuse*, The McGraw-Hill Inc., New York, 2003.

Tebbutt, T.H.Y. and Christoulas, D.G. (1975) Performance relationships for primary sedimentation. *Water Research*, 9 (4):347-356.

Tessier, A.; Campbell, P.G.C. And Bisson, M. (1979) Sequential Extraction Procedure For The Speciation Of Particulate Trace-Metals. *Analytical Chemistry*, 51 (7): 844-851.

Toumi, A.; Nejmeddine, A. and Belkoura, M. (2003) The Fate Of Heavy Metals (Zn, Cu, Pb, Cd And Cr) In An Integrated Wastewater Treatment Plant: Two Phase Anaerobic Reactor (Rap) - High Rate Algal Pond (HRAP). *Environmental Technology*, 24 (2): 153-159.

Tuccillo, M.E. (2006) Size Fractionation Of Metals In Runoff From Residential And Highway Storm Sewers. *Science Of The Total Environment*, 355 (1-3): 288-300.

Tyagi, R.D. and Couillard, D. (1987) Bacterial Leaching Of Metals From Digested Sewage-Sludge. *Process Biochemistry*, 22 (4): 114-117.

Upadhyay, A.R.; Mishra, V.K.; Pandey, S.K. and Tripathi, B.D. (2007) Biofiltration Of Secondary Treated Municipal Wastewater In A Tropical City. *Ecological Engineering*, 30 (1): 9-15.

Ure, A.M.; Quevauviller, P.; Muntau, H. and Griepink, B. (1993) Speciation Of Heavy-Metals In Soils And Sediments - An Account Of The Improvement And Harmonization

Of Extraction Techniques Undertaken Under The Auspices Of The Bcr Of The Commission-Of-The-European-Communities. *International Journal Of Environmental Analytical Chemistry*, 51 (1-4): 135-151.

US Public Health Service (1965) *Manual of Septic Tank Practice*. USPHS, Washington, DC.

Van Veen, E.; Burton, N.; Comber, S. and Gardner, M. (2002) Speciation Of Copper In Sewage Effluents And Its Toxicity To Daphnia Magna. *Environmental Toxicology And Chemistry,* 21 (2): 275-280.

Wang, C.; Hu, X.; Chen, M.L. and Wu Y.H. (2005) Total Concentrations And Fractions Of Cd, Cr, Pb, Cu, Ni And Zn In Sewage Sludge From Municipal And Industrial Wastewater Treatment Plants. *Journal Of Hazardous Materials*, 119 (1-3): 245-249.

Wendt, H.and Kreysa, G. G′enie ′ Electrochimique—Principes et Proc′ed′es, Dunod, Paris, France, 2001, 386 p.

Wilkinson, J.F. (1958) The Extracellular Polysaccharides Of Bacteria. *Bacteriological Reviews,* 22 (1): 46-73.

Wong, J.W.C. and Selvam, A. (2006) Speciation of heavy metals during co-composting of sewage sludge with lime. *Chemosphere,* 63 (6): 980-986.

Wong, J.W.C.; Fang, M.; Li, G.X. and Wong M.H (1997) Feasibility Of Using Coal Ash Residues As Co-Composting Materials For Sewage Sludge. *Environmental Technology*, 18 (5): 563-568.

Wong, P.T.S; Chau, Y.K. And Luxon, P.L. (1975) Methylation Of Lead In Environment. *Nature,* 253 (5489): 263-264.

Wu, J.L.; Yang, Y.S. and Lin, J.H. (2005) Advanced Tertiary Treatment Of Municipal Wastewater Using Raw And Modified Diatomite. *Journal Of Hazardous Materials*, 127 (1-3): 196-203.

Zhai, Y.; Wei, X.; Zeng, G.; Zhang, D. and Chu, K. (2004) Study of adsorbent derived from sewage sludge for the removal of Cd^{2+}, Ni^{2+} in aqueous solutions. Separation and *Purification Technology*, 38 (2):191-196.

Zhang, P. and Wu, Z.C. (2005) Municipal Sludge As Landfill Barrier Material. *Journal Of Environmental Sciences-China*, 17 (3): 474-477.

Zhang, S.; Qu, Y.; Liu, Y.; Yang, F.; Zhang, X.; Furukawa, K. and Yamada, Y. (2005) Experimental study of domestic sewage treatment with a metal membrane bioreactor. *Desalination,* 177(1-3):83-93.

Zorpas, A.A.; Constantinides, T.; Vlyssides, A.G.; Haralambous, I. and Loizidou, M. (2000) Heavy metal uptake by natural zeolite and metals partitioning in sewage sludge compost. *Bioresource Technology*, 72 (2):113-119.

Zufiaurre, R.; Olivar, A.; Chamorro, P.; Nerin, C. and Callizo, A. (1998) Speciation Of Metals In Sewage Sludge For Agricultural Uses. *Analyst,* 123 (2): 255-259.

In: Sewage Treatment: Uses, Processes and Impact
Editor: Anna Stephens and Mark Fuller
ISBN 978-1-60692-959-9

Chapter 2

EFFECTS OF STABILIZATION PROCESSES ON SEWAGE SLUDGE´S N AND P FERTILIZING VALUE

Rodrigo Studart Corrêa[1]
The University of Brasília, Department of Forestry
Caixa Postal 04.401 CEP 70910 - 900 Brasília/DF, Brazil

ABSTRACT

High concentrations of N and P in sewage materials are one of the strongest reasons for their agronomic use. Such residues have been reported to increase yields and supply plant nutrients. However, digested sewage sludges still present highly putrescible organic matter and pathogens and therefore they usually need further sanitation and stabilization. Sanitation processes such as composting, liming, thermal-drying and solar-irradiation are frequently required to make sewage sludge safe enough for its agricultural use. For an appropriate sludge management and due to environmental concerns, it is essential to understand side effects of stabilization processes on sewage sludge's N and P.

This chapter discusses the common ways to stabilize domestic sewage sludge and consequences on N and P contents and availability in the produced biosolids. Stabilization of digested sewage sludge has been reported to affect its N and P fertilizing value. There is not a stabilized sewage material (biosolid) that is totally similar to another or to the fresh sewage sludge. The process that most alters N and P content in digested sludge is composting. Compared to other biosolids, composted biosolid remains as the poorest potential source of N and P for plants. Heat-drying at 250 °C and solar-treatment are the only processes which do not cause significant losses of total-N and total-P in sewage sludge. Nutrient contents and availability have a profound effect on agronomic value of biosolids. Therefore, potential markets should be assessed prior to deciding for a stabilization process of sewage sludge.

---

[1] Correspondence to: rodmanga@yahoo.com.br
Bolsista de Produtividade em Pesquisa do CNPq

1. Introduction

Wastewater collection and treatment systems are designed and managed to protect human and environmental health (Muga and Mihelcic, 2008). It is done by removing solids, pathogens, nutrients, and contaminants from collected sewages before the treated water is returned to the environment. Mechanical sewage treatment plants have been built in urban areas all around the world for such task. But treatments systems generate sewage sludge, a by-product which is the concentrated bioactive residue of mostly organic clay-sized particles derived from wastewater treatment processes (O'Kelly, 2008).

Water quality legislation in many countries has increased the standards regarding nutrient removal in order to overcome eutrophication problems in receiving waters. The establishment of secondary treatment of wastewater prior to effluent discharge in the United States, for example, has raised quantities of sewage sludge requiring disposal from 4.6 million dry tonnes in 1972 to 6.9 million dry tonnes in 1998 (USEPA, 1999). It accounts for a 50% increase in sludge tonnage for a 29% population growth over the same period (CEA, 1999). Therefore, more intensive methods of water cleaning have created problems of sewage sludge disposal (Bitton et al., 1980; Pastor et al., 2008).

Sewage sludge tonnage is still increasing and it is expected a production of 10 - 13 million dry tonnes in the United Sates by 2010 (USEPA, 1999; Lemainski and Silva, 2006). Figures for European countries are similar: the amount of sludge dry solids increased from 1.7 to 2.5 million dry tonnes over 15 years (1975 - 1990) in Germany, and from that to 3.2 million dry tonnes in 1991 after implementation of wastewater secondary treatment (Frank, 1998). Production of municipal sewage sludge increases each year in the European Union due to the implementation of the Urban Waste Water Treatment Directive (91/271/EEC). Sewage sludge generation reached 10 million dry tonnes in 2005 (Ramírez et al., 2008). Yet, it is predicted that by 2010, 40 - 50% of the wastewater generated in China will be collected and treated. This indicates that a dramatic increase in sewage sludge generation will soon occur, and China will have to face the problem of how to dispose it (Pengcheng et al., 2008).

Sewage sludge disposal is an expensive operation (Pengcheng et al., 2008) and physical, chemical and biological characteristics of dewatered sewage materials are of concern for its management (O'Kelly, 2008). Landfilling may result in environmental pollution due to sludge's harmful constituents. Consequently, safe and economical methods for sludge disposal shall consider the whole plethora of risks and benefits from such material (Pengcheng et al., 2008). Actually, the ultimate disposal of sewage sludge has become one of the most difficult and expensive problems in the field of wastewater engineering (Metcalf and Eddy, 2003; Farnet et al., 2008).

Environmental legislation has been tightened worldwide (Farnet et al., 2008) and options for sludge disposal have decreased. Landfills are filling up and new permits are becoming more expensive. Several countries have banned landfilling, incineration and ocean dumping (Snow, 1996). The greatest economic impacts for mechanical systems arise from their high capital, operation and management (Muga and Mihelcic, 2008). Once more, handling and final disposal of sewage materials are currently the most problematic and costly part in the operation of wastewater treatment plants (Spellman, 1997; Plachá et al., 2008).

However, sludge from municipal wastewater treatment represents a valuable source of organic matter, nitrogen (N), phosphorus (P), and some trace elements (Plachá et al., 2008).

The idea of feeding organic matter, primarily containing N and P, back into the natural cycle forms the basis of using sludge on land. Such a management in the western World began in the United States and in Europe, but it has quickly spread through many other countries (Frank, 1998). There is general agreement on the positive effects of sewage sludge as a soil amendment and on soil chemical properties (Pengcheng et al., 2008). Thus, the use of digested sewage sludge in agriculture is considered an acceptable solution in medium and long terms, provided the waste complies with sanitary and environmental requirements. Nowadays, about 50% of the sludge produced in the European Union is used as soil fertilizer, 25% is disposed in landfills and the rest is incinerated. (Ramírez et al., 2008).

In order to relieve sewage sludge disposal problems by sustainable means, the term biosolids has been coined to boost the use of sewage sludge as soil fertilizer and conditioner. Biosolids is defined as the organic product yielded by municipal wastewater treatment process that can be beneficially used (USEPA, 1999). Sustainable sludge management can be defined as a method that meets requirements of efficient recycling of resources without generating harmful substances to humans or to the environment (Ramírez et al., 2008). Biosolids are reputed for improving soil characteristics, releasing plant nutrients, having beneficial effects on plant growth, and are reported to yield more biomass where they are applied (Crush et al., 2006; Walker et al., 2008). As a result, a wide spectrum of agricultural products referred as biosolids has been manufactured from sewage sludge (Hu et al., 1996, Mbarki et al., 2008; Ostos, et al., 2008; Plachá et al., 2008).

Recycling has become increasingly important in societies with high consumption standards (Barton et al., 2008). Population growth in urban areas has resulted in large transfers of plant nutrients from agricultural soils to urban wastes (Frossard and Morel, 1995). As an example, approximately 60% of the total amount of P and 70% of N ingested by humans via foods are flushed into domestic sewers after being metabolized (Kvarnstrom and Nilsson, 1999). Then, the recycle of agricultural nutrients involves the use of sewage sludge. The increasing interest in waste recycling itself is another cause to advocate the use of organic wastes as soil amendments (Ostos et al., 2008). Transformation of wastes into agricultural products can create alternative sources of nutrients and avoid detrimental effects in the environment (Nóvoa-Muñoz et al., 2008).

Conservation of natural resources has been necessary. Phosphate rock is a finite resource thereby an efficient phosphorus recovery process should be promoted (Pastor et al., 2008). Natural reserves of P are expected to be depleted in another five hundred years. There is an annual consumption of 10 million tonnes of P worldwide only for food production. If all the produced sewage sludge were applied to land, the amount of P used through artificial fertilizers could be reduced by more than 40% (Frank, 1988). Quotations for nutrient sources have been soaring in the international market. Prices of N fertilizers increased 54% and P prices raised 84% in the last couple of years.

The current fertilizers' market situation represents an extra burden on food production in developing countries (Nishanth and Biswas, 2008). However, it also gives an extra argument for recycling nutrients from sewage sludge. Cost calculations have shown that direct application of residues and wastes on land is a more economical alternative than other kinds of possible sludge disposal (Frank, 1988; Bridle and Skrypski-Mantele, 2000). Soil application is actually one of the most efficient ways of sludge disposal and plant nutrient recovery (Nóvoa-Muñoz et al., 2008).

In addition, soil amendment with municipal solid wastes represents a management strategy that could counteract the depletion of organic matter in soils (Mbarki et al., 2008). Soil conservation is an issue all around the World. Inappropriate technologies have resulted in soil quality deterioration, leading to soil organic matter losses and degradation. Thus, the application of organic wastes with high organic matter content has become a common environmental practice for soil restoration. It can maintain soil organic matter, reclaim degraded soils, and supply plant nutrients (Tejada et al., 2008).

However, benefits from sludge application on croplands have to be weighed against the potential hazards associated with certain sludge-borne constituents (Pengcheng et al., 2008). Land application of sewage sludge can cause deleterious effects such as the increase of nitrate leaching rate, contaminate crops with metals, promote nutritional disorders in crops, increase soil salinity, contaminate watertable with pesticides, hormones and pharmaceuticals, and decrease the stability of native soil organic matter (Bertoncini et al., 2008; Mbarki et al., 2008). Yet, pathogens in the biosolids can have negative impacts on human health, either directly or indirectly through contamination of food products and water (Crush et al., 2006).

Metals, transmission of diseases and environmental hazards related to the sludge's high N and P content stand out among problems arising with soil application of sewage materials (Sparkes, 1990). Therefore, health safety standards, maximum rates and frequency of application have been set according to the concentrations of metals, pathogens, N and P in sewage materials.

Concerns on metals are related to risks of their entry into the food chain or water supply. Metal content may both limit unit loading and the design application period for a particular site (Pengcheng et al., 2008). They are often the controlling design parameter for long-term beneficial land use land of biosolids (Reed et al., 1995). The USEPA, for instance, allows application rates of 10 t dry solids/ha/year for 100 years if metal concentrations in sewage sludge are within safety limits. Pathogens, like metals, may also limit both the unit loading and the design application period for a particular site. They are divided among bacteria, protozoa, worms, and viruses and a broad range of them have been isolated even from digested sludge.

Pathogens, however, are mostly a short-term concern on the beneficial use of sewage sludge. Different methods of sludge stabilization have been developed to reduce concentrations of human pathogens in biosolids (Plachá et al., 2008). The USEPA Regulation 40 CFR Part 503 has addressed potential hazards represented by the disposal and beneficial use of biosolids. It is a risk-based rule to protect public health and the environment from adverse effects of pollutants that may be present in these materials (USEPA, 1995). It also describes sludge stabilization processes, considering their capability of pathogen removal, and also determines the possible beneficial uses of each stabilized product.

Dewatered sewage sludges typically have high water content (75 - 90%), which increases transport costs and their physical properties make them difficult to manage (Crush et al., 2006). Digested sludges contain putrescible organic matter as well. Actually, there are few instances where the application of sewage sludge to land is straightforward. Economic constraints, environmental and health risks are usually in conflict with sludge disposal and beneficial use (Plachá et al., 2008). Therefore, sludge stabilization processes also aim to dehydrated, make sewage sludge less offensive to humans and the environment and turn it into a more costly effective product (Corrêa et al., 2007).

Different characteristics of biosolids have given rise to debate whether they are slow release organic fertilizers or soil amendments/conditioners. Whatever biosolids might be, N and P concentrations have the strongest appeal for their use, compared to manures and other organic matter sources (Pengcheng et al., 2008). Products made of sewage sludge (biosolids) vary in their physical and chemical characteristics, but all of them contain N (< 0.1 - 17.6%) and P (< 0.1 - 14.6%) in their dry solids (Cameron et al, 1996).

2. Sewage Sludge Stabilization

Stabilization refers to a number of processes that reduce the volume, pathogen levels, odor and volatile solids content in sewage sludge (USEPA, 1999). Sewage sludge has to be stabilized to some extent before being used in agriculture, landscaping, gardening, and forestry or even disposed into the environment without major problems (Ostos et al., 2008). Economics, use options and safety of disposal are strongly influenced by the degree of sludge stabilization and contents of water, pathogen, organic matter, hazardous chemicals, and plant nutrients in biosolids. One of the primary factors in selecting a stabilization process is the final option for disposal or use of biosolids, especially focusing on pathogen reduction (Snow, 1996).

Although stabilization is a relative concept, sludge stabilization processes falls into two categories:

1) Those that define how much improvement in any of one or more sludge characteristics has to be achieved.
2) Those that specify a treatment which will commonly attain some significant improvement in one of these characteristics.

Reducing pathogens to a given level is an example of the first category. In this sense, the USEPA Regulation 40 CFR Part 503 defines different pathogen levels for different uses of biosolids on land (Classes A, B and C). A sludge disposal guideline that states what kind of treatment the sludge has to undergo prior to disposal or use, such as long-term storage, is an example of the second category. Again, the USEPA Regulation 40 CFR Part 503 specifies stabilization processes which are able to achieve different levels of pathogenic removal.

The selection of a stabilization process demands the knowledge of what the ultimate use will be and what sludge characteristics are required. If a land is far away from civilization, for instance, pathogen destruction or odor reduction would be questionable. Disposal on strip mine land would require dewaterability to minimize transport costs. For this last particular case, stabilization should produce a product that does not greatly generate nitrate due to risks of groundwater contamination, as water tables are usually shallow in such places.

Common ways to eliminate nuisance conditions and pathogens in sewage sludge by means of stabilization are biological reduction or chemical oxidation of volatile matter, addition of chemicals to render sludge unsuitable for pathogen survival or application of heat to disinfect or sterilize sludge. The most common sewage sludge stabilization processes are aerobic and anaerobic digestion. A number of post-treatment processes have bee also applied to reduce sludge water content and biological contamination. (Ramírez et al., 2008). Current post-treatment technologies include composting, alkaline treatment, thermal-drying, and sludge radiation (Barton et al., 2008; Metcalf and Eddy, 2003; Plachá et al., 2008). Pathogenic content in sewage sludge is strongly related to the stabilization process and concentrations of pathogens in the original wastewater. Some stabilization processes are better on specific target pathogens than others.

The USEPA Regulation 40 CFR Part 503 has defined biosolids standards according to pathogen concentrations present in them:

Class A Biosolids: undergo advanced treatment to reduce pathogen concentrations to below detectable levels. They are often sold in bags and can be used without pathogen-related restrictions.

Class B Biosolids: pathogens are reduced to levels protective of human health and the environment, but not to undetectable levels. Class B biosolids require crop harvesting and site restrictions that minimize the potential for human and animal contact until further natural attenuation of pathogens has occurred.

Class C Biosolids: pathogenic concentration has to be 1.5 times lower than the original pathogen concentration in the sewage sludge. They have more restrictions on crops and harvesting than Class B biosolids, when applied to arable lands.

Processes listed in Table 1 have to disinfect sewage sludge to concentrations lower than detectable levels if Class A biosolids is to be achieved. Thermal-drying certainly achieves it (Ramírez et al., 2008), whilst composting and alkaline treatment may produce pathogen-free biosolids. Class A biosolids are usually marketable products, but they may also be produced in special cases when sludge from communities known to contain carriers of serious infectious diseases will be used.

Table 1. Stabilization processes to reduce pathogens in sewage sludge and biosolids potential use

Process	Requirements
Aerobic or Anaerobic Digestion	They produce biosolids used as soil amendment and organic fertilizer; additional treatment is advisable. It may achieve Class A biosolids.
Composting	Minimum temperature of 55 °C for at least 5 days. A post period of 2 months for maturation is advisable. It produces highly organic, soil-like biosolids with conditioning properties for horticultural, nursery, and landscape uses. If temperature requirements are met, it can achieve Class A biosolids.
Alkaline treatment	Liming to reach pH 12 for 12 hours, then limed-sludge is stored for 2 weeks at pH > 11. It produces biosolids useful for land application and daily landfill cover. It usually achieves Class B biosolids.
Thermal-drying	Heating for a minimum 30 minutes at a minimum temperature of 70 °C. It produces biosolids for fertilizers generally used at lower rate because of higher cost and high nutrient content. It always produces Class A biosolids.
Solar-irradiation	Digested sludge is stored for 2 weeks under solar-irradiated ambient. It produces biosolids similar to digested sludge, although higher in solid content and lower in pathogen concentration. It often ends as Class B or Class C biosolids.

Composting biologically converts wastes' organic components to a state in which it can be handled, stored, and applied to land without adversely affecting the environment (Saludes et al., 2008). Composting processes are often depicted in terms of time-temperature relationship and therefore temperature is usually the monitoring parameter (Epstein, 1997). Temperatures high enough for long enough are crucial for sanitation of piles containing sewage materials.

Builder's lime [$Ca(OH)_2$] and burnt lime (CaO) are the two most common products employed for alkaline treatment of sewage sludge. They are applied at rates of 20, 30, 40 or 50% of lime to sludge's dry solids (Imhoff and Imhoff, 1986; Metcalf and Eddy, 2003). Application rates depend on lime efficiency to raise sludge pH. For thermal-drying process, temperatures range from 100 to 350 °C. There is an inverse relationship between temperature and detention time and the process is run until dry pellets (< 20% moisture) are yielded. A similar time-temperature relation applies to solar irradiation process. This process is convenient for tropical countries. However, solar radiation may sterilize sludge surface but can not penetrate deep layers of sludge mass.

USEPA standards for biosolids have been set on the basis of public health concerns. However, some stabilization processes end with products better for soil fertility, whilst others may serve as stable organic matter for the rehabilitation of physically degraded soils (Mbarki et al., 2008). Thus, a better knowledge of stabilization effects on biosolids agronomic characteristics is essential for their more effective use. Hence, environmental concerns have resulted in legislation and policies that require more intensive management of biosolids N and P in order to avoid deleterious impacts of these nutrients on water quality (Maguire et al., 2000).

3. Sewage Sludge Composting

3.1. Composting Performance

It is well known the contribution of composting to organic matter stabilization, sanitation regarding weeds and pathogens, deodorization, improvement of handling, and safe storage and transportation before agronomic or horticultural use (Gil et al., 2008; Robin et al., 2008). Among various raw materials for composting, municipal sewage sludges are considered of ones of the great use due to their chemical composition (Barreira et al., 2008).

However, digested sludge is too dense, wet and presents an inappropriate C/N ratio to compost alone (Table 2). Several studies have already emphasized the importance of adding lignocellulosic bulking agents for composting sewage sludge. Sawdust can be used as carbon source, water-absorbing and bulking agent in order to achieve C/N ratios of 20 - 30:1, moisture between 60 - 70% (weight/weight) and to increase porosity to 30 - 40% in composting mixtures. A growing demand for traditional bulking agents such as sawdust, straw, and woodchip has recently shifted interest in using alternative bulking agents from other sources, such as wallboard paper (Saludes et al., 2008).

Municipal sewage sludge composting develops temperatures high enough for the due time to achieve Class A biosolids as long as a carbon source and bulking agent take part in the mixture (Figures 1 and 2). Temperature starts to raise from 2nd day after mixing and reaches values higher than 55 °C from day 4th onwards (Table 3). The highest temperatures in composting sludge piles have been measured in the first quarter of the process (Saludes et al.,

2008; Geisler, 2000). High concentrations of available nutrients in sewage materials stimulate microbial activity and metabolic heat production (Joshua, 1996). Temperature, however, has to be controlled to avoid premature exhaustion of organic carbon. It can be achieve by turning piles frequently (Esptein, 1997).

Table 2. Some characteristics of sewage sludge at dry basis (means ± standard errors)

Property	Digested sludge
Total-N (%)	6.5 ± 0.02
Mineral-N (mg kg^{-1})	624 ± 54
Total-P (%)	7.2 ± 0.13
Available-P (mg kg^{-1})	268 ± 17
Total-C (%)	37.8 ± 0.03
C/N ratio	5.8 ± 0.1
Potassium (g kg^{-1})	4.0 ± 2.8
Aluminum (g kg^{-1})	3.0 ± 1.0
Arsenic (mg kg^{-1})	0,6 ± 0.1
Cadmium (mg kg^{-1})	5.0 ± 1.4
Lead (mg kg^{-1})	10.5 ± 0.7
Copper (mg kg^{-1})	95.5 ± 12.0
Chromium (mg kg^{-1})	19.5 ± 2.1
Mercury (mg kg^{-1})	3.8 ± 0.7
Molybdenum (mg kg^{-1})	6.9 ± 0.7
Nickel (mg kg^{-1})	5.5 ± 0.7
Selenium (mg kg^{-1})	1.3 ± 0.3
Zinc (mg kg^{-1})	164 ± 7.1
pH (1:5 water w/v)	6.4 ± 0.09
Gravimetric water [$\%_{(w/w)}$]	87.8 ± 0.2
Dry solids [$\%_{(w/w)}$]	12.2 ± 0.2
Volatile solids [$\%_{(w/w)}$]	74.7 ± 0.2
Bulk Density (t m^{-3})	1.2 ± 0.1

Compost temperature is a consequence of microbial activity, which is generated during organic matter decomposition and growth of indigenous microorganisms. The process is divided into four stages: (1) mesophilic, (2) thermophilic, (3) mesophilic, and (4) maturing. Acid-forming bacteria and fungi appear during the initial mesophilic stage. Mesophilic microorganisms utilize soluble and readily degradable carbohydrates, increase microbial activity and raise temperature favorable to thermophilic microorganisms.

As temperature increases to 40 °C, degradation is taken over by thermophilic bacteria, actinomycetes, and thermophilic fungi. At 65 - 70 °C, spore-forming bacteria and actinomycetes become dominant and organic matter degradation speeds up during the thermophilic stage of composting. Temperature falls down when most of the easily degradable materials are consumed and the rate of heat generation becomes less than rate of heat loss. At this stage, mesophilic bacteria and fungi reappear (Saludes et al., 2008). The passage through the four phases is important for sanitation and stabilization of sewage materials.

When piles run appropriately heat is generated and temperature is used as sanitation parameter and indication of composting efficiency.

Table 3. Some properties of piles during composting of sewage sludge (dry matter basis, except where stated)

Time (days)	Temperature (ºC)	Moisture (% wet basis)	pH_{water} (1:5w/v)	Total-N (%)	NH_4^+-N ($mg\ kg^{-1}$)	NO_3^--N ($mg\ kg^{-1}$)	Total-C (%)	C/N ratio	Total-P (%)	Available-P ($mg\ kg^{-1}$)
0	13.1 a	68.4 a	6.0 a	1.83 a	183 a	0.0	47.7 a	24.1 a	2.02 a	329 ac
2nd	19.3 b	59.9 b	6.2 ad	1.43 bd	160 b	0.0	46.8 ab	25.6 ab	1.98 a	339 ac
4th	59.1 cdef	59.4 b	8.0 b	1.20 c	157 b	0.0	44.7 bc	37.2 c	2.02 a	391 b
6th	64.6 c	59.3 b	8.3 b	1.23 bc	291 c	0.0	42.0 cd	34.1 cd	2.14 b	346 a
8th	55.6 d	60.3 b	8.2 b	1.27 c	283 c	0.0	40.4 d	31.8 de	2.26 c	322 c
10th	56.0 defg	59.6 b	8.0 b	1.32 bcd	334 d	0.0	40.0 de	30.3 e	2.31 cd	336 ac
12th	56.6 d	63.3 bc	8.4 b	1.35 bc	484 e	2.1 a	39.5 e	29.2 ef	2.36 d	341 ac
14th	55.5 d	62.4 bc	8.3 b	1.37 bcd	468 f	3.0 ab	29.5 e	28.5 f	2.40 de	325 c
16th	57.9 de	63.8 c	8.2 b	1.36 bcd	426 g	2.8 ab	39.1 e	28.4 f	2.41 e	326 c
18th	57.3 de	64.9 ac	6.8 cd	1.39 bc	249 h	5.0 b	39.0 e	27.2 bf	2.45 ef	336 ac
20th	59.7 e	64.2 c	7.2 c	1.43 bd	292 c	3.5 ab	38.6 edf	26.3 af	2.47 f	332 ac
22nd	54.6 f	61.0 b	7.0 c	1.47 bd	287 c	5.0 ab	37.9 ef	25.5 af	2.47 f	350 bd
24th	52.0 fg	59.0 bd	6.9 cd	1.46 bd	208 i	6.3 bc	37.6 f	25.4 af	2.48 f	367 bd
26th	49.3 g	57.2 bd	6.5 de	1.50 d	216 i	7.9 bc	37.4 f	24.8 a	2.50 f	369 bd
28th	51.5 fg	55.9 d	6.5 de	1.51 d	204 i	6.6 bc	37.1 f	24.5 a	2.49 f	366 bd
30th	40.6 h	53.6 de	6.5 de	1.49 d	176 aj	4.6 a	37.3 f	25.0 af	2.49 f	371 bd
32nd	34.1 i	50.9 e	6.4 e	1.51 d	171 abj	3.6 a	36.8 f	24.4 a	2.45 f	377 b
34th	31.2 i	50.3 e	6.6 de	1.53 d	167 bj	9.4 c	36.8 f	24.0 a	2.44 f	379 b
94th	21.3 g	55.1 d	6.1 ae	1.59 d	180 aj	97.7 d	36.7 f	23.1 a	2.42 f	377 b

Means (n=3) with same letters within each column are not significantly different by test of Tukey at $p < 0.05$.

Thermal extermination of pathogens is an important via of waste sanitation (Saludes et al., 2008), but nutrient changes and microbial competition also play an important part in the composting hygiene process. Pathogenic organisms survived longer when exposed only to high temperatures than when they were placed in contact with composting organisms (Golueke, 1975). Sewage sludge contains its own microbial population, the pathogens. They are a numerically insignificant fraction of the total microbial population in well run composting piles. An established saprophytic community is therefore essential to avoid pathogenic re-colonization and the development of nuisance conditions in composted sludge.

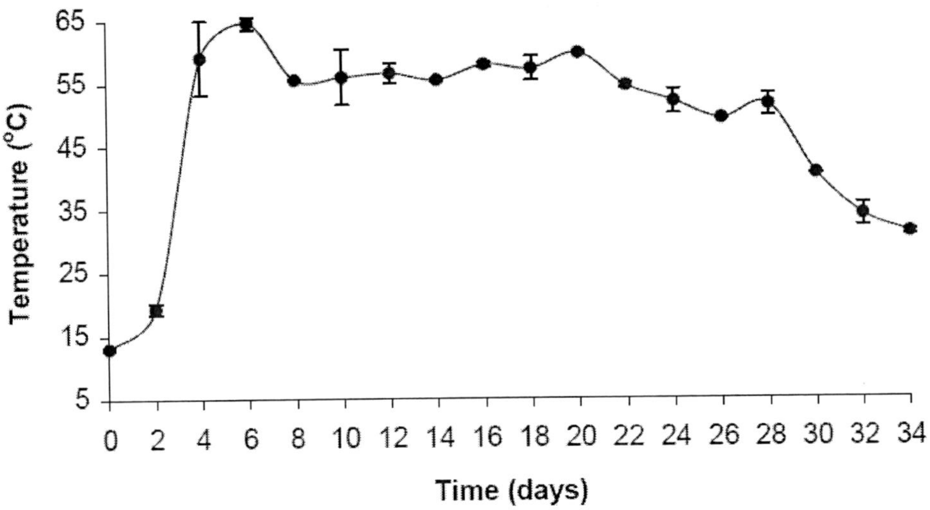

Figure 1. Temperature during composting of sewage sludge.

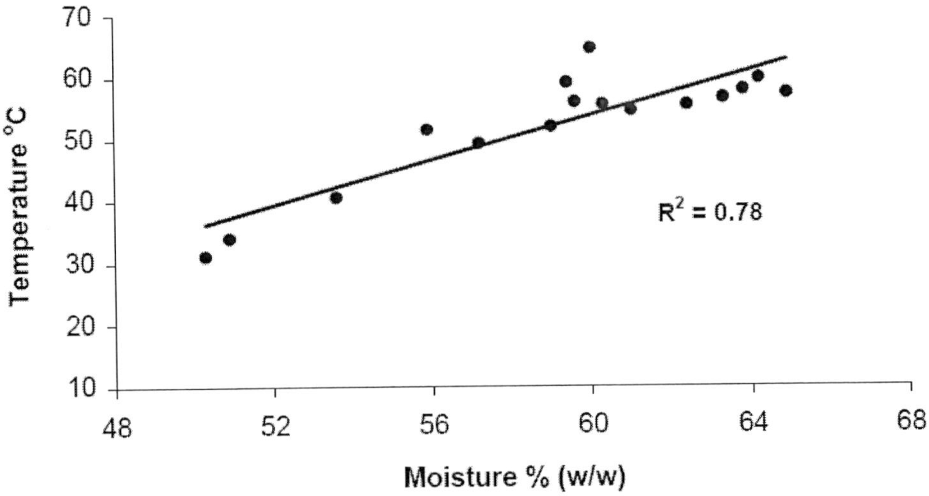

Figure 2. Correlation between temperature and moisture during composting of sewage sludge.

Composting is also a dewatering process. Piles containing sewage sludge start at 65 - 70% moisture to end the process at 50 - 55% water content. This represents 30% loss of water present in the initial mixture and a solid concentration of 60% on a wet weight basis. Initial sludge composting mixtures are wet, sticking, smelly and heavy. Dewatering and other transformations during the process reduce pile size to approximately one-third of the original volume, as commonly observed for the composting of various organic matter sources (Robin et al., 2008). The composted product is an odorless, humus-like material easy to handle.

However, there is a major decrease in organic matter content during the first stage of composting, after which it remains practically constant until the end of the process. This is caused by mineralization of labile organic compounds which mainly occurs during the thermophilic phase. It is expected that organic matter content of compost decreases during the processes, as microorganisms feed on organic matter. Finally, humification prevails over mineralization during the maturation stage afterwards (Gil et al., 2008; Saludes et al., 2008).

Microbial breakdown of organic matter releases CO_2 to the air and water into the composting medium. Moisture correlates with temperature during the process (Figure 3) as microbial respiration releases significant amounts of water into composting mixture. However, many authors defend the cost advantage of composting compared to other stabilization processes because of its capability for stabilization and dewatering sludges in one step (Barreira et al., 2008; Epstein, 1997). The mixture of sawdust to sludge at 3:1 ratio and the composting process result in more than 40% water loss and 4-fold solid concentration compared to the initial moisture in digested sewage sludge.

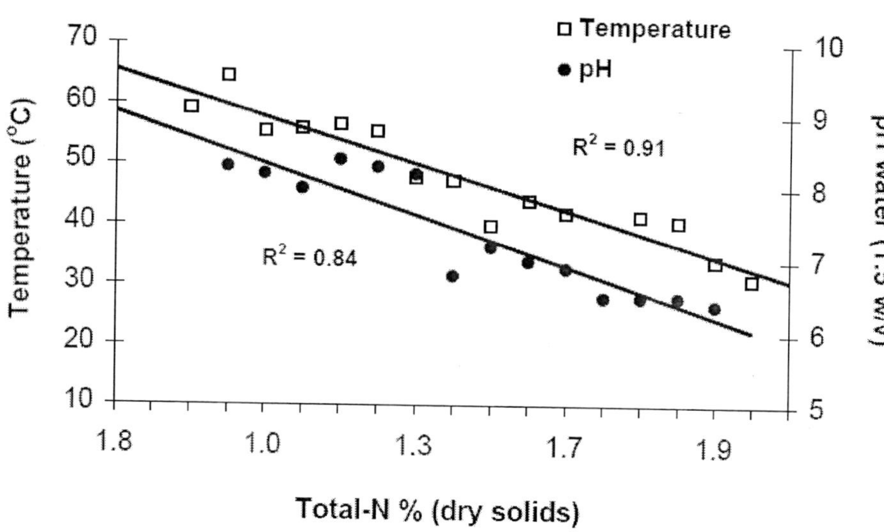

Figure 3. Correlation between temperature and pH and total-N during composting of sewage sludge.

Piles containing sewage sludge increase pH from 6.0 to 8.0 when composting process reaches thermophilic temperatures - Table 3 (Gil et al., 2008). Numerous studies have shown relationship between pH changes and nitrogen transformations during composting. Organic-N is transformed in NH_4^+-N and OH^- through ammonification, which is an alkalizing reaction. In the earlier stages of composting, odor compounds like ammonium-N (NH_4^+-N) and hydrogen sulphide (H_2S) are abundant due to hydrolysis of quaternary ammonium compounds. They partially volatilize and oxidize biologically and then, such compounds tend to disappear as the process progresses (Citterio et al., 1986; Saludes et al., 2008).

Ammonification of organic-N from sewage sludge explains pH increases in composting piles. The release of ammonia (NH_4^+-N) into the mixture results in raises of pH. On the other hand, high pH condition converts non-volatile ammonium ions to volatile ammonia, and high temperature condition promotes volatilization (Saludes et al., 2008). The production of NH_4^+-N needs an organic-N source and there is a decrease in total-N concentrations from sewage sludge during the composting process. Consequently, total-N concentrations negatively correlate with both temperature and pH.

Maximum ammonia emission occurs on the first days of composting and begins to decline rapidly thereafter. Ammonia volatilization is undesirable because of the N loss in the final product and odor generation. N volatilization is affected by several factors, such as ammonia concentration, pH, temperature, moisture content, and aeration rate (Esptein, 1997).

Values of pH decrease in the third phase of sludge composting (Table 3). This can involve N-volatilization, nitrification, and immobilization. Volatilization of NH_4^+-N is a pH neutral reaction as the H^+ ions released from NH_4^+-N neutralize the OH^- ions produced during ammonification of organic-N, until pH is stabilized around neutral values (White, 1997). Nitrification can follow ammonification, producing 2 H^+ for each NH_4^+-N nitrified (Saludes et al., 2008). Further decomposition of organic matter and accumulation of organic acids also contribute to the decline in pH. Finally, synthesis of bacterial structures can immobilize mineral-N in compost.

When NH_4^+-N concentration peaks, approximately two weeks after the process start, nitrate (NO_3^--N) can be detected in sludge composting mixtures. Although not at proportional levels, decreases in NH_4^+-N concentrations are accompanied by increased levels of NO_3^--N (Table 3) and consequent decay in pH. However, nitrification normally takes place during the maturing stage when temperatures are below 40 °C and aeration conditions are favorable (Saludes et al., 2008). NO_3^--N produced by nitrification corresponds to approximately 3% of NH_4^+-N lost during the composting of sewage sludge. Therefore, volatilization and immobilization of NH_4^+-N remain as the major paths of mineral-N losses in sewage sludge composting.

The composting process simultaneously transforms carbon and nitrogen by means of heterotrophic microorganisms (Hotta and Funamizu, 2007). Therefore C/N ratio is a major factor affecting the process as C and N play a paramount part in the process of microbial growth and activity. Carbon is the principal source of energy, and N is needed for cell synthesis. Phosphorus, sulfur and micronutrients are also important, but less is known about their role in composting (Epstein, 1997). Usually organic wastes provide enough of all other nutrients for the composting microorganisms, as long as an appropriate C/N ratio is achieved.

Ideal C/N ratios for composting range between 20 - 30 although composting mixtures with C/N ratios between 20 and 50 are frequently reported to work (Handreck, 1993). At C/N ratios exceeding 50, the composting process slows because of rapid cell growth and depletion of available N. C/N ratios lower than 20 result in NH_3^- volatilization, particularly under alkaline conditions (Handreck and Black, 1994). Piles containing sewage sludge increase C/N in the beginning of composting to decrease it soon after as a consequence of the release of CO_2 from microbial respiration (Figure 4).

Increases in C/N ratios are not result of C gains but rather caused by large losses of N in the first days of trial (Gil et al., 2008). Sánchez-Monedero et al. (2001) reported N losses up to 41% of the total-N in piles where sewage sludge was added to the composting mixture.

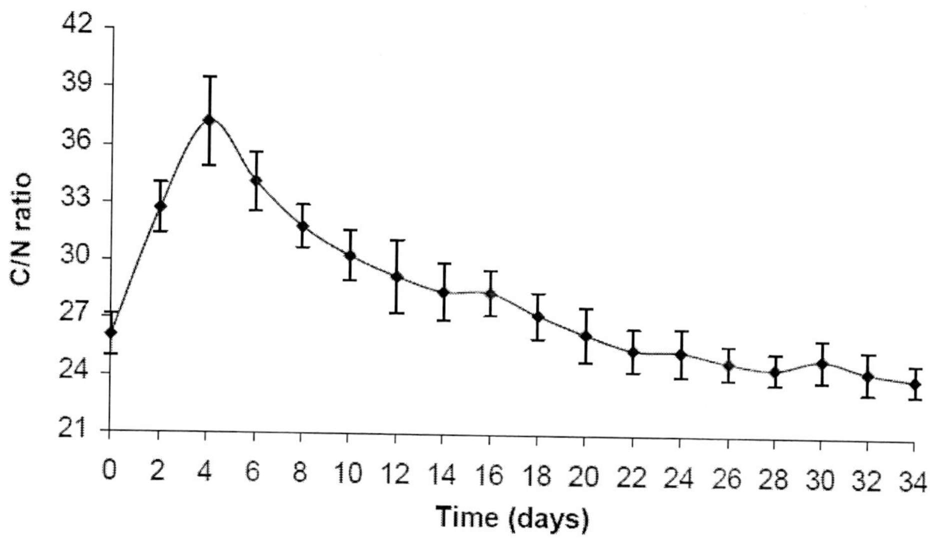

Figure 4. C/N ratios during composting of sewage sludge.

N-volatilization leads to high C/N ratios in the first week of composting trials, when temperature and pH also reach the highest values. In an experiment run with digested sludge, only 6% total-C was consumed for a 39% total-N loss in the first week of composting (Corrêa, et al., 2006). From that, losses of C became higher than losses of N and C/N ratios started to drop. Such a pattern is frequently reported when composting piles contain sewage sludge (Epstein, 1997; Saludes et al., 2008).

3.2. Total N and P

The addition of sawdust or other C source to meet suitable composting conditions dilutes N and P to one third of their original concentrations in digested sludge dry solids. Nitrogen volatilization and the use of mineral-N by microorganisms further reduce total-N to approximately 25% of the initial sludge N content (Nishanth and Biswas, 2008). However, the dewatering effect of composting concentrates N in the bulked product (wet basis). Adding sawdust (10% moisture) to digested sludge (88% moisture) results in a mixture with 68% moisture. The composting process can further decrease water content to 55% on wet basis. As a result, composted sludge's dry solids are fourfold concentrated relative to digested sludge.

Temperature is well related to nutrient changes from organic to inorganic forms, while pH plays the main role on the availability of N and P. As soon as composting process starts, N and P follow different fates. There are significant losses of total-N and a relative increase in total-P concentration (Figure 5). Unlike P, a fairly large loss of N occurs when raw materials are converted to compost and almost all of it is a result of NH_3^- volatilization (Gil et al., 2008; Hotta and Funamizu, 2007; Katayama et al., 1986).

Amounts of total-N lost are bigger than amounts of organic-N mineralized. Amounts of mineral-N that remain in the compost's dry matter are far bellow the quantity of organic-N mineralized during composting because N volatilization occurs during the process. It is common in open air composting systems for an initial decrease in N to be followed by a partial N recovery (Figure 5).

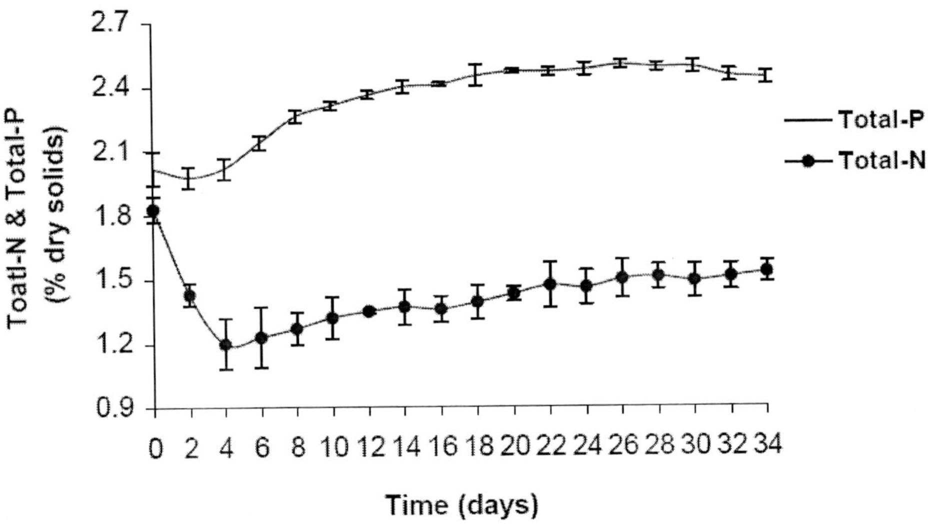

Figure 5. Total-P and N during composting of sewage sludge.

Increased total-N concentrations are sometimes reported as a result of sewage sludge composting (Joshua, 1996). Seekins (2000) reported 25% N losses in windrows starting at C/N = 20. When the process started at C/N = 45, there was 30% N accumulation in the final composted product. The author suggested bacterial fixation as the way of increase N concentration in high C/N ratio piles. However, microorganisms capable of fixing N from the air have not been reported in composting mixtures.

Alternatively, building of N in composting piles may be explained in terms of relative nutrient concentration (Andreoli et al., 1999; Corrêa et al., 2006; Hotta and Funamizu, 2007). A composting study showed that 66% of the initial fecal N and 85% of the initial fecal C were biodegradable in the process (Hotta and Funamizu, 2007). When C is consumed, C/N ratio drops and total-N increases (Figure 5). C/N ratios continue to drop because of the bigger losses of total-C relatively to total-N. It alters the proportion of elements in dry solids. Losses between 10 and 20% of organic-C content have been reported during thermophilic phases of sludge composting. Overall, approximately 60% of organic-C is lost during the process (Robin et al., 2008).

Taking C loss into account, every 100 kg of dry solids in initial composting mixture renders 92 kg of composted biosolid dry solids. Then, 1.83 kg of total-N (Table 3) present in initial 100 kg mixture would reach 1.99% total-N in the final product if volatilization did not occur. But composting of sewage sludge causes loss of approximately 25% volatile solids. Such a reduction in volatile solids could have increased total-N to 2.3%. Consequently there is, in fact, 0.8% total-N lost during the composting process or 40% relative to the digested sludge's initial total-N content (Corrêa, 2001; Sánchez-Monedero et al. 2001).

Epstein (1997) measured 25% decrease in total-N caused by poor aeration of a static pile. He could reduce losses by increasing bulking agent ratio from 1:1 to 3:1. The higher the N concentration in the mixture, the greater is the N loss potential (Seekins, 2000). But the adoption of higher C/N ratios would dilute N even more. Some ammonia loss is practically inevitable in composting and hence, with higher C/N ratios total-P content starts the process with lower concentrations. Therefore, bulking mixtures more than necessary is not a measure of nutrient conservation for composting sewage sludge.

Unlike N, the composting of sewage sludge does not result in P losses. On the contrary, total-P concentration increases in about 20% during composting trial (Table 3). Magnification of total-P can be explained through the release of C as CO_2, NH_3^- volatilization and losses of volatile solids. Total-P increases mainly in the first week of composting trial, when C and N losses occur at the highest levels (Figure 5). The masses of non-volatile elements (P, K, Mg, Ca, Fe, Cu, Zn, and Mn) are quite stable during composting, while there are losses of volatile elements such as C, N (Robin et al., 2008). As a result, some elements appear diluted in the composted product and others appear in concentrated forms.

3.3. Mineral-N (NH_4^+-N + NO_3^--N) and Available-P

Not only total concentrations but also the availability of nutrients in composted materials can vary greatly (Figure 6). When sewage sludge undergoes composting, more than 10% of ammonia (NH_4^+-N) is lost to the air even before the process reaches thermophilic temperatures (Table 3). When temperature achieves thermophilic levels, concentrations of ammonia double and continue to increase until peak at the middle course of the process (Figure 6). From that, ammonia concentration returns to the same initial levels because volatilization and immobilization occur (Hotta and Funamizu, 2007). Although NH_4^+-N concentration at the end of the process is lower than the initial value (Table 3), the account for NO_3^--N brings mineral-N (NH_4^+-N + NO_3^--N) to amounts similar to initial values measured in composting mixtures (Table 3). NH_3^- volatilization in the first stage of composting and immobilization at the end of the process are the most reported mechanisms of ammonia (NH_4^+-N) loss from composting piles (Epstein, 1997).

Nitrate (NO_3^--N) significantly increases only during the maturing phase of composts. Some changes in heaps take place only at low temperatures and in well decomposed organic matter (Gil et al., 2008; Rynk et al., 1992). NO_3^--N usually becomes noticeable only at end of the second mesophilic phase. As a result, the presence of NO_3^--N has been traditionally related to mature composts (Diaz-Burgos et al., 1991; Hotta and Funamizu, 2007).

Nitrifying microorganisms take on composting piles during the maturing phase, when conditions are favorable to their colonization. Nitrification is intermediated by bacteria of the genera Nitrosomonas and Nitrobacter, which derive energy from oxidation of NH_4^+ (White, 1997). Furthermore, Nitrobacter organisms are inhibited at pH 8 or higher, and because of the inappropriate temperature and pH during composting it seems reasonable that NO_3^--N significantly appears only during the maturing phase.

The fate of available-P in composting piles containing sewage sludge is simpler than that of mineral-N (Figure 6). Available-P increases nearly 20% in piles as soon as temperature reaches thermophilic levels in the first days of composting (Figure 6) as a consequence of organic matter mineralization. Then, available-P returns to concentrations similar to that initially measured in piles by the end of the process (Table 3, Figure 6). Reduction of P availability follows elevations of pH from values of 8.0 to higher ones. Then, P-availability increases again following pH decays to values around 7.0. Concentration of available-P is approximately 20% higher at the end of sludge composting than in the beginning of it (Table 3). The best factor that explains changes of P availability in composting piles is pH. It seems that part of the P released from organic pool becomes immediately unavailable due to high pH in the first half of the composting process.

Some authors state that nutrients in composts are more plant-available than in the organic matter from which the compost originated (Parkinson, 1996). However, others have a different opinion as they report lower availability of nutrients in composted wastes than in their original feedstock (Barreira et al., 2008; Gil et al., 2008; Sims, 1990; Stentiford, 1996). A mature composted biosolid takes three months to be produced and presents 278 mg kg^{-1} of mineral-N (Table 3) against 624 mg kg^{-1} present in its original feedstock (Table 2). Available-P appears in composted biosolid at 377 mg kg^{-1} against 268 mg kg^{-1} in the fresh feedstock. Therefore, the composting process reduces mineral-N by 55% and increases available-P by 40% relative to the digested sludge from which compost is originated.

The account for nutrient availability on wet weight basis shows mature composted biosolid can readily deliver 125 mg kg^{-1} of mineral-N and 169 mg kg^{-1} of available-P while the fresh feedstock accounts for 76 mg N kg^{-1} and 33 mg P kg^{-1} respectively. Therefore, there are 5.2 times more available-P and 1.6 more mineral-N in a tonne of composted biolsolid (55.1% moisture) than in a tonne of digested sludge (87.8% moisture) from which compost is originated. Nevertheless, compost can not work as effectively as chemical fertilizers because its mineralization time is generally unknown and therefore, the availability of nutrients to plants is also uncertain (Gil et al., 2008). As source of plant nutrients, composted sewage sludge is closer to manure than to chemical fertilizers. This causes available N and P to be released slowly and steadily. For that reason, compost rather works as soil conditioner than as fertilizer. Composted sewage sludge improves a number of soil quality parameters when applied to land such as soil water content, bulk density, porosity, structure, and many others (Pengcheng et al., 2008).

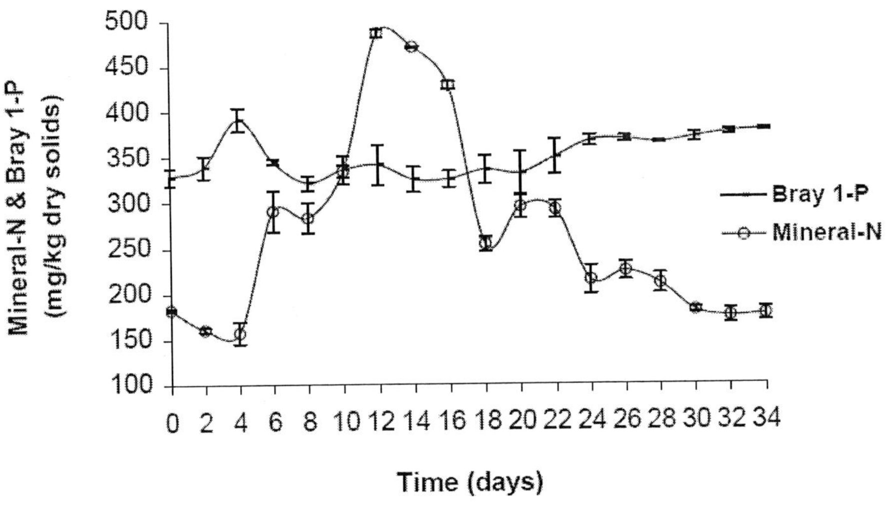

Figure 6. Changes in mineral-N and available-P during composting of sewage sludge.

From the practical point of view, the dewatering effect of composting overrides changes in nutrient availability during the process. Many factors affect composting and nearly all of them are interrelated. Temperature significantly correlates with several composting properties (Table 3). Only NO_3^--N and available-P does not show direct relations to composting temperature. Therefore, temperature seems very appropriate as a process´s monitoring factor. Composting of digested sludge comes to the end without the existence of any limiting factor, even without C/N ratio limitation for the process to continue (Table 4). The most important factors that have to be controlled for successful composting are aeration, moisture, and C/N ratio, as digested sludge contains all other necessary nutrients.

Curing phase of compost begins when piles non-longer reheat after turning. This stage is characterized by much lower rates of metabolic activity, although further aerobic decomposition of resistant compounds and organic acids and reduction of volatile solids still occur. There is also formation of organic complexes such as humus and the build up of organisms that play an essential role in minimizing the growth of undesirable pathogens in the product (Robin et al., 2008).

The main change after curing refers to N transformations (Table 3). There is a small increase in NH_4^+-N concentration and a ten times increase in NO_3^--N concentration. Nitrification reduces pH and piles reabsorb some moisture from air but without any raise of temperature. Some countries use N drawdown index to reflect the potential of composted products to either act as N sinks or sources in soil.

Physical and chemical properties of initial and final compost can be summarized in Table 4. Chemical analysis shows that total-N, P, K, and Na concentrations in the final compost are slightly higher after composting. Raises in nutrient concentrations can be attributed to the net loss of compost dry mass associated with organic matter degradation, which lead to the concentrating effect on some nutrients (Saludes et al., 2008).

Table 4. Summary of some changes during composting of sewage sludge

Time (days)	Microbial stage	Temperature (°C)	pH_{water} (1:5w/v)	C/N ratio (dry weight)
0 - 4th	Mesophilic	13 → 59	6 → 8	24 → 37
4th - 28th	Thermophilic	49 - 65	8 - 8.4 → 6.5	37 → 25
28th - 34th	Mesophilic	51 → 31	6.4 - 6.6	24 - 25
34th - 94th	Curing	31 → 21	6.6 - 6.1	23

4. Alkaline Treatment

4.1. Liming Performance

Alkaline treatment has some advantages over the other sludge stabilization processes such as the simplicity of the device for lime application, it is easy to operate and highly efficient for sanitation (Li et al., 2008). Liming treatment aims to kill pathogens by the rise of pH to at least 12 and by keeping nearly it for 14 days or longer (Plachá et al., 2008) (Table 1). High pH causes protein to loose their natural shapes, saponification of lipids and hydrolysis of RNA. Then, cells are exposed to environment with extremes of pH thereby can not keep the appropriate turgor pressure, are disrupted and the inner matters are released. Therefore, alkaline treatment can solubilize sludge and release inner water (Li et al., 2008).

The strength of liming materials depends on their ability to remove H^+ from colloid surfaces. This ability decreases in the order oxides > hydroxides > carbonates and therefore different performances on liming sewage materials usually happen. Sludge moisture is another variable which strongly affects pH values during alkaline treatment of sewage materials.

The main disadvantage of using chemical disinfectants is the negative effect of these agents upon the living environment. An exception refers to the substances used to disinfect sludge, which do not load the environment, are accessible and even capable of improving the sludge fertilizing quality. For this purpose lime in the form of lime hydrated (calcium hydroxide) or quicklime (calcium oxide) is used. The latter, if added to sludge, reacts with water and produces $Ca(OH)_2$ and heat (exothermic reaction). The heat acquired and the alkalic pH are decisive disinfecting factors (Plachá et al., 2008).

Sufficient quantity of OH^- has to be added to sludge in order to combine with all free carbonic acid of the bicarbonates to raise pH. Mechanisms of pH decay involve chemical and biological processes, depending on the pH level. Decreases of pH caused by NH_3^- volatilization from sewage sludge down to levels where biological activity can re-start is frequently reported (Costa and Krohling, 1998).

Digested sludges (87.8% moisture) have to receive from 30 to 50% lime [CaO or $Ca(OH)_2$] to sludge dry solids (weight/weight) to reach pH 12 (Table 5). There is no significant difference between CaO and $Ca(OH)_2$ at 30 - 50% treatments soon after liming. However, the process is run for two weeks and only those samples holding pH higher than 11 are considered appropriate for alkaline treatment (Table 5). $Ca(OH)_2$ applied at 40 - 50% and CaO applied at 30 - 50% sludge dry solids are able to meet USEPA Regulation 40 CFR Part 503 criteria on lime stabilization. Alkaline treatments at 20% lime/sludge present pH decays before the end of the process (Figure 7), which indicates inappropriate sanitary conditions. To lime digested sludge with $Ca(OH)_2$ at 30% lime/sludge may be not enough, since pH drops too early (Figure 7). Keeping pH higher than 11 for at least two weeks is critical to stopping biological activity in limed-sludge (Metcalf and Eddy, 2003).

Table 5. Sludge pH responses to alkaline treatment

Biosolids	$Ca(OH)_2$-sludges				CaO-sludges			
Lime rate								
0%	20%	30%	40%	50%	20%	30%	40%	50%
pH at day 0								
6.2 a	11.8 b	12.2 cg	12.2 cg	12.3 d	11.9 b	12.2 cg	12.3 c	12.3c
pH at day 14th								
7.4 d	8.2 e	10.3 f	11.6 b	11.8 b	8.5 e	11.9 bg	12.1 cg	12.1cg

Means (n=3) of same letter are not statistically different by test of Tukey ($p < 0.05$).

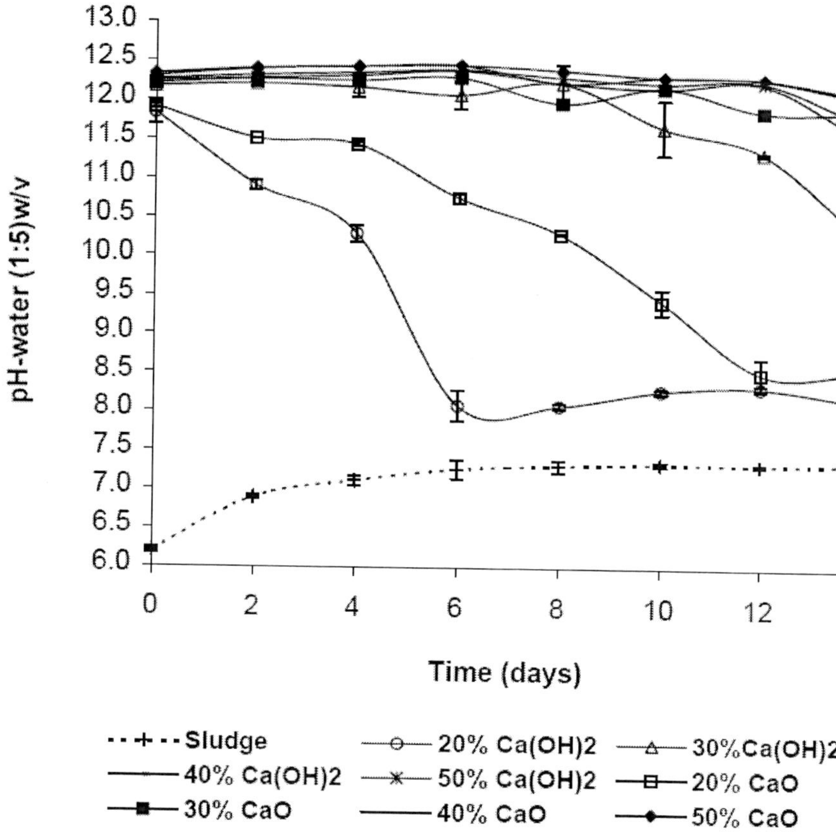

Figure 7. pH of sewage sludge treated with CaO and Ca(OH)2 during 14 days of trial.

Application of $Ca(OH)_2$ at 40% and 50% and CaO at 30 - 50% can achieve criteria for sewage sludge sanitation (Tables 1 and 5). Results from limed-sludge at 50% $Ca(OH)_2$ are similar to lime it at 30% CaO-sludge (Table 6). Therefore, the use of CaO can be considered as 1.7 more effective than $Ca(OH)_2$ to appropriately stabilize sewage sludge. Calcium oxide (CaO) is taken as the most effective of all the liming materials commonly employed in sanitary engineering (Tisdale and Nelson, 1971). It is usually used to burn off high volatile contaminants and fix non-volatile ones (such as metals) in soil particles and other materials. Although lime is referred to in a general manner only $Ca(OH)_2$ and CaO are able to meet USEPA pH criteria for sewage sludge sanitation.

Calcium oxide (CaO) shows a better performance on alkaline treatment of sewage sludge than calcium hydroxide [$Ca(OH)_2$] when they are applied at similar rates. The use of CaO at 30% represents the saving of up to 20% resource compared to application rates of 40 and 50% of CaO or $Ca(OH)_2$. Subiaco Wastewater Treatment Plant in Perth (Australia) has been successfully achieving Class B biosolids with the application of CaO at 20% to sludge dry solids (Bridle et al., 2000). Health matters are the main concern in sewage sludge management and mixtures up to 50% lime to sludge dry solids may not be a waste of resources. This is especially true to avoid the onset of putrefying conditions if sludge will be kept for periods longer than two weeks. Taking sludge dry solids into account, two effects can be noticed on liming process: an immediate increasing in solid content, usually proportional to the amounts of lime applied, and a further increase of it due to water losses through time (Table 6).

Table 6. Alkaline treated sludge´s moisture content and total water losses

Biosolids	$Ca(OH)_2$-sludges				CaO-sludges			
Lime rate								
0%	20%	30%	40%	50%	20%	30%	40%	50%
Water content at day 0 (%)								
87.8 a	85.4 b	84.7 b	83.0 c	82.4 ce	84.7 b	83.6 c	82.0 e	80.8 f
Water content at day 14th (%)								
80.1 f	78.0 g	77.3 g	75.7 hi	75.0 h	77.2 g	76.3 i	74.8 h	73.5 j
Total water loss (kg water/t sludge 87.8% moisture)								
388 a	338 b	325 b	301 cd	294 d	374 a	308 c	286 d	273 e

Means (n=3) with same letter are not statistically different ($p < 0.05$) by test of Tukey.

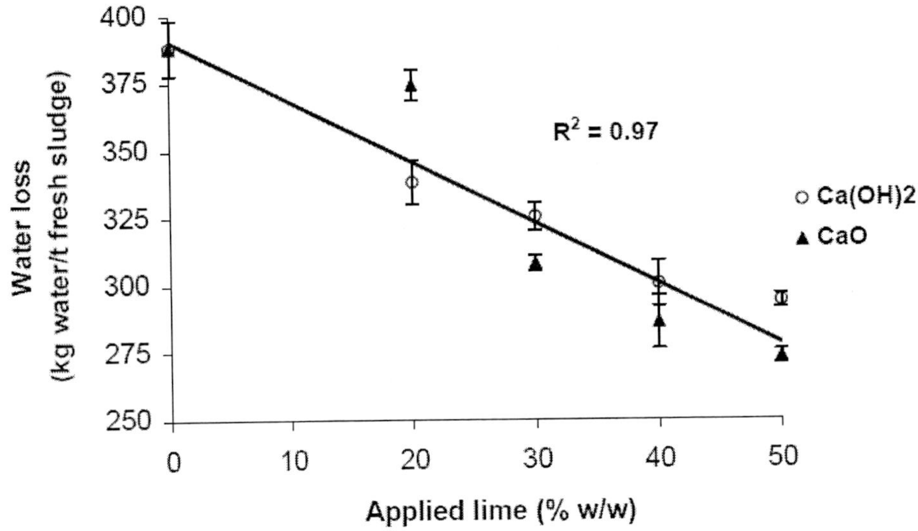

Figure 8. Relation between water lost and lime application rates to sewage sludge after two weeks of resting period.

Total water losses in two weeks ranges from 27% (weight/weight), for the 50% CaO-sludge, to 39% (w/w) for non-limed digested sludge (Table 6). Water losses in the first week account for 94% of the total moisture losses in the process. Water loss in limed-sludge becomes insignificant from second week onwards. Volume reduction of sludge materials is very important for their management. Lime application during wastewater treatment easies sludge dewatering ability by mechanical means (Li et al., 2008).

However, the higher the rate of lime applied to mechanically dewatered sludge the lower the dewatering effect on it afterwards (Figure 8). At an initial digest sludge's moisture of 87.8% and bulk density of 1.2 t m^{-3}, water loss accounts for a volume reduction to nearly half for the digested sludge and to two-thirds for limed-sludge. Two weeks resting period increases more sludge dry solids, as a result of water loss, than the initial solid concentration caused by the adding of lime (Table 6). Tertiary digested sludge can loose 40% of its initial 87.8% moisture in two weeks of resting to end up with 20% dry solids. Sewage sludge limed at 50% CaO can loose less than 30% of its water content over the same period to finish at 27% dry solids, of which approximately half is sludge and half is lime.

Water is the major constituent of sewage sludges and it has a profound impact on the economics, disposal, and reuse options. As more water is released during stabilization as easier and more profitable is the biosolids management. Liming the sludge at 30% CaO seems the best option not only for the sake of lime savings but also for reduction of water content in the biosolids. Limed-sludge at 30% CaO looses more than 300 kg water/t digested sludge against values between 273 - 286 kg water/t digested sludge limed at 50% and 40% CaO respectively (Table 6).

4.2. Total N and P

Amounts of total-N in sludges alkaline-treated vary according to amounts and forms of lime applied to them. Dilution of nutrients is more pronounced in sludges treated with CaO than with $Ca(OH)_2$ (Figure 9). Calcium oxide (CaO) is more aggressive to organic matter and causes more NH_3^- volatilization than $Ca(OH)_2$ at similar application rates. Hence, part of the $Ca(OH)_2$ applied for alkaline stabilization is converted into water, which is driven off during drying works. It leaves more sludge dry solids and consequently more N in the stabilized biosolids. Among alkaline treatments that achieve USEPA pH-criteria, CaO at 30% and $Ca(OH)_2$ at 40% (weight/weight) preserve 60% of the total-N present in digested sludge dry solids. More than half of digested sludge's total-N is lost if CaO is applied at 40 and 50% rate (Figure 10).

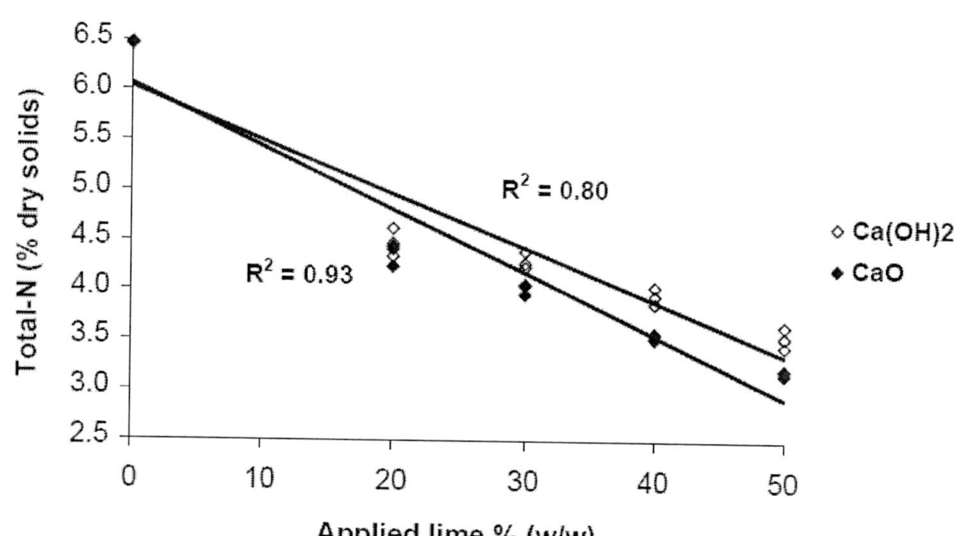

Figure 9. Relation between total-N and lime application rates on day 0 of the alkaline treatment of sewage sludge.

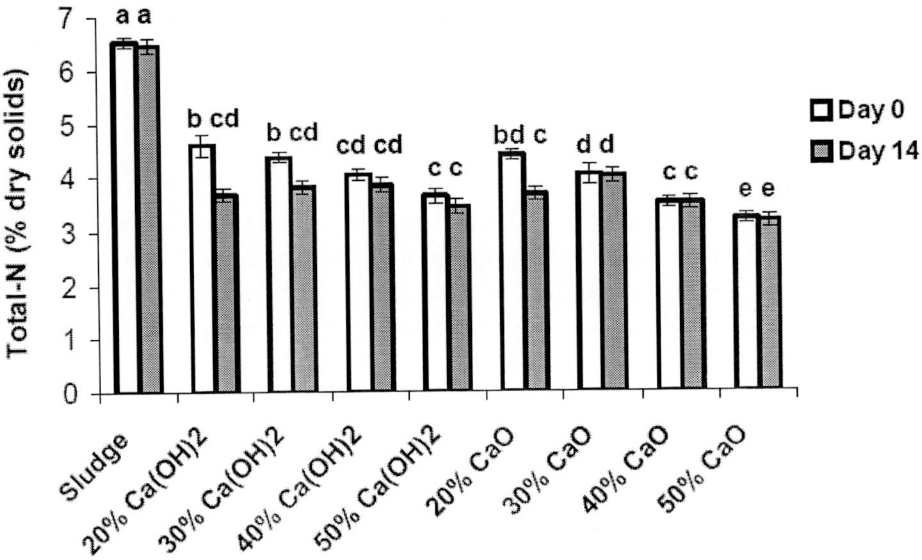

Figure 10. Total-N on days 0 and 14th of alkaline treatment of sewage sludge.

Liming sludge at either 30% or 40 - 50% CaO rates accounts for a difference of 5,000 - 8,000 mg N (kg dry solids)$^{-1}$ which is highly significant for plant growth. There are also N losses through volatilization during the two weeks resting period determined by USEPA criteria (Table 1). Such losses reach 13 - 20% of the total-N in limed-sludge dry solids and such figures represent 6,000 - 9,000 mg N (kg dry solids)$^{-1}$. Nitrogen losses through NH_3^- volatilization occur if pH decreases and it allows the re-start of ammonification in lime-sludge mixtures. At pH 6 - 7 essentially NH_4^+-N is present as mineral-N in sewage sludge. On the other hand, at pH > 8.5 NH_3-N dominates the mineral-N pool (Reed et al., 1995). Some N is also drained by leaching from limed-sludge piles, but it accounts for a minor part of total N losses.

Lime is applied to sludge chiefly to stop biological activity (Girovich 1996a) and ammonification should be also inhibited. N losses are high when ammonification is not inhibited. Digested sludge applied at 20% $Ca(OH)_2$ rate can lose up to 20% of its total-N when pH drops to values near 8.0. If CaO is applied at the same 20% rate, losses are kept under 15% for the same period because CaO is more efficient to raise and keep pH high. As long as pH is kept near 11, there are not significant losses of N in sludge undergoing alkaline treatment. Consequently, liming sludge at 30 - 50% CaO and 40% - 50% $Ca(OH)_2$ application rates stabilize organic matter and conserve N.

Mechanisms of N conservation in limed-sludge rely on the suppression of mineralization. Where NH_4^+-N is absent, there is not volatilization of NH_3^- gas. Low lime application rates initially conserve more N in sludge dry solids, since dilution is less pronounced. However, NH_3^- volatilization soon follows and limed-sludges end the alkaline process poor in N content. Losses of N and reset of a nuisance environment in limed-sludges indicate poor sanitation conditions. Under these circumstances, higher doses of lime or a better product shall be applied.

Concentrations of total-P are also reduced soon after liming due to dilution of sludge dry solids (Figure 11). But, after this initial dilution of total-P concentrations remain the same throughout the trial. There is some loss of P through sludge drained water. Losses of 0.2 - 1.4 mg P L^{-1} of leachate are not significant in agronomic terms but they may be environmentally significant as tiny P concentrations may trigger water eutrophication.

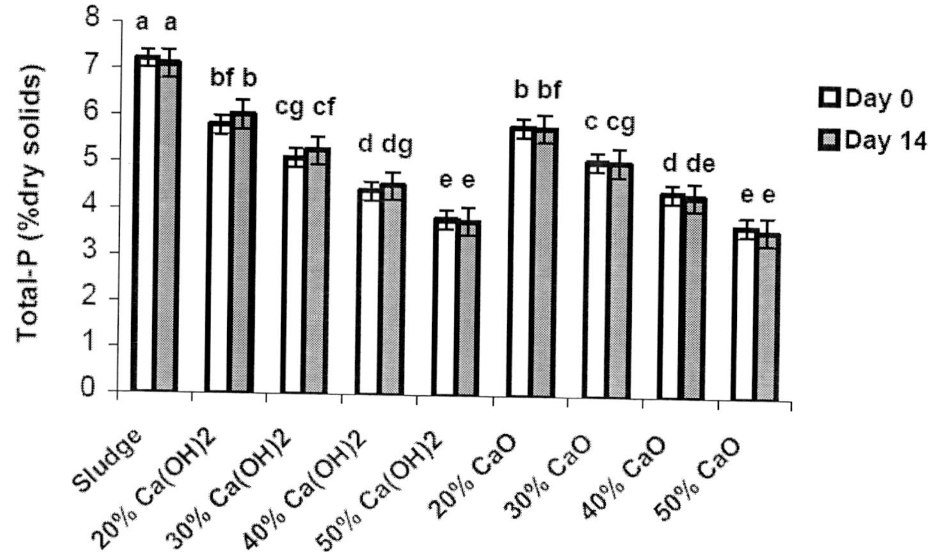

Figure 11. Total-P on days 0 and 14th of alkaline treatment of sewage sludge.

Liming treatments that cause significant losses of total-N (Figure 10) tend to increase total-P concentrations (Figure 11). Release of CO_2 through ammonification and NH_3^- volatilization alter the balance of nutrients remaining in limed-sludges dry matter (Plachá et al., 2008). Digested sludge treated at 30% CaO and 40% $Ca(OH)_2$ keep the highest total-P concentrations among limed-sludges that achieve USEPA Regulation 40 CFR Part 503 criteria (Table 1; Figure 7). These two treatments maintain respectively 70 and 60% of the original digested sludge´s total-P. The other treatments that achieve USEPA Regulation 40 CFR Part 503 criteria (Figure 7) dilute total-P to approximately half the amount present in digested sludge dry solids.

The USEPA Regulation 40 CFR Part 503 criteria for alkaline treatment (Table 1) rely on pH levels. Liming digested sludge at 30% CaO and 40% $Ca(OH)_2$ is enough to raised and maintain pH at 12 level and keep the highest concentrations of total-N and P in their dry solids. Applications of 50% CaO or $Ca(OH)_2$ to digested sludge decrease total-N to less than half of the original amount (Figure 10) and total-P to approximately half of the original concentration (Figure 11). Such rates do not improve pH performance compared to results achieve with 30% CaO or 40% $Ca(OH)_2$. Hence, application of 30% CaO saves 10% lime relative to 40% $Ca(OH)_2$ for similar results on pH and contents of N and P.

4.3. Mineral-N (NH_4^+-N + NO_3^--N) and Available-P

Amounts of mineral-N in digested sludge can be taken entirely as NH_4^+-N because NO_3^--N concentrations are neglected under the anaerobic conditions that prevail in sludges.

Immediately after liming, NH_3^- volatilization causes NH_4^+-N concentrations to drop to less than half the value measured in digested sludge (Figure 12). Mineral-N content in sludge treated with CaO are always lower than in sewage materials treated with $Ca(OH)_2$ at similar application rates. Higher liming rates also imply lower amounts of NH_4^+-N in alkaline treated sludge.

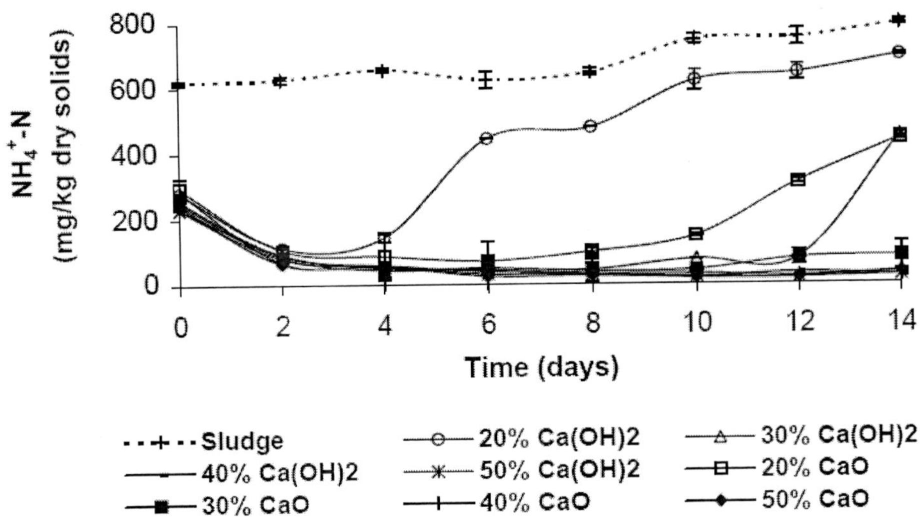

Figure 12. Changes in NH_4^+-N concentrations during alkaline treatment of sewage sludge.

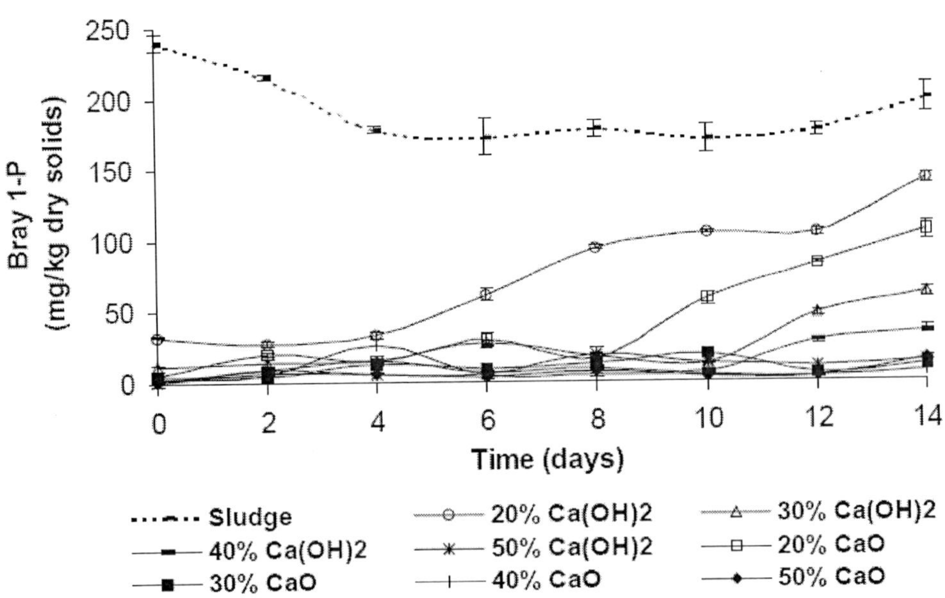

Figure 13. Available-P during alkaline treatment of sewage sludge.

Concentrations of NH_4^+-N continue to decrease during the process up to values close to null (Figure 12). But, pH drops result in the raise of NH_4^+-N concentrations (Figure 12) and some N loss may occur, since NH_4^+-N in high pH medium leads to NH_3^-volatilization.

Dilution of dry solids and NH_3^- volatilization remove all mineral-N originally present in digested sludge. However, such N loss corresponds to approximately 0.05% of the total-N in sludge dry solids. Thereafter, alkaline treatments that kept pH high show neither increase of NH_4^+-N during the process (Figure 12) nor significant losses of total-N (Figure 10). Losses of mineral-N throughout the stabilization process (0.05%) are neglected compared to losses of total-N caused by dilution of sludge dry solids, as long as pH is kept at high values. NH_3^- volatilization accounts for 0.05% of the total-N in sewage sludge and dilution accounts for 20 - 50% of it.

Phosphorus availability drops to nearly null values soon after the alkaline treatment due to high pH environment. As soon as pH drops to values around 8.0, concentrations of available-P recover (Figure 13). However, concentrations of available-P do not reach original values because optimum pH range for P-availability is 6 - 7 (White, 1997). Phosphorus do not volatilize from lime-stabilized sludge. Its availability increases after mixing lime-treated sludge with non-alkaline soils as calcium phosphates dissolve under acidic soil conditions (Smith et al., 2001). Therefore, besides P dilution caused by application of lime, alkaline treatment do not cause further losses of P.

Alkaline treated-biosolids contain residual lime that can increase pH of acid soils and save agricultural lime (Plachá et al., 2008). Some works have reckoned limed-sludges as 68% effective as $CaCO_3$ (99.9% pure) in raising soil pH (Willett, 1984). In that sense, the neutralizing value of digested sludge treated at 50% $Ca(OH)_2$ is similar to sludge treated at 30% CaO (Table 7). For similar amounts of lime applied to sewage sludge, CaO presents as much as twice residual neutralizing value compared to $Ca(OH)_2$.

Table 7. Residual neutralizing value (%) of alkaline-treated biosolids

Biosolids	$Ca(OH)_2$-sludges				CaO-sludges			
0%	20%	30%	40%	50%	20%	30%	40%	50%
-3.8 a	10.0 b	14.3 c	19.4 d	22.9 e	12.5 c	21.8 e	26.3 f	46.9 g

Means (n=3) with same letter are not statistically different by test of Tukey ($p < 0.05$).

5. Thermal-Drying of Sewage Sludge

5.1. Thermal-Drying Performance

A promising approach for reducing volume, offensive odors, and facilitating storage of sewage sludge is drying and pelletizing. Thermal-drying greatly reduces water content, saves transport costs, and increases environmental acceptability and financial value of wastes (Ramírez et al., 2008). The reduced moisture content of pelletized organic wastes makes them easier to store, transport, and apply to land (López-Mosquera et al., 2008).

Thermal-drying is the process of removing water and volatile substances by heating to yield a product with less than 20% moisture content. The process is highly effective to eliminate pathogens, demands short time and does not depend on sludge chemical and biological characteristics. Heat-drying treatment is especially recommended for biological sludge, where water is mostly contained in cells and their destruction is necessary to decrease moisture content. Sufficient heat has to be applied for enough time to destroy cells and allow the bound water to escape.

Sludge particle surface contains free water that evaporates at constant rate in the first stage of drying. After that, moisture has to be transported by capillary and diffusion forces from the interior of particles to their surface, when drying rates fall. As more water is extracted from sludge, higher energy and longer time are demanded to take out the remaining moisture (Girovich, 1996b).

Dryers are evaporators, classified according to methods of heat transfer (conduction or indirect drying, convection or direct drying, radiant, dielectric), type of drying hardware (rotary drum, fluidized bed, spray, tray, conveyor), mode of operation (continuous, batch), or residence time. In such a stabilization process only two variables are involved: temperature and time.

At commercial operations sewage sludge is subjected to 110 - 115 °C for at least 75 min, increasing the solids concentration to over 90% and producing pellets that are approximately 3 mm in diameter (Kalogo and Bagley, 2008). Sewage sludge can be also heated in pressured vessels to temperatures between 180 and 280 °C for an hour (Girovich, 1996b), which is enough to kill pathogens (López-Mosquera et al., 2008). Drying sewage sludge at 350 °C would take only 20 min, but energy consumption at such a high temperature may be not economically affordable. Thermal-stabilization process is recognized as the most expensive method prior to agricultural use of biosolids, whilst direct land application has been the cheapest way to beneficially use sewage sludge (Bridle and Skrypski-Mantele, 2000).

Energy demand to reach 350 °C is too high and eight hours of resident time at 100 °C is too long because the losing of economic effectiveness (Girovich, 1996b). A reasonable choice would be to process digested sludge at 200 - 250 °C that corresponds to the range where higher temperatures would not significant reduce resident time (Figure 14). Sludge can be dried in 1 - 2 hours at that temperature range. However, commercial dryers have either some mechanical shaker or pressured system that can shorten resident time.

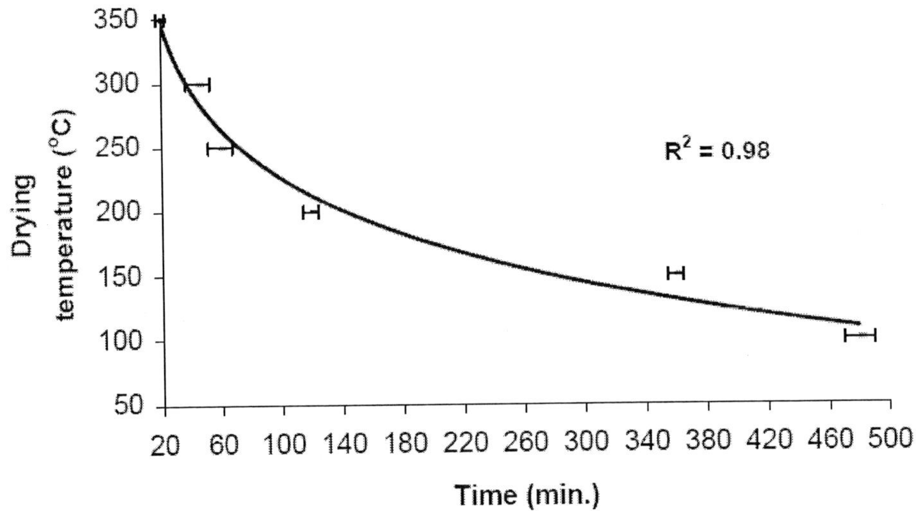

Figure 14. Relation between temperature and resident time in thermal-drying stabilization of sewage sludge.

Commercial plants do not extract water beyond 5% moisture due to energy consumption, which is an economic matter. Some thermal-drying plants in Europe have opted to dry sludge at 200 °C, 1.8 MPa pressure for 45 min (Pommel et al., 1991). Other plants also in Europe apply 200 °C, 1.6 MPa for an hour (Jaeger and Mayer, 2000). A plant in Western Australia has chosen to operate a furnace at 275 °C for approximately half an hour. The semi-dried material is then conveyed into a rotary drum and pelletized (1 - 4mm) at 100 °C.

5.2. Total N and P

There is a common sense that thermal-drying stabilization does not diminish fertilizer value of sewage sludge (Girovich, 1996b). However, there may be changes in nutrient content when sludge is stabilized by heating at certain temperatures. Nitrogen losses of 30% and 14% can occur when digested sludge is dried at 100 and 150 °C respectively. Hence, approximately 6% of total-N is lost when sewage sludge is dried at temperatures ≥ 350 °C (López-Mosquera et al., 2008). On the other hand, drying temperatures between 200 - 300 °C does not cause significant losses total-N and P in sewage sludge (Figure 15).

Digested sewage sludge has a high latent heat that keeps its temperature at approximately 75 °C when drier is at 100 °C (Figure 16). Water is evaporated at 75°C until sewage sludge dries out. Meanwhile, mineralization of organic-N is boosted and NH_3^- volatilizes. Heating stimulates the hydrolysis of proteinaceous materials, which results in cell destruction and release of soluble organic compounds to the sludge medium and NH_3^- to the air (Metcalf and Eddy, 2003). Volatilization of NH_3^- at 100 °C drying temperature results in relative increment of P in thermally dried bisolids (Figure 15). If higher concentrations of P than N are required in biosolids, the 100 °C treatment may be an option.

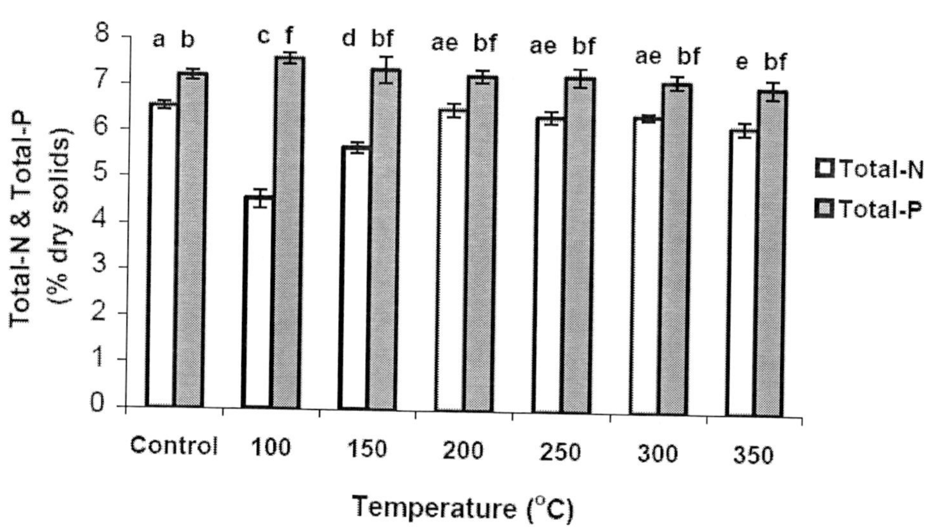

Figure 15. Total-N and P contents in thermal-dried biosolids at different temperatures.

In spite total-N losses (Figure 15), NH_4^+-N concentrations are higher in biosolids dried at 100 and 150 °C than in sewage sludge, but lower than in biosolids stabilized at 200 - 350 °C. Mineralization occurs in sewage sludge at any drying temperature because there are increments of available-P in all heat-dried biosolids. Available-P concentrations are usually higher than NH_4^+-N in thermally stabilized biosolids because P is not lost during the drying process (Figure 17).

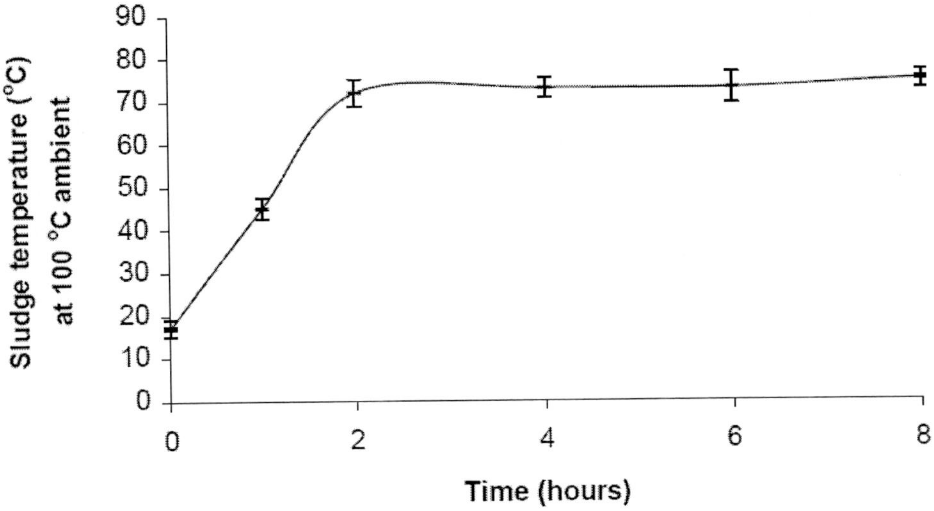

Figure 16. Temperature of sewage sludge under 100 °C drying temperature.

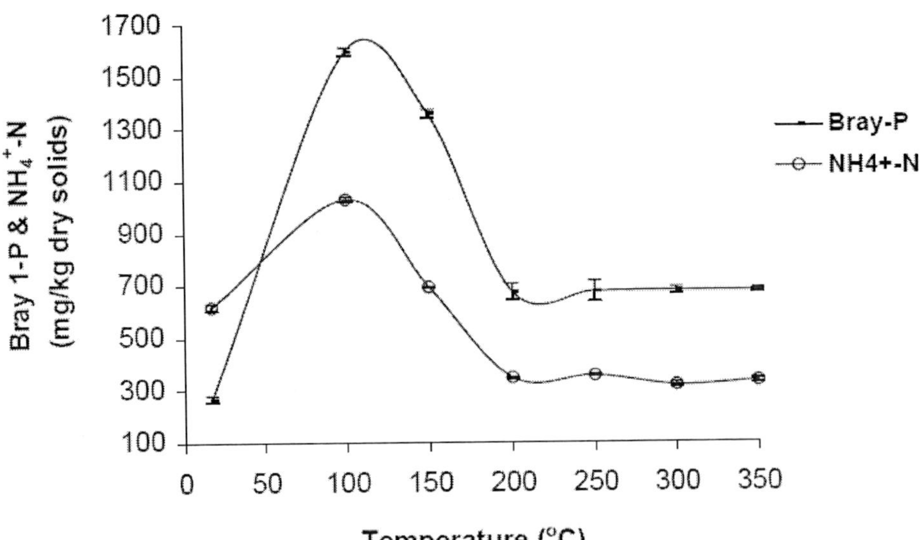

Figure 17. Available-P (Bray 1-P) and NH_4^+-N in sewage sludge at different drying temperatures.

Biosolids stabilized at 350 °C present lower concentration of total-N compared to the sewage sludge used as raw material (dry basis). Employment of excessively high temperatures drives off most of volatile solids from sewage materials (Jaeger and Mayer, 2000). The heat-dried biosolids would have lower concentrations of nitrogen and sulfur compared to its feedstock, and little to mineralize as the most of organic content had gone. Sewage sludge and heat-dried biosolids at 200 - 250 °C present both approximately 75%

volatile solids in their dry solids. Drying sludge at 300 - 350 °C reduces volatile solids to nearly 70%. Then, the processing of a tonne of digested sludge at 300 - 350 °C drives off 25 - 32 kg (t dry solids)$^{-1}$ of mineralizable organic matter, while drying it at 200 - 250 °C would save all of it. Bridle et al. (2000) report huge amounts of potassium and phosphate left as ashes in furnace when 275 °C is employed to dry sewage sludge from Perth, Australia. They also stressed that a temperature of 200 °C is more appropriate to dry sewage sludge if nutrient conservation is desirable.

Sludge products from heat-drying are marketable fertilizers and soil conditioners, of which the value depends primarily on N concentration (Girovich, 1996b). Nitrogen concentrations vary greatly among different sewage sludges and even in similar sewage materials treated alike. Blending various sewage sludges and dried pellets is a way to produce a consistent 6% total-N commercial product in USA, because total-N contents differ greatly in pellets originated from different feedstocks and batches (Girovich, 1996b). Phosphorus concentrations are more stable although they depend on the original P content in sewage material and on drying temperatures to avoid ash formation (Bridle et al., 2000).

Despite losses of N and volatile solids at certain temperatures, thermal stabilization process concentrates nutrients (López-Mosquera et al., 2008): a tonne of 5% moisture heat-dried bisolids can contain 60 kg of total-N and 73 kg of total-P, whilst a tonne of 88% moisture sewage sludge contains 8 kg total-N and 9 kg total-P. Therefore, there is 7.5-fold concentration factor for total-N and 8.3-fold for total-P when sewage sludge is transformed in pellets by thermal means. If sewage sludge is dried at 200 - 250°C, there is a consistent 8-fold concentration factor for both N and P.

5.3. Mineral-N (NH_4^+-N + NO_3^--N) and Available-P

Thermal-drying process of sewage sludge affects concentrations of both mineral-N and available-P. Drying temperatures of 100 - 150 °C can not cease mineralization of organic matter until sewage sludge dries off. As result, biosolids dried at 100 - 150 °C present higher concentrations of NH_4^+-N than its original feedstock (Figure 17). If sludge is dried at a minimum 200 °C temperature, mineralization slows down quickly. However, some NH_3^- volatilization occurs and the biosolids end the process having lower concentrations of NH_4^+-N than the sewage sludge from which they were originated - dry solids basis (Figure 17).

Mechanisms of organic-N mineralization and subsequent NH_3^- volatilization in sewage sludge have to rely on temperature and time because pH is little affected in the thermal process besides the amount of NH_4^+-N released. Some mineralization of organic matter seems unavoidable even at high drying temperatures (Figure 17). The highly putrescible nature of sewage sludge and its high enthalpy, which keeps temperature lower than drier's, boost mineralization of sludge's organic matter. The raise of P availability in the thermal process can better display the extent of mineralization under different drying temperatures, as there is not volatilization of this nutrient from driers (Figure 17). Available-P concentrations increase 2.5-times when sewage sludge is stabilized by heating.

From the agronomic point of view, the availability of nutrients and the mineralizable part of the total organic pool over the time have to be considered in order to beneficially apply sewage materials to land (Reed, 1995). This can avoid nutrient shortages or excesses. The heat-drying process is capable to produce different stabilized biosolids regarding N and P availability and total contents. Drying digested sewage sludge at 200 - 250 °C yields biosolids with increased concentrations of mineral-N and available-P. They also present are great potential for releasing nutrients in soils since there are not significant losses of nutrients and volatile solids at such drying temperature.

6. Solar-Irradiation of Sewage Sludge

6.1. Solar-Irradiation Performance

In solar-irradiation process sludge is placed on an open surface (lagoons, sand-drying beds) and left to dry up to a desired solids content. Open air solar drying has been practiced for many years, but sludges are dried to a level that depends upon uncontrollable conditions (Girovich, 1996b). The process is more appropriate for hot dry climates and the main effect is desiccation, which can result in a considerable reduction of volume. Shrinkage of biosolids during dehydration is beneficial because it reduces the volume of the residual material, and hence transport and spreading costs. Growing plants in biosolids will accelerate dehydration rates compared to fallow biosolids. Although raw biosolids have very high water content, some of this water is tightly bound and not readily accessible to plants (Crush et al., 2006).

Digested sewage sludge at 88% moisture reduces its water content to approximately 80% after two weeks under sunny tropical conditions (Figure 18). More than 95% of sludge water is released in the first week. Solar-irradiation of sludges works more efficiently to reduce moisture up to 80%, after which rates of water release are too small to justify the process (Lima et al., 1998). Trials conducted in Perth (Australia) could not yield material drier than 70% moisture after eight months of solar drying process (Videon et al., 1996). Such findings suggest that solar-irradiation process is suitable for the drainage of sludge's free-water, but not for the dehydration of sludge bound-water (Girovich, 1996b).

A difference of 8% between initial and final sewage sludge's water accounts for approximately 390 L water loss per tonne of wet sludge that resulted in a volume reduction by half (Figure 18). Various authors stress the importance of water loss for sludge management, because economics and handling characteristics are improved after sludge desiccation (Kukenberger, 1996). Sewage sludge with 99% reduces its volume by half if solids concentration increases from 1% to 2%. The increase of solids concentration to further 5% results in a volume reduction to 20%. Sludge dewatering from 90% to 80% results in volume reduction by half as well.

Desiccation also plays an important role in sludge sanitation, although ultra-violate (UV) radiation is the biggest cause of pathogen die off in this process. A study conducted in Australia with 40 m^3 sludge piles assessed the effect of this stabilization process on giardia cysts and bacteria during autumn and mid-winter conditions (Videon et al., 1996). Giardia cysts decreased their original concentration to between one-sixth and one-fourth and the material was fully disinfected of bacterial pathogens. However, as soon as the sludge re-wetted in the rainy season, concentrations of all pathogens soared to values higher than the original ones in the fresh sewage sludge. Therefore, sludge desiccation is a key factor in pathogen decay, discarding UV radiation as the main action in pathogenic die off in stockpiled sludge (Videon et al., 1996).

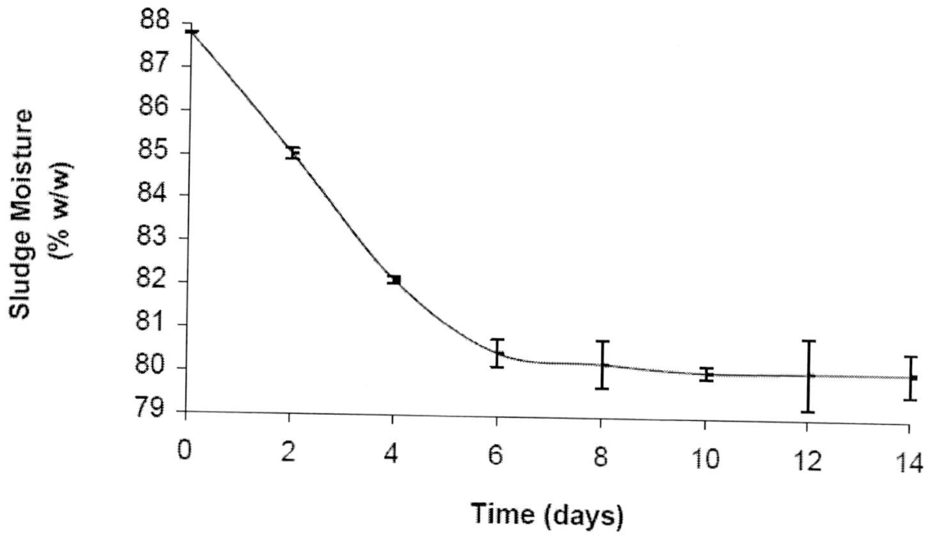

Figure 18. Sewage sludge water content during solar-irradiation stabilization process.

Another major sludge change during solar-irradiation process is the raise of sludge pH. Digested sewage sludge left under open natural conditions increases pH 6.2 to 7.4 as a result of organic-N ammonification. Sludge's pH significantly increases in the first week of solar-irradiation process and keeps it high until the last stabilization day. The raise of pH within a week indicates the quick mineralization pattern of digested sludge, which also indicates its high putrescible characteristic.

6.2. Total N and P

Solar-irradiation process does not affect concentrations of total-N present in sewage sludge dry solids. Some authors claim sludge dewatering removes significant amounts of soluble nutrients, making it less beneficial for land application (Munn, 1995). There is in fact high concentrations of N in the sludge's leachate, but the total amount of N leached (concentration x volume of leachate) is insignificant compared to total amount of N present in sludge dry solids. Organic-N is the predominant form of N in sewage sludges and losses of soluble N through leaching are slight (Girovich, 1996a). However, a leachate containing 782 mg N L^{-1} and 5.62 mg P L^{-1} from digested sewage sludge may be a source of pollution on a large-scale operation, especially considering the release 390 L leachate (t wet sludge)$^{-1}$ occurring mostly within a week.

There is not significant NH_3^- volatilization during solar-irradiation process of sewage sludge. Some studies on NH_3^- volatilization from fresh sludge in three plants in Sydney (Australia) concluded small amounts of N volatilized in the three first weeks of solar stabilization, and no further detectable loss thereafter (Robinson and Polgase, 2000). Therefore, changes in total amounts of nutrients caused by solar-irradiation are restricted to dehydration effect. Sludge dehydration from 88% to 80% moisture by solar-irradiation means increases concentrations of total-N and P in 1.6 times on wet basis (Figure 19). Such increments are proportional and concomitant to the losses of water from sewage sludge.

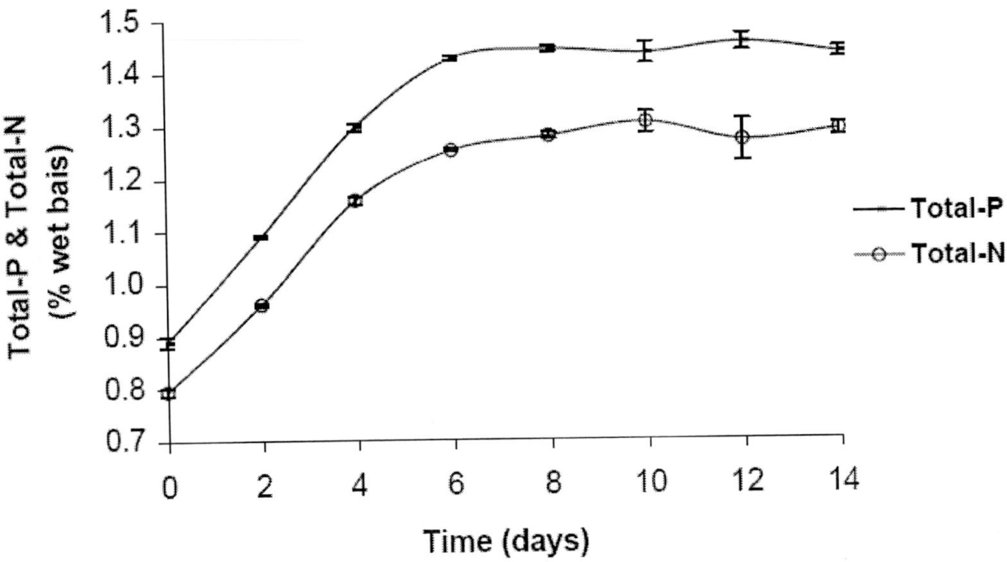

Figure 19. Total-P and N during solar-stabilization of sewage sludge (wet basis).

The biggest hassle for sewage sludge utilization has been handling and application due to high water content (Wisniewski et al., 1996). Transportation and application costs depend strongly on the material's water content. Transport distance from treatment plant to the point of discharge or land application is a critical factor and must be included in the final decision to use it (Reed at al., 1995). Some authors consider economically feasible to transport sludge at 92% moisture for a distance of 16 km from the source. Others state that press-dewatered sludge can be economically used within 100 km distance from the treatment plant (Lemainski and Silva, 2006). Economic analysis is mainly done based on sludge's total-N and P contents. Therefore, nutrient increments allow transportation of solar-irradiated biosolids to longer distances compared to the fresh feedstock. As long as transportation cost is the concern, there is a double effect on nutrients caused by sludge dehydration from 88 to 80% moisture: nutrients appeared 1.6 more concentrated and due to volume reduction 1.9 more sludge volume can be carried per load.

6.3. Mineral-N (NH_4^+-N + NO_3^--N) and Available-P

Digested sewage sludge contains nearly all its mineral-N in NH_4^+-N form (Figure 20). The highly anaerobic environment in sewage sludges is not appropriate for nitrification to occur. Then, nitrate (NO_3^--N) concentrations make only 0.2 - 1.1% of sludge mineral-N. When sewage sludge is left under solar-irradiation conditions, a slight part of its NH_4^+-N nitrifies during the process. However, ammonification of organic-N prevails over other biochemical process in piles under solar-stabilization process.

NH_4^+-N concentrations in stocked sludge piles start at high levels (620 mg kg^{-1}) and increase even more (790 mg kg^{-1}) during solar-irradiation treatment (Figure 20). Ammonification of organic-N could double NH_4^+-N concentrations in sludge piles if NH_3^- volatilization did not occur. It is reckoned 40 - 50% losses of NH_4^+-N through NH_3^-

volatilization when sludges are left for solar-drying (Imhoff and Imhoff, 1986). Other authors claim that 75% of the NH_4^+-N from sludges can volatilize within a week of solar-irradiation treatment (Robinson and Polgase, 2000). However, such amounts of NH_3^- volatilized are insignificant as such losses account for less than 0.2% of sewage sludge's total-N.

Meanwhile, available-P is present in sewage sludge at 240 mg kg^{-1} (dry basis) and such concentration decreases in 20 - 30% as the solar-irradiation process follows its course (Figure 20). This is an opposite behavior compared to thermal-drying process, in which NH_4^+-N concentrations decrease and P availability increases in sludge's dry solids (Figure 17). Ammonification increases sludge's pH (6.2 → 7.4), which makes P partially unavailable for plant uptake. There is a negatively linear regression between sludge's pH and available-P, especially in the first week of solar-irradiation process.

Besides the 20 - 30% lower concentrations o available-P in solar-irradiated biosolid's dry solids, dehydration of digested sludge increases available-P concentrations in 1.4 times and NH_4^+-N concentration doubles on wet weight basis after two weeks of solar-stabilization. Nitrogen is the main parameter for sludge land application design, and the amount of mineral-N in biosolids is considered an important factor limiting land application due to N nitrification and losses to groundwater (Loehr, 1979).

However, to determine application rate based on N is a complex matter because of the various forms of it in sewage sludge and the various pathways N can take after soil incorporation (Tomer et al., 1998). Land application rate of solar-irradiated biosolids can be halved relative to the freshly belt-pressed sewage sludge, since mineral-N concentration doubles on wet basis during the stabilization process. This would limit available-P to half the original content present in the fresh sewage sludge.

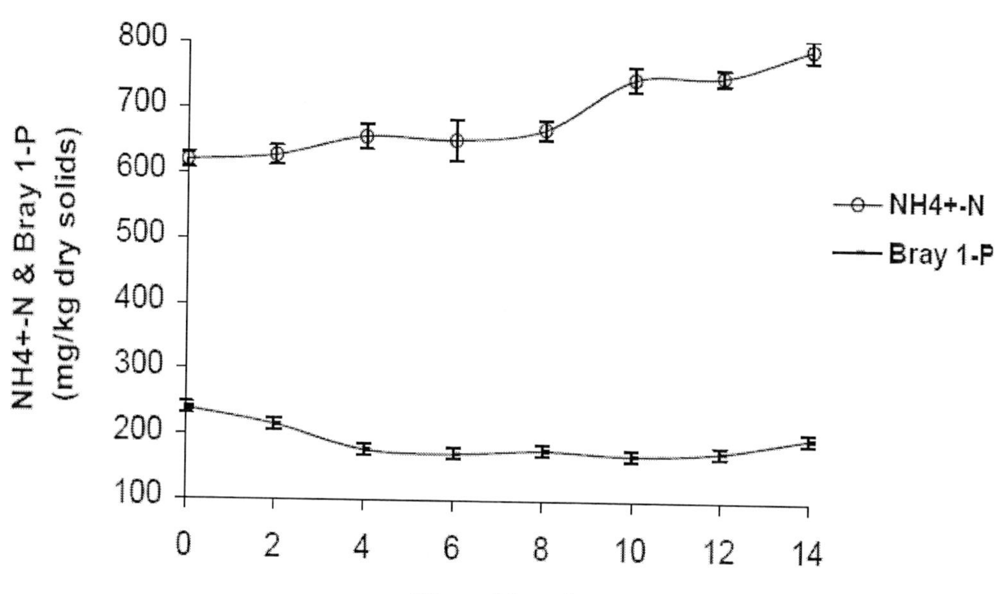

Figure 20. NH_4^+-N and available-P during solar-stabilization of sewage sludge (dry basis).

CONCLUSIONS

The selected USEPA stabilization processes (Table 1) works without major problems on sanitation of digested sewage sludge. Composting is the most demanding process as it requires control of many parameters: moisture, aeration, pore space, C/N ratio, nutrients, pile size, turning frequency, temperature and time. However, adding a carbon source and bulk agent in correct amounts and turning piles frequently are sufficient to meet all the requirements to start up and keep the process running. The gathering of an appropriate carbon source and bulking agent is a key factor to achieve a quality composted biosolid. The final product has a strong market appeal, even for nursery growing media (Ostos et al., 2008).

Thermal-drying process is the second in demanding control as temperature-time relation has to be carefully managed. Low drying temperatures combined with a long resident time causes huge losses of total-N. On the other hand, temperatures too high may cause losses of P as ash. Alkaline treatment and solar-irradiation do not require further management after the processes are set up. Biological activity stops in sewage sludge as long as pH is raised and kept higher than 11. Such pH level is essential to stop ammonification and, therefore, essential for N conservation. Alkaline stabilization goals are achieved as long as an appropriate lime product is applied at right rate. Finally, solar-irradiation process refers mostly to drain out sludge's free-water. Most of dehydration and N ammonification occur within a week. It is difficult to imagine that ultra-violate radiation can penetrate the thick and dense mass of stockpiled sewage sludge. Stabilization by the means of solar-irradiation seems more appropriate if sewage sludge is spread on land surface; otherwise desiccation has to be the main process of pathogen decay.

Composted and the heat-dried biosolids may be long-term stored and sold without nuisance problems. There are significant changes in digested sewage sludge after it undergoes composting and thermal-drying processes. Both stabilized biosolids are loose, easy to handle, dry and free of strong smells. Biosolids heat-dried at 250 °C are surely free of pathogens and mature composted biosolid does cause nuisance problems even after being re-wetted.

Alkaline-stabilized biosolids still present the original sticky and smelly features of digested sewage sludges. Solar-irradiation process makes digested sludge even stickier and stronger in smell. However, this process results in a significant volume reduction and a probable decay of pathogenic concentrations. Alkaline and solar treatments are suitable soon before applying them to land. Health safety, regarding liming, and gain of economic strength for transportation, regarding solar-drying, may be strong appeals for running such processes. Otherwise, if sanitation and nuisance problems are the concern, limed and solar-dried biosolids seem too similar to digested sewage sludge to justify treatment efforts.

All the stabilization processes affect in some way sewage sludge´s N and P fertilizing value. From the agronomic point of view, there is not a stabilized biosolid similar to another or to the digested sewage sludge from which they originated (Table 8). The stabilization process that most alters contents of sludge's total-N and total-P is composting. Compost treatment of digested sewage sludge decreases its volatile solids, total-N and mineral-N concentrations (Corrêa et al., 2007). However, available-P concentration increases during composting. Compared to other stabilized biosolids, compost remains as the poorest source of N and P on dry basis. Composted wastes usually are poor sources of N due to volatilization during the processing (Pascual et al., 1997).

However, mineral-N, available-P (Bray 1-P), and volatile solids contents in composted biosolid are bigger than in biosolids stabilized with CaO at 30% rate (Figure 21). Alkaline stabilization causes dilution of dry solids and significant losses of mineral-N by volatilization. Hence, available-P is bound as calcium phosphate ($CaHPO_4$) in lime-treated biosolids.

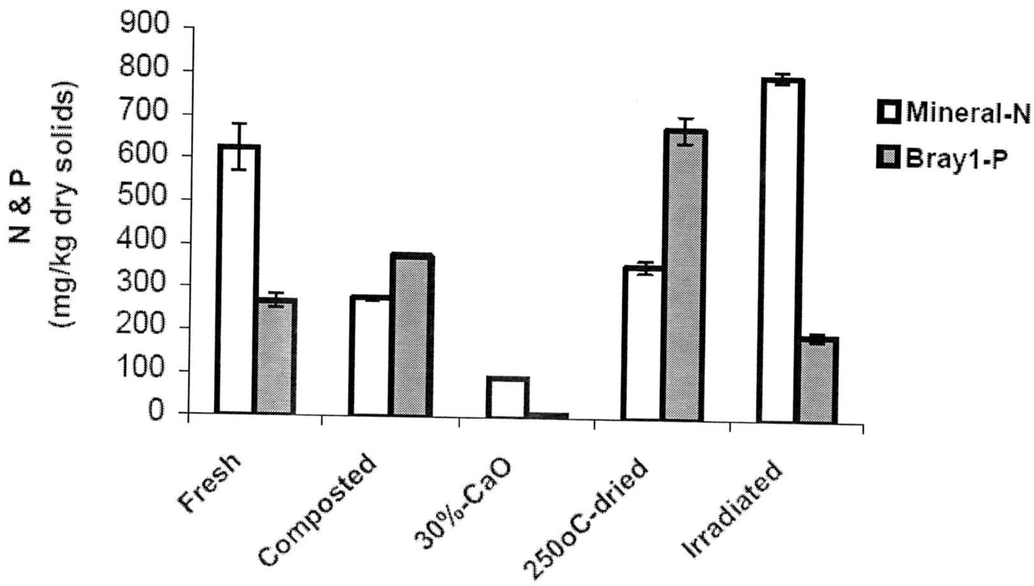

Figure 21. Mineral-N and available-P (Bray 1 - P) in stabilized biosolids.

As result, alkaline-stabilized biosolids present the lowest concentrations of mineral-N, available-P and volatile solids among all biosolids (Figure 21, Table 8). Total amounts of N and P in dry solids stand as the second lowest. Low content of volatile solids in limed-sludges also indicate a poor potential for the release of nutrients in soils through time. However, considering total amounts of nutrients, limed-sludge (30% CaO) presents approximately twice the potential of composted biosolid to deliver N and P.

Thermal-drying (250 °C) and solar-irradiation processes do not cause significant losses of total-N, total-P, and volatile solids in sewage sludge dry solids. However, mineral-N and available-P are affected by all the processes (Figure 21). All treatments cause losses of mineral-N, even thought solar-irradiation process is able to increase and keep higher concentrations of mineral-N in dry solids compared to digested sewage sludge (Figure 21). Solar-irradiated biosolid has the highest concentration of mineral-N, which has to work as the upper limiting loading factor for land application. The other stabilized biosolids keep between 15% and 57% of the original mineral-N content in their dry solids.

Phosphorus is generally not lost during the stabilization processes (Table 1). It can be temporally immobilized in some treatments. Concentrations of available-P increase by 40% in composted biosolid and 2.5-fold in thermal-dried biosolid (Figure 21). Alkaline and solar-irradiation processes cause changes in pH. As a result, concentrations of available-P in these two biosolids are reduced compared to original sewage sludge (Figure 21, Table 8).

Overall, heat-dried biosolid combines high concentrations of total and available nutrients with good stabilized physical characteristics. It has four times more total-N, three times more total-P, 1.8 times more available-P and 1.3 times more mineral-N than the other long-term stable product, the composted biosolid. These two products present very different N and P nutritional values and consequently different potentials to be used in soils. Total-N and N-mineralization rate in soils are the most important parameters to discriminate sludges that undergone different treatments (Johansson et al, 1999).

Figures on wet weight basis show a different scenario. Except heat-dried biosolid, the biggest part of all stabilized biosolids is water (Table 8). The high water content diminishes the nutrient value of most biosolids, but makes composted biosolid worthier on comparative basis. There is more total-P, mineral-N, and available-P in compost wet weight than in the digested sludge at 88% moisture. However, concentration of total-N remains the lowest among all biosolids (Table 8).

Lime-treated and solar-irradiated biosolids still present high water content (77 - 80%). Solar-dried biosolid presents 37% more total-N and total-P, seven times more mineral-N and fourteen times more available-P than alkaline-treated bisolid on wet weight basis. Alkaline stabilization is the treatment that mostly affects sewage sludge's volatile solids and availability of nutrients due to pH changes. However, limed-sludges also have neutralizing value that may be worth for raising soil pH. But, if a quick pathogenic decay is not necessary before land application, liming achieves only a short-term stabilization and does not justify the huge losses of nutrients in sewage sludges.

Table 8. Some agronomic characteristics of stabilized biosolids

Parameter	Digested sewage sludge	Composted biosolid	30%-CaO biosolid	250°C-dried biosolid	Solar-dried biosolid
Dry weight basis					
Total-N (%)	6.51a ±0.02	1.59b ±0.07	4.01c ±0.17	6.48a ±0.03	6.53a ±0.06
Total-P (%)	7.21a ±0.13	2.42b ±0.03	5.06c ±0.15	7.25a ±0.21	7.26a ±0.08
Mineral-N (mg kg^{-1})	624a ±54	277b ±5	93.8c ±4.6	356d ±15	803e ±13
Bray 1-P (mg kg^{-1})	268a ±17	377b ±1	11.9c ±0.2	678d ±31	199e ±10
Volatile solids (%)	74.7a ±0.2	68.4b ±0.7	54.6c ±1.7	74.6a ±0.1	74.9a ±0.2
Wet weight basis					
Water content (%)	87.8a ±0.2	55.1b ±1.2	76.3c ±0.2	3.4d ±0.1	80.1e ±0.2
pH water (1:5 w/v)	6.4a ±0.1	6.1ab ±0.3	11.9c ±0.2	5.8b ±0.1	7.4d ±0.1
Total-N (%)	0.79a ±0.01	0.71b ±0.03	0.95c ±0.04	6.26d ±0.05	1.30e ±0.01
Total-P (%)	0.88a ±0.02	1.09b ±0.01	1.20c ±0.04	7.00d ±0.20	1.45e ±0.02
Mineral-N (mg kg^{-1})	76.1a ±6.6	124.5b ±2.2	22.2c ±1.1	345d ±15	159.8e ±2.6
Bray 1-P (mg kg^{-1})	32.7a ±2.1	169.1b ±0.4	2.8c <±0.1	655d ±29	39.5a ±6.9

Means (n=3) with same letter within rows are not statistically different by test of Tukey ($p < 0.05$).

REFERENCES

Andreoli, C. V.; Lara, A. I. and Fernandes, F. (1999). Reciclagen de biosólidos: transformando problemas em soluções. Sanepar. Curitiba-PR, Brazil.

Barreira, L. P.; Philippi Junior, A., Rodrigues, M.S. and Tenório, J. A. S. (2008). Physical analyses of compost from composting plants in Brazil. *Waste Management*, v. 28, n. 8, p. 1417 - 1422.

Barton, J. R.; Issaias, I. and Stentiford, E. I. (2008) Carbon - Making the right choice for waste management in developing countries. *Waste Management*, v. 28, n. 4, p. 690 - 698.

Bertoncini, E. I.; D'Orazio, V.; Senesi, N. and Mattiazzo, M. E. (2008). Effects of sewage sludge amendment on the properties of two Brazilian Oxisols and their humic acids. *Bioresource Technology*, v. 99, n. 12, p. 4972 - 4979.

Bitton, G.; Damron, B. L. and Davidson, J. M. (1980). *Sludge-health risks of land application*. Arbor Science Publishers Inc., Michigan, USA.

Bridle, T. R.; Molinari, L.; Skrypski-Mantele, S.; Ye, D. P. and Mills, J. (1999). Commisioning the Subiaco ENERSLUDGEtm Plant Part 2: sludge drying, energy recovery and gas cleaning. In: *18th Federal Convention of the AWWA*, 8 p. Adelaide, Australia.

Bridle, T. R.; Molinari, L.; Skrypski-Mantele, S.; Ye, D. P. and Mills, J. (2000). Start-up of the Subiaco Enersludgetm plant. *Water Science and Technology: sludge management for the 21st Century,* v. 41, p. 31 - 36. Australia.

Cameron, K. C.; Di, H. J. and Mclaren, R. G. (1996). Is the soil an appropriate dumping ground for our wastes? In: *Australian and New Zealand National Soils Conference 1996. Plenary papers*, p. 31 - 53. The University of Melbourne, Australia.

CEA (1999). Economic report of the president: 1999. Council of Economic Advisors, US Government Printing Office. 85 p. Washington DC, USA.

Citterio, B.; Civilini, A.; Rutili, A; and Pera, A; (1986). Control of a composting process in bioreactor by monitoring chemical and microbial parameters. In: *Compost: production, quality, and use* (Eds. de Bertoli, M.; Ferranti, P.; L'Hermite, P. and Zucconi, F.), p. 633 - 642. Elsevier Applied Science. Udine, Italy.

Corrêa, R. S. (2001). *Beneficial use of biosolids based on their N and P fertilising value*. The University of Melbourne, ILFR. Ph.D. Thesis, 305p. Melbourne, Australia.

Corrêa, R. S.; Fonseca, Y. M. F. and Corrêa, A. S. (2007). Production of agricultural biosolid by composting and vermicomposting sewage sludge. *Brazilian Journal of Agricultural and Environmental Engineering*, v. 11, n. 4, p. 420 - 426.

Corrêa, R. S.; White, R. E. and Weatherley, A. J. (2006). Effect of compost treatment of sewage sludge on nitrogen behavior in two soils. *Waste Management*, v. 26, n. 6, p. 614 - 619.

Costa, A. N. and Krohling, B. (1998). Efeito do processo de higinenizaçao do lodo de esgoto com cal virgem no desenvolvimento do cafeeiro *Coffea canephora*. In: *I Seminário sobre gerenciamento de biossólidos do Mercosul*. Gráfica Editora, p. 285 - 289. Sanepar, Abes-PR. Curitiba/PR, Brazil.

Crush, J. R.; Sarathchandra, U. and Donnison, A. (2006). Effect of plant growth on dehydration rates and microbial populations in sewage biosolids. *Bioresource Technology*, v. 97, n. 18, p. 2447 - 2452.

Diaz-Burgos, M. A.; Ceccanti, B. and Polo, A. (1991). Changes in the inorganic nitrogen during sewage sludge composting. In: *Treatment and use of sewage sludge and liquid agricultural wastes* (Ed. L'Hemite, P.), p. 513-519. Elsevier Applied Science. Brussels, Belgium.

Epstein, E. (1997). *The science of composting*. Technomic Publishing Company Inc, Lancaster-Pennsylvania, USA.

Farnet, A. M.; Prudent, P.; Cigna, M. and Gros, R. (2008). Soil microbial activities in a constructed soil reed-bed under cheese-dairy farm effluents. *Bioresource Technology*, v. 99, n. 14, p.6198 - 6206.

Frank, R. (1998). The use of biosolids from wastewater treatment plants in agriculture. *Environmental Management*, v. 9, p. 165 - 169.

Frossard, E. and Morel, J. L. (1995). Assessment of phosphate fertilizing value of urban sewage sludges. In: *Soil management in sustainable agriculture* (Eds. Cook, H. F. and Lee, H. C.), p. 226 – 230. Wye College Press, Wye College. University of London, UK.

Geisler, M. (2000). Composting biosolids and food wastes. In: *Compost 2000 Downunder Conference*. Day 2, Stream 1, 9 pages. DMP Conference Organisers. Carlton Crest Hotel, Melbourne, Australia.

Gil, M. V.; Carballo, M. T. and Calvo, L. F. (2008). Fertilization of maize with compost from cattle manure supplemented with additional mineral nutrients. *Waste Management*, v. 28, n. 8, p.1432 - 1440.

Girovich, M. J. (1996a). Biosolids characterization, treatment and use: an overview. In: *Biosolids treatment and management - processes for beneficial use* (Ed. Girovich, M. J.), p. 1 - 45, Chapter 1. Marcel Dekker Inc. New York, USA.

Girovich, M. J. (1996b). Thermal treatment. In: *Biosolids treatment and management - processes for beneficial use* (Ed. Girovich, M. J.), p. 271 - 341 Chapter 6. Marcel Dekker Inc. New York, USA.

Golueke, C. G. (1975). *Composting - a study of the processes and its principles*. Rodale Press Inc. Book Division, Emmaus, Pensylvania USA.

Handreck, K. A. (1993). Composting - making soil improver from rubbish. In: *CSIRO Division of Soils,* 19 p. Auckland, NZ.

Handreck, K. A. and Black, N. D. (1994). *Growing media for ornamental plants and turf*. University of New South Wales Press.

Hotta, S. and Funamizu, N. (2007). Biodegradability of fecal nitrogen in composting process. *Bioresource Technology*, v. 98, n. 17, p. 3412 - 3414.

Hu, C. J.; Gibbs, R. A.; Mort, N. R.; Hofsted, H. T.; Ho, G. E. and Unkovich, I. (1996). Reuse dominates sludge options. *Waste Management and Environment*, v. 7, p. 15 - 23.

Imhoff, K. and Imhoff, K. (1986). *Manual de tratamento e águas residuárias*. Edgard Blucher Ltda, São Paulo Brasil.

Jaeger, M. and Mayer, M. (2000). The Noell Conversion Process - a gasification process for the pollutant-free disposal of sewage sludge and the recovery of energy materials. *Water Science and Technology: sludge management for the 21st Century*, v. 41, p. 37 - 44.

Johansson, M.; Stenberg, B. and Torstensson, L. (1999). Microbial and chemical changes in two arable soils after long-term sludge amendments. *Biology and Fertility of Soils*, v. 30, p. 160 - 167.

Joshua, R. S. (1996). Biorecovery of food effluent residues. La Trobe University, School of Microbiology, Faculty of Science and Technology. Ph.D. Thesis, 400 p. Bundoora, Victoria, Australia.

Kalogo, Y. and Bagley, D. M. (2008). Fermentative hydrogen gas production using biosolids pellets as the inoculum source. *Bioresource Technology*, v. 99, n.3, p. 540 - 546.

Katayama, A.; Ker, K. C.; Hirai, M.; Shoda, M. and Kubota, H. (1986). Stabilization process of sewage sludge compost in soil. In: *Compost: production, quality and use* (Eds.: de Bertoli, M., Ferranti, P., L'Hemite, P. and Zucconi, F.), p. 341 - 350. Elsevier Applied Science, Udine Italy.

Kukenberger, R. J. (1996). Conditioning and dewatering. In: *Biosolids treatment and management - processes for beneficial use* (Ed. Girovich, M. J.). p. 131-164, Chapter 3. Marcel Dekker Inc. New York, USA.

Kvarnstrom, E. and Nilsson, M. (1999). Reusing phosphorus: engineering possibilities and economic realties. *Journal of Economic Issues*, v. 33, p. 393 - 341.

Lemainski, J. and Silva, J. E. (2006). Agronomic and economic evaluation of biosolid application on soybean production. *Brazilian Journal of Agriculture - PAB*, v. 41, n. 10, p. 1477 - 1484.

Li, H.; Jin, Y.; Mahar, R. B.; Wang, Z. and Nie, Y. (2008). Effects and model of alkaline waste activated sludge treatment. *Bioresource Technology*, v. 99, n. 12, p. 5140 - 5144.

Lima, M. R. P.; Müller, P. S. G. and Gonçalves, R. F. (1998). Desidratação de lodo de lagoas anaeróbicas de estabilização em leitos de secagem na Região Sudeste do Brasil. In: I Seminário sobre gerenciamento de biossólidos do Mercosul (Ed. Finep/Praosab/Sanepar/ Abes-PR), p. 211 - 222. Caiobá Editora, Curitiba/PR, Brazil.

Loehr, R. C. (1979). *Land application of wastes*. Van Nostrand Reinhold Company. New York USA.

López-Mosquera, M. E.; Cabaleiro, F.; Sainz, M. J.; López-Fabal, A. and Carral, E. (2008). Fertilizing value of broiler litter: effects of drying and pelletizing. B*ioresource Technology,* v. 99, n. 13, p. 5626 - 5633.

Maguire, R. O.; Sims, J. T. and Coale. F. J. (2000). Phosphorus solubility in biosolids-amended farm soils in the Mid-Atlantic Region of the USA. *Journal of Environmental Quality,* v. 29, p. 1225 - 1233.

Mbarki, S.; Labidi, N.; Mahmoudi, H.; Jedidi, N. and Abdelly, C. (2008). Contrasting effects of municipal compost on alfalfa growth in Clay and in sandy soils: N, P, K, content and heavy metal toxicity. *Bioresource Technology*, v. 99, n. 15, p. 6745 – 6750.

Metcalf and Eddy (2003). *Wastewater engineering: treatment, disposal and reuse*. McGraw-Hill, Boston.

Muga, H. E. and Mihelcic, J. R. (2008). Sustainability of wastewater treatment Technologies. *Journal of Environmental Quality*, v. 88, n. 3, p. 437 - 447.

Munn, K. J. (1995). Effects of metal-contaminated sewage sludge on biological nitrogen fixation and mineralization of legume nitrogen. The University of Melbourne, School of Agriculture and Forestry. 232 p. Melbourne, Australia.

Nishanth, D. and Biswas, D. R. (2008). Kinetics of phosphorus and potassium release from rock phosphate and waste mica enriched compost and their effect on yield and nutrient uptake by wheat (*Triticum aestivum*). *Bioresource Technology*, v. 99, n. 9, p. 3342 - 3353.

Nóvoa-Muñoz, J. C.; Simal-Gándara, J; Fernández-Calviño, D; López-Periago, E. and Arias-Estévez, M. (2008). Changes in soil properties and in the growth of *Lolium multiflorum* in an acid soil amended with a solid waste from wineries. *Bioresource Technology*, v. 99, n. 15, p. 6771 - 6779.

O'Kelly, B. (2008). Effect of biodegradation on the consolidation properties of a dewatered municipal sewage sludge. *Waste Management*, v. 28, n. 8, p. 1395 - 1405.

Ostos, J. C.; López-Garrido, R.; Murillo, J. M. and López, R. (2008). Substitution of peat for municipal solid waste and sewage sludge-based composts in nursery growing media: effects on growth and nutrition of the native shrub *Pistacia lentiscus* L. *Bioresource Technology,* v. 99, n. 6, p. 1793 - 1800.

Parkinson, R.; Fuller, M.; Jury, S. and Groenohof, A. (1996). An evaluation of soil nutrient status following the application of (I) co-composted MSW and sewage sludge and (ii) greenwaste to maize. In: *The science of compost* (Eds. de Bertoli, M.; Sequi, P.; Lemmes, B. and Papi T), p. 469 - 476. Blackie Academic and Professional Publisher. Glasgow, Scotland.

Pascual, J. A.; Ayuso, M.; Garcia, C. and Hernández, T. (1997). Characterization of urban wastes according to fertility and phytotoxicity parameters. *Waste Management and Research*, v. 15, p. 103 - 112.

Pastor, L.; Marti, N.; Bouzas, A. and Seco, A. (2008). Sewage sludge management for phosphorus recovery as struvite in EBPR wastewater treatment plants. *Bioresource Technology*, v. 99, n. 12, p. 4817 - 4824.

Pengcheng, G.; Xinbao, T.; Yanan, T and Yingxu, C. (2008). Application of sewage sludge compost on highway embankments. *Waste Management*, v. 28, n. 9, p. 1630 - 1636.

Plachá, I.; Venglovský, J.; Makova, Z. and Martinéz, J. (2008). The elimination of *Salmonella typhimurium* in sewage sludge by aerobic mesophilic stabilization and lime hydrated stabilization. *Bioresource Technology*, v. 99, n. 10, p. 4269 - 4274.

Pommel B., Billiet R., Gandilhon A. and Roche R. (1991). Ability of a thermic sludge to supply crops with phosphorus. In: *Treatment and use of sewage sludge and liquid agricultural wastes* (Ed. L'Hemite P), p. 288 - 292. Elsevier Applied Science. Brussels, Belgium.

Ramírez, W. A.; Domene, X.; Ortiz, O. and Alcañiz, J. M. (2008). Toxic effects of digested, composted and thermally-dried sewage sludge on three plants. *Bioresource Technology,* v. 99, n. 15, p.7169 - 7175.

Reed, S. C.; Crites, R. W. and Middlebrooks, E. J. (1995). *Natural systems for waste management and treatment*. 2nd ed. McGraw-Hill Inc. New York, USA.

Robin, R.; Ablain, F.; Yulipriyanto, H.; Pourcher, A. M.; Morvan, T.; Cluzeau, D. and Morand, P. (2008). Evolution of non-dissolved particulate organic matter during composting of sludge with straw. *Bioresource Technology*, v. 99, n. 16, p. 7636 - 7643.

Robinson, M. B. and Polgase, P. J. (2000). Volatilization of nitrogen from dewatered biosolids. *Journal of Environmental Quality*, v. 29, p. 1351 - 1355.

Rynk, R.; van de Kamp, M.; Willson, G. B.; Singley, M. E.; Richard, T. L.; Kolega, J. J.; Gouin, F. R.; Laliberty, L. J.; Kay, D.; Murphy, D. W.; Hoitink, H. A. J. and Brinton W. (1992). *On-farm composting handbook.* Northeast Regional Agricultural Engineering Service and Cooperative Extension. Ithaca, NY - USA.

Saludes, R. B.; Iwabuchi, K.; Miyatake, F.; Abe, Y. and Honda, Y. (2008). Characterization of dairy cattle manure/wallboard paper compost mixture. *Bioresource Technology*, v. 99, n. 15, p. 7285 - 7290.

Sánchez-Monedero, M. A., Roig, A., Paredes, C. and Bernal, M. P. (2001). Nitrogen transformation during organic waste composting by the Rutgers system and its effect on pH, EC and maturity of the composting mixtures. *Bioresource Technology*, v. 78, n. 2, p. 301 - 308.

Seekins, B. (2000). Determining nutrient changes in a farm windrow composting system. In: *Composting 2000 Downunder Conference*, Day 2, Stream 2, 4 pages. DMP Conference Organisers, Carlton Crest Hotel, Melbourne, Australia.

Sims J.T. (1990). Nitrogen mineralization and elemental availability in soils amended with co-composted sewage sludges. *Journal of Environmental Quality*, v.19, n.4, p. 669 - 675.

Smith, D. R.; Moore, Jr. P. A.; Griffis, C. L.; Daniel, T. C.; Edwards, D. R. and Boothe, D. L. (2001). Effects of alum and aluminium chloride on phosphorus runoff from swine manure. *Journal of Environmental Quality*, v. 30, p. 992 - 998.

Snow, K. J. (1996). Digestion. In: *Biosolids treatment and management - processes for beneficial use* (Ed. Girovich MJ), p. 165 - 191. Marcel Dekker Inc. New York, USA.

Sparkes, J. (1990). Environmental implication of the use of sewage products. In: *The Agricultural and Horticultural Use of Sewage Sludge* (Eds. Elmes, D. and Mallen-Cooper, J.), Paper Number 8. AIAS. The Australian Institute of Agricultural Science and New South Wales Branch, University of Western Australia. Hawkesbury Campus, Australia.

Spellman, F. R. (1997). *Dewatering biosolids*. Technomic Publishing Company, Lancaster USA.

Stentiford, E. I. (1996). Composting control: principles and practice. In: *The Science of Composting* (Eds. de Bertoli, M.; Sequi, P.; Lemmes, B. and Papi, T.), p. 49 - 59. Blackie Academic and Professional Publisher. Glasgow, Scotland.

Tejada, M.; Gonzalez, J. L.; García-Martínez, A. M. and Parrado, J. (2008). Application of a green manure and green manure composted with beet vinasse on soil restoration: effects on soil properties. *Bioresource Technology*, v. 99, n. 12, p. 4949 - 4957.

Tisdale, S. L. and Nelson, W. L. (1971) *Soil fertility and fertilizers*. 2nd ed. The MacMillan Company. New York, USA.

Tomer, M.; Smith, T.; Charleson, T.; Barton, L.; Gielen, G.; Thorn, A.; Peacock, A. and Hopkins, K. (1998). Nitrogen and phosphorus removal in an effluent-irrigated forest plantation in New Zealand. In: National Soils Conference. *Environmental Benefits of Soil Management,* p. 79-85. Brisbane, Australia.

USEPA (1995) A guide to the biosolids risk assessments for the EPA Part 503 rule. United States Environmental Protection Agency. *Office of Wastewater Management*. 143 p. Washington DC, USA.

USEPA (1999) Biosolids generation, use, and disposal in the United States. United States Enviromental Protection Agency, Municipal and Industrial Solid Waste Division, *Office of Solid Waste*. 75 p. Washington DC, USA.

Videon, G.; Catherine, H.; Gibbs, R.; Mort, H. H. and Ho, G. (1996). Reuse dominates sludge options. *Asia-Pacific Magazine of Environmental Business and Technology*, v. 7, p. 15 - 23.

Walker, P. M.; Kelley, T. R. and Smiciklas, K. D. (2008). Evaluation of pulverized trommel fines for use as a soil amendment. *Bioresource Technology*, v. 99, n. 16, p.7848 - 7858.

Willett, I. R.; Jakobsen, P.; Malafant, K. W. J. and Bond, W. J. (1984). Effects of land disposal of lime-treated sewage sludge on soil properties and plant growth. In: p. 56. *CSIRO Institute of Biological Resources*, Division of Water and Land Resources. Canberra, Australia.

Wisniewski, C.; Neto, J. A. M.; Pereira, A. M.; Radomski, M. I. and Sessegolo, G. C. (1996). Uso do lodo de esgoto da ETE - Belém na recuperação de áreas degradadas por mineração de calcário. *Sanare,* v5, p. 76 - 85.

White, R. E. (1997). *Principles and practices of soil science: the soil as a natural resource.* 3rd ed. Blackwell Science Ltd. Victoria, Australia.

In: Sewage Treatment: Uses, Processes and Impact
Editor: Anna Stephens and Mark Fuller
ISBN 978-1-60692-959-9

Chapter 3

Fate of Cadmium, Copper, Lead and Zinc in Soils after Application of Different Treated Sewage Sludge in Soils of the Pampas Region, Argentina

Silvana Irene Torri* and Raúl Silvio Lavado
Cátedra de Fertilidad, Facultad de Agronomía, Universidad de Buenos Aires,
Avda San Martín 4453, Buenos Aires (C1417 DSE), Argentina

Abstract

The accumulation and availability of PTE after application of sewage sludge on agricultural land has become a world wide critical problem. This is due to the risk of environmental problems. The risks of human, crop and/or environmental toxicity posed by these elements are a function of their availability and therefore on their soil speciation. PTE commonly found in sewage-sludge include Cd, Cu, Pb and Zn. The toxicity and mobility of these metals strongly depend on their specific chemical forms and on their binding state in the sludge. Alternatively, soil physical–chemical properties, especially pH, CEC, organic matter and clay content, are also likely to assume great importance in determining PTE behavior. This chapter emphasizes on studies performed on representative soils of the Pampas Region, Argentina, showing the effects of soil application of raw sewage sludge, anaerobically digested sewage sludge and a mixture of sewage sludge and its own incinerated ash. The inorganic fraction became the most prevalent form for sludge-borne PTE, whereas sludge-borne Cd was only found in the residual fraction. The chemical behavior of sludge-borne Pb did not depend on soil physico-chemical characteristics. On the contrary, availability of sludge-borne Cu and Zn depended on soil pH and not in soil texture. The results obtained in this study provide supporting evidence for the protection theory, which hypothesizes that mineral components may compensate for any loss of metal retention capacity caused by mineralization of organic components. We also considered soil application of soluble salts of the four studied metals. The relationship between the soil chemical forms of the four studied metals and plant absorption was also studied. In pot experiments, plant

* E-mail: torri@agro.uba.ar

absorption depended on PTE availability. However, field experiments showed very little PTE mobility, availability or accumulation in crops. Under controlled conditions, the use of sewage sludge as a soil amendment did not pose a significant risk of contaminating soil, water or plants. Moreover, the mixture of sewage sludge with its own incinerated ash also offers a potential viable utilization of this organic waste as a soil amendment.

INTRODUCTION

The Disposal of Sewage Sludge

The accumulation sewage sludge (SS) poses a growing environmental problem. At present, there are different disposal methods for processed sludge, such as agricultural application as a soil amendment, land filling and incineration (Bridle and Skrypski-Mantele 2000). Composting is also recognized as a recycling option in some countries. Although no documented scientific evidence has been found that sewage sludge regulations have failed to protect public health (Darvodelsky and Fien, 2005), these disposal methods are still under scrutiny (Laturnus et al., 2007). Nevertheless, agricultural application of SS has become the most widespread method of disposal in most countries and is generally the most economical outlet. This practice is considered a very good option because it offers the possibility of recycling plant nutrients and organic matter. Numerous studies have indicated that the use of this waste as a source of organic matter improves the chemical and physical properties of agricultural soils, decreasing bulk density, increasing pore size, soil aeration and root penetrability, water holding capacity and biological characteristics as stimulating soil microorganism activity (Chambers et al., 2002; García-Orenes et al., 2005; Tejada and Gonzalez, 2007), resulting in an increase in crop yields (Antolín et al, 2005; Lavado 2007). However, one of the main concerns regarding soil application of SS is the presence of heavy metals or other potentially toxic elements (PTE). Nowadays, the content of PTE in SS is generally below the maximum permissible concentration established in current Argentinean regulations. However, PTE are persistent environmental contaminants, and continuous sludge application to agricultural lands may pose a serious risk for contamination of the food chain. PTE accumulation in soils (Kidd et al., 2007; Singh and Agrawal., 2007) and, in some cases, in crops (Jamali et al., 2007) have been reported when sludge application or irrigation with recycled residual water have been carried out for a long time. Such concerns, together with surface-water pollution that may arise from runoff, have motivated and maintain a high degree of interest in the subject (Ashworth, Alloway 2007; Antoniadis et al., 2008).

Land filling has shown to be the cheapest means of eliminating sludge, in term of exploitation and capital costs (Metcalf, Eddy 2003). However, this option not only takes up valuable land space, but also causes air, water and soil pollution, discharging chemicals and pesticides into the earth and groundwater. In Europe, the Directive 99/31strongly limits this disposal outlet. Sludge incineration results in a large reduction of sludge volume, which accounts for 10% of the volume of mechanically dewatered sludge; the thermal destruction of toxic organic constituents and pathogens, and the recovery of the flue gas energy (Werther, Ogada 1999). The ash can also improve soil physical properties because of its silt-size nature (Saikia et al., 2006) and can be an effective liming agent (Zhang et al., 2002). Public acceptance of incineration is, however, hampered by concerns about potential adverse environmental impact, mainly due to non-volatile hazardous constituents that are concentrated in the ash and potentially limit the extent of its land application. Moreover, land filling and incineration have negative climate effects, for methane or carbon dioxide is produced. Another important issue of sludge combustion is the emissions of chlorinated dibenzo-p-dioxins (PCDDs) and chlorinated dibenzofurans (PCDFs), NO_x, N_2O, SO_2, as well as HCl, HF and C_xH_y. (Werther, Ogada 1999). In fact, how to manage and treat sewage sludge in a sustainable manner and its safe disposal is one of the major environmental concerns throughout the world and has recently been a central issue of debate. Most experts in wastewater treatment and sludge handling consider that direct land application is the most sustainable method (Ødegaard et al., 2002; Guibelin, 2002; Lundin et al. 2004). Combined use of sewage sludge and its incinerated ash may prove to be a beneficial means of disposal., improving soil quality and crop production. As an organic amendment, SS improves physical., chemical., and microbiological properties of soils, whereas its stable organic matrix may act as both a source of and sinks for metals (Corey et al., 1987; Smith, 1996). On the other hand, soil addition of silt-size particles present in incinerated ash promotes better aeration, percolation and water retention capacity. Even though SS tends to increase soil acidity as a result of proton release from organic matter decomposition and mineralization, oxides formed during incineration would buffer pH decrease (Torri 2001).

The Study Area

In terms of population, Buenos Aires City is the third largest city in Latin America, and the eighth largest city in the world. From 1975 to 2000, the capital city of Argentina had an average annual population increase of 1.28% (United Nations 2004). The federal district of Buenos Aires alone consists of about three million inhabitants (INDEC 2001), representing the 34% of the total population. Buenos Aires City and its outskirts annually produce about 1.800.000 metric tons of sewage sludge. At present, sludge products are first aerobically stabilized in waste water treatment plants and, due to regulation and jurisdiction problems are discarded in landfarming or, to a minor extent, as a soil amendment on lawns or land filling. However, as in the rest of the world, agricultural soils application of sludge may be a common practice in Argentina in the future. A logistic aspect to be considered in this case is the distance between the place where sludge is produced and the agricultural zones where this waste could be potentially applied. The nearby agricultural area of Buenos Aires City, The Pampas Region, is a wide plain with more than 52 million ha of lands suitable for cropping and cattle rearing. Combined use of sewage sludge and its incinerated sludge is not yet practiced in Argentina, but it is thought to fit the local situation, which combines localized high-population and a large cropping area (Torri, 2001). Taking into account that the agricultural region of feasible land application of sewage sludge comprises the most productive soils of the country, the chemical availability of potentially toxic elements present in sewage sludge should be carefully studied in order to prevent transferability to the food chain. This Chapter addresses the dynamics of some trace elements of concern found in sewage sludge from Buenos Aires City when sewage sludge or a mixture of sewage sludge and its own incinerated ash are applied to soils of the Pampas Region, Argentina.

Characteristics of the Soils of the Pampas Region

The Pampas Region is one of the largest temperate field cropland areas of the Southern Hemisphere and is located between 32° to 39°S and 56 to 67°W. This zone covers about 30 Mha of agriculturally useful land, the remaining being either marginally suitable or unsuitable for cropping, mainly as a result of slight differences in relief.

The Aeolian sediments from which the soils of the Pampas have developed were brought from the south west, resulting in a progressively finer texture from south-west to north-east. This, combined with a gradient in rainfall which increases in the same direction, has produced a geographic sequence of Mollisols, with Entic Haplustolls (US Soil Taxonomy, USDA, 1999) at the western limit of the region, and the progressive appearance of Entic Hapludolls, Typic Hapludolls, Typic Argiudolls and Vertic Argiudolls towards the east (Soriano, 1992). These soils are associated with minor proportions of Aquolls in the drainage ways. The mineralogy of the soils is dominated by nearly unweathered volcanic material such as plagioclases, glasses and lithic fragments; minor components include quartz and orthoclase. In the clay-fraction, the predominant mineral is illite, some montmorillonite and a lesser proportion of interstratified illite/ montmorillonite make up the reminder (Soriano, 1991). The climate of the region is subtropically humid, characterized by long warm summers and mild winters, an average air temperature of 11 °C in July and 25.5 °C in January, and a mean annual precipitation of 1147 mm. These weather conditions allow good development and production of forage and crop species typical of temperate regions.

Figure 1. Map of the Pampas Region, Argentina.

The Pampas Region was under grassland vegetation up to about 150 years ago. Soils of this region are moderately acid and low in available P. Total organic carbon levels increase from southwest to northeast, in the same direction as the textural classes and are inversely related with sand content (Díaz-Zorita, Grosso 2000), with organic matter content between 9.9 g kg^{-1} and 85.0 g kg^{-1} (Hevia et al., 2003; Aparicio, Costa 2007).

In a first research carried out in the area, several heavy metals (Cd, Co, Cr, Cu, Ni, Pb and Zn) were found in lower concentrations in the open field than in Buenos Aires city (Lavado et al., 1999). A second and more compete work, showed that the top soils of the eastern pampas had concentrations and dispersion values of some elements (As, B, Ba, Cd, Co, Cr, Cu, Pb, Mn, Hg, Mo, Ni, Ag, Se and Zn) similar to other non-contaminated soils of the world. These results did not depend on taxonomical position or land use. PTE concentrations increased with depth or showed a maximum concentration at the Bt horizon (US Soil Taxonomy, USDA, 1999). This was related to parent material and to pedogenetic processes, and not to recent contamination (Lavado et al., 2004). In addition, PTE concentrations from subsoil samples were far below threshold values for contaminated soils (Lavado and Porcelli, 2000). From this information, and taking into account that there are no mines or factories in the region, the obtained data was taken as the baseline of PTE concentration in this region, corroborating the lack of soil contamination in rural areas of the Pampas Region.

Factors Affecting PTE Availability in Soils

Most sewage sludge metal regulations around the world are based on the use of total metal concentrations (MAFF, 1993; NSW EPA, 1997; US EPA, 1997, Chaney et al., 1999 and EU, 2000). Even though the evaluation of total PTE concentration may be useful as a global index of contamination, it provides limited information on the specific bioavailability or mobility of PTE in sludge-amended soil (McBride, 1995, Torri 2002). It is now widely recognized that the toxicity and mobility of these pollutants strongly depend on their specific chemical forms and on their binding state in the soil (i.e. precipitated with primary or secondary minerals, complexed by organic or inorganic agents, competing cations and so on). Soil physical-chemical properties are also likely to assume great importance in determining PTE behavior (Smith, 1994, Chaudri et al., 2000; McBride et al., 2000), especially at low sludge application rates. It has been postulated that, at very high application rates, sewage sludge properties dominate PTE chemistry and bioavailability in the zone of sludge incorporation in the short to medium term (Logan et al., 1997; Zufiaurre et al., 1998). With time, it is thought that sludge properties will have progressively less influence over metal behavior and that soil characteristics will ultimately control metal speciation, irrespective of the speciation in the original sludge (Smith, 1996). The time frame over which this occurs may greatly vary, from 12 weeks to over 8 years (Parkpain et al., 1998; Smith, 1996).

Organic matter dynamics and its role on PTE availability has been extensively investigated. However, there is still continuing controversy over the likely long-term effects of sludge applications on metal uptake, especially with regard to the effects of sludge adsorption properties on metal bioavailability (the sludge "protection" hypothesis), and the role of organic matter decomposition following cessation of applications (the "time bomb" hypothesis). It has been argued that in temperate climates, where organic matter decomposition is not particularly fast, the protective role of organic matter remains unaltered decades after sludge application. Therefore, PTE would be sorbed onto the remaining non-decomposable fraction of organic matter and also onto the inorganic matrix, for inorganic sorption phases are not altered in the time-scale of a few decades (Antoniadis et al., 2008), in good agreement with the sludge "protection" hypothesis. Consequently, land application of sewage sludge may cause an increase of organic matter in degraded soils through its stable organic matter pool (Torri et al., 2003).

Another important soil parameter affecting heavy metal availability is soil pH (Sauvé et al. 2000, McBride et al., 2004). Land application of sewage sludge has been reported to initially increase soil pH (Torri, Lavado, 2002). Afterwards, sludge´s organic matter mineralization lowers soil pH, leading to a decrease in PTE adsorption (Richards et al., 2000; Sukreeyapongse et al., 2002). Texture and adsorption strength of the soil matrix also affect the bioavailability of PTE. Results indicate that crops grown on sandy, low organic matter status soils are likely to have a greater uptake of certain elements (in particular Zn and Cd) compared with crops grown on soils with higher clay and organic matter contents (Alloway, 1990). Likewise, Richards et al. (2000) reported that mobility of Ni, Cd, and Zn was greater in fine sandy loam soil than in a silt loam soil at similar pH levels. However, Sauerbeck (1991) reported that soil metal concentration and soil pH are more important than organic matter and texture in regulating the availability of Cd, Cu, Ni, and Zn to several plant species. In this way, there are many soil, climate and sludge characteristics that are known to have a direct or indirect effect on PTE behavior in soils, so sludge-borne PTE availability in soils is difficult to predict and has to be carefully studied.

Soil Speciation

Quite a lot of work has been done to find a method that can reliably estimate bioavailability of PTE in sludge amended soils and thereby predict their impact on the soil ecosystem. The most important methods include: single batch extraction of soil samples with salt solutions (Adriano and Weber, 2001); sequential extraction with increasingly strong extractants designed to dissolve metals bound to different solid phases (Pierzynski 1998); and column leaching experiments (Paramasivam et al., 2006).

Chemical speciation can be defined as the process of identifying and quantifying different species, forms or phases present in a material. Sequential extraction is a useful technique for separating elements into various operationally defined fractions. Such information is potentially valuable for predicting bioavailability, metal leaching rates and transformations between chemical forms in agricultural and polluted soils (Tsadilas et al., 1995). In this technique, the soil is subjected to a series of chemical reagents of increasing reactivity, with phytoavailability and mobility of PTE decreasing in the order of the sequential extraction step. In most protocols, PTE are divided into the following physicochemical forms: (1) simple or complexed ions in solution and exchangeable ions; (2) bound to organic matter; (3) bound to carbonates; (4) bound to iron and manganese oxides and hydroxides; and (5) in the mineral lattice of silicates or residual fraction (McLaren, Crawford 1973; Tessier et al., 1979; Emmerich et al. 1982; Sims, Kline 1991; Morabito 1995). The terms of the fractions are more likely to be operationally, rather than chemically defined. However, each extractant in the sequentially selective procedure mostly targets one major solid-phase component. In this way, water-soluble and exchangeable fractions are considered to be bioavailable; oxide-, carbonate- and organic matter bound fractions may be potentially bioavailable; while the residual fraction is mainly not available to either plants or microorganisms (Ma, Rao 1997; He et al., 2005)

Studies on Pampas Soils

Sewage Sludge Used

Sewage sludge from Buenos Aires City was provided by the local water operator Aguas Argentinas S.A. The aerobically stabilized sludge used in these experiments was previously dried in holding pools in the sewage sludge treatment plant.

For the experiments described later, the sludge was oven-dried at 60°C, ground and sieved (<2 mm) (Figure 2 A), and then split into two portions. One portion was incinerated at 500° C, and the ash obtained (AS, Figure 2 B) was mixed with a portion of the previously sieved sewage sludge, resulting in a new mixed waste containing 30% DM as ash (SSA). Moisture content was determined by drying subsamples at 105°C to a constant weight. Analytical data (dry mass basis) for SS and SSA is presented in Table 1. The main crystalline component of SS was quartz, with a trace of plagioclase. Incineration of the sludge had little influence on the overall mineralogy of the sludge components (Torri, Lavado 2008 a).

Cadmium, Cu, Pb and Zn were studied because they are generally the PTE of greatest concern and concentration in sewage sludge (Wang et al., 2008; Antoniadis et al., 2008; Egiarte et al., 2008). Concern for Cd arises from its possible entry into the food chain. Kidney disorders are the first biochemical signs of Cd toxicity. Although Cu and Zn are known to be essential for the normal growth of plants or animals, high soil availability of both elements can cause phytotoxicity prior to produce plant concentrations that would be toxic to humans. Phytoxicity is thus a more relevant issue than risks of food chain contamination for these elements (Thomas et al., 2005). Copper is more phytotoxic than Zn and is therefore of greater concern for the environment. Although lead is not an essential element for plant growth, it is easily absorbed and accumulated in different plant parts organs (Khatik et al., 2006)

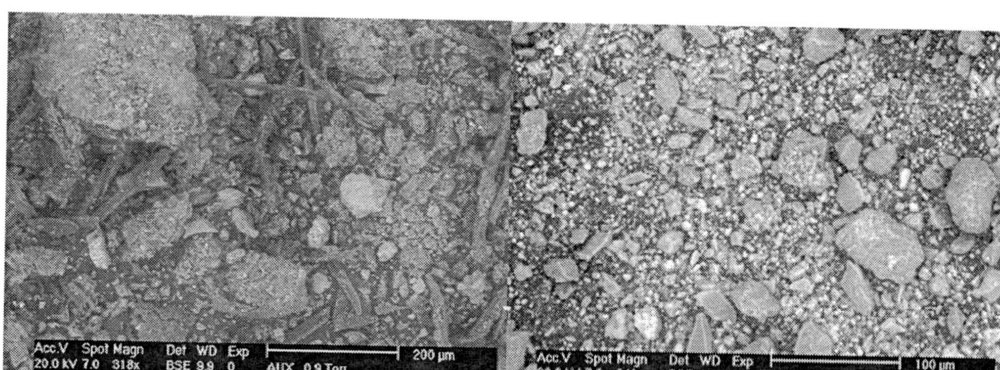

A B

Figure 2. SEM-EDS Images of sewage sludge (A) and incinerated sewage sludge at 500° C (B).

Table 1. Selected properties of sewage sludge (SS) and 70:30 DMW mixture of sewage sludge and sewage sludge ash (SSA)

	SS	SSA
pH	5.82	6.17
Moisture content (%)	5	4.5
Total organic carbon (mg g^{-1})	251	176
Total N (mg g^{-1})	19.3	22.5
Total P (mg g^{-1})	0.052	0.086
Electrical conductivity (dS m^{-1})	0.90	0.89
Cation exchange capacity ($cmol_{(c)}$ kg^{-1})	11.95	nd
Ca (mg g^{-1})	22.5	nd
Mg (mg g^{-1})	5.6	nd
K (mg g^{-1})	10.7	nd
Total Cd (mg kg^{-1})	10.08	13.08
Total Cu (mg kg^{-1})	750.8	894.7
Total Pb (mg kg^{-1})	334.2	365.9
Total Zn (mg kg^{-1})	2500	3150

nd = not determined.

The contents of Cd, Cu, Pb and Zn in the SS and SSA of this study did not exceed ceiling concentrations for land application recommended by Argentine regulation (S.A.D.S. 2001, Res. 97/01), whose values are similar to USEPA´s limits (USEPA, 1993). Moreover, Cd, Cu and Zn concentrations were within the numerical standards for biosolids not subject to cumulative pollutant loading rates (CPLRs) permitted by the USEPA regulations (Table 2).

Because of high Pb concentration, this sludge would not be allowed for use in agriculture without maintaining records of cumulative applications, 300 kg/ha for Pb. Total Pb loading used in this study for SS and SSA complies with the Argentine and U.S. regulations on cumulative loadings for sludge-treated soils.

Sequential chemical fractionation (McGrath and Cegarra, 1992) of Cd, Cu, Pb and Zn in SS and SSA samples is shown in Table 3. Cd was only recovered from the residual fraction. These results differ from other reports, where Cd in sludge was principally distributed in the oxidizable and residual fractions or between the easily assimilatable, exchangeable and reducible fraction (Fuentes et al., 2004 a; Wang et al., 2005).

Table 2. Argentine and USEPA acceptable standards of Cd, Cu, Pb and Zn in sludge applied to soils (USEPA 1993)

	Cd	Cu	Pb	Zn
			mg kg^{-1}	
SS	10.08	750.8	334.2	2500
SSA	13.08	894.7	365.9	3150
Argentina	85	4300	840	7500
USEPA - PC Biosolid [1]	39	1500	300	2800
USEPA – MAMC [2]	85	4300	840	7500
			kg ha^{-1}	
CPLRs [3]	39	1500	300	2800

[1] PC = Pollutant Concentration Biosolid.
[2] MAMC = Maximum Allowable Metal Concentrations.
[3] CPLRs = Cumulative Pollutant Loading Rates.

Incineration reduced the availability of Cu, Pb and Zn, increasing the percentage of residual fractions, which agrees with the results found by Obrador et al. (2001). Cu, Pb and Zn were predominantly in the residual fraction of both SS and AS, indicating that the elements were probably occluded in primary minerals. Only 17-20% of Cu was recovered as OM-Cu, differing from other studies (Zufiaurre et al., 1998, Scancar et al., 2002; Fuentes et al., 2004 b). The high proportion of Pb recovered from inorganic and residual fraction should also be mentioned. Different results have been reported in other studies (Scancar et al., 2002; Wang et al., 2005), where Pb was distributed between the organic and the residual fraction. Between 14.7-21.9% of Zn was recovered as OM-Zn in SS or AS, a higher percentage than other studies (Amir et al., 2005; Obrador et al., 2001), whereas relatively low amounts of Zn were found in the water-soluble and exchangeable fractions.

These results confirm that the mode by which the sludge is processed or the different origin of PTE in the sludge may afterwards affect individual metal association with soil components (Merrington et al., 2003; Smith, 1996). Sludge properties may affect PTE availability both directly and indirectly. Direct effects include PTE concentration and the sorptive capacity of sludge inorganic and organic components (Jing, Logan 1992, Xiao et al., 1999; Zhou, Wong 2001). Other sludge properties such as pH, EC and ionic strength indirectly affect the availability of sludge-borne elements (Hooda, Alloway, 1993; Weggler-Beaton et al., 2000).

Table 3. Distribution of Cd, Cu, Pb and Zn in sewage sludge (SS) and sewage sludge plus 30% DM ash (SSA), among water-soluble and exchangeable fraction (EXCH), organic matter bound fraction (OM), inorganic precipitate fraction (INOR) and residual fraction (RES) (mean ± SD, n = 4)

		SS	AS
		mg kg^{-1}	
Cd	RES-Cd	10.80 ± 1.75	13.08 ± 3.74
Cu	EXCH-Cu	8.45 ± 0.49	9.58 ± 1.05
	OM-Cu	157.7 ± 9.8	159.10 ±4.7
	INOR-Cu	108.9 ± 6.1	94.00± 7.9
	RES-Cu	475.8 ± 5.4	632.0 ± 6.6
Pb	EXCH-Pb	ND	ND
	OM-Pb	31.2 ± 4.4	14.2 ± 0.4
	INOR-Pb	157.9 ± 6.5	127.5 ± 8.9
	RES-Pb	145.0 ± 4.5	224.2 ± 8.0
Zn	EXCH-Zn	232.75 ± 9.17	159.5 ± 8.4
	OM-Zn	548 ± 6.3	399 ± 9.4
	INOR-Zn	888.2 ± 5.9	660.75 ± 9.8
	RES-Zn	773 ± 4.6	1223.5 ± 9.5

ND = not detectable (below analytical detection limit).

Dynamics of Cd, Cu, Pb and Zn in Sludge Amended Soils

This study includes a greenhouse incubation experiment during a period of one year. The soils selected for this experiment were Typic Hapludoll, Typic Natraquoll and Typic Argiudoll, sampled near Carlos Casares (35° 37' S - 61° 22' W), Pila (36° 1' S - 58° 8' W) and San Antonio de Areco (34° 15' S - 59° 29' W) towns respectively. Relevant soil properties are presented in Table 4.

Both sludge amendments were homogeneously mixed with each of the three soils at proportions equivalent to a field application rate of 150 t DM ha^{-1}. Soil moisture was maintained at 80% of water holding capacity during the experiment.

The application of the SS and SSA amendments to the three soils caused a significant build up of total Cd, Cu, Pb and Zn. The initial speciation of Cd, Cu, Pb and Zn in the three amended soils was governed by the chemical state in which these elements occurred in the sludge (Zufiaurre et al., 1998).

Table 4. Selected properties of the A horizon (0-15 cm)

	Typic Hapludoll	Typic Natraquoll	Typic Argiudoll
Clay (%)	19.2	27.6	30.3
Silt (%)	23.2	43.0	53.6
Sand (%)	57.6	29.4	16.1
pH	5.12	6.21	5.72
Organic carbon (g kg^{-1})	28.6	35.31	24.5
Electrical conductivity (dS m^{-1})	0.61	1.18	0.7
Cation exchange capacity ($cmol_{(c)}$ kg^{-1})	20.3	22.3	24.5
Exchangeable cations			
Ca^{2+} ($cmol_{(c)}$ kg^{-1})	10.2	9.1	12.6
Mg^{2+} ($cmol_{(c)}$ kg^{-1})	2.0	5.4	4.3
Na^{+} ($cmol_{(c)}$ kg^{-1})	0.3	2.1	0.2
K^{+} ($cmol_{(c)}$ kg^{-1})	2.8	1.6	2.1
Total Cd (mg kg^{-1})	nd	nd	nd
Total Cu (mg kg^{-1})	22	11	16
Total Pb (mg kg^{-1})	18	9	13
Total Zn (mg kg^{-1})	55	47	59

Cadmium

Cadmium (Cd) is a toxic element with high mobility in the environment. It can accumulate in some plant species without showing symptoms of phytotoxicity. Cadmium toxicity especially affects humans rather than animals, through its bioaccumulation in the food chain (Guo et al., 2007; Sun et al., 2008).

Domestic sewage sludge usually contains Cd because it is present in several goods washed down drains or toilets: it is found in cigarette butts flushed down toilets, or it is given off from rubber when car tires run over streets, and after a rain, it is washed into sewage systems where it collects in the sludge (Kirkham 2006).

In our studies, Cd was only extracted from the residual fraction from the SS or SSA amended soils, which is considered mainly to be highly crystalline Fe oxides and silicate minerals. EXCH-Cd, OM-Cd and INOR-Cd remained below analytical detection limits in the three amended soils during a year's period. On the contrary, some authors mentioned that exchangeable forms of Cd increased significantly under sludge application (Berti, Jacobs 1996; Vaca-Paulín et al., 2006), while others have reported that the greater percentage of Cd was in the inorganic fraction (Walter and Cuevas, 1999).

Copper

Copper (Cu) is an essential element for plant growth. However, its soil concentration in quantities lower or greater than optimal amount can adversely affect plant growth (Alva et al., 2000).

In our experiments, Cu was mainly extracted as RES-Cu (36.7-65%) from either SS or SSA amended soils, indicating tightly bonded forms - presumably corresponding to Cu associated to lattices of primary minerals, Fe-Mn oxides, and sulphides (Miller et al., 1986; Vaca-Paulín et al., 2006). A lower percentage of Cu was found to be associated with the organic (20.7-34.7%) and inorganic (12.9-27.6%) fractions, whereas EXCH-Cu was below analytical detection limit, suggesting extremely low initial levels of Cu availability.

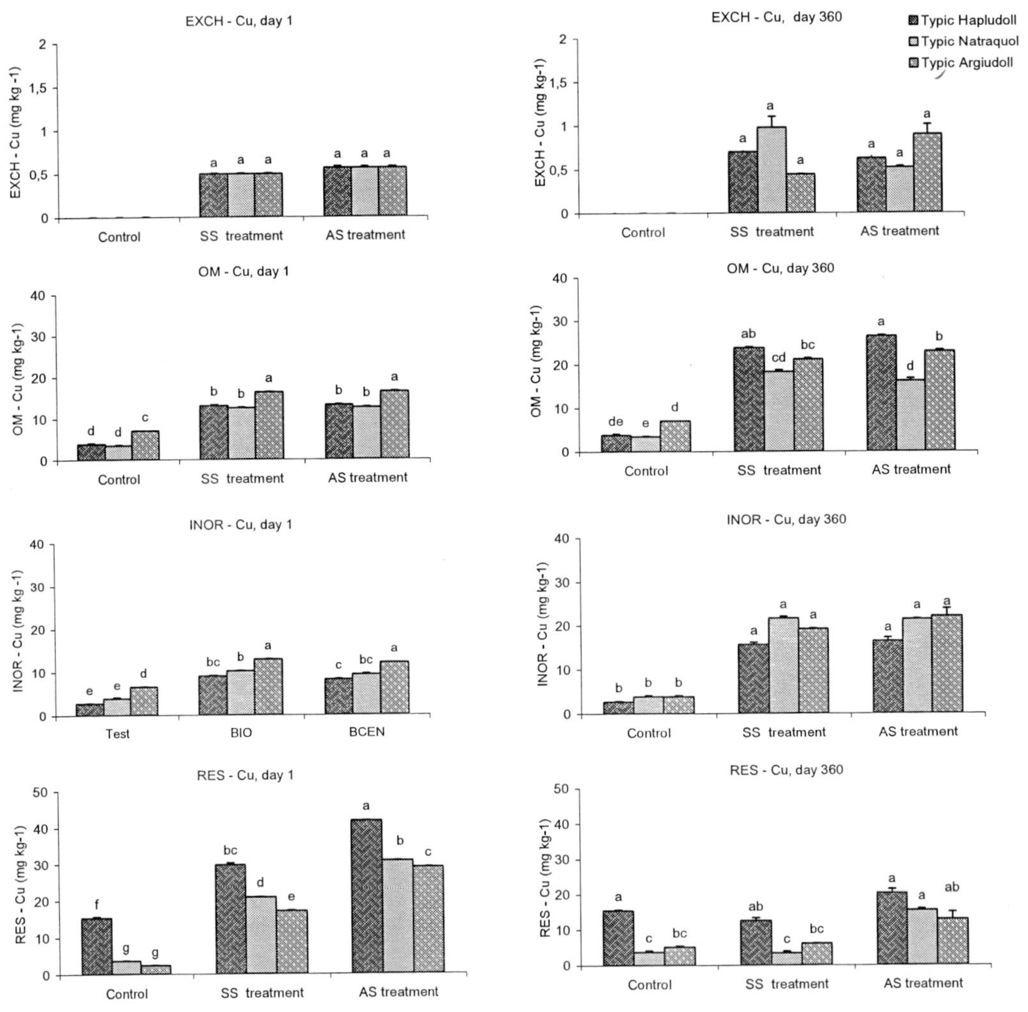

Figure 3. Distribution of Cu among soil fractions in a Typic Hapludoll, Typic Natraquoll and Typic Argiudoll of the Pampas region, Argentina, amended with pure sludge (SS), and the 70:30 (w/w) mixture of sewage sludge and sludge ash (AS). For each date, different letters indicate significant differences (Tuckey, $p<0.05$).

With the passage of time, an increase in OM-Cu and INOR-Cu fractions along with a decrease in the RES-Cu was observed in all sludge amended soils. These results are similar to findings reported by Berti and Jacobs (1996). Other research has indicated that most of the Cu was associated with organic forms (Su, Wong 1994), or organic and residual forms (Nyamangara 1998). At the end of the year, the magnitude of the increase in the OM-Cu fraction in the amended soils depended on soil chemical characteristics, and followed the sequence: Hapludoll > Argiudoll > Natraquoll, with no significant differences between SS and SSA treatment (Figure 3). These results indicate that Cu incorporated to soils through sewage ash did not adsorb onto sludge organic matter. Copper is well known to be predominantly associated with soil organic matter (Irving and Williams, 1953; Walker et al., 2003). However, the proportion of OM-Cu did not correlate with the organic carbon content of the amended soils: SOM was higher in the Natraquoll (38.46 mg C g^{-1} soil), compared to the Hapludoll (31.3 - 33.9 mg C g^{-1} soil) or the Argiudoll (25.53 - 28.22 mg C g-1 soil) (Torri and Lavado, 2002). The soil factor most closely associated with the distribution of OM-Cu in the three soils was soil pH. There was a negative correlation between OM-Cu fraction and soil pH at the end of the experiment, indicating that OM-Cu increased as soil pH decreased (Figure 4). Conversely, a positive correlation was observed between INOR-Cu and soil pH in this date (Figure 5).

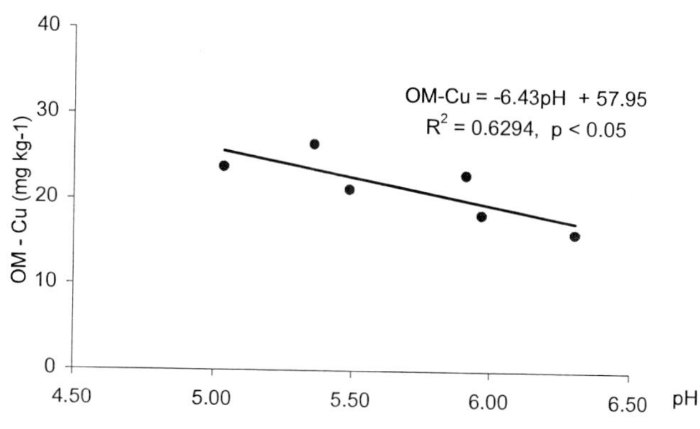

Figure 4. Relation between OM-Cu and soil pH in the amended soils.

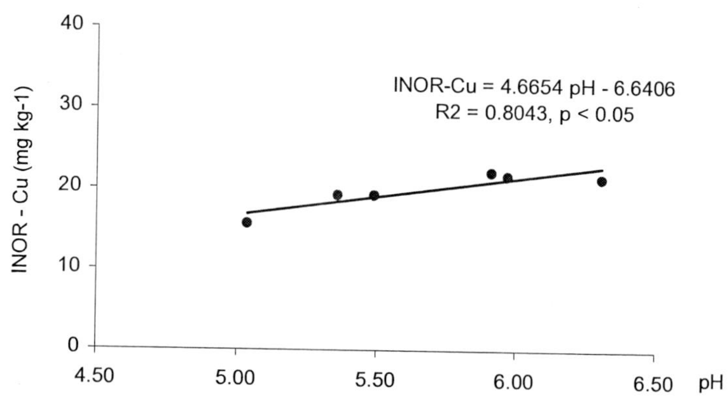

Figure 5. Relation between INOR-Cu and soil pH in the amended soils.

A small percentage of EXCH-Cu was detected a year after sludge application in the amended soils (0.95 - 2.21%). It may be expected that Cu would show an increase in the labile fraction over time due to the appreciable amount of organic carbon mineralization over the incubation period. In these soils, mineralization constants, calculated as the fraction mineralized per time unit (day^{-1}) were found to be between 0.03 - 0.071 (Torri et al., 2003)

Several studies on the relationship between PTE and dissolved organic matter (DOM) showed that organic molecules were, in many instances, responsible for the availability of PTE in sludge amended soils. Hsu and Lo (2000) observed an increase in Cu availability with increasing pH and a concurrent increase in DOM. They attributed these results to the irreversible dissolution of organic matter with organically bound Cu at high pH during natural weathering of the sludge in the soil. However, in our studies, EXCH-Cu was below analytical detection limits in the period of intense mineralization of sewage sludge organic matter. Martínez and Motto (2000) studied the relationship between PTE and pH, and found an approximate threshold pH value at which EPT solubility increased in amended soils. This pH value for Cu was 5.5 in non-calcareous soils, concluding that lower pH values may enhance Cu biological availability. In our studies, soil pH of SS and SSA amended soils ranged from pH = 5.04 - 6.30 at the end of the year. Although the lowest pH value was below the above mentioned threshold, EXCH-Cu was below analytical detection limits.

Lead

Lead has been widely studied due to its long persistence in soil and high toxic effects on both crop production and human health. Most of the Pb taken up by plants is restricted to root and very small amount is transported to the shoots (Reddy et al., 2005). However, plant leaves may show symptoms of Pb toxicity through inhibition of photosynthesis (Islam et al., 2008). Lead uptake capacity was found to correlate with soil pH and soil organic matter content (Gao et al., 1997 and Hooda and Alloway, 1998),

Most of the Pb in our sludge amended soils was initially found as RES-Pb (49- 68 %) and INOR-Pb (23-46 %), and a minor proportion as OM-Pb (1.8 - 5 %). Concentrations of EXCH-Pb remained below the levels of analytical detection.

With the passage of time, an increase in INOR-Pb along with a decrease in RES-Pb was observed in the three SS and SSA amended soil samples (Figure 6).

The increase of Pb bound to Fe-Mn oxides at the expense of the residual fraction usually indicates anthropogenic contamination, and has been reported in contaminated sites (Jensen et al. 2006). At the end of the year, most of the Pb in our soils was found as INOR-Pb, possibly precipitated as mineral phases (Morin et al., 2001). It is well known that phosphate reduces Pb mobility by ionic exchange and precipitation of pyromorphite-type minerals. The newly formed minerals have a very low solubility and bioaccessibility (Hettiarachchi et al., 2001; Scheckel and Ryan, 2003).

Although soils of the Pampas region are low in available P, phosphorous is found in large amounts in other different chemical forms (Gutierrez Boem et al, 2008). The results obtained herein for Pb are in good agreement with other studies (Wong et al., 2001, Walker et al., 2003; Jensen et al., 2006).

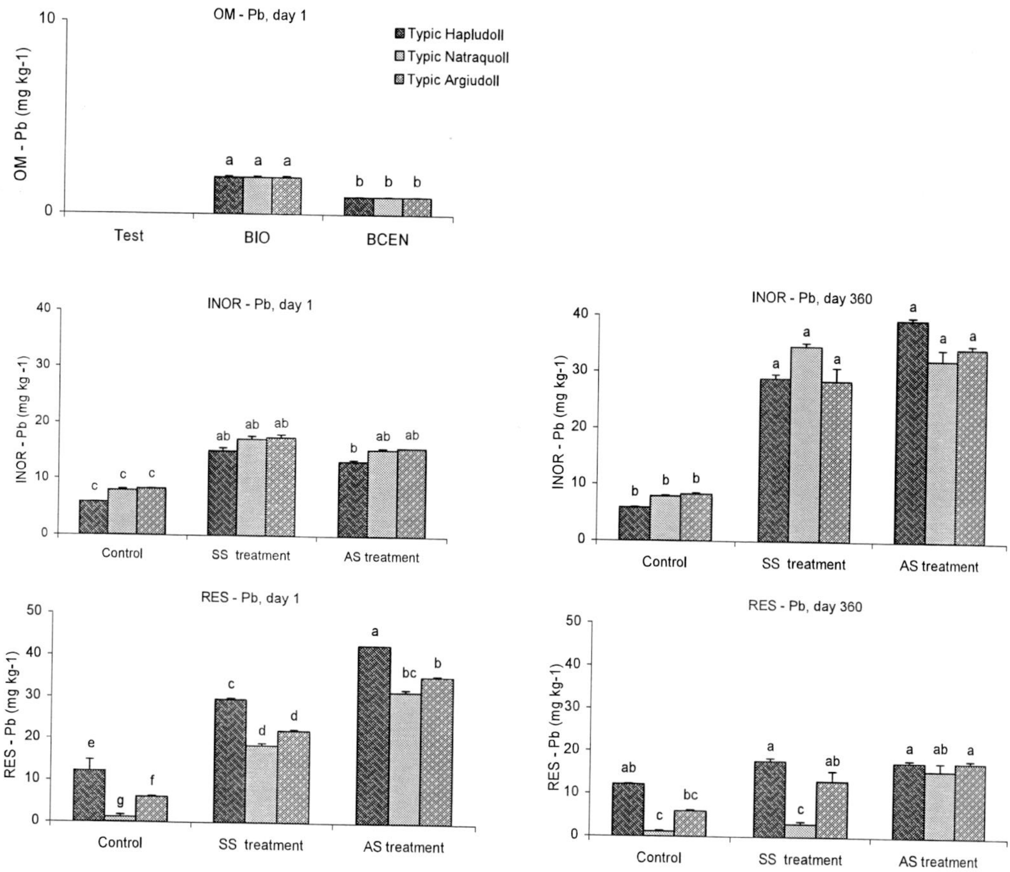

Figure 6. Distribution of Pb among soil fractions in a Typic Hapludoll, Typic Natraquoll and Typic Argiudoll of the Pampas region, Argentina, amended with pure sludge (SS), and the 70:30 (w/w) mixture of sewage sludge and sludge ash (AS). For each date, different letters indicate significant differences (Tuckey, $p<0.05$).

No significant differences in INOR-Pb were observed between SS or SSA treatments for the same soil or among amended soils, indicating that differences in soil texture or soil pH did not appear to influence Pb redistribution to the inorganic fraction.

Lead is reported to form very stable complexes with humic substances, reducing its availability (Hooda, Alloway 1998; Strawn, Sparks 2000). However, OM-Pb in this study was below analytical detection limits and differs from other reports (Strawn et al., 2000; Sánchez-Martín et al., 2007). Studies on Pb sorption and desorption kinetics revealed that soil organic matter increased the adsorption and impeded desorption of Pb from soil (Strawn and Sparks, 2000). However, our studies showed that the stable organic matter present in the SS or SSA amended soils had a weak tendency to form stable organic complexes with Pb.

A possible reason for this is that the inorganic matrix of both SS and SSA amendments enhanced the formation of inorganic Pb compounds, regardless soil pH. The formation of thermodynamically stable minerals minimizes Pb bioavailability (Arnich et al., 2003). The fact that EXCH-Pb was below analytical detection limits in all sampling dates, suggest that solubilization reactions were negligible.

Zinc

Zinc is an ubiquitous element in nature, though it is believed that approximately 96% of its release into the global environment is a result of anthropogenic activities (Landner and Reuther, 2004). Vegetable crops are generally sensitive to high soil levels of zinc, while grasses are usually tolerant. Several field studies have shown yield reduction related to Zn phytoxicity on sludge amended soils (Jarausch-Wehrheim et al. 1999; Bhogal et al. 2003). De Brouwere and Smolders (2006) reported that maize Zn concentrations increased linearly during the first year of all sludge treatments with increasing sludge application rate without reaching a plateau, indicating a potential pathway of exposure to Zn. In our assay, EXCH-Zn was initially significantly higher in SS than in AS treatment for all soils. An increase in this fraction was observed in all sludge-treated soils over the time frame of the experiment (Figure 7), depending on soil characteristics. No significant differences in EXCH-Zn were observed comparing SS and AS treatments for the same soil. These results indicate that Zn incorporated into the soils as a mixture of sewage sludge and ash did not increase Zn availability compared to the pure sewage sludge treatment. Conversely, other studies reported increases in bioavailable Zn in soils amended with sludge ash (Bierman and Rosen, 1994; Saikia et al., 2006). Oxidizing conditions such as incineration are postulated to change organic Zn into oxides (Chang et al., 1999), which may furthermore form chloride compounds (Belevi and Moench, 2000), increasing Zn bioavailability in ash amended soils. Xiao et al. (1999) also reported that ash/sludge mixtures have elevated concentration of dissolved organic matter that increased PTE bioavailability. However, in our study, the amounts of EXCH-Zn in the AS treatment were initially significantly lower than in the SS treatment. The significant increase of EXCH-Zn in all sludge-treated soils is partially due to the mineralization of sludge's organic matter. This is consistent with several other studies (McGrath et al., 2000 and Nyamangara, 1998). However, the concentration of EXCH-Zn depended on soil characteristics. Soil particle size distribution and soil pH are usually considered to play important roles in controlling trace metal availability. In this study, the concentration of EXCH-Zn was not related to clay content but with soil pH (Figure 8). Similar results were found in other studies (Basta and Sloan, 1999).

Organic matter bound Zn was initially significantly higher for the SS treatment compared to the AS treatment at the beginning (Figure 2), but no significant differences were observed among both treatments for each soil at the end of the year. The intense mineralization of the labile organic matter pool of sludge treated-soils (Torri et al., 2003) resulted in a decrease in OM-Zn. The Zn released increased EXCH-Zn as well as the inorganic fraction of the amended soils, in agreement with other studies (Shuman, 1999 and Xiao et al., 1999), and despite the decrease of soil pH with time.

Shuman (1999) reported that some organic waste materials, such as spent mushroom compost and humic acid, lowered the potential availability of Zn by redistributing it from the exchangeable to the less soluble fractions like manganese oxide or organic matter fractions. We conclude that the transformation of raw organic matter to stable humic substances with time favored the complexation of sludge borne Zn onto stable organic forms regardless soil pH.

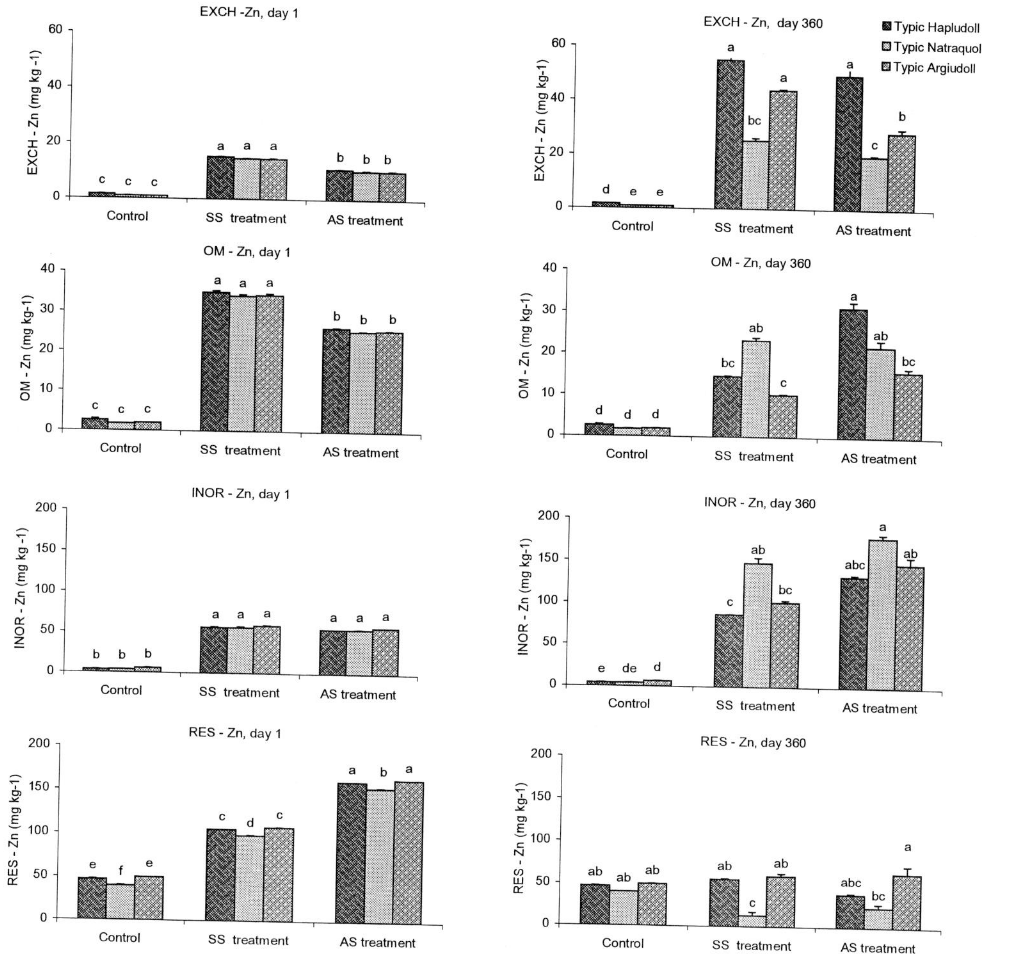

Figure 7. Distribution of Zn among soil fractions in a Typic Hapludoll, Typic Natraquoll and Typic Argiudoll of the Pampas region, Argentina, amended with pure sludge (SS), and the 70:30 (w/w) mixture of sewage sludge and sludge ash (AS). For each date, different letters indicate significant differences (Tuckey, $p<0.05$).

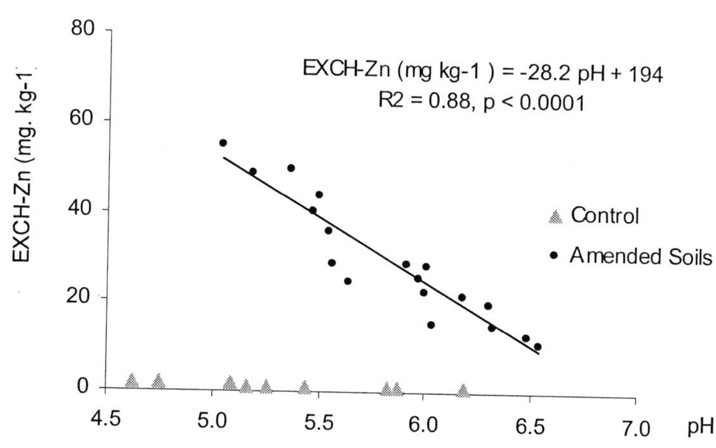

Figure 8. Relation between EXCH-Zn and soil pH in the amended soils, day 360.

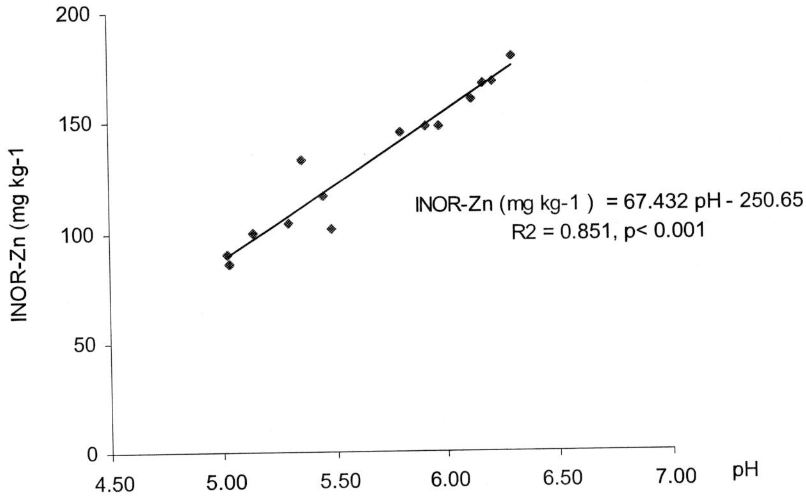

Figure 9. Relation between INOR-Zn and soil pH in the amended soils, day 360.

Inorganic bound Zn increased in both sludge treatments at the end of the studied period, representing the most abundant fraction, in good agreement with other reports (Qiao et al., 2003; Walter and Cuevas, 1999). Zinc has a relatively high affinity for sorption on the surfaces of Fe/Mn/Al-hydroxides (Meima and Comans, 1999), which are usually present in large amounts in municipal solid waste incinerator ash (Stipp et al., 2002). These processes are enhanced by increasing soil pH (Figure 9) in good agreement with other reports (Alloway and Jackson, 1991; Luo and Christie, 1998; Morera et al., 2002). Again, no significant differences were observed between SS and AS amended soils.

Conclusions

The results obtained in these experiments suggest that Cd was strongly adsorbed by reactive sorptive solid and inorganic sludge components or was incorporated in neoformed clay-like minerals. These results reflect the low reactivity and bioavailability of sludge-born Cd in soils amended with sewage sludge from Buenos Aires city and its outskirts.

On the other hand, Cu originally bound to the organic fraction of SS or SSA was retained by other sludge components instead of being released to water-soluble or exchangeable fractions. The kind of retention of Cu depended on soil pH. Cu had a stronger affiliation with the resistant, non-readily decomposable organic materials of SS or SSA at lower pH. The results obtained in our studies indicated that the proportion of OM-Cu / INOR-Cu is governed by soil pH. When soil pH is low, the association of Cu with the organic fraction predominates. As soil pH increases, there is a shift of Cu towards inorganic species. Conversely, the chemical behavior of Pb in different sludge amended soils may be similar, regardless soil characteristics.

A dynamic equilibrium of Zn forms in soils was observed. The increase of water-soluble and exchangeable Zn with time in both sludge-treated soils indicates that these amendments are an important short-term source of available forms of Zn. A negative and significant correlation between exchangeable Zn and soil pH was found. Redistribution towards the inorganic fraction was observed, representing the most abundant fraction in the three soils a year after sludge application. At the end of the year, inorganic Zn was positively and significantly correlated with soil pH.

Other Experiments

A set of large field experiments were carried out on commercial farms, in the North and West of Buenos Aires province, Argentina. The soils selected for this experiment were Typic Argiudolls, Vertic Argiudolls and Typic Hapludolls, located in the Rolling and Flat Pampas. The study was performed on i) nine field experiments with maize (*Zea mays* L.). Treatments included controls and different biosolids doses. The experiments were repeated for three years; ii) Four field experiments with wheat (*Triticum aestivum* L.) with similar treatments. The experiments were repeated for two years. Soil total PTE were determined after a digestion with a mixture of concentrated nitric and hydrochloric acid in a volumetric ratio of 1:3 respectively, whereas available PTE were extracted with aqueous solutions of ethylene diamine tetra-acetic acid (EDTA).

Few significant relationships between PTE soil content and PTE plant concentration were found in both crops. Total Cu, Zn, and Pb showed significant correlation with Zn concentration in wheat grains and shoots; EDTA extracted Cu significantly correlated with Cu concentration in maize grains (Table 5). Comparing mild (EDTA) and strong (concentrated hydrochloric, perchloric and nitric acids) extractions, PTE extraction using EDTA do not represent better PTE availability to crops than total extraction (Lavado et al., 2006). This is in agreement with above stated results.

Other study quantified the availability of PTE in soils and its uptake by maize (*Zea mays* L.) in soils which had received digested or non digested biosolids. Evidences were found that soils which received non-digested biosolids exhibited higher PTE crop availability and its accumulation in maize tissues in general terms. This could be caused by a higher level of interaction between EPT and the organic and inorganic biosolid matrix during digestion. Cd, one of the most mobile PTE, with high environmental impact, was the most remarkable element: it was the only PTE which was significantly higher in grains in non-digested biosolids treatment (Lavado et al., 2005).

Finally, in an experiment with soybean it was found that the application of the high biosolid doses caused significant increases in bioavailable Cu, Pb and Zn, as related with the check concentrations However there was not any accumulation of them either in soybean leaves or in grains. In the other trace elements studied (Cd), the plant concentration did not differ among treatments. A linear correlation between the Zn concentrations in leaves and grains and the bioavailable Zn soil concentration was found (Zn in grains ($mg.kg^{-1}$) = 0.69 Zn in soil ($mg.kg^{-1}$) x 36.58, R^2= 0.728). The high correlation suggests that the crop absorbed most Zn from the biosolid.

Table 5. Correlation between total PTE and EDTA extractable content in soils (mg kg^{-1}) and PTE concentration in maize and wheat shoot (mg kg^{-1})

Crop	Independent variable	Dependent Variable	Regression Equation	R^2
Maize	Total Mn in soils	Mn in shoot	Y=-43.45 + 0.15 X	0.40^{**}
	EDTA extractable Cu in soils	Cu in grains	Y=-1.30+ 2.54 X	0.82^{***}
wheat	Total Pb in soils	Pb in shoot	Y=0.43 + 0.13 X	0.87^{***}
	Total Zn in soils	Zn in shoot	Y = 10.65 * (ln X) – 24.60	0.70^{***}
	Total Cu in soils	Cu in grains	Y = 0.63 – 0.06 X	0.84^{***}
	Total Pb in soils	Pb in grains	Y=0.40 + 0.14 X	0.85^{***}
	Total Zn in soils	Zn in grains	Y = 15.73 * (ln X) – 8.01	0.68^{***}

***: Significant at the 0.1 % level.
**: Significant at the 1 % level.

General Conclusions

The largest proportion of Cd, Cu, Pb and Zn in the three sludge amended soils of the pampas region were initially found in the residual fraction. Although PTE associated with this fraction is considered not available to plants or microorganisms, a year after sludge application a shift of Cu, Pb and Zn from the residual to other stable fractions considered unavailable for plant uptake was observed.

The chemical partitioning of Cu and Pb into unreactive forms in sludge amended soils indicates that Cd and Pb bioavailability are not of major concern for these soils. Moreover, the results obtained in our studies for sludge –borne Cd and Pb indicate that the chemical behavior of these elements in soils with different physicochemical-characteristics may be similar. However, a high pH dependence of Cu and Zn soil fractions in sludge amended soils was observed.

Overall, the studies performed over three soil samples of representative soils of the Pampas Region, Argentina, showed that the use of a mixture of sewage sludge containing 30% DM of its own incinerated ash as a soil amendment did not pose a significant risk of soil, water or plants being contaminated. Furthermore, for the same chemical form, there were no significant differences between the concentration of each studied PTE in sludge or sludge-ash amended soils, in spite of the high rates applied. Based on these results, land spreading of sewage sludge with its own ash may be similar to sewage sludge disposal in terms of Cu. Pb or Zn mobility. Nevertheless, these results cannot be directly extrapolated on long-term field conditions. The long-term fate of sludge-added PTE after cessation of sludge application, then, need to be investigated.

The set of field experiments with different crops in the Pampas region indicated that, although application of biosolids can increase PTE concentration in soils, this does not imply a correlative increase of PTE concentration in plants. We conclude that in areas without other sources of PTE, the application of sewage sludge following local regulations does not imply a risk of PTE accumulation in crops.

REFERENCES

Adriano D.C. and J.T. Weber, 2001. Influence of fly ash on soil physical properties and turfgrass establishment, *J. Environ. Qual.* 30 (), pp. 596-601.

Alloway BJ and Jackson AP . 1991. The behaviour of heavy metals in sludge amended soils. *Science of the total Environment* 100, 151 - 176.

Alva, A.K., B.Huang and S.Paramasivam. 2000. Soil pH affects copper fractionation and phytotoxicity. *Soil Sci. Soc. Am. J.* 64: 955-962.

Amir S., Hafidi M, Georges Merlin G and J. Revel. 2005. Sequential extraction of heavy metals during composting of sewage sludge. *Chemosphere,* 59: 801-810.

Antolín M. C, Pascual, I. Carlos García C., Polo, A; Sánchez-Díaz M. 2005. Growth, yield and solute content of barley in soils treated with sewage sludge under semiarid Mediterranean conditions. *Field Crops Research*, Volume 94, Issues 2-3, 15 November 2005, Pages 224-237.

Antoniadis V., Robinson J.S. and Alloway B.J.. 2008. Effects of short-term pH fluctuations on cadmium, nickel, lead, and zinc availability to ryegrass in a sewage sludge-amended field. *Chemosphere*, 71: 759-764.

Aparicio V, Costa J. L. 2007. Soil quality indicators under continuous cropping systems in the Argentinean Pampas. *Soil and Tillage Research*, Volume 96, Issues 1-2, October 2007, Pages 155-165.

Arnich, N., M.-C. Lanhers, F. Laurensot, R. Podor, A. Montiel, and D. Burnel. 2003. In vitro and in vivo studies of lead immobilization by synthetic hydroxyapatite. *Environ. Pollut.* 124:139–149.

Ashworth, D. J., Alloway, B. J. 2007. Complexation of copper by sewage sludge-derived dissolved organic matter: Effects of soil sorption behaviour and plant uptake. *Water Air and Soil Pollution,* 182, 187–196.

Basta,N.T. and J.J.Sloan.1999. Bioavailability of heavy metals in strongly acidic soils treated with exceptional Quality biosolids. *J. Environ.Qual.* 28: 633-638.

Belevi, H., Moench, H., 2000. Factors determining the element behavior in municipal solid waste incinerators. 2. Laboratory experiments. *Environmental Science and Technology* 34, 2507–2512.

Berti, W.R. and L.W.Jacobs. 1996. Chemistry and phytotoxicity of soil trace elements from repeated sewage sludge applications. *J. Environ. Qual* 25: 1025-1032.

Bhogal, A., Nicholson, F.A., Chambers, B.J. and Shepherd, M.A. 2003. Effects of past sewage sludge additions on heavy metal availability in light textured soils: implications for crop yields and metal uptakes. *Environmental Pollution*, 121, 413–423.

Bierman, P.M., Rosen, C.J. 1994. Sewage sludge incinerator ash effects on soil chemical properties and growth of lettuce and corn. *Commun Soil Sci and Plant Anal* 25 (13-14) , 2409-2437.

Bridle and Skrypski-Mantele, 2000 T. Bridle and S. Skrypski-Mantele, Assessment of sludge reuse options: a life cycle approach, *Water Science and Technology* 41 (8) (2000), pp. 131–135.

Chambers, B., S. Royle, S. Hadden and S.Maslen, 2002. The use of biosolids and other organic substances in the creation of soil-forming materials. *Journal of the Chartered Institution of Water and Environmental Management* 16, 34-39.

Chaney, R.L., Brown, S.L., Stuczynski, T.I., et al., 1999. In-situ remediation and phytoextraction of metals from hazardous contaminated soils. Presented at US-EPA's Conference: Innovative Clean-Up Approaches: Investments in Technology Development, *Results and Outlook for the Future*. Nov. 2–4, Bloomingdale, IL.

Chang, Y.M., Chang, T.C., Lin, J.P., 1999. Effect of incineration temperature on lead emission from a fixed bed incinerator. *Journal of Chemical Engineering of Japan* 32, 626–634.

Chaudri et al., 2000 A.M. Chaudri, C.M.G. Allain, V.L. Barbosa-Jefferson, F.A. Nickolson, B.J. Chambers and S.P. McGrath, A study of the impacts of Zn and Cu on two rhizobial species in soils of a long-term field experiment, *Plant Soil.* 221 (2000), pp. 167–179.

Corey, R.B. King L.D., Leu-Hing C., Fanning D.S., Street J.J. and J.M. Walker, 1987.Effects of sludge properties on accumulation of trace elements by crops. In: A.L. Page, T.J. Logan and J.A. Ryan, Editors, *Land Application of Sludge: Food Chain Implications*, Lewis Publishers, Chelsea, MI (1987), pp. 25–51.

Darvodelsky P, Fien M (2005): The health impact of biosolids use on land. *Water* 21 12, 20–22.

Díaz-Zorita, M.. Grosso G. A. 2000. Effect of soil texture, organic carbon and water retention on the compactability of soils from the Argentinean pampas. *Soil and Tillage Research*, Volume 54, Issues 1-2, 121-126.

Egiarte et al, 2008. G. Egiarte, M. Pinto, E. Ruíz-Romera, M. Camps Arbestain. Monitoring heavy metal concentrations in leachates from a forest soil subjected to repeated applications of sewage sludge. *Environmental Pollution*, In Press, Corrected Proof, Available online 3 July 2008.

Emmerich, W.E, L.J.Lund, A.L.Page and A.C.Chang, 1982. Movement of heavy metals in sewage sludge-treated soils. *J. Environ.Qual.*11: 174-178.

EU, 2000. EU, 2000. *Working document of sludge*, 3rd Draft. European Union, Brussels. EnV.E.3/LM.

Fuentes Ana, Mercedes Lloréns, José Sáez, Antonio Soler, M Isabel Aguilar, Juan F. Ortuño, Victor F. Meseguer. 2004 a. Simple and sequential extractions of heavy metals from different sewage sludges. *Chemosphere,* 54: 1039-1047.

Fuentes Ana, Mercedes Lloréns, José Sáez, Ma Isabel Aguilar, Juan F. Ortuño and Victor F. Meseguer. 2004 b. Phytotoxicity and heavy metals speciation of stabilised sewage sludges. *Journal of Hazardous Materials*, 108: 161-169.

Gao, S.A., W.J. Walker, R.A. Dahlgren, and J. Bold. 1997. Simultaneous sorption of Cd, Cu, Ni, Zn, Pb, and Cr on soils treated with sewage sludge supernatant. *Water Air Soil Pollut.* 93:331–345.

García-Orenes F., C. Guerrero, J. Mataix-Solera, J. Navarro-Pedreño, I. Gómez, J. Mataix-Beneyto. 2005. Factors controlling the aggregate stability and bulk density in two different degraded soils amended with biosolids. *Soil and Tillage Research*, Volume 82, Issue 1, May 2005, Pages 65-76.

Guibelin, E. 2002 Sustainability of thermal oxidation processes: strength for the new millennium, *Water Science and Technology* 46 (10) (2002), pp. 259–267.

Guo B., Y.C. Liang, Y.G. Zhu and F.J. Zhao, 2007. Role of salicylic acid in alleviating oxidative damage in rice roots (Oryza sativa) subjected to cadmium stress, *Environ. Pollut.* 147 (2007), pp. 743–749.

He et al., 2005. Z.L. He, X.E. Yanga and P.J. Stoffellab, 2005Trace elements in agroecosystems and impacts on the environment. *Review, J. Trace Elem. Med. Biol.* 19: 125–140.

Hettiarachchi G.M., G.M. Pierzynski and M.D. Ransom, 2001 In situ stabilization of soil lead using phosphorus, *Journal of Environmental Quality* 30: 1214–1221.

Hevia, G.G., Buschiazzo, D.E., Hepper, E.N., Urioste, A.M., Antón, E.L.. 2003. Organic matter in size fractions of soils of the semiarid Argentina. Effects of climate, soil texture and management. *Geoderma* 116 (3-4), pp. 265-277

Hooda P.S. and B.J. Alloway, 1993. Effects of time and temperature on the bioavailability of Cd and Pb from sludge-amended soils. *J. Soil Sci.* 44 (1993a), pp. 97–110.

Hooda P. S., B. J. Alloway. 1998. Cadmium and lead sorption behaviour of selected English and Indian soils. *Geoderma*, 84: 121-134.

Hsu, J.-H.; S.L. Lo. 2000. Characterization and extractability of copper, manganese, and zinc in swine manure composts, *Journal of Environmental Quality* 29: 447-453.

Irving and Williams, 1953 H. Irving and R.J. Williams, Stability of transition metal complexes, *Journal of Chemical Society* (1953), pp. 3182–3210.

Islam E., Dan Liu, Tingqiang Li, Xiaoe Yang, Xiaofen Jin, Qaisar Mahmood, Shengke Tian, Junying Li. 2008. Effect of Pb toxicity on leaf growth, physiology and ultrastructure in the two ecotypes of *Elsholtzia argyi. Journal of Hazardous Materials*, 154: 914-926.

Jamali M. K., Kazi T. G., Arain M. B., Afridi H. I., Jalbani N., Memon A. R. 2007. Heavy metal contents of vegetables grown in soil, irrigated with mixtures of wastewater and sewage sludge in Pakistan, using ultrasonic-assisted pseudo-digestion. *Journal of Agronomy and Crop Science* 193: 218–228.

Jarausch-Wehrheim, B., Mocquot, B. and Mench, M. 1999. Absorption and translocation of sludge-borne zinc in field-grown maize (Zea mays L.). *European Journal of Agronomy*, 11, 23–33.

Jensen et al., 2006 P.E. Jensen, L.M. Ottosen and A.J. Pedersen, Speciation of Pb in industrially polluted soils, *Water, Air, and Soil Pollution* 170 (1–4) (2006), pp. 359–382.

Jing and Logan, 1992. J. Jing and T.J. Logan, Effects of sewage sludge cadmium concentration on chemical extractability and plant uptake. *J. Environ. Qual.* 21 (1992), pp. 73–81.

Khatik, S.K. Thakur, R. Sharma, G.D. Lead: The heavy metal in soil water and plant environment. *Journal of Industrial Pollution Control*. Volume 22, Issue 2, 2006, Pages 233-244

Kidd P.S., Domínguez-Rodríguez M.J, Díez J. and C. Monterroso. 2007. Bioavailability and plant accumulation of heavy metals and phosphorus in agricultural soils amended by long-term application of sewage sludge. *Chemosphere,* 66: 1458-1467.

Kirkham, M.B. 2006. Cadmium in plants on polluted soils: Effects of soil factors, hyperaccumulation, and amendments. *Geoderma* 13: 19-32.

Landner, L. and Reuther, R. (2004). *Metals in the society and in the environment*. Dordrecht: Kluwer Academic Publishers.

Laturnus F, von Arnold K, Grøn C. 2007. Organic contaminants from sewage sludge applied to agricultural soils. False alarm regarding possible problems for food safety? False alarm regarding possible problems for food safety? *Environmental Science and Pollution Research.* Volume 14, Number 1 / mayo de 2007, 53-60.

Lavado R. S., Porcelli C. A., Alvarez R. 1999. Concentration and distribution of extractable elements in a soil as affected by tillage systems and fertilization. *The Science of The Total Environment,* 232: 185-191.

Lavado R.S. ,and C.A. Porcelli. 2000. Contents and main fractions of trace elements in Typic Argiudolls of the Argentinean Pampas. *Chemical Speciation and Bioavailability.* 12(2): 67-70.

Lavado R.S., M. Rodríguez, R.Alvarez, M.A. Taboada and M.S. Zubillaga. 2006. Transfer of potentially toxic elements from biosolid-treated soils to maize and wheat crops. *Agriculture, Ecosystems and Environment* 118: 312-318.

Lavado, R., M. Zubillaga, R. Alvarez and M. Taboada. 2004. Baseline levels of potentially toxic elements in pampas soils. *Soil and Sediment Contamination: an International Journal.*, 13 (5) : 329-339.

Lavado, R.S. 2007. Concentration of potentially toxic elements in field crops grown near and far from cities of the Pampas (Argentina). *Journal of Environmental Management* 80 (2006) 116–119.

Lavado, R.S., M. B. Rodríguez and M. A. Taboada. 2005. Treatment with biosolids affects soil availability and plant uptake of potentially toxic elements. *Agriculture, Ecosystems and Environment* 109: 360-364.

Lavado, R.S., M. Zubillaga, Ralvarez and M.A. Taboada.2004. Baseline levels of potentially toxic elements in pampas soils. *Soil and Sediment Contamination: an International Journal.,* 13 (5) : 329-339.

Logan,T.J., B.J.Lindsay, L.E.Goina and J.A.Ryan. 1997. Field assessment of sludge metal bioavailability to crops : sludge rate response. *J. Environ. Qual* 26: 534-550.

Lundin et al., 2004 M. Lundin, M. Olofsson, G.J. Pettersson and H. Zetterlund, Environmental and economic assessment of sewage sludge handling options, *Resources, Conservation and Recycling* 41 (4) (2004), pp. 255–278.

Luo YM, Christie P. 1998. Bioavailability of copper and zinc in soils treated with alkaline stabilized sewage sludges. *J. Environ.Qual.* 27: 335-342.

Ma, L.Q. and G.N.Rao. 1997. Chemical fractionation of cadmium, copper, nickel and zinc in contaminated soils. *J. Environ. Qual.* 26: 259-264.

MAFF, 1993. MAFF, 1993. Review of the Rules for Sewage Sludge Application to Agricultural Land: *Soil Fertility Aspects of Potentially Toxic Elements.* PB1561. Her Majesty's Stationary Office, London.

Martínez and Motto, 2000 C.E. Martínez and H.L. Motto, Solubility of lead, zinc and copper added to mineral soils, *Environmental Pollution* 107 (2000), pp. 153–158.

McBride M, Martinez C, Topp E, Evanis L. 2000. Trace metal solubility and speciation in a calcareous soil 18 years no-till sludge application. *Soil Sci.* 165: 646-656.

McBride, M.B. 1995. Toxic metal accumulation from agricultural use of sludge: Are USEPA regulations protective? *J. Environ. Qual.*24:5-18.

McBride, M.B., Richards, B.K., Steenhuis, T.S. 2004. Bioavailability and crop uptake of trace elements in soil columns amended with sewage sludge products. *Plant and Soil* 262: 71–84.

McGrath, S.P. and J.Cegarra. 1992. Chemical extractability of heavy metals during and after long-term applications of sewadge sludge to soil. *Journal Soil Sci.*43: 313-321.

McGrath, S.P., Zhao, F.J., Dunham, S.J., Crosland, A.R., Coleman, K., 2000. Long-term changes in the extractability and bioavailability of Zn and Cd after sludge application. *Journal of Environmental Quality* 29, 875–883.

McLaren, R,G, and D.V.Crawford. 1973. Studies on soil copper: 1. The fractionation of copper in soils. *J. Soil Sci.* 24: 172-181.

Meima J., Comans R. 1999. The leaching of trace elements from municipal solid waste incinerator bottom ash at different stages of weathering. *Applied Geochemistry*, 14: 159-171.

Merrington, G., Winder, L. and Green, I., 1997. The bioavailability of Cd and Zn from soils amended with sewage sludge to winter wheat and subsequently to the grain aphid Sitobion avenae. *Sci. Total Environ.* 205, pp. 245–254.

Metcalf and Eddy, 2003 *Wastewater engineering: treatment, disposal, and reuse* (fourth ed.), by Inc. Metcalf and Eddy, McGraw-Hill Publishing Company Ltd., New York (2003).

Miller, W.P., Martens, D.C., Zelazny, L.W. and Kornegay, E.I. 1986 Forms of solid phase copper-enriched manures. *Journal of Environmental Quality* 15, 69–72.

Morabito R. (1995 Speciation of organotin compounds in environmental matrices. *Microchem. J.*, 51, 198-206.

Morera, M., Echeverrı´a, J., Garrido, J., 2002. Bioavailability of heavy metals in soils amended with sewage sludge. *Canadian Journal of Soil Science* 81, 405–414.

Morin, G., Juillot, F., Ildefonse, P., Calas, G., Samama, J.C., Chevallier, P., Brown, G.E., 2001. Mineralogy of lead in a soil developed on a Pb mineralized sandstone (Largentiere, France). *American Mineralogist* 86, 92–104.

NSW EPA, 1997. Environmental Guidelines: Use and Disposal of Biosolids Products. New South Wales Environment Protection Authority, Sydney.

Nyamangara, J. 1998. Use of sequential extraction to evaluate zinc and copper in a soil amended with sewage sludge and inorganic metal salts, *Agriculture, Ecosystems and Environment* 69 (1998), pp. 135–141.

Obrador A., M. I. Rico, J. M. Alvarez and J. Novillo. 2001. Influence of thermal treatment on sequential extraction and leaching behaviour of trace metals in a contaminated sewage sludge. *Bioresource Technology*, 76: 259-264.

Ødegaard, H. , B. Paulsrud and I. Karlsson, 2002. Wastewater sludge as a resource: sludge disposal strategies and corresponding treatment technologies aimed at sustainable handling of wastewater sludge, *Water Sci. Technol.*, 46: 295–303.

Paramasivam S., K. S. Sajwan and A. K. Alva. 2006. Incinerated Sewage Sludge Products as Amendments for Agricultural Soils: Leaching and Plant Uptake of Trace Elements, *Water, Air, and Soil Pollution*, 171: 273–290.

Parkpain et al., P. Parkpain, S. Sirisukhodom and A.A. Carbonell-Barrachina, 1998. Heavy metal and nutrient chemistry in sewage sludge amended Thai soils. *J. Environ. Sci. Health* 33 (1998), pp. 573–597.

Pierzynski G.M., 1998. Past, present, and future approaches for testing metals for environmental concerns and regulatory approaches, *Commun. Soil Sci. Plant Anal.* 29: 1523–1536.

Reddy, A.M.; S.G. Kumar, G. Jyonthsnakumari, S. Thimmanaik and C. Sudhakar, Lead induced changes in antioxidant metabolism of horsegram (Macrotyloma uniflorum (Lam.) Verdc.) and bengalgram (Cicer arietinum L.), *Chemosphere* 60 (2005), pp. 97–104.

Richards, B.K., Steenhuis, T.S., Peverly, J.H. and McBride, M.B., 2000. Effect of sludge-processing mode, soil texture and soil pH on metal mobility in undisturbed soil columns under accelerated loading. *Environmental Pollution* 109, pp. 327–346.

Rodríguez, M.B. and R.S. Lavado. 2004. Uptake and distribution of trace elements by soybean from a physically degraded soil treated with biosolids. *Agrochimica.* 48: (1-2): 1-10.

Saikia N., Kato S, Kojim T . 2006. Compositions and leaching behaviours of combustion residues. *Fuel,* 85: 264-271.

Sánchez-Martín M.J., M. García-Delgado, L.F. Lorenzo, M.S. Rodríguez-Cruz, M. Arienzo. 2007. Heavy metals in sewage sludge amended soils determined by sequential extractions as a function of incubation time of soils. Geoderma, Volume 142, Issues 3-4, 15 December 2007, Pages 262-273.

Sauerbeck, D.R. 1991. Plant, element and soil properties governing uptake and availability of heavy metals derived from sewage sludge. *Water, Air and Soil Pollution* 57-58: 227-237.

Sauvé, S., W. Hendershot, and H.E. Allen. 2000. Solid–solution partitioning of metals in contaminated soils: Dependence of pH, total metal burden, and organic matter. *Crit. Rev. Environ. Sci. Technol.* 34:1125–1131.

Scancar J., R. Milacic, M. Strazar and O. Burica, 2002. Total metal concentrations and partitioning of Cd, Cr, Cu, Fe, Ni and Zn in sewage sludge. *Sci. Total Environ.* 250: 9–19.

Scheckel and Ryan, 2003 K.G. Scheckel and J.A. Ryan, In vitro formation of pyromorphite via reaction of Pb sources with soft-drink phosphoric acid, *Science of the Total Environment* 302 (2003), pp. 253–265.

Shuman, L.M., 1999. Organic waste amendments effect on zinc fractions of two soils. *Journal of Environmental Quality* 28, 1442–1447.

Sims, J.T.and J.S.Kline. 1991. Chemical fractionation and plant uptake of heavy metals in soils amended with co-composed sewage sludge. *J. Environ. Qual.* 20: 387-395.

Singh R.P., Agrawal M. 2007. Effects of sewage sludge amendment on heavy metal accumulation and consequent responses of *Beta vulgaris* plants. *Chemosphere.* 67: 2229-2240.

Smith, S.R. 1994. Effect of soil pH on availability to crops of metals in sewage sludge treated soils. 2. Cd uptake by crops and implications for human dietary intake. *Eniron. Pollut.* 86:5-13.

Smith, S.R. 1996. *Agricultural Recycling of Sewage Sludge and the Environment*, CAB International, Wallingford (1996).

Soriano, A. 1991. Temperate subhumid grasslands of South America. In: Temperate subhumid grasslands. R.T. Coupland (ed.) Ecosystems of the World, Volume 8A, Natural Grasslands. With sections by R.J.C. León (Geographic Limits, Geomorphology and Geology, Regional Subdivisions and Vegetation), O.E. Sala (Structure and Function),

R.S. Lavado (Soils), J.H. Lemcoff (Climate), A. Soriano, V.A. Deregibus and R.S. Lavado (Land use), M.A. Cahuepé, C.A. Velazquez and O.A. Scaglia (Fauna). Elsevier Scientific Publishing Company, Amsterdam. 367-407 pp.

Soriano, A., León, R.J.C., Sala, O.E., Lavado, R.S., Deregibus, V.A., Cahuepé, M.A., Scaglia, O.A., Velázquez, C.A., Lemcoff, J.H. 1992. Río de la Plata grasslands: In: Coupland, R.T. (ed.) *Ecosystems of the world 8A. Natural grasslands. Introduction and western hemisphere*. Elsevier, New York, pp. 367-407.

Stipp, S., Hansen, M., Kristensen, R., Hochell, M., Bennedsen, L., Dideriksen, K., Balic-Zunic, T., Le´onard, D., Mathieu, H., 2002. Behaviour of Fe-oxides relevant to contaminant uptake in the environment. *Chemical Geology* 190, 321–337.

Strawn D.G. and D.L. Sparks, Effects of soil organic matter on the kinetics and mechanisms of Pb(II) sorption and desorption in soil, *Soil Sci. Soc. Am. J.* 64 (2000), pp. 144–156.

Su D.C. and J.W.C. Wong, , 2004. Chemical speciation and phytoavailability of Zn, Cu, Ni and Cd in soil amended with fly ash-stabilized sewage sludge, *Environment International* 29 (2004), pp. 895–900.

Sukreeyapongse, O., Holm P. E., Strobel B W., Panichsakpatana S, Magid J and H C. Bruun Hansen.2002. pH-Dependent Release of Cadmium, Copper, and Lead from Natural and Sludge-Amended Soils. *Journal of Environmental Quality* 31:1901-1909

Sun, Y., Li, Z., Guo, B., Chu, G., Wei, C., Liang, Y. 2008. Arsenic mitigates cadmium toxicity in rice seedlings. Environmental and Experimental Botany. Available from: http://www.sciencedirect.com, in press.

Tejada, M., Gonzalez, J.L. 2007. Application of different organic wastes on soil properties and wheat yield. *Agronomy Journal* 99: 1597-1606.

Tessier, A, P.Campbell and M.Bisson. 1979. Sequential extraction procedure for the speciation of particulate trace metals. *Anal. Chem.* 51:844-850.

Thomas S., D. Mahammed, M. Clairotte M.R Benedetti, M. Castrec-Rouelle', F. Persin% P. Feu^, J. Martinez^, and P. Hinsinger. 2005. Bioavailability and extractability of copper and zinc in a soil amended with pig slurry: Effect of iron deficiency in the rhizosphere of two grasses. Chapter 11. In *Biogeochemistry of Trace Elements in the Rhizosphere*. P.M. Huang and G.R. Gobran (Editors) 337. © 2005. Elsevier B.V..

Torri S, Alvarez R, Lavado R. 2003. Mineralization of Carbon from Sewage sludge in three soils of the Argentine pampas. *Commun Soil Sci and Plant Anal* 34 (13-14): 2035-2043. ISSN 00103624.

Torri, S. 2001. Distribucio´n y biodisponibilidad de Cd, Cu, Pb y Zn en suelos fertilizados con bioso´ lidos. (Distribution and availability of Cd, Cu, Pb y Zn in sewage sludge amended soils). M Sci. Dissertation. University of Buenos Aires, Faculty of Agronomy, Argentina.

Torri, S.I., Lavado, R.S. 2002. Distribution and availability of potentially toxic elements in representative soils of Buenos Aires province as a result of biosolid application | [Distribución y disponibilidad de elementos potencialmente tóxicos en suelos representatitvos de la Provincia de Buenos Aires enmendados con biosólidos]. Ciencia del Suelo 20 (2), pp. 98-109.

Torri, S.I., Lavado, R.S. 2008 a. Dynamics of Cd, Cu and Pb added to soil through different kinds of sewage sludge. *Waste Management*, Volume 28, Issue 5, 2008, Pages 821-832

Torri, S.I., Lavado, R.S. 2008 b. Zinc distribution in soils amended with different kinds of sewage sludge. *Journal of Environmental Management*, Volume 88, Issue 4, September 2008, Pages 1571-1579.

Tsadilas, C.D., T.Matsi, N.Barbayiannis and D.Dimoyiannis.1995. Influence of sewage sludge application on soil properties and on the distribution and availability of heavy metal fractions. *Commun. Soil Sci.Plant Anal.* 26 (15and16), 2603-2619 .

Tsadilas, C.D., T.Matsi, N.Barbayiannis and D.Dimoyiannis.1995. Influence of sewage sludge application on soil properties and on the distribution and availability of heavy metal fractions. *Commun.Soil Sci.Plant Anal.* 26 (15and16), 2603-2619 .

USEPA 1993. U.S. Environmental Protection Agency. 1993. 40 CFR Part 503 Standards for the use and disposal of sewage sludge subpart B Land Applications (503.13-Pollutant limits) 58FR9387. Washington, D.C.

USEPA, 1997. Tesy methods for evaluating solid waste, physical/chemical methods. OSWER, SW-847.

Vaca-Paulín, R., Esteller-Alberich, M.V., Lugo-De La Fuente, J., Zavaleta-Mancera, H.A. 2006. Effect of sewage sludge or compost on the sorption and distribution of copper and cadmium in soil. *Waste Management* 26 (1), pp. 71-81.

Walker et al., 2003 D.J. Walker, R. Clemente, A. Roig and M.P. Bernal, The effects of soil amendments on heavy metal bioavailability in two contaminated Mediterranean soils, *Environ. Pollut.* 122 (2003), pp. 303–312.

Walter, I. y G. Cuevas. 1999. Chemical fractionation of heay metals in a soil amended with repeated sewage sludge application. *The Sci. of the Total Environment.* 113-119.

Wang C, Hu X, Chen M, Wu Y. 2005. Total concentrations and fractions of Cd, Cr, Pb, Cu, Ni and Zn in sewage sludge from municipal and industrial wastewater treatment plants. *Journal of Hazardous Materials*, 119: 245-249.

Wang X, Tao Chen, Yinghua Ge, Yongfeng Jia. 2008. *Studies on land application of sewage* sludge and its limiting factors.

Weggler-Beaton et al., 2000. K. Weggler-Beaton, M.J. McLaughlin and R.D. Graham, Salinity increases cadmium uptake by wheat and Swiss chard from soil amended with biosolids. *Aust. J. Soil Res.* 38 (2000), pp. 37–45.

Werther J. and T. Ogada, 1999. Sewage sludge combustion, *Prog. Energy Combust. Sci.* 25 (), pp. 55–116.

Wong J.C., K. Li, M. Fang and D.C. Su, 2001. Toxicity evaluation of sewage sludges in Hong Kong, *Environ. Int.* 27 (2001), pp. 373–380.

Xiao, C., Ma, L.Q. and Sarigumba, T., 1999. Effects of soil on trace metal leachability from papermill ashes and sludge. *J. Environ. Qual.* 25: 321–333.

Zhang, F.S., Yamasaki, S., Nanzyo, M., 2002. Waste ashes for use in agricultural production: I. Liming effect, contents of plant nutrients and chemical characteristics of some metals. *Sci. Total Environ.* 284, 215-225.

Zhou and Wong, 2001. L.Z. Zhou and J.W.C. Wong, Effect of dissolved organic matter from sludge and sludge compost on soil copper sorption. *J. Environ. Qual.* 30 (2001), pp. 878–883.

Zufiaurre et al., 1998. R. Zufiaurre, A. Olivar, P. Chamorro, C. Nerín and A. Callizo1998, Speciation of metals in sewage sludge for agricultural uses. *Analyst* 123: 255–259.

In: Sewage Treatment: Uses, Processes and Impact
Editor: Anna Stephens and Mark Fuller

ISBN 978-1-60692-959-9

Chapter 4

TREATMENT OF SEWAGE FOR USE IN AGRICULTURE

José Tavares de Sousa[1], Wilton Silva Lopes[1], Shiva Prasad[2] and Valderi Duarte Leite[1]

[1]Department of Chemistry, Universidade Estadual de Paraíba, Campina Grande, PB, Brazil

[2]Department of Chemical Engineering, Universidade Federal de Campina Grande, PB, Brazil

ABSTRACT

Our environmental sanitation group developed research into the treatment of municipal sewage with the objective of producing effluents for use in agriculture during the period 1999 to 2005. A low-cost pilot-scale treatment system was installed and monitored at Experimental Station for Biological treatment of Sewage (EXTRABES) in Campina Grande City, northeast of Brazil. The system comprised an UASB (Upflow Anaerobic Sludge Blanket Reactor) with a hydraulic flow rate of 20 m^3 day^{-1}, followed by three post-treatment options: a constructed wetland, a gravel bed system and polishing ponds. The effluents were analyzed for physical and chemical parameters as well as for thermotolerant coliform and helminthes eggs. The mean values for nitrogen, phosphorus and potassium in the effluents produced were analyzed and expressed as kgN, kgP_2O_5, kgK_2O per hectare and the effluents used to irrigate cotton and sesame. The results showed that the polishing pond effluent was free of helminth eggs and contained <1000 FCU/100mL of thermotolerant coliforms and was thus suitable for unrestricted irrigation as per WHO guidelines (1989). Due to high pH values, which varied between 8.4 and 9.8, considerable removal of phosphorus and nitrogen occurred, reducing the concentrations of nutrients available for fertirrigation. There was no statistical difference ($p>0.05$) between thermotolerant coliform removal between the planted and unvegetated systems with mean effluent concentrations in both systems of 2500 FCU/100 mL. Thus according to the WHO guidelines these effluents are suitable for fertirrigation of cereals and crops which are not consumed raw. The removal efficiencies of carbonaceous material expressed as COD and VSS in the three systems was considerable (72 to 80 mg/L of COD and 10 mg/L of VSS) giving effluent concentrations between 72 and 80 mg/L for COD and 10 mg/L of VSS, with no significant difference ($p>0.05$) between the treatments. Finally after five years of research the technology was passed to farming families of low-income communities.

1. Introduction

Humanity is faced with different types of environmental, economic and social problems which are reflected in the degradation of natural resources, such as water resources. These acquire special importance due to the increasing demands, associated with the accelerating impacts of population increase, demanding with this, greater attention to the necessities of use of water for the multiple activities. These necessities charge their tributes as much in quantitative as in qualitative terms, and become evident principally in regions of urban and agricultural development (Sousa and Leite 2003).

Through the pressing necessity of alternative sources to increase the water supply, one of the alternatives pointed out as a possible solution is the reuse of water, which is an important instrument of environmental management of hydraulic resources. The reuse in agriculture has shown great importance, among the principal forms of utilization of wastewater. Various studies have evidenced its efficiency and have concluded that this practice promotes an increase in agricultural productivity and a reduction in costs of fertilizers (Leon and Cavallini, 1996; Monte and Sousa, 1992; Vazquez-Montiel *et al.*, 1996; Sousa and Leite, 2003). Therefore, this modality of reuse can be an effective solution for the water supply in irrigation since it is a sector which demands more than 70% of the available water (Bastos *et al.*, 2003).

The planned reuse of treated wastewater for agriculture as practiced in developed countries such as the United States, Germany, Israel and Australia, is increasing significantly in arid and semi-arid regions of developing countries including Chile, India, Mexico and Peru and the reuse is regulated by specific legislation. In Chile, all the sewage of Santiago is utilized for irrigation of areas close to the city. In drought periods, the direct reuse is normally practiced for irrigation of vegetables.

Current water demand in agriculture for food production is at a critical unsustainable level and innovative management practices are essential in the short term to reverse this situation. However the reversal of this situation cannot be accomplished merely by establishing priorities and control mechanisms such as approval and recovery in relation to water supply. The treatment and reuse of poor quality water is an important component in the overall process of sustaining national water resources (Lima, 2004).

The application of raw sewage to agricultural soils to refurbish nutrient levels is a very ancient practice, indeed this has occurred in China for over 2000 years. In Europe, this practice was initiated in 1550 with the development of sewage farms in Germany and subsequently in 1700 in the United Kingdom. Sewage use became legal in London, Boston and Paris in 1815, 1833 and 1880, respectively. The awareness that water quality and sanitation were key factors in the preservation of public health led to the process of water filtration initiated in 1848, with the purpose of preventing outbreaks of typhoid fever in the United Kingdom (1860). Some thirty years later, Pasteur, in France, demonstrated the presence of contaminating bacteria in drinking water, and the use of sodium hypochlorite to disinfect water supplies started in UK in 1895. Next came the chlorination of surface water in 1910 as a form of disinfection (Asano and Levine, 1996).

A significant contribution in relation to sewage treatment arose in England, in 1914, with the development of the activated sludge process. This process, in its original and modified forms is currently in use to treat domestic and industrial wastewaters throughout the world.

The World Health Organization (WHO) first suggested health protection criteria for different uses of treated sewage in 1970. In Brazil, the São Paulo section of the Brazilian Association of Environmental and Sanitary Engineering (ABES), published a document that discussed the question of the reuse of waste water in 1992.

According to Santos *et al.* (1992), this document established the following recommendations for the scientific society:

i. Inclusion of disciplines, in graduate programs, which deal with management of hydraulic resources, principally aiming at the recycling and reutilization of waste water;
ii. Creation of research centers for the development of advanced processes of sewage treatment, with the aim of producing effluents suitable for reuse;
iii. Development of inter-institutional activities among research scientists, aimed at the exchange of experiences in the field of reuse of domestic waste water;
iv. Municipal sewage, when suitably treated, can be used for potable and non-potable uses. However, the reutilization of domestic wastewater for direct human consumption poses the question of what potentially toxic substances that could constitute a potential public health risk might still be present in the effluent even after the sewage treatment process. These could include a range of synthetic substances such as contraceptive drugs, antibiotics and hormones. On the other hand, there is also difficulty in relation to utilization of low cost advanced technologies that come for effective control of water quality from microbiological point of view, with potable quality standard.

However, treated municipal sewage can be safely used for a range of non-potable uses and there is now awareness by a large part of the population that water should be consumed in an economical and responsible way. The chemical, physical and biological characteristics of treated wastewater effluents will determine the type of reuse. Some types of reutilization of water and its respective purposes are presented in Table 1.

In the semi-arid region of northeast Brazil, the annual rainfall varies between 300 to 800 mm of precipitation, distributed over a three month period, exceeding evapo-transpiration occurs, and the plants suffer hydro-deficiency. Thus, irrigation becomes the safest practice to guarantee agricultural production in the region; however, to meet this water demand, effective management of the local hydrographic basins is necessary and also a rational policy of wastewater treatment in the big urban centers, with objective of producing treated effluents suitable for reuse in agriculture. Wastewater reuse not only helps overcome water shortages in agriculture but also provides valuable plant nutrients, thus, contributing directly to the sustainable development of the region (Sousa *et al.*, 1998; Sousa *et al.*, 2001).

The utilization of treated municipal sewage effluent in Northeast Brazil, particularly for irrigation, can be a feasible alternative to the use of surface and groundwater. Presently, the reuse of waste water of domestic origin has a negative connotation for a major part of the population that has no perception of the treatment process despite the fact that well designed and operated treatment systems can produce good quality effluents (Sousa *et al.*, 1998).

Table 1. Types of non-potable reuse and their respective objectives

Areas of Reuse	Application Examples
Agriculture	Irrigation of regional crops Irrigation of pasture areas
Livestock production	Watering of animals
Aquiculture	Production of fishes and shrimps Production of aquatic plants
Recreational and public areas	Irrigation of parks Irrigation of sports fields Charging of ornamental lakes
Industry	Cooling towers Heat exchangers Boilers Process waters
Sewage dilution	Maintenance of minimum wastewater flow conditions at the treatment plant headworks
Domestic	Sanitary discharges Car washes Green areas of condominiums
Aquifer recharge	Prevention of salt water intrusion in coastal regions

The usual irrigation techniques are inundation, furrow, subsurface and localized drip systems and micro-aspersion. The first two techniques are those most commonly used with sewage effluents. The advantages and disadvantages in the use of sewage effluents in agriculture are presented in Table 2.

Table 2. Advantages and disadvantages of sewage use in agriculture

Advantage	Disadvantage
Permanent availability of water	Contamination of the crops by pathogenic microorganisms
Availability of nutrients	Bioaccumulation of toxic chemicals
Improvement in soil structure	Increase in soil salinity
Higher productivity	Disequilibrium of nutrients in the soil
Larger cultivated area	Encrustation of Soils

The utilization of treated sewage in agriculture brings benefits three ways:

1) Social

i. Reduction of water borne diseases;
ii. Conservation of precious water resources, especially in semi-arid regions;
iii. Improvement in the quality of life of the population by preventing surface water contamination.

2) Economic

i. Reduction in medical costs;
ii. Decrease in the use of artificial fertilizers;
iii. Increase in irrigation area and fish production;
iv. Soil conservation by the application of biologically treated sludge.

3) Environmental

i. Decrease in the organic pollution of surface waters;
ii. Decrease in microorganisms thrown in the atmosphere;
iii. Maintenance and regeneration of the ecosystems.

From the chemical point of view, certain parameters are of fundamental importance when evaluating water quality for irrigation notably: pH, dissolved salts concentrations, the concentrations of organic material and suspended solids, toxic ions, traces of heavy metals and the sodium adsorption. The use of sewage in agriculture in the fertirrigation of crops and the replenishment of soil nutrients is an important practice. The planned reuse of wastewater in agricultural leads to the control of water pollution, economy in chemical fertilizer use, an increase in productivity, economy in the use of water of better quality in irrigation, and the promotion of organic agriculture. It also addresses the issues of sustainable development, and the provision of a better quality of life for the population. However, when the management of agricultural activities is inadequate, the negative impacts on the environment can be considerable notably on the quality of surface and ground water, as shown in Table 3.

Table 3. Agricultural activities and their consequences

Agricultural activity	Impact on Surface water	Impact on Groundwater
Land plowing	Loading of phosphorus and of pesticides adsorbed on to sediment particles	Leaching of recalcitrant material
Mineral fertilization	Presence of nutrients, mainly phosphorus, promotes eutrophication, causing taints and odors in public water supplies	Nitrate leaching to underground waters, threatening public health
Organic fertilization	Contamination by pathogens, metals, phosphorus and nitrogen, causing eutrophication	Contamination, mainly by nitrates
Pesticide application	Contamination of biota present in water bodies, affecting the fishes which consume it. The pesticides are carried to a long distance	Leaching of pesticides, reaching the aquifer and contaminating people
Pastoral activities	Contamination of surface water due to the presence of pathogens and nutrients	Contamination by nitrate
Irrigation	Water contamination by fertilizers, pesticides and salts. Bioaccumulation of contaminants in fishes	Enrichment of ground by salts and nutrients

Source: Ongley (1996).

2. Characteristics of Municipal Sewage

The characteristics of sewage are such as to require its treatment to eliminate those constituents that are known to be harmful to aquatic life and cause human disease. These diseases include bacillary dysentery, polio, cholera, gastroenteritis, infectious hepatitis, leptospirosis, salmonellosis, giardiasis, amoebic dysentery, trachoma, and scabies, among others. To treat sewage is not to let eutrophize ground water and conserve the environment; it is guarantee of quality of life. It is a right of citizen and a duty of the state.

From a technical point of view, the purpose of sewage treatment is to reduce the solid constituents present in wastewater that degrade the water quality. The principal constituents of an organic and mineral nature, which should be removed or reduced, are presented in Table 4. The reduction or removal of pathogenic organisms usually occurs as part of the tertiary treatment process by chlorination, ozonation, UV treatment or by application of effluent to soil or containment in maturation or polishing ponds.

Table 4. Constituents present in waste water and their respective consequences

Constituents	Consequences
Suspended solids	Decrease in photosynthetic activity and concentration of dissolved oxygen; production of turbidity and color
Biodegradable organic material	Decrease in dissolved oxygen
Nutrients (N and P)	Excessive increase in algal blooms; eutrophication
Heavy metals, generally coming from commercial and industrial sources	Cardiovascular disease; cerebral damage; carcinogenesis
Refractory organic material: Detergents and their derivatives; phenols and pesticides	Harm to human health and to environment
Dissolved inorganic solids: Ca^{2+}, Na^{+}, SO_4^{2-}	Causes corrosion, alteration in taste and soil salinity
Bacteria, Helminthes, Virus and Protozoa	Epidemic and endemic diseases

2.1. Nutrients Present in Municipal Sewage

Irrigation with municipal sewage effluents provides sufficient nutrients for the development of agricultural crops without the need for the application of commercial fertilizers and the organic component of the effluent increases the soil's capacity for storing nitrogen and retaining water.

Analysis of the chemical components of plant tissues demonstrates that of the sixty or more elements present, only sixteen are important namely: carbon, oxygen, hydrogen, nitrogen, phosphorus, potassium, calcium, magnesium, sulphur, manganese, boron, zinc, chlorine, copper, molybdenum and iron. The plant utilizes the first nine in relatively large quantity and these elements are abundant in sewage. The approximate quantities of the principal plant nutrients produced daily by an individual are presented in Figure 1.

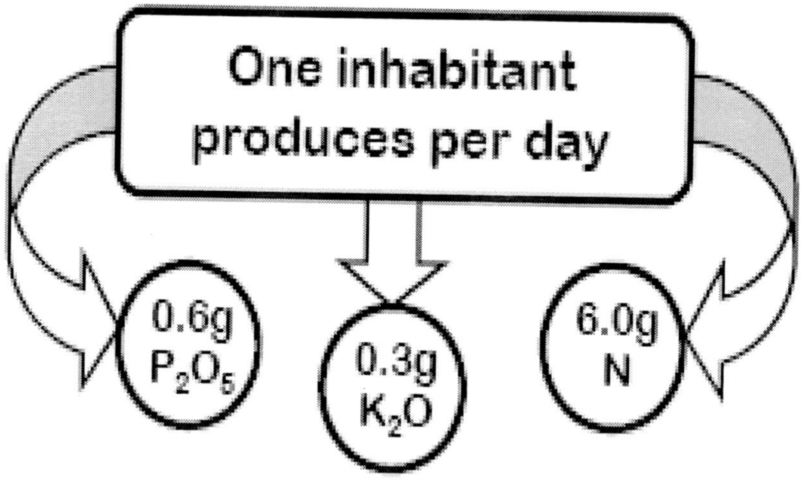

Figure 1. Nitrogen, phosphorus and potassium produced per capita in sewage of Campina Grande.

The macro nutrients nitrogen, phosphorus and potassium are present enough in municipal sewage. Each person resident in a medium sized city produces about 100 liters of sewage per day and analyses performed in the laboratory of the Program of Research in Basic Sanitation (PROSAB), located in the city Campina Grande–PB, Northeast of Brazil, demonstrated that each inhabitant generates 6.0, 0.6 and 0.5 g daily of total Kjeldahl nitrogen, total phosphorus and potassium, respectively.

The reuse of treated sewage effluents in agriculture avoids the use of mineral fertilizers and, consequently, presents at least three advantages: 1) It dispenses with the need for mineral fertilizers, 2) Diminishes the contamination of surface and groundwater by fertilizers, and 3) Reduces the scarcity of water for domestic uses (Sousa, *et al.*, 2001).

A person produces, on average, 200 g feces and 1 kg urine daily. Municipal sewage contains 81 to 100 g $inhab^{-1}.day^{-1}$ of dry solids, of these, 60 to 85% is organic material and 15 to 40% are mineral salts, in which 5 to 10% Nitrogen (N), 2.5 to 4.5% phosphorus ($P_2 0_5$), 3.0 to 4.5 % potassium ($K_2 0$) and 4.5 to 6.0% calcium (Andrade Neto, 1997).

2.2. Microbiological Aspects

It is known that sanitary sewage, when used untreated, can contaminate the environment with bacteria, parasites and virus, causing disease and serious problems of public health particularly amongst agricultural workers in contact with the soil, the population living close to the cultivated areas and the consumers of the irrigated crops (Leon and Cavallini, 1996; Metcalf and Eddy, 1991). The wastewater reused in irrigation should be of good physico-chemical and sanitary quality. The principal constituents that should be evaluated are: the salts present in the effluent and soil, which reduce water availability to the plant; high concentrations of sodium and conversely low concentrations of calcium, which decrease the rate of water infiltration; the toxicity of specific ions (sodium, chloride and boron); the potential excess of nutrients, and the concentrations of pathogenic organisms (Sousa *et al.*, 2000).

The greatest drawback to the use of municipal sewage effluents in irrigated agriculture has been the sanitary aspects. The principal biological contaminants of importance for public health are viruses, bacteria, protozoa and helminthes. The sanitary quality of sewage depends on its level of treatment. The possibility of a person contracting a disease by the utilization of municipal sewage in irrigation depends principally on the following factors:

i. The number of pathogenic microorganisms present in the effluent capable of provoking an infection. (Infective dose).
ii. Latency – That is the time interval between the excretion of the pathogen and the time taken to become infective.
iii. Survival capacity of the microorganism in the environment (persistence).
iv. Susceptibility of the individual to the disease.
v. Capacity for multiplication of the pathogen in the human body.

Aspects relative to public health and the transmission of diseases via the water route, as a consequence of the use of municipal sewage in agriculture, are treated in the specialized literature by Feachem *et al.*, (1983); Shuval *et al.*, (1986); and Bastos *et al.*, (2003). Thus the quality of treated municipal sewage has to be established to guarantee its safe use in irrigation. In Brazil, no norms or criteria exist for the reuse of inferior quality water, despite the use of municipal sewage in agriculture to be a millenary practice realized in all the continents.

Table 5. Principal biological contaminants and their respective infections

Biological contaminants	Epidemiological characteristic	Infection
Bacteria	Non-latent Medium or high Infectious dose Moderate Persistent Capable of multiplying	Typhoid and paratyphoid Salmonellosis bacillary dysentery Cholera Diarrhea Compylo-bacterial enteritis
Helminthes	Latent Persistent	Ascaridiasis Tricuriasis Ancylostomiasis Esquistossomosis
Viruses	Non-latent Persistent	Poliomyelitis Meningitis Diarrhea Respiratory diseases Hepatitis Fever
Protozoa	Non-latent Low infectious dose	Giardiasis Amoebiasis Balantidiasis

In absence of norms, the World Health Organization (WHO, 1989) guidelines for unrestricted irrigation, recommends less than 1 helminth egg per liter and less than or equal to 1000 fecal coliforms per 100 mL of wastewater. These recommendations seem to be very rigorous, even when irrigating crops eaten raw, besides this, it omits any recommendation in relation to protozoa and viruses.

Treatment of sewage is necessary for its adequate use in irrigation, besides guaranteeing the hygienic quality, for correcting some undesirable characteristics such as: high concentration of solids and fermentable organic matter.

To avoid that pathogenic microorganisms are maintained in the fields, the municipal sewage should be treated with efficient methods before being used in irrigation, because under the conditions of utilization of these sewages in agriculture, the groups exposed to the infection, such as rural workers and consuming public of the agricultural products, could enter in contact with the pathogenic microorganisms.

WHO published in 1989 sanitary recommendations for the use of sewage in agriculture, based on only two microbiological indicators: number of helminthes eggs per liter and concentration of fecal coliform.

Table 6. WHO Guidelines for the agricultural use of treated wastewaters (1989-2005)

Category	Conditions of use	Exposed Group	Intestinal nematodes (eggs/L) (1)	Fecal coliform (CFU/100 mL) (2)	Treatment process
A	Raw consumed crops, sports fields, public gardens	Workers, consumers and public in general	≤ 1	≤ 1000	Stabilization lakes in series, or equivalent treatment in terms of pathogenic removal
B	Crops of cereals, fodder plants, meadows and fructiferous trees (3)	Workers	≤ 1	No any norm recommended	Stabilization lakes with 8 to 10 days of time of detention or removal of thermotolerant coliform and helminthes
C	Category B, without workers and public being exposed	None	Not applicable	Not applicable	Pre-treatment according to the irrigation method, for minimum primary sedimentation.

(1) Human intestinal nematodes: *ascaris*, *trichinella*, *necator* and *ancylostoma*; arithmetical mean during the irrigation period. (2) Geometric mean during the irrigation period. (3) In case of fructiferous trees, the irrigation should be terminated two weeks before the harvest and no any fruit should be picked from the ground; sprinkling irrigation should not be employed.

Source: WHO (1989).

From these indicators, WHO (1989) classifies the variety of crops, their corresponding contact mediums and the required treatment in the three categories (A, B, C).

The crops consumed raw and also the sportive fields and gardens, where public can have direct contact with sewage are included in category A. In this case, it is recommended that for utilization of treated sewage may contain up to 1 egg of intestinal nematodes (*ascaris, trichinellas* and *ancylostomas*) per liter, and fecal coliform up to 1000 per 100 mL.

The crops of cereals, the fodder plants, the meadows and the fructiferous trees are included in category B. Here, the workers who handle water in irrigation process and participate in other cultural activities may have contact with sewage. The recommendation is for have up to 1 egg of intestinal nematodes per liter. The irrigation should stop two weeks before the harvest. The collected fruits should not have contact with the soil.

Category C, is included in category B, except that it does not include the direct exposure of the workers and the public. For this case, there is no any recommendation relative to the two microbiological indicators presented by WHO.

Since the publication of the WHO guidelines in 1989, different studies have been conducted to evaluate the potential risks of using treated wastewaters and helminthes have emerged as the principal risk. Blumenthal *et al.* (2000) raised the following important points:

i. Validation of the bacteriological standard (10^3 thermotolerant coliforms per 100 mL) for unrestricted irrigation.
ii. Establishment of a stricter parasitological standard of (< 0.1 egg.L^{-1}) in some situations, for example the exposure of children.
iii. Establishment of a stricter parasitological standard for restricted irrigation (10^4 - 10^5 thermotolerant coliform per 100 mL).
iv. Establishment of epidemiological standards for viruses and protozoa.

The WHO's guidelines reuse of wastewater for irrigation had been updated with basis mainly in Disability ad justed life years (DALYs) and Quantitative microbial risk assessment (QMRA) (WHO, 2006).

3. Municipal Sewage Treatment

Wastewater treatment processes can be classified as physical, chemical and biological. The physical processes are characterized by physical removal of constituents such as coarse, floating and decanted solids. The chemical processes are generally characterized by the addition of chemical products. The most common chemical processes are flocculation, precipitation and oxidation. Countless discussions about chemical treatment processes have been raised, because these processes produce significant amounts of sludge, and because of the concentrations of these chemicals in the wastewater being treated.

The biological treatment processes result in transformation of the wastewater constituents into more simple and stable molecules. The organic material present gets oxidized, thus transforming into simpler molecules structure and of lower energy content.

The biological oxidation process can occur by two ways: aerobic and anaerobic. The aerobic way occurs by means of bacteria that breathe oxygen of air; where as the anaerobic occurs by intervening of the bacteria that utilize other type of electron receptor. However, by both the ways, the prevalent mechanism for the removal of the organic material is the bacterial metabolism, which takes care of enzymatic transformations of the material and of the energy, starting from simple substances and reaching to synthesis of cellular material.

The bacterial metabolisms are chemical processes that occur simultaneously in the cell, include well defined two steps: catabolism and anabolism. The catabolism is the step that utilizes the organic material as source of energy. In this step, the degradation of the organic material occurs or even the transformation of nutrients of the cell itself in the final oxidized products. Like this, it liberates free energy which is generally utilized in syntheses of the cellular material.

Van Haandel and Lettinga (1994) showed that, in the bacterial metabolism, the catabolic fraction subdivides in two processes: 1) Oxidative catabolism (breathing) signifies the oxidation of the organic material by an oxidant present in the liquid phase, which can be molecular oxygen; in this case, the breathing is denominated as aerobic. However, when the electronic receptor is nitrate or sulfate, occurs the anaerobic breathing. *2*) The fermentative catabolism which involves a process in absence of oxidant; consists of "intramolecular" transfer of electrons, which means, a rearrangement of electrons in a molecule. The process is characterized by occurrence of successive fermentations that could finally result in formation of, in minimum, two products. The anaerobic digestion can be presented as example, which produces methane and carbon dioxide as the final products.

For better comprehension of oxidative catabolism, the principal donors and acceptors of electrons in reactions mediated by aerobic and anaerobic microorganisms are shown in Table 7. It is observed in Table 7 that, besides nitrate and sulfate, considered in oxidative catabolism as acceptors of electrons, carbon dioxide (CO_2) should also be included, which is converted to methane, by the reaction mediated by hydrogenotrophic methanogenic bacteria that utilize inorganic substrate as energy source (electron donor).

There occurs biochemical transformation of substrate (catabolism), by metabolism of organic material by microorganisms, forming more stable products with liberation of energy. Simultaneously to this process of catabolism there occurs a process of anabolism, which is synthesis or assimilation of a new cellular mass. The source of material for synthesis of new cells can be organic (see Table 7), the microorganism is called heterotrophic. Where as, when the source of carbon is CO_2, the microorganism is called autotrophic. Lawrence and Mccarty (1970) presented kinetic models considering the endogenous breathing and admitted the first order kinetics, in relation to the sludge concentration in terms of volatile suspended solids (VSS). In reality, they understood that not all the sludge (microorganism mass) present in the biological process is oxidized. From this on, there appeared new models, by considering kinetics of first order, but in relation to concentration of biodegradable sludge.

Table 7. Classification of microorganisms by donor and acceptor of electrons, source of carbon and final products

Types of bacteria	Reactions	Carbon source	Electron donor	Electron acceptor	Products
Aerobic heterotrophic	Aerobic oxidation	Organic compound	Organic compound	O_2	CO_2, H_2O
Aerobic autotrophic	Nitrification	CO_2	NH_4^+ NO_2^-	O_2	NO_2^-, NO_3^-
	Sulphur oxidation	CO_2	H_2S, S^0 $S_2O_3^{-2}$	O_2	SO_4^{2-}
Facultative heterotrophic	Denitrification	Organic compound	Organic compound	NO_2, NO_3^-	N_2, CO_2, H_2O
Anaerobic heterotrophic	Acidic fermentation	Organic compound	Organic compound	Organic compound	VFA, acetate, propionate, butirate
	Reduction of sulfate	Organic compound	Organic compound	SO_4^{2-}	H_2S, CO_2, H_2O
	Methanogenic	Organic compound	VFA	CO_2	Methane

Source: Metcalf and Eddy, 2003.

In previous studies with activated sludge systems without sludge return and feed with soluble substrate, it was found that the rate of endogenous respiration fell and the sludge was not fully oxidized. The small fraction remaining which only partially degraded was called endogenous waste (Washington and Hetling, 1965; Mccarty and Brodersen, 1962).

There is possibility of measuring the metabolism occurred in the system after knowing the concentration of the removed affluent organic material. In the case of catabolism, the measurement can be done through the consumption of oxidant, when the process is aerobic (oxidative process), or through the production of methane in the methanogenic anaerobic processes. On the other hand, the anabolism is measured by the concentration of volatile suspended solids (VSS). The sludge production coefficient depends on catabolic process. In case of activated sludge (oxidative catabolism), the production of free energy is much higher than of the anaerobic digestion (fermentative catabolism). Thus, in activated sludge, the bacteria should catabolize little organic material, due to large quantity of energy available for anabolism, where as that, in anaerobic digestion, the bacteria should catabolize higher quantity of organic material, to guarantee the required energy for anabolism. Like this, the sludge production coefficient (Y) in anaerobic digestion should be quite smaller, when compared with that of the aerobic sludge production coefficient (Y) (van Haandel and Lettinga, 1994).

Henzen and Harramöes (1983) showed experimental results in anaerobic culture (acidogenic and methanogenic bacteria) in which the sludge production coefficient (Y) was 0.18 mg VSS.(mg COD)$^{-1}$; where as Marais and Ekama (1976) determined a $Y = 0.45$ mg VSS.(mg COD)$^{-1}$ for activated sludge. Thus, it is necessary to know the relation between COD and VSS to be able to determine the synthesized fraction as cellular material. The production of sludge (synthesized cellular mass) is proportional to the removed affluent COD (metabolized organic material). Therefore, the cellular production coefficient is defined as follows:

$$Y = -(\frac{dX}{dS}) \tag{1}$$

where: Y: cellular production coefficient, mg VSS.L^{-1} (mg COD.L^{-1});
X: concentration of volatile suspended solids, mg VSS.L^{-1};
S: concentration of organic material, mg COD.L^{-1}.

3.1. Biological Systems for Sewage Treatment

3.1.1. Description of Technology for UASB Reactor

The UASB (upflow anaerobic sludge blanket) is a reactor unfurnished with artificial agitation. The affluent to be digested is distributed in lower part of the reactor; it crosses a sludge blanket with methanogenic activity, and discharged in its upper part. In the reactor, the biodegradable organic material of affluent comes in contact with the sludge blanket, and is stabilized in methane and carbon dioxide. In their rising trajectory, the bubbles meet a liquid-gas interface in separator of phases. This interface is maintained by an external hydro seal. In the interface, the gas bubbles become part of the gas accumulated in separator of phases (gas chamber), and the sludge flakes, eventually aggregated to the bubbles, tend to sediment, returning to the lower part of the reactor. It is observed that the hydro seal is an essential device for adequate functioning of the ascending flux reactor.

The liquid phase, in its ascending trajectory, passes through openings in the separator of phases and through the decantation zone, where the ascending velocity of the liquid gradually decreases due to progressive increase in transversal area. Therefore, the dragging force of sludge particles, which are eventually carried by the liquid, decreases with the approximation of the liquid to the upper level of water (free surface). When the gravitational force on a particle becomes greater than the dragging force of the liquid, this particle tends to decant, depositing on the separator of phases. The inclination of the separator enables slipping of particles deposited on it, which return to the lower part of the reactor. Thus, the effluent will essentially be free from suspended particles. Other particularity of the ascending flux reactor is in the sludge which grows. After a given period of operation (50 to 200 days), the formation of a granulated sludge is observed. The sludge granules have a high mechanical resistance and a high density which permits the formation of a blanket of sludge in the lower part of the reactor, with a very high concentration of total solids: 40 to 100 $g.L^{-1}$, depending on the substrate.

Constructive Aspects

UASB reactor is a simplified system of sewage treatment of low implantation cost (US$12 to 24 per capita) and of elevated sustainability, related to a little dependence on the energy use, and is of simple operation and control. It occupies small areas, presents flexibility in relation to future expansion, and has little problem with a disposition of sludge produced at the station. It presents small scale application possibility, can function in decentralized system with little dependence on big interrupters. The subproducts: effluents and sludge can be recuperated in irrigation and in fertirrigation of agricultural crops.

The construction of UASB system:

i. Requires a little availability of area for construction;
ii. The land utilized can be irregular and steep; and
iii. Requires impermeable soil.

3.1.2. Description of Constructed Wetland Technology

Naturally moist lands are areas where the soil surface remains saturated, at times swampy containing vegetation, predominantly regional. Generally, it has the following characteristics: the substrate consists of little drained soil or accumulated sediments always populated by hydrophytes, generally covered by water for a certain period of time during the year.

Constructed moist lands are artificially projected systems for utilizing macrophytes (aquatic plants) in substrates as sand, gravel or other inert material, where proliferation of biofilms that aggregate varied populations of microorganisms occurs, which, by means of biological, chemical and physical processes, treat waste waters (Coraucci Filho *et al.*, 2001).

Constructive Aspects

Studies and research works in laboratory and also at real scale have demonstrated that the utilization of wetland is technically and economically feasible. These systems present real capacity for removal of BOD, suspended solids, nitrogen, phosphorus and metals. The removal of amounts of these constituents is the result of actions of different mechanisms of sedimentation, precipitation, sorption and interaction of chemical and biochemical reactions.

Constructed moist land system exposed to environment has a fundamental importance, such as: 1) Climatic effects; promotes improvement in global heating of the earth, fixes carbon and maintains the balance of CO_2, influences the microclimates and maintains the rain quantity. 2) Biodiversity; maintains the scenery and diversification of ecosystem with diversity of microbiological populations and activity. 3) Habitat; maintains fauna, biodiversity as well as ground water. 4) Water quality; occurs nitrogen stripping, biodegradation of toxic compounds, reduction of pathogenic agents, accumulation of heavy metals and biodegradation of carbonaceous material. 5) Hydrological; protects erosion, maintains hydrodynamics of plains and dynamics of hydrographic basins, promoting drainage and recharging of ground water.

Following aspects should be observed for construction of these systems:

i. Existence of impermeable soil;
ii. Downward slop of land between 0 to 3%;
iii. Existence of rivers, lakes, reservoirs or other water bodies in the proximities;
iv. Availability of areas;
v. Distance of inundation plain from water reservoirs.

3.1.3. Description of Gravel Bed System Technology

In a gravel bed system the influent sewage comes into contact with a fixed microbial film attached to the surfaces of the gravel media which is responsible for degrading the complex molecules of the sewage into simpler ones.

Generally the effluent from the gravel bed is of low turbidity, with a low concentration of organic material. It resists shock loads in affluent, has reduced loss of biological solids, does not necessitate initial seeding to start the process and is simple to construct and operate. The principal limitations of it are obstruction risks of gravel bed (clogging of the interstices) and of relatively large volume, due to the space occupied by the supporting medium (Sousa *et al.*, 2005)

The gravel which functions as filling material is very important in relation to its purpose in the system:

i. to permit large accumulation of biomass, maintaining thus a greater retention time of the cells;
ii. to provide better contact of the affluent and biological solids contained in the system;
iii. to act as retainer of solids, impeding that these are carried out of the system; and
iv. to help in promoting greater uniformity of drainage from the reactor.

Constructive Aspects

Results of research work realized have demonstrated that utilization of gravel bed is successfully applicable. These systems present real capacity for removal of BOD, suspended solids and eggs of helminthes. The great disadvantage is that the gravel bed does not show sustainability, there exists a real condition of vacant space initially occupied by sanitary sewage being occupied by the accumulated sludge with passage of operation time of the system.

3.1.4. Description of Technology of Polishing Ponds

Polishing ponds are lentic water bodies, are not mechanized, the treatment occurs in natural form by means of physical, chemical and biological processes. The principal organisms present in the ponds are bacteria, alga and protozoa. They have rectangle or square configurations, constructed by shallow excavations, surrounded by land talus, receive effluents of anaerobic reactors of UASB type. The post-treatment in these ponds is realized for additional removal of organic material, nutrients and pathogenic organisms (thermotolerant coliforms and helminthes eggs), principally with the objective of reuse in irrigation, thus serves for general health protection.

Polishing ponds are usually dimensioned with low depth for favoring the criteria that promote the fall of pathogenic organisms, and maintain solar radiation, temperature, pH, among others. The fall of microorganisms occurs slowly, this being the factor that determines the hydraulic detention time, which varies from 10 to 12 days. During this time the removal of helminthes eggs and protozoa cysts will occur.

For the removal of organic material and of suspended solids the principal biological processes are: photosynthesis, oxidation of organic material by aerobic bacteria and fermentation by anaerobic digestion process. The utilization of pond for effluent treatment with low concentration of organic material; will produce reduced amount of CO_2. Besides this fact, in the tropics of high insolation on the surface of the accumulated water in the pond promotes intense photosynthetic activity, increasing thus consumption of carbon dioxide gas, which increases the pH of the medium, hence favoring the fall of pathogenic bacteria (Cavalcanti, *et al.*, 2000; van Haandel and Lettinga, 1994). This increase in pH is fundamental for sanitization of the effluent, but promotes, concomitantly, the reduction of nutrients. The pH higher than 9.0 creates disequilibrium of the phosphate species, resulting in precipitation of phosphorus, generally in the forms hydroxylapatite and calcium phosphate. The ammonical nitrogen, also present in the medium, can be removed in the form of ammonia gas, by physical process of adsorption, besides absorption by the phytoplantonic biomass, in rapid growth (Cavalcanti, *et al.*, 2000; Sousa *et al.*, 2006).

Constructive Aspects

The hydraulic characteristics of polishing pond should be taken in to consideration, such as: the geometric form, the conditions between length and breadth, to design shallow ponds of depth less than one meter, and of large length and small breadth seeking thus less dispersion for maintaining better efficiency.

The bottom of the pond should be made impermeable, even knowing that the generation of sludge in the pond will tend to make it impermeable. This becomes necessary when the rate of infiltration is higher than 10% of the feeding rate; materials such as clay, bitumen, asphalt or plastic have been utilized in coating the bottom of the pond (Silva and Mara, 1979).

It is necessary that the entrance and the exit of the polishing pond should be located at opposite points. This practice will minimize the possibility of occurrence of hydraulic short circuits. It is advised to construct ponds with entrance and exit in diagonal position to avoid the short circuit induced by wind.

4. EXPERIMENT OF TREATMENT AND APPLICATION OF THE EFFLUENT

Department of Chemistry, Universidade Estadual da Paraíba (UEPB), together with research workers of Research Program for Basic Sanitation (PROSAB) and professors of Graduate Program in Development and Environment (PRODEMA), installed in area belonging to Water and Sewage Company of Paraiba State (CAGEPA), where Biological Treatment Station for Sewage (EXTRABES) is situated, has been developing studies on treatment of municipal sewage by means of biological processes (Figure 2) for its utilization in agriculture. The studies investigated effluents produced from the following municipal sewage treatment systems: upflow anaerobic sludge blanket (UASB) reactor, constructed wetland systems (planted gravel bed and un-vegetated gravel bed) and polishing pond (Ceballos et al.,2001; Figueiredo, 2005; Meira, 2004; Lima, 2003; Lima, 2004; Sousa and Leite, 2003; Sousa et al., 2003; Sousa et al., 2005).

High rate anaerobic reactors, for example UASB reactor, followed by post-treatment units are being largely used presently for municipal sewage treatment. In this context, a study was realized on post-treatment of anaerobic effluent in three different systems: constructed wetland, gravel bed systems and polishing ponds, for the purpose of evaluating and comparing the quality of the effluents produced to use it for irrigation.

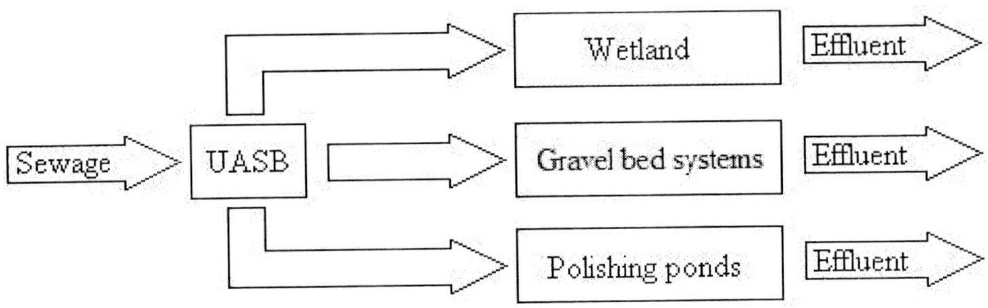

Figure 2. Sewage treatment systems, Sousa et al., (2005).

Table 8. Physical and operational characteristics of the experimental units

Unit	Dimension		HDT	Flow
	Depth (m)	Volume (m³)	(day)	(m³/day)
UASB	2.50	5.00	0.25	20
Wetland	0.60	2.28	7	0.325
Gravel bed systems	0.60	2.88	7	0.411
Polishing ponds	0.60	30.00	12	2

4.1. Upflow Anaerobic Sludge Blanket – UASB

The UASB reactor, shown in Figure 3, was fabricated in PVC with useful volume of 5 m^3, functioning with hydraulic detention time (HDT) of 6 hours, was supplied with raw sewage collected from inspection well of municipal sewage system of Campina Grande City, PB, Brazil.

Figure 3. Photo of UASB reactor (PROSAB, UFCG).

4.2. Gravel Bed Systems

The gravel bed was constructed in masonry, with 10 m^2 area (10 m length and 1m breadth) filled with gravel of granulometry varying from 15 to 53 mm, forming a layer of 0.60 m height and with 48% vacant space as shown in Figure 4. Sub-surface level loading with a flow rate 0.411 m^3/day, which corresponded to HDT of 7 days, was maintained.

Figure 4. Photo of gravel bed systems (PROSAB, UFCG).

4.3. Constructed Wetland

The wetland shown in Figure 5 was constructed in masonry and coated with impermeable mass. It was also with 10 m^2 area (10 m length and 1m breadth) filled with sand of granulometry varying from 2.8 to 4.8 mm, forming a layer of 0.60 m height and with 38% vacant space. At the entrance and at exit a layer of gravel of 0.40 m breadth and 0.60 m height was deposited with the objective of improving the flow distribution. The wetland was vegetated with macrophytes of *Juncus ssp* type, presenting a density of 25 vegetative propagules per square meter. The feeding rate and HDT were 0.325 m^3/day and 7 days, respectively.

Figure 5. Photo of constructed wetland (PROSAB, UFCG).

4.4. Polishing Ponds

Five polishing ponds (Figure 6) were functioning in series; each one of them was having 10 m length, 1 m breadth and 0.60 m useful depth. A UASB system followed by polishing pond in series has been functioning for the last 3 years. During the functioning period, the feeding rate of the ponds was 2m^3/day and total HDT of the pond was 15 days.

While the wetland and un-vegetated gravel bed systems were operated with sub-surface level flow, the polishing ponds were interconnected with pipeline by connection known as top-bottom, the effluent comes at the top part and through a 20 mm tube it gets distributed in the bottom.

Figure 6. Photo of polishing ponds (PROSAB, UFCG).

4.5. Methods

The physico-chemical, bacteriological and parasitological variables were investigated with the purpose of evaluating quality of the final effluent and its feasibility for utilization in agriculture, of contamination of the soil and of the irrigated products. The methods used are shown in Table 9. The microbiological evaluations of the effluents were performed by using filtration membrane technique for thermotolerant coliforms and the modified Bailenger method, 1979, for helminthes eggs.

Table 9. Analytical methods used for physico-chemical and microbiological examinations

Physico-chemical analysis	Analytical methods
N-ammonical (mgN-NH_4.L^{-1})	Semi-micro Kjeldahl (APHA, 1998)
TKN (mgNTK.L^{-1})	Semi-micro Kjeldahl with digestion (APHA, 1998)
Total phosphorus (mgP.L^{-1})	Spectrophotometric, ascorbic acid and digestion (APHA, 1998)
Soluble ortho-phosphate (mg P-PO_4^{-3}.L^{-1})	Spectrophotometric, ascorbic acid (APHA, 1998)
Potassium (mgK.L^{-1})	Photometric (APHA, 1998)
COD (mg.L^{-1})	Redox titration with dichromate (APHA, 1998)
TSS (mg.L^{-1})	Gravimetric (APHA, 1998)
VSS (mg.L^{-1})	Gravimetric (APHA, 1998)
Conductivity (dS. m^{-1})	Instrumental method (direct reading) (APHA,1998)
Total alkalinity (mg $CaCO_3$ L^{-1})	KAPP method by Buchauer (1998)
pH	Potentiometric (APHA, 1998)
Microbiological examination	Analytical methods
Irrigation water Thermotolerant Coliforms (UFC/100mL)	Filtration Membrane (APHA, 1998)
Soil Thermotolerant Coliforms *Escherichia coli* (NMP/100g)	Multiple tubes and confirmation with defined substrate, MUG (APHA, 1998)
Culture (lettuce) Thermotolerant Coliforms Escherichia coli (NMP/100g)	Multiple tubes and confirmation with defined substrate, MUG (APHA, 1998)
Parasitological examination	Analytical methods
Irrigation water Helminthes eggs (eggs.L^{-1})	Modified Bailenger (WHO, 1979)
Soil Helminthes eggs (eggs.L^{-1})	Modified Meyer (MEYER, 1978)
Culture (lettuce and sweet pepper) Helminthes eggs (eggs/L^{-1})	Modified Bailenger (WHO, 1979)

4.5.1. Characteristics of the Produced Effluents

The results of physical, chemical and microbiological analysis of the treated effluents during the experimental period are presented in Table 10; the mean values and standard deviation of forty determinations of concentrations of macronutrients: (N-ammonical, potassium, phosphorus), chemical oxygen demand and electrical conductivity, solids and their fractions present in effluent of Upflow Anaerobic Sludge Blanked (UASB), in effluent of un-vegetated wetlands (UVWL), in effluent of planted wetlands (PWL) and of effluent of polishing ponds (PP), besides the values of only one determination of concentration of micronutrients are given.

Table 10. Mean values of quality parameters of the four treated effluents

Parameter	Effluent of UASB[1]	Effluent of UVWL[2]	Effluent of PWL[3]	Effluent of PP[4]
pH	6.7-7.1	7.0-7.2	7.0 -7.8	8.2 – 9.0
EC (dS.m^{-1})	1. 9±0.3	1.63±0.37	1.56±0.09	1.0±0.2
COD (mg.L^{-1})	220±50	78±32	60 ±17	164±58
TKN (mg.L^{-1})	49.9±11	34.9±6.0	28.0±9.0	15.5±4.1
N-ammonical (mg.L^{-1})	38.8±8.5	30.8±7.7	23.5±8.9	8.9±3.6
Phosphorus (mgP.L^{-1})	6.9±1.2	4.99±1.5	3.0±1.2	4.7±1.3
Orthophosphat (mgP-PO_4L^{-1})	5.8±1.0	3.0±2.4	2.6±1.1	3.1±1.4
Potassium (mgK.L^{-1})	22±3.1	22±0.3	25.0 ±3.4	26±2.0
Total Solids (mg.L^{-1})	788±178	377±209	102±59	806±92
TVS (mg.L^{-1})	231±108	134±47	49±21	336±223
SS (mg.L^{-1})	98±57	15±2.0	6.0±4.0	21±15
VSS (mg.L^{-1})	75±51	10±2.0	4.0±3.0	21±15
Copper (mg.L^{-1})	<0.02	<0.02	-	0.02
Zink (mg. L^{-1})	0.1	0.1	-	0.02
Iron (mg.L^{-1})	1.6	1.6	-	0.13
Manganese (mg.L^{-1})	0.06	0.06	-	0,04
Sodium (mg.L^{-1})	125	127	-	130
Boron (mg.L^{-1})	0.12	0.13	-	0.17
Molybdenum (mg.L^{-1})	<1	<1	-	<1
Nickel (mg.L^{-1})	0.02	0.02	-	<0.02
Chloride (mg.L^{-1})	178±33	180±29	198±32	179±39
Thermotolerant coliforms	9.8x10^{6}	9.8x10^{3}	7.9x10^{3}	9.2x10^{2}

[1] Upflow Anaerobic Sludge Blanked.
[2] Gravel bed systems.
[3] Constructed w etland.
[4] Polishing ponds.

4.5.2. Removal of Nutrients

The removal of nutrients is not desired when the treatment objective is utilization in irrigation. However, nitrogen in ionized ammonia form can be harmful to the plants when the concentration exceeds 5 mg/L (Ayers and Westcot, 1991). The effluents of UASB reactor and of un-vegetated wetlands maintained high values of ammonical nitrogen (38.8 and 30.8 mg/L, respectively), where as the effluents of vegetated wetlands and of polishing ponds maintained the average 23.5 and 8.9, respectively, as shown in Table 10. In polishing ponds, when the pH is established above 8.5, the liberation of ammonia gas occurs, thus loosing nitrogen for the atmosphere. Besides this, high pH can precipitate phosphorus in the form of insoluble salts (estruvite, vivianite and hydroxyapatite). This phenomenon only takes place if there exist considerable concentrations phosphate species ($H_2PO_4^{2-}$, HPO_4^{-2}, PO_4^{-3}) and calcium and magnesium ions and high pH in the medium (van Haandel and Lettinga, 1994).

The removal of nutrients in vegetated wetland systems does not occur in linear form, supposedly, the variations should apply to vegetative cycle of the aquatic plants. The removal of phosphorus occurs by chemical precipitation, sorption, assimilation of macrophytes and bacterial biofilms formed on substrate and on root system of the aquatic plants (Ceballos, *et al.*, 2001). The absorption of nutrients by the root system of the plant depends on pH, water used in irrigation, soil and the vegetable specie. The adequate range of pH for irrigation varies from 6.5 to 8.4 (Ayers and Westcot, 1991), therefore, as shown in Table 10, only the effluent of polishing pond is away from this range, thus able to cause nutritional disequilibrium of the plant.

The average values of nitrogen, phosphorus and potassium made available per hectare, from the effluents of UASB reactor, gravel bed and constructed wetland systems and from the final effluent of polishing ponds are shown in Table 11. The amount of available nitrogen, phosphorus and potassium, was estimated considering that 600 mm of well distributed rain are sufficient to guarantee the production of most of the regional cultures of the semi-arid, such as cotton, bean and corn that, generally, have vegetative cycle of 100 days; thus, the volume of 6000 m^3 of water per hectare will be used by each cycle of the culture.

Knowing the concentration of nutrients ($g.m^{-3}$) and the volume of applied effluent, and that present in each treatment system, as shown in Table 11, the amount of nitrogen, phosphorus (P_2O_5) and potassium (K_2O), per hectare can be calculated.

It can be seen in Table 11 that the effluents do not present in a constant form sufficient concentrations of nutrients to supply all the necessity of the plant. According to Malavolta *et al.*, (2000), 200 kg N/hectare/cycle of crop will be necessary for irrigating graminea in soils with organic material less than 2.5%; therefore, the effluent of polishing pond does not possess enough nitrogen to guarantee the fertirrigation of crops with these characteristics.

However, the effluent produced by shallow polishing pond is poor in macronutrients, mainly in nitrogen, phosphorus and potassium, it presents good sanitation quality (concentration of thermotolerant coliform lower than 1000 FCU/100mL), absence of helminthes eggs, and protozoa cysts, such as *crytosporidium*, *entamoeba* and *giardia*. It has relatively high concentration of sodium, which should be taken in to consideration at the time of application of certain type of fertilizer.

Table 11. Average values of nitrogen, phosphorus and potassium contained in 6000 m3 of effluents of different treatment systems

Systems	N (Kg ha^{-1})	P_2O_5 (Kg ha^{-1})	K_2O (Kg ha^{-1})
UASB	230	80	175
Gravel bed systems	185	41	181
Polishing ponds	53	42	108
Constructed wetland	141	37	176

Fonte: Sousa et al., (2005).

It can also be observed in Table 11, that the effluent produced by UASB reactor contain sufficient nutrients for the plants during the vegetative cycle, if there is a decrease in effluents produced by planted and un-vegetated constructed flooded systems.

At the time of entrance in wetlands the velocity of liquid decreases by the physical barrier effect of the substrate (sand, stone, carbon, soil, etc.) and of the rhizosphere, consequently the sedimentation velocity of the particles increases, thus resulting in decrease of water and sewage turbidity. A part of the organic material is degraded by the adhered biota on the roots (microbial biofilm) and of the system itself.

The death of microorganisms succeeds by autogenic activity among the populations, by predation, by excretion of microbiocide substances by roots and due to sedimentation and posterior biodecomposition. The growth of vegetables contributes sufficiently with the removal of the nutrients in the effluent along with the microbial growth in the biofilm. The combination produces useful biomass and, simultaneously, accentuated improvements in effluent water quality which can be used again, for example, in unrestricted irrigation. The system is easily constructed at low cost and therefore it is recommended for agricultural families.

4.5.3. Removal of Thermotolerant Coliforms

Although the indirect reuse is a much used practice in Brazil, but lamentably it is being developed without sanitary control. Today, the knowledge of water-generated diseases is quite popularized but, the "emerging" and "reemerging" diseases are also being discussed besides of zoonoses, for example, of cryptosporidiosis and gastroenteritis. The first disease is caused by protozoa and the second by bacteria as *Campylobacter* (Bastos *et al., 2003).*

The utilization of treated sewage in agriculture is a sustainable alternative that has economic, social and environmental feasibility, always when the effluent used in fertirrigation has sanitary quality. Otherwise, it can be tragic, since it may facilitate dissemination of hydro-vehiculation of diseases such as salmonellosis, shigellosis, cholera, hepatitis, giardiasis, amebiasis and esquistossomiasis among many others, some of them already present endemic state, for example, in Northeast of brazil.

As shown in Table 12, only the effluent of polishing ponds attends the requirements of World Health Organization for thermotolerant coliforms. Therefore, the use of shallow polishing ponds for production of effluents destined for fertirrigation shows itself as environmentally correct technology for producing effluents for unrestricted irrigation. However, the same can not be affirmed about its chemical quality, since nutritionally the effluent obtained is chemically unbalanced.

The microbiological results of the planted and un-vegetated wetland effluents showed similar behavior. The effluent of planted wetland maintained the mean value $7.8x10^3$ UCF/100 mL and that of the un-vegetated system remained at 10^4 UCF/100 mL. This small decrease in number of coliforms between the planted and un-vegetated wetland effluents, is statistically insignificant ($p>0.05$). Meira (2004) using vegetated wetlands in treatment of waste waters, obtained effluent with concentration of thermotolerant coliforms varying from 10^3 to 10^4 UCF/100 mL , therefore, having similar values.

The evidence of higher removal of coliforms in the planted wetlands occurs due to the existence of the ecosystem formed by vegetation roots adhered to the inert substrate on which there increases a dense biofilm. Thus, the microorganisms are eliminated by physical, chemical and biological processes. Even so, the wetlands effluents, with hydraulic detention time of about 10 days, should be used only for restricted irrigation.

Table 12. Frequency of protozoa cysts (n=12) in raw sewage and in effluents of UASB, UVWL and PP, during the period April-June, 2004

Protozoa cysts	RS		UASB		UVWL		P P	
	N°	%	N°	%	N°	%	N°	%
*Entamoeba**	12	40.0	25	47.2	0	0.0	0	0.0
*Entamoeba***	08	26.6	14	26.4	0	0.0	0	0.0
Iodamoeba sp.	01	3.3	03	5.7	0	0.0	0	0.0
Giardia sp.	02	6.7	03	5.7	0	0.0	0	0.0
Endolimax sp.	05	16.7	05	9.3	0	0.0	0	0.0
Balantideo sp.	02	6.7	03	5.7	0	0.0	0	0.0
Total	30	100	53	100	0	0.0	0	0.0
Mean	5.0	-	8.8	-	0	-	0	-
Standard deviation	4.3	-	9.0	-	0	-	0	-

* *Entamoeba* with up to 4 nucleus (characteristic of *Entamoeba histolytica, pathogenic*).
** *Entamoeba* with up to 8 nucleus (characteristic of *Entamoeba coli, non-pathogenic*).
Source: Figueiredo et al. (2005).

4.5.5. Removal of Helminthes Eggs and Protozoa Cysts

The occurrence and the removal of helminthes eggs and protozoa cysts in the four effluents produced have been verified. The examination performed on raw sewage and on effluent of UASB reactor, evidenced a considerable concentration of helminthes, 354 and 230 eggs/L, respectively.

In the effluent of UASB reactor the specie *Ascaris lumbricoides* prevailed over the others, followed by *Trichuris trichiura* (21.5%), *Enterobius sp* (4.4%,) *Ancilostomatideos* (8.7%) and *Hymenolepis sp* (3.9%). The effluents originated from un-vegetated constructed wet lands and from polishing ponds maintained free from helminthes eggs (Sousa *et al.*, 2005).

The performance data for removal of protozoa cyst in the studied systems are shown in Table 12. The average frequency of appearance of *Entamoeba*[1] cysts was 40% and 47%, respectively in raw sewage and in the effluent of UASB reactor. *Giardia sp* cysts were also found in the raw sewage (6.7%) and in the effluent of UASB reactor (5.7%).

The removal of protozoa cysts and of helminthes eggs occur by discreet sedimentation that favors the accumulation of these organisms in the sediment; the sedimentation process depends on rate of surface hydraulic application ($m^3.m^{-2}.day^{-1}$) and it only transfers helminthes eggs and cysts, which does not mean death of these organisms. Cavalcanti *et al.*, (2000) succeeded removal of 100% of helminthes eggs in polishing ponds operated with surface hydraulic charge of 0.20 $m^3.m^{-2}.day^{-1}$.

The sedimentation velocities for *Ascaris* eggs and for *Giardia* cysts 65 $cm.h^{-1}$ (Shuval et al., 1986) and 1 $cm.h^{-1}$ (Bouhoum et al., 2000), respectively, are sufficiently differentiated.

[1] *Entamoeba* with up to 4 nucleuses (characteristic of *Entamoeba histolytica*).

5. Application of Treated Sewage in Fertirrigation

Even though four types of effluents in the systems of sewage treatment were produced in this work, only three effluents (effluents of UASB reactor, of gravel bed system and of polishing pond) were applied for fertirrigation of two distinct crops: sesame and herbaceous cotton. The municipal sewage treatment was realized with a low cost system, functioned at the installation site at a family level agriculture farm. The preliminary treatment by sand box and equalization tank was initially performed, which was followed by a UASB reactor and then the effluent produced was post-treated in polishing pond.

Cultivation of Sesame and Cotton

Utilization of duly treated municipal sewage in agriculture, observing the standards recommended by World Health and Sanitary Vigilance Organization can become an ecologically sustainable, economically feasible and sanitarily safe agricultural practice. This study analyses some results of agricultural application of municipal sewage treated by biological processes, on two different crops, in Northeast of Brazil near Campina Grande City. The objective was to evaluate the effect of irrigation with sewage treated in UASB reactor, wetland system and stabilization ponds on sesame and cotton crops, and to compare its productivity with that of cotton irrigated by well water with and without mineral fertilization.

Experimental Plan

The variety of crop used in the study was herbaceous cotton (*Gossypiu hirsutum L.r.latifolium hutch*). The irrigation technique applied was the system of furrows, in area of 10 m^2, with a spacing of 0.80 m between rows and 0.20 m between plants. Regarding the water management, the system functioned with watering shift of 4 days, with the application of a blade depth of 24 mm. The second crop experimented was of sesame G-4, observing the spacing of 15 cm between plants and 60 cm in rows. The experimental design for fertirrigation was of the randomized blocs, with five treatments, according to the details shown in Table 13. The used area was divided in 20 portions. Each portion for cultivating G-4 sesame had an area of 10 m^2, where as, for the cotton, each portion measured 12 m^2.

Table 13. Experimental design for sesame and cotton cultures

Treatment	Description
T1	Irrigation with well water
T2	Irrigation with well water and inorganic fertilizer
T3	Irrigation with effluent of polishing pond
T4	Irrigation with effluent of gravel beds
T5	Irrigation with effluent from UASB reactor

Preparation and Fertilization of Soil

The soil sample for characterization was prepared as follows. Ten soil samples, up to the depth of 20 cm, were collected from each portion of the land. Then the samples collected from all the twenty portions were mixed up and homogenized. This soil was used for all the chemical and physical characterizations. The inorganic fertilization of treatment T2 was carried out according to the procedure recommended by Malavolta (1965).

The chemical and physical characteristics of the soil used as substrate for planting the studied crops are presented in Table 14.

Table 14. Chemical and physical characteristics of the soil before installation of the experiment

	Treatments				
	T1	T2	T3	T4	T5
Chemical characteristics					
pH	7.5	6.75	7.05	6.95	7.02
Calcium ($cmol_c/dm^3$)	4.89	4.70	7.20	4.44	5.0
Magnesium ($cmol_c/dm^3$)	3.16	2.98	5.92	2.96	3.4
Sodium ($cmol_c/dm^3$)	0.86	0.89	1.62	1.83	0.98
Potassium ($cmol_c/dm^3$)	0.21	0.25	0.45	0.34	0.32
Nitrogen ($cmol_c/dm^3$)	1.2	1.7	2.4	1.3	1.3
Total base ($cmol_c/dm^3$)	9.12	9.07	15.88	9.57	9.79
Hydrogen ($cmol_c/dm^3$)	0.00	0.66	0.16	0.00	0.00
Aluminum ($cmol_c/dm^3$)	0.00	0.00	0.00	0.00	0.00
Cattion exchange capacity ($cmol_c/dm^3$)	9.12	9.73	16.04	9.57	9.79
Organic carbon (g/kg)	12.9	17.5	24.7	13.4	13.4
Organic material (g/kg)	22.2	30.2	426	23.1	23.1
Assimilable phosphorus (g/kg)	0,125	0,125	0,125	0,125	0,105
Electrical conductivity (ds/m) Water-soil suspension	0,4	0,26	0,28	0,25	0,25
Physical characteristics					
Granulometry (g/kg)					
Sand	709.2	707.5	701.2	703.1	709.2
Silt	131.5	130.2	131.8	129.3	130.5
Clay	159.	156.8	154.1	158,8	157.5
Apparent density (g/cm^3)	1.41	1.42	1.40	1.41	1.41
Real density (g/cm^3)	2.60	2.60	2.60	2.60	2.60
Porosity (%)	45.79	45.86	45.96	45.74	45.76
Natural moisture (%)	1.11	1.11	1.11	1.11	1.11
Field capacity (0,33 atm)	14.09	14.99	14.90	14.89	14.69
Withering point (15 atm)	5.30	5.35	5.35	5.35	5.35
Available water	8.79	9.64	9.64	9.64	9.64

Foliar Area and Productivity of the Crops: Sesame and Cotton

The average values of foliar area and productivity, for sesame and cotton crops are presented in Table 15. The foliar area measures the plant growth, which is directly proportional to the photo-synthetically active surface. About 90% of the dry material accumulated by the plants during their growth comes from photosynthetic activity

(Benincasa, 1988). It can be observed in Table 15 that the average foliar area of sesame did not show significant difference in relation to treatments 1 and 3 ($p > 0.05$), treatment 1: soil without fertilizer and irrigation with well water; treatment 3: soil without fertilizer and effluent of polishing pond. It means that the effluent of polishing pond being poor in nutrients had a performance similar to that of the treatment with well water and soil with fertilizer.

Also there were no significant differences ($p > 0{,}05$) observed among the variables of the treatments 2, 4 and 5, irrigation with well water and soil with mineral fertilizer, irrigation with well water and soil with effluent of wetland, and soil irrigated with effluent of UASB reactor, respectively. Therefore, the effluent of UASB reactor presented performance similar to those with the treatments with soils fertilized as much with complete mineral nutrients as with the effluent of the wetland system.

It is also observed in Table 15 that the average foliar area and productivity of cotton did not show significant difference in relation to treatments 2, 3 and 4 ($p > 0.05$). Therefore, the analysis of growth and productivity of cotton among the treatments: irrigation with effluent of polishing pond, soil with mineral fertilization and soil with effluent of wetland, did not show statistically significant difference. In relation to treatment 1, irrigation with well water and unfertilized soil presented the lowest average, where as treatment 5, irrigation with effluent of UASB, maintained the highest average. Thus, treating sanitary sewage for utilization in agriculture is fundamental, because it functions as fertilizer, besides fulfilling for water demand.

Table 15. Foliar area and productivity of the crops fertirrigated with effluents of UASB reactor, of wetland system and polishing ponds, during the period 2001 to 2003

Treatment	Sesame		Cotton	
	Foliar area (cm^2)	Productivity (kg ha^{-1})	Foliar area (cm^2)	Productivity (kg ha^{-1})
T1	3000 b	750 a	500 c	1500 b
T2	5100 a	1250 c	1850 ab	2800 a
T3	3350 b	980 ab	1980 a	2900 a
T4	4780 a	1250 c	1450 b	1980 a
T5	4850 a	1000 bc	700 c	1600 b

Where: Values in a column followed by same letter do not differ at the probability level of 0.5 Tukey test. T1: irrigation with well water and un-fertilized soil; T2: irrigation with well water and mineral-fertilized soil; T3: irrigation with polishing pond effluent; T4: irrigation with wetland system effluent and T5: irrigation with UASB reactor effluent.

Source: Lima (2003).

The system composed of UASB reactor followed by polishing pond functions adequately since it produced effluent with sanitary quality standard for unrestricted irrigation. It can be installed in small rural properties who desire to practice the planned reuse of waste water.

The effluent of UASB reactor detain high concentration of nutrients, however, it preserves high concentration of thermotolerant coliforms and helminthes eggs. It does not present adequate sanitary quality as per World Health Organization recommendations (WHO,1989) for unrestricted reuse in agriculture, since it presents potential risk to the groups exposed to the infection, such as rural workers and consumer of the irrigated products.

Figueiredo et al. (2005) examined effluent originated from UASB reactor and evidenced great diversity of helminthes eggs. The study performed on twelve samples of UASB effluent revealed on an average 176 eggs of helminthes per liter, of which 65% were *Ascaris lubricoides* (114 eggs/L) followed by *Trichuris trichiura* with 22% (39 eggs/L) and *Enterobius vermiculares, Hymenolepis sp* and *Amcilostomi.deo* with 6.4 and 3%, respectively.

Sanitary Quality of the Fertirrigated Products

The fruits harvested irrigated by effluent of polishing ponds and with well water and soil with mineral fertilizer presented thermotolerant coliforms varying from $7.21x10^0$ to $2.7x10^1$ MPN/g. Where as, which were irrigated with wetland effluent and UASB reactor effluents presented concentrations $2.9x10^3$ MPN/g and $1.16x10^4$ MPN/g, respectively. According to the standard established by WHO the crops that presented contamination by thermotolerant coliforms higher than 100 MPN/g are improper for human consumption.

The data of *E. coli,* thermotolerant and total coliforms present in irrigated crops with the effluents and with well water are shown in Table 16. *Escherichia coli* is presently the best indicator of fecal contamination being exclusive of intestinal tract of homoeothermic animals

Table 16. Concentration of microorganisms present in the crops

Treatment	Total coliforms ($MPN.g^{-1}$)	Thermotolerant coliforms ($MPN.g^{-1}$)	*E. coli* ($MPN.g^{-1}$)
T1	$3.2x10^1$	$4.2x10^1$	$3.6x10^0$
T2	$2.52x10^1$	$2.7x10^1$	$2.7x10^1$
T3	$4.5x10^0$	$2.9x10^0$	$4.5x10^0$
T4	$8.7x10^2$	$7.21x10^3$	$1.6x10^2$
T5	$6.0x10^4$	$9.16x10^4$	$9.3x10^2$

Cost and Investment for Treating Municipal Sewage

Sewage detains considerable amounts of nutrients. Studies performed by Monte and Sousa (1992); Vazquez-Montiel et al. (1996); Mota et al. (1997) and Sousa and Leite (2003) showed that the productivity (t/ha) of irrigated crops with treated sewage was higher (15 to 30%) than those of crops irrigated with supply water and soil fertilized with NPK, demonstrating the feasibility of use of sludge in irrigation.

Municipal sewage treatment should be considered as a right of citizen and a duty of the state. It can be considered as an integrated general necessity for exercising the rights and duties of citizenship. The sewage treatment reduces diseases related to eutrophication of ground water, conserves environment, thus guarantying better quality of life to the population.

Technically, sewage treatment consists of three levels: (a) primary treatment involves removal of material in suspension, crude solids, using grids, sand retention boxes or sedimentors, (b) secondary treatment, which means removal of carbonaceous material and (c) tertiary treatment that removes nutrients and pathogenic organisms. The final effluent produced in the polishing ponds presented good sanitary quality: low concentration of thermotolerant coliforms (less than 1000 FCU/100 mL) and absence of helminthes eggs, which according to WHO (1989) can be used in unrestricted irrigation. In case of sanitary sewage, the reduction of pathogens occurs, by chloration, ozonization, application of effluent in soil, maturation and polishing ponds.

The investment costs shown in Table 17 are based on estimated calculations for the northeast region of Brazil. Each region has its specificity, such as in utilization of land area for construction of polishing ponds.

As can be observed from Table 17 that the treatment of sewage of a population of ten thousand inhabitants requires US$195,000, which means it requires only 19 dollars/inhabitant. This amount is relatively very low, considering its social, economic and environmental impact.

The area that can be irrigated with the volume of sewage produced by the population depends on climate, type of crop, soil and on the application method. Considering that 1 mm of rain corresponds to 10 m^3 per hectare, the 100 L of sewage produced by one person corresponds to 0.01 mm/day of rain, which means, 3.65 mm of rain per year. Considering further that, for semi-arid region, 600 mm of rain in a well distributed year have guaranteed the crop of cotton, bean and sesame, the regional crops regarded as family agriculture commonly developed in countryside of northeast of Brazil. Following this reasoning, the sewage produced per person during one year will give for irrigation of 0.006 hectares.

Considering that, there should be necessity of 600 mm of rain for irrigation of cotton during the vegetation cycle of 100 days; the following estimation can be made. 600 mm of rain makes 6000 m^3 for irrigation of one hectare of cotton during the cycle of 100 days, which means 60 m^3 dia^{-1} per hectare are necessary. Considering an intense evaporation in semi-arid region, 100 $m^3.dia^{-1}$ per hectare can be adopted in place of 60. Using the sewage volume of a population of ten thousand inhabitants will give 1000 m^3 per day (100 $L.inhab^{-1}.day^{-1}$), which will be sufficient to irrigate ten hectares of cotton per day.

Table 17. Investment for sewage treatment of a population of ten thousand inhabitants

System	Volume m^3	Cost (US$) per m^3	Total cost (US$ 1,000)
UASB	334	132	45
Polishing ponds	15,000	10	150
Total investment			195

Brazil has 144 million inhabitants in the urban zone out of whom 95.3% have water service. Almost half of the Brazilian urban population does not possess a sanitary sewage line. According to the Brazilian government, it would have to invest about US$ 285 billion for universalization of the water supply service and for collection and treatment of sewage.

Considering that this benefit should be realized during the period of 20 years, an investment of US$ 14.2 billion/year will be required. An investment of US$ 7.2 billion/year was planned for the period 2004 to 2007. If this utopia becomes reality, in the next twenty years, Brazil will be able to serve half of its population that necessitates this service.

CONCLUSION

Although UASB reactor does not remove significantly the pathogenic organisms and nutrients, the reactor with HDT of 6 hours, presented a removal performance of COD and VSS higher than 60 and 70%, respectively. This considerable removal represents a significant diminution of organic load, favoring, therefore, the post-treatment in systems such as wetlands, polishing ponds and gravel beds. Therefore the effluent produced by UASB reactor presents itself as a good alternative for reuse from the nutritional point of view. In relation to sanitary aspects, it does not follow the WHO directives for its unrestricted use in fertirrigation. Therefore, effluents of UASB reactor necessarily require a post-treatment for producing waste water of sanitary quality for attending these directives. In crops of lower sanitary risk and under controlled conditions of access to production areas and of handling of the application system, its use can become feasible.

Due to its poor hygienic quality, effluents of UASB reactors should only be used in restricted irrigation. But UASB reactors followed by polishing ponds have been shown to be an efficient treatment unit for sewage treatment. The final effluent produced in polishing ponds presented good sanitary quality: low count of thermotolerant coliforms (1000 FCU/100 mL) and absence of helminthes eggs, which can be used for unrestricted irrigation according to WHO (1989).

Wetland and un-vegetated gravel bed systems, operated under the conditions presented in this study, produced effluents free from helminthes eggs, where as they presented thermotolerant coliforms concentrations higher than 1000 FCU/100 mL. These effluents do not attend the WHO (1989) directives for unrestricted irrigation use.

The fertiliztion value of the evaluated effluents was, in general, high; except the polishing pond effluent, which does not have sufficient nutrients for good development of crops, in spite of presenting excellent sanitary characteristics. The high level of treatment of municipal sewage harmed the chemical quality of the effluent.

The fertirrigation treatment with UASB reactor effluent presented foliar area similar to that obtained in treatment of irrigation of the crop with well water, where the soil received complete mineral fertilizer.

The effluents produced in constructed wetland systems indicated nutrient amounts and pH adequate for fertirrigation but also presented high concentrations of thermotolerant coliforms (10^3 < UFC/100 mL < 10^4). These effluents can be used for fertirrigation of industrial crops, such as cotton, fructiferous tree, pastures and forages, according to WHO directives.

The crop fertirrigated with effluent produced from UASB reactor presented greater phyto-mass, consequently, higher productivity, but due to elevated number of thermotolerant coliforms and helminthes eggs, this effluent should not be used for unrestricted irrigation and should not be applied by sprinkling.

The fertilization with anaerobically pre-treated sewage and with mineral fertilizer, provided equal performance in reference to productivity, evidencing the importance of organic material and of the nutrients contained in primary effluent.

The treatment techniques which proportionate the maintenance of. fertilizers in the effluents, enabling the substitution of all or part of the mineral fertilizer, are the most perfectly applicable techniques for the modality of family-level agriculture, and deserve greater attention from government research workers.

With the exception of the final effluent, originating from polishing ponds, the amount of macronutrients contained in the other three effluents is sufficient for most of the regional crops cultivated in the semi-arid region northeast of Brazil.

REFERENCES

Andrade Neto, C. O. (1997). *Simple systems for sanitary sewage treatment: Brazilian experience* (In Portuguese). Rio de Janeiro, RJ: ABES. 301 p.

APHA. (1998). *Standard methods for the examination of water and wastewater.* (15th ed.) American Public Health Association. American Water Works Association, Water Pollution Control Federation: Washington, DC. 1134 p.

Asano, T. and Levine, A. D. (1996). Wastewater reclamation, recycling and reuse: Past, present, and Future. *Water Science and Technology*, v.33, n.10-11, p 1-14.

Ayers, R. S. and Westcoot, D. W. (1991). *A qualidade de água na agricultura.* Transl. Gheyi, H.; Medeiros, J. F.; Damaceno, F. V. Campina Grande, UFPB, 218 p.

Bailenger, J. (1979). Mechanisms of parasitological concentration in coprology and their practical consequences. *Journal of American Medical Technology*, v.41, p 65-71.

Bastos, R. K. X.; Bevilacqua, P. D. and Keller, A. (2003). *Pathogenic organisms and effects on human health.* In: *Desinfection of sanitary effluents* (In Portuguese). PROSAB 3. Rio De Janeiro. RJ.: ABES. p. 27-88.

Benincasa, M. M. P. (1988). *Analysis of growth of plants: basic notions* (In Portuguese). Jaboticabal-SP: Editora da UNESP. 128 p.

Blumenthal, U. J.; Peasey, A.; Ruiz-Palacios and Mara, D. D. (2000). *Guidelines for wastewater reuse in agriculture and aquaculture: Recommended revisions based on new research evidence.* London: WELL Study. 67 p.

Bouhoum, K.; Amahmid, O. and Asmama, S. (2000). Occurrence and removal of protozoan cysts and helminth eggs in waste stabilization ponds in Marrakech. *Water Science and Technology*, v. 42, n. 10-11, p. 1597-1604.

Buchauer, K. A. (1998). A comparison of two simple titration procedures to determine volatile fatty acids in effluents of wastewater and sludge treatment process. *Water SA*, v. 24, n. 1, p. 49-56.

Cavalcanti, P. F. F.; Van Haandel, A. C. and Lettinga, G. (2000). Polishing ponds for post treatment of digested sewage: Sequential batch ponds. *Proc. of the VI Latin-American Workshop and Seminar of Anaerobic Digestion*. Recife, Brazil, p. 352-359.

Ceballos, B. S. O.; Rolim, H. O.; Meira, C. M. B. S. and Sousa, J. T. (2001). River water quality improvement by natural and constructed wetland systems in the tropical semi-arid region of Northeastern Brazil. *Water Science and Technology*, v. 44, n. 11-12, p. 599-605.

Coraucci Filho, B.; Andrade Neto, C. O.; Souza Melo, H. N; Sousa, J. T. and Nour, E. A. A. (2001). *Post-treatment of effluents of anaerobic reactors for soil application systems.* In: Chernicharo, C. A. L. *Post-treatment of effluents of anaerobic reactors* (In Portuguese) (1st ed.) Belo Horizonte, MJ: ABES. P. 35-90.

Feachem, R. G.; Bradley, D. J.; Garelick, H. and Mara, D. D. (1983). *Sanitation and disease - health aspects of excreta and wastewater management.* Washington, D.C. USA: John Wiley and Sons. 534 p.

Figueiredo, A. M. F. (2005). *Treatment, sanitary quality and the use of municipal effluent in irrigation of okra* (In Portuguese). Campina Grande, PB: Universidade Federal da Paraíba/ Universidade Estadual da Paraíba – PRODEMA, MS Dissertation. 105 p.

Figueiredo, A. M. F.; Ceballos, B. S. O.; Sousa, J. T. and Araújo, H. W. C. (2005). Effect of fertirrigation of sanitary quality treated municipal sewage (In Portuguese). *Revista Brasileira de Engenharia Agrícola e Ambiental* (Online), v. 9, n. supplement, p. 322-327.

Henze, M. and Harremoës, P. (1983). Anaerobic treatment of wastewater in fixed film reactors: A literature review. *Water Science Technology*, v. 15, n. 8-9, p 1-101.

Lawrence, A. W. and McCarty, P. L. (1970). Unified basics for biological treatment design and operation. *Journal of the Sanitary Engineering Division*, v. 89, n. 5, p 757-775.

Léon S. G. and Cavallini, J. M. (1999). *Tratamento e uso de águas residuárias.* Transl. Gheyi, H. R.; König, A.; Ceballos B. S. O. and Damasceno, F. A. V. Campina Grande, PB: Editor UFPB. 110 p.

Lima, E. P. C. (2003). *Sanitary sewage treatment and utilization in fertirrigation of sesame: A proposal for family-level agriculture* (In Portuguese). Campina Grande: Universidade Federal da Paraíba/Universidade Estadual da Paraíba – PRODEMA, MS Dissertation. 160 p.

Lima, S. M. S. (2004). *Treatment and utilization of municipal sewage in family-level horticulture* (In Portuguese). Campina Grande: Universidade Federal da Paraíba/Universidade Estadual da Paraíba – PRODEMA, MS Dissertation.131 p.

Malavolta, E. (1965). *Métodos para la determinación de deficiencia. In: Fitopatologia – Curso Moderno.* A. A. Sarasola and M. A. R. de Sarasola (Ed.). Buenos Aires: Editorial Hemisfério Sul. p. 244-247.

Malavolta, E.; Pimentel-Gomes, F. and Alcarde, J. C. (2000). *Fertilizers and Fertilization* (In Portuguese) . São Paulo, SP: Editora Nobel. 200 p.

Marais, G. V. R. and Ekama, G.A. (1976). The activated sludge process - Part. I: State Behavior. *Water SA*, v. 2, n. 4, p. 163-200.

McCarty, P. L. and Brodersen, C. F. (1962). Theory of extended aeration activated sludge. *Journal Water Pollution Control Federation*, v. 34, n. 11, p 1095-1102.

Meira, C. M. S. (2004). *Utilization of wetlands in treatment of polluted surface waters: Influence of the type of beds and of macrophytes.* Campina Grande, PB: Universidade Federal da Paraíba, PhD Thesis. 171p.

Metcalf and Eddy. Inc. (1991). *Wastewater engineering treatment disposal reuse.* 3rd ed. NewYork: McGraw-Hill Book. 1334p.

Metcalf and Eddy. Inc. (2003). *Wastewater engineering treatment disposal reuse.* 4th ed. NewYork: McGraw - Hill Book. 1815 p.

Meyer, K. B. and Kaneshiro, E. S. (1978). Recovery of ascaris eggs from sludge. *Journal of Parasitology*, v. 64, n. 2, p. 380-383.

Monte, M. H. F; Sousa, M. S. (1992). Effect of irrigation by facultative pond effluent on the crop. *Water Science and Technology*, v. 26, n. 7-8, p 1603-1613.

Mota, S; Bezerra, F. C. and Tomé, L. M. (1997). Evaluation of performance of crops irrigated by treated sewage (In Portuguese). In: 19ª Congresso Brasileiro de Engenharia Sanitária e Ambiental, Foz do Iguaçu: ABES, p. 20-25.

Ongley, E. D. (1996). *Control of water pollution from agriculture, irrigation and drainage*. Rome, Italy: FAO. 101p.

Santos, H. F.; Morita, D. M.; Grull, D.; Rodrigues, J. M. C.; Pimentel, J. S. and Botelho, M. H. C. (1992). Reuse of water (In Portuguese). *Revista DAE*, v. 167, n. 5, p. 23-32.

Shuval, H.; Adin. A.; Fattal, B. (1986). *Wastewater irrigation in developing countries. Health effects and technical solutions*. Paper Number 51, Integrated Resource Recovery Projects series number GLO/80/004, Washington, DC: World Bank Technical. 324 p.

Silva, S. A. and Mara, D. D. (1979). Biological treatment of wastewater: Stabilization ponds (In Portuguese). Rio de Janeiro: ABES. 140p.

Sousa, J. T.; Araújo, H. W. C.; Catunda, P. F. C. and Florentino, E. R. (1998). Treatment of sanitary sewage by slow filter, for producing effluent to reuse in agriculture (In Portuguese). In: VIII *Simpósio Luso-Brasileiro de Engenharia Sanitária e Ambiental*. João Pessoa, PB: ABES, v.1, T.1, p. 317-327.

Sousa, J. T. and Leite, V. D. (2003). Treatment and utilization of municipal sewage in agriculture (In Portuguese). Campina Grande, PB: EDUEP. 135 p.

Sousa, J. T.; Leite, V. D.; Dantas, J. P. and Dionisio, J. A. (2000). Reuse of sanitary sewage effluent in rice crop (In Portuguese). In: *IX Simpósio Luso-Brasileiro de Engenharia Sanitária e Ambiental*. Porto Seguro, BA: ABES, p. 1058-1063.

Sousa, J. T.; Leite, V. D. and Luna, J. G. (2001). Development of rice crop irrigated with already treated sanitary sewage. *Revista Brasileira de Engenharia Agrícola e Ambiental*, v. 5, n. 1, p. 107-110.

Sousa, J. T.; van Haandel, A. C.; Cavalcanti, P. F. F. and Figueiredo, A. M. F. (2005). Sewage treatment for reuse in northeastern semi-arid agriculture (In Portuguese). *Engenharia Sanitária e Ambiental*, v. 10, n. 3, p. 260-265.

Sousa, J. T.; van Haandel, A. C. and Guimarães, A. V. A. (2003). Performance of constructed wetland systems treating anaerobic effluents. *Water Science and Technology*, v. 48, n. 6, p. 295-299.

van Haandel, A. C. and Lettinga. G. (1994). *Tratamento Anaeróbio de Esgoto: um manual para a região de clima quente*. Campina Grande-PB: Epgraf. (Undefined numbers).

Vasquez-Montiel, O.; Horan, N. J. and Mara, D. D. (1996). Management of wastewater for reuse in irrigation. *Water Science and Technology*, v. 33, n. 10-11, p. 355-362.

Washington, D. R. and Hetling, L. J. (1965). Volatile sludge accumulation in activated sludge plants. *Journal Water Pollution Control Federation*, v. 37, n. 4, p. 499-507.

WHO. World Health Organization (1989). *Health guidelines for the use of wastewater in agriculture and aquaculture*, Technical Report Series, N° 778, Geneva: World Health and Organization. 72 p.

WHO. World Health Organization (2006). *Guidelines for the safe use of wastewater, excreta and greywater*. Volume 2: *Wastewater and excreta use in agriculture*. Geneva: World Health and Organization. 85 p.

In: Sewage Treatment: Uses, Processes and Impact
Editor: Anna Stephens and Mark Fuller

ISBN 978-1-60692-959-9

Chapter 5

ISOLATION OF POLLUTANT-DEGRADING MICROBES FROM A SEWAGE TREATMENT PLANT

Katsuhiko Fujii*
Department of Agriculture, Yamaguchi University,
Yoshida, Yamaguchi, 7538511 Japan

ABSTRACT

Activated sludge method is widely used for the biological sewage treatment. Activated sludge contains aerobic microbes including bacteria, fungi, amoeba, and protozoa, which play a critical role for efficient treatment process. Since domestic and industrial wastewaters are known to contain various pollutants, activated sludge is therefore thought to be acclimated to pollutants, and might be an attractive source for isolation of pollutant-degrading microbes. Although a microbial community of activated sludge is recently analyzed using culture-independent methods, enrichment and isolation works still play an important part to study biochemical and physiological characters and examine an application potential of microbes.

The author has isolated several pollutant-degrading microbes including a nonylphenol-degrading bacterium, estradiol-degrading bacterium, and benzophenone-degrading yeast from sewage treatment plants. We assimilate the pollutants as a sole carbon source.

A bioreactor using the immobilized nonylphenol-degrading bacterium worked properly and removed nonylphenol in wastewater, suggesting that a biotreatment system using a pollutant-degrading microbe can contribute to the preservation of aquatic environments. The bacterial strains isolated in this study were found sphingomonads. Sphingomonads are well known for various pollutant-degrading activities and were recently reported as ubiquitous microbes in activated sludge. Considering the fact that almost of microbes in the activated sludge have not been isolated yet, an employment of suitable culture conditions can bring a finding of novel pollutant-degrading microbes in an activated sludge.

* Phone/fax: +81-(0)839335835; Email: kfujii@yamaguchi-u.ac.jp

Introduction

Sewage treatment is a process for removing contaminants, mainly organic ones, from a wastewater, and is indispensable to maintain a public health. Several kinds of sewage treatment processes are used nowadays, and the most common one is the activated sludge method. In this method, organic matters in a wastewater are mineralized by the activated sludge, a brownish flocs formed by aerobic bacteria, fungi, and protozoa [1-3].

A microbial community of the activated sludge is studied using molecular biological methods such as ribosomal RNA/DNA [4-6] or quinone [7, 8] analysis, and found that more than 99% of microbes in the activated sludge have not been isolated or cultivated yet by microbiologist [9-11].

On the other hand, recent urban sewage contains various industry-derived pollutants. Among them, endocrine disrupter chemicals (EDCs) have lately attracted considerable attention. EDCs are naturally occurring or man-made chemicals that act like an endogenous hormone and interfere with the endocrine system of wildlife and human. In fact, widespread sexual disruption or "feminization" has been observed in riverine fish in several developed nations and is thought to be due to EDCs in the STP effluent [12-15].While pollutants such as EDCs are generally resistant to biodegradation, an acclimation of activated sludge microbes to pollutants is reported [16, 17], suggesting that a sewage treatment plant (STP) is one of sources for isolation of pollutants-degrading microbes. Considering the fact that more than 99% of microbes in the environment remain to be isolated as described above, novel pollutant-degrading microbes can be found by further researches. In this chapter, characteristics of EDC-degrading microbes isolated from sewage treatment plant and their potential for bioremedial applications were discussed.

Strategy for Isolation of EDC-Degrading Microbes

An outline of the experimental method for a search of EDC-degrading microbes is shown in Figure 1. Since almost of EDCs are organic compounds, the basic strategy for obtanining EDC-degrading microbes is the enrichment cultivation technique using a minimal salt medium containing EDC as a carbon source. Yeast Nitrogen Base without amino acids (YNB) from Beckton Dickinson (Franklin Lakes, NJ) supplemented with EDC is used for searching EDC-degrading microbes in my laboratory, although YNB was originally developed and is used for classifying yeasts based on their carbohydrate requirement. As shown in Table 1, YNB consists of $(NH_4)_2SO_4$ as a nitrogen source, other salts (KH_2PO_4, $MgSO_4$, NaCl, and $CaCl_2$), trace metals, and a very small amount of vitamins. No other carbon sources are present, and EDC is thus considered to be nearly the sole carbon source in the medium. In this medium, therefore, microbes having EDC-degrading activity can grow dominantly, because EDCs are generally resistant to biodegradation and a microbial growth found in this culture condition implies the existence of EDC-degrading microbes.

In order to enrich EDC-degrading microbes, we collected various environmental samples including river water, river sediment, field soil, wastewater, or activated sludge and inoculated directly into the medium. The samples were then cultivated at 25°C, rotating at 150 rpm/min.

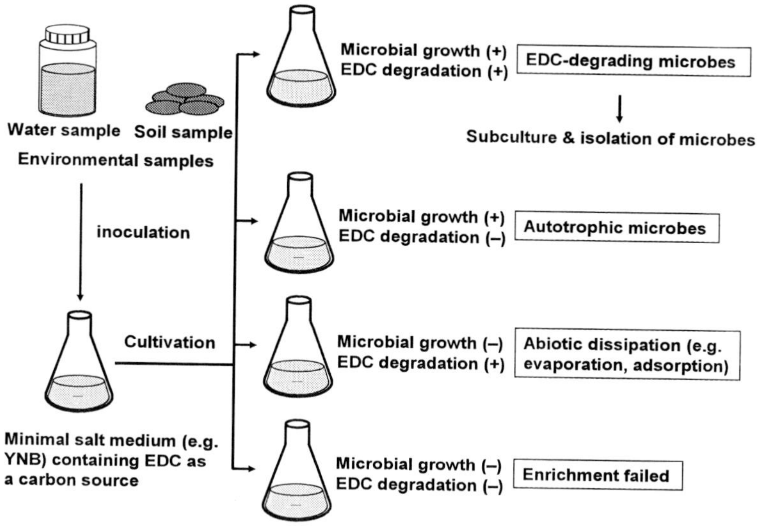

Figure 1. An experimental scheme for a search of EDC-degrading microbes.

Table 1. Medium composition for cultivation of EDC-degrading microbes used in this study

Constituents		Final concentration
EDC	Carbon source	1 g/l
Ammonium sulfate	Nitrogen source	5 g/l
Potassium phosphate monobasic	Salts	1 g/l
Magnesium sulfate		0.5 g/l
Sodium chloride		0.1 g/l
Calcium chloride		0.1 g/l
Boric acid		500 μg/l
Copper sulfate		40 μg/l
Potassium iodide		100 μg/l
Ferric chloride		200 μg/l
Manganese sulfate		400 μg/l
Sodium molybdate		200 μg/l
Zinc sulfate		400 μg/l
Biotin	vitamins	2 μg/l
Calcium pantothenate		400 μg/l
Fplic acid		2 μg/l
Inositol		2,000 μg/l
Niacin		400 μg/l
p-Ainobenzoic acid		200 μg/l
Pyridoxine hydrochloride		400 μg/l
Riboflavn		200 μg/l
Thiamine hydrochloride		400 μg/l

Microbial growth (turbidity) and EDC degradation in the culture was monitored during cultivation. EDC degradation was monitored by a high performance liquid chromatograph (HPLC) or a gas chromatograph-mass spectrometer GCMS. Cultures showing both microbial growth and EDC degradation were subcultured several times and then streaked on agar media to isolate EDC-degrading microbes. The agar medium was incubated at 25°C and emerging single colonies were isolated by a sterile toothpick onto a fresh agar medium. The agar-grown isolates were then examined for their EDC-degrading activity in a pure culture (Figure 2). If microbial growth and EDC degradation were observed in the isolate pure culture, the isolate is suggested to possess EDC-degrading activity. If microbial growth was observed but EDC was not degraded, the isolate might be an autotroph. If neither microbial growth nor EDC degradation was observed, the isolate seems to assimilate metabolites of true EDC-degraders or a member of EDC-degrading microflora.

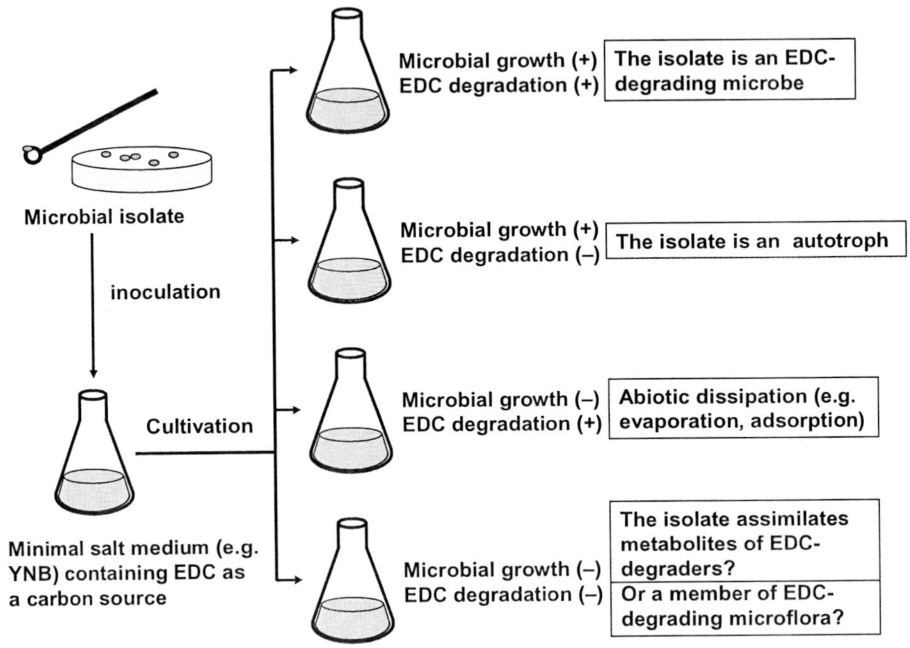

Figure 2. An experimental scheme for studying EDC-degrading activity of the isolates.

Nonylphenol-Degrading Microbe

Nonylphenol (NP, Figure 3a) is known as an important intermediate in the production of many commercial and industrial materials. The major use of NP is for the production of nonylphenol polyethoxylate, one of non-ionic surfactants used widely in many industrial applications. However, NP is also known as a ubiquitous pollutant in a wastewater, and there is growing evidence that NP is an EDC. Therefore we tried to isolate NP-degrading microbes to extend our knowledge of microbial degradation of NP [18].

A HPLC system with an octadecyl silica (ODS , C_{18}) column was employed to determine NP concentrations in a mixed culture or isolate pure culture.

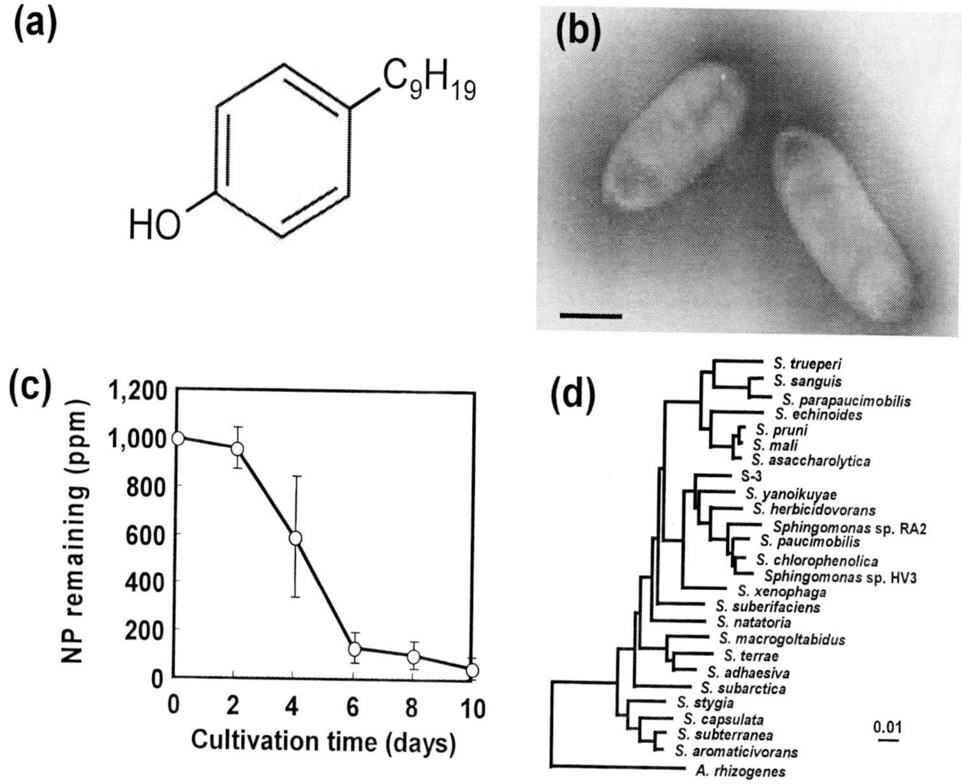

Figure 3. Nonylphenol (NP) and NP-degrading strain isolated in this study. (a) Molecular structure of NP. (b) Transmittion electron micrograph of the NP-degrading strain S-3. Bar, 1 μm. (c) Typical time courses for the NP degradation by the strain S-3. (d) Phylogenetic tree constructed by the neighbour-joining method based on 16S rDNA sequences of the strain S-3 and related bacteria. Scale bar represents an evolutionary distance (Knuc) of 0.01.

For the mobile phase (flow-rate 1.0 ml/min), an acetonitrile/water mixture (90/10 vol%) was used. NP was detected by UV absorption at 270 nm. As a result of enrichment cultivation of various environmental samples in YNB containing 1,000 ppm (mg/l) NP (NP-YNB medium), culture of domestic sewer obtained in a STP showed microbial growth. Approximately 90 % of NP was degraded in the 15-day culture.

After three times subculture, an aliquot of the subculture was inoculated on NP-YNB agar medium, cultivated statically at 25°C. A gram-negative and rod-shaped bacterial strain was isolated and named as S-3 (Figure 3b).

The NP-degrading activity of S-3 pure culture was examined [19]. A single colony of S-3 from an agar plate was inoculated into fresh NP-YNB medium and cultivated at 25°C while being rotated at 150 rpm. In independent assay with an initial concentration of 1,000 ppm, NP began to be degraded within 2 days and a complete degradation was accomplished within 10 days (Figure 3c). The total biomass of the isolate in liquid culture was found to increase significantly after 10 days of cultivation, indicating that S-3 utilized NP as a sole carbon source. Although UV absorption at 270 nm was used for monitoring an aromatic part of NP in the HPLC analysis of the culture, we did not find any aromatic compounds other than NP. Therefore, a gas chromatograph-mass spectrometer (GCMS) was used to identify metabolites derived from NP degradation. S-3 was cultivated in NP-YNB medium, and chloroform and hexane extracts of 0 to 10-day-old cultures were analyzed with a GCMS under the following conditions: column, CBP-20M capillary column (25 m × 0.25 mm i.d., Shimadzu, Kyoto, Japan); carrier gas, helium; temperature gradient, 50°C for 3 min, 50-150°C at 5°C/min, 50-220°C at 10°C/min, and 220°C for 20 min. The GCMS analysis revealed that the major metabolite of NP degradation was nonanol (C_9H_{19}-OH). Various structural isomers of nonanol through the branched nonyl-group exist, suggesting that nonanols are derived from the nonyl-group of NP. Nuclear magnetic resonance (^{1}H-NMR) spectrometer was also employed for the metabolite analysis and found that a phenolic part of NP was degraded.

The strain S-3 was studied phenotypically, genotypically and phylogenetically [20]. Analysis of the 16S ribosomal DNA sequence, the G+C content of the DNA, the isoprenoid quinone composition, and the whole-cell fatty acid profile revealed that the strain is a member of the genus *Sphingomonas*. However, 16S rDNA similarity with that of known *Sphingomonas* spp. was found to be at most 96% (Figure 3d), and the results of DNA–DNA hybridization experiments clearly indicated that the isolate represents a new *Sphingomonas* species, for which the name *Sphingomonas cloacae* was proposed.

ESTRADIOL-DEGRADING MICROBE

Natural estrogens entering the environment through the excretions of humans, domestic or farm animals, and wildlife are also thought to be EDCs [21-24]. Among the natural estrogens, 17β-estradiol (E2, Figure 4a) is the most potent and is found ubiquitously in many water systems [25-28].

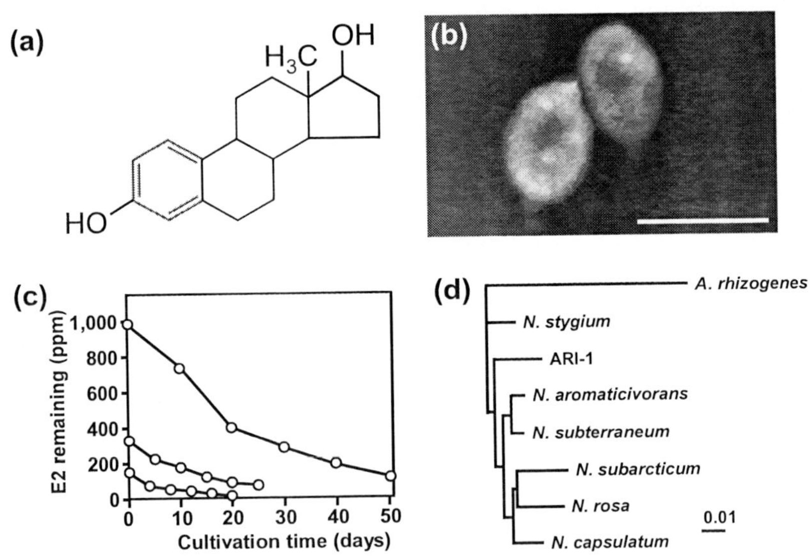

Figure 4. Estradiol (E2) and E2-degrading strain isolated in this study. (a) Molecular structure of E2. (b) Transmittion electron micrograph of the E2-degrading strain ARI-1. Bar, 1 μm. (c) Typical time courses for the E2 degradation by the strain ARI-1. (d) Phylogenetic tree constructed by the neighbour-joining method based on 16S rDNA sequences of the strain ARI-1 and related bacteria. Scale bar represents an evolutionary distance (Knuc) of 0.01.

It has been suggested previously that natural steroidal estrogens, including E2, are poorly degraded during STP processing [29, 30], and the proportion of steroidal estrogens remaining in the effluent is still capable of producing estrogenic effects [31]. Therefore, it is thought that E2 is also responsible for the estrogenic effects found for STP effluent. These prospects led us to search for microbes with an E2-degrading activity.

A HPLC system with an ODS column was used to determine E2 concentrations in a mixed culture or isolate pure culture. For the mobile phase (flow-rate 1.0 ml/min), an methanol/water mixture (80/20 vol%) was used. E2 was detected by UV absorption at 210 nm. As a result of enrichment cultivation of various environmental samples in YNB containing 330 ppm E2 (E2-YNB), culture of activated sludge in a STP showed microbial growth [32]. In this sample, 60% of E2 was degraded in 14 days. After three times subculture, an aliquot of the subculture was inoculated on nutrient agar medium, cultivated statically at 25°C. A gram-negative and oval-shaped bacterial strain was isolated and named strain ARI-1 (Figure 4b).

The E2-degrading activity of ARI-1 pure culture was examined [32]. A single colony of ARI-1 from the agar plate was inoculated into fresh E2-YNB medium and cultivated at 25°C while being rotated at 150 rpm. In independent assay with an initial E2 concentration of 160-1,000 ppm, ARI-1 steadily degraded E2 (Figure 4c). Plate counts on nutrient agar showed that the cell density in 30-day-old ARI-1 pure culture (1,000 ppm E2) was about 0.5×10^9 CFU/ml, while it was 0.8×10^3 CFU/ml in 0-day-old culture, suggesting that ARI-1 utilizes E2 as a carbon source. We examined whether ARI-1 could degrade other estrogens, including estron (E1), estriol (E3), and ethynylestradiol (EE2). The experimental conditions used in the E2 degradation assay were adopted. ARI-1 was found to degrade E1 and E3, but not EE2. Although UV absorption at 210 nm was used for monitoring aromatic parts of E2 in the HPLC analysis of the culture, we did not find any aromatic compounds other than E2.
Therefore, a GCMS was used to identify metabolites derived from E2 degradation. ARI-1 was cultivated in E2-YNB, and chloroform and hexane extracts of 0-, 5-, 10-, 15-, 20-, 25-, and 30-day-old cultures were analyzed with a GCMS under the following conditions: column, BPX-5 capillary column (25 m × 0.22 mm i.d., SGE, Melbourne); carrier gas, helium; temperature gradient, 50°C for 3.4 min, 50-360°C at 10°C/min, and 360°C for 10 min. However we could not find any metabolites in extracts of 0- through 30-day-old cultures. To examine whether the aromatic part of E2 is degraded by ARI-1, chloroform extracts of cultures were analyzed by a ^{1}H-NMR. The ^{1}H-NMR signal for the protons of both the aromatic and nonaromatic parts of E2 was found to have disappeared almost completely in culture, which was consistent with those obtained from GCMS. Therefore it was strongly suggested that E2 was degraded by ARI-1 to compounds with very low molecular mass (for instance, CO_2) or simple organic acids which may not be extractable with the conditions used.

The strain ARI-1 was studied phenotypically, genotypically and phylogenetically [33]. Analysis of the 16S ribosomal sequence, the G+C content of the DNA, the isoprenoid quinone composition, and the whole-cell fatty acid profile revealed that the strain is a member of the genus *Novosphingobium*. However, 16S rDNA similarity with that of known *Novosphingobium* spp. was found to be at most 97% (Figure 4d), and the results of DNA–DNA hybridization experiments clearly indicated that the isolate represents a new *Novosphingobium* species, for which the name *Novosphingobium tardaugens* was proposed.

Benzophenone-Degrading Microbe

Benzophenone (BPH, Figure 5a) is a well-known, important intermediate in the production of medicines and cosmetics.

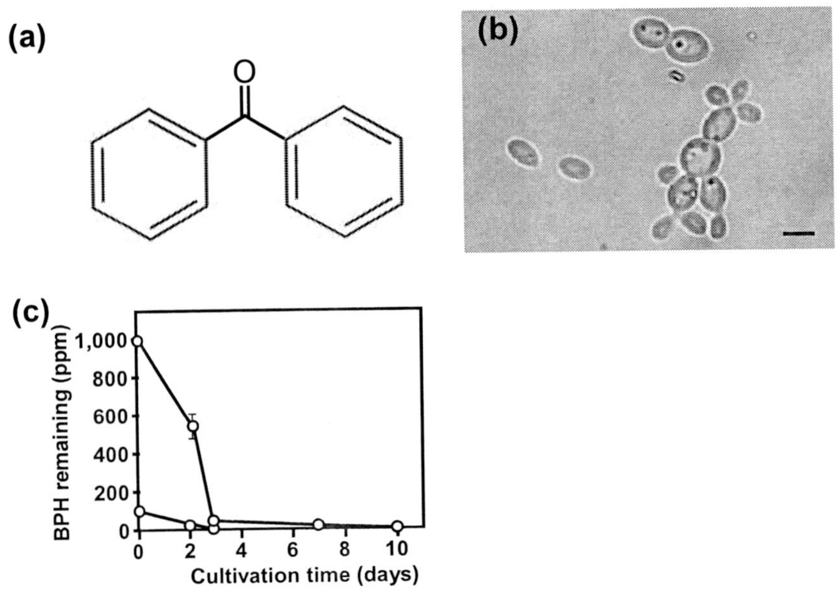

Figure 5. Benzophenone (BPH) and BPH-degrading strain isolated in this study. (a) Molecular structure of BPH. (b) An optical micrograph of the BPH-degrading strain MU-1. Bar, 10 μm. (c) Typical time courses for the BPH degradation by the strain MU-1.

However, BPH is also known as a ubiquitous pollutant in the urban aquatic environment, found in both sediment and surface water [34-36]. Only partial removal of BPH has been observed in a municipal wastewater treatment process [35], and it was recently reported that BPH has a certain endocrine disrupting activity in mammals [37-39]. However, BPH is characterized by recalcitrance, and there has been no previous report of the microbial degradation of BPH or isolation of BPH-degrading microbes. These prospects led us to search for BPH-degrading microbes.

A GCMS system was employed for monitoring BPH degradation and studying the degradation metabolites. Mixed culture or an isolated BPH-degrading microbe was cultivated in YNB containing 1,000 ppm BPH (BPH-YNB medium), and hexane and chloroform extracts were analyzed under same conditions used in the E2 metabolite analysis. As a result of cultivation of 10 environmental samples, including 3 activated sludge and 7 soil samples, significant BPH-degrading activity was observed in the culture of one activated sludge sample obtained from a STP. In this sample, approximately 90% of 1,000 ppm BPH was degraded through 30-day cultivation. After five times subcultures, an aliquot of the culture was streaked onto a YPD agar and incubated at 25 °C to isolate the BPH-degrading microbes. Gray colony was isolated and named the strain MU-1 (Figure 5b).

The BPH-degrading activity of MU-1 pure culture toward various concentrations of BPH was studied. A single colony of MU-1 from the agar plate was inoculated into fresh BPH-YNB medium and cultivated at 25°C while being rotated at 150 rpm. The pure culture of

MU-1 degraded 100 ppm of BPH completely within 3 days, and even 1000 ppm of BPH within 1 week (Figure 5c). Subsequently, hexane and chloroform extracts of the culture were analyzed by a GCMS. Although no peaks of metabolites were detected in the hexane extract, one peak of a metabolite was found in the chloroform extract. The metabolite was identified as phenol (C_6H_5-OH). The appearance of the phenol peak was transient, and it was decreased by further cultivation, suggesting that BPH was degraded to hydrophilic compounds with low molecular mass via conversion to phenol. Whether MU-1 could degrade other aromatic compounds (benzene, toluene, benzenesulphonate, benzoic acid, styrene, biphenyl, and diphenyl ether) was also examined, but MU-1 could not degrade any of those compounds. Therefore it was suggested that MU-1 has the BPH-degrading activity, but does not have a wide range of aromatic compound-degrading activity.

The strain MU-1 was studied phenotypically, genotypically and phylogenetically [40]. The cell size and shape of MU-1 suggested that the strain was not a bacterium but a yeast or a fungus. Hyphae and pseudohyphae with blastspores were observed by microscopic analysis. 18S ribosomal analysis suggested the nearest candidate for MU-1 was *Rhinocladiella aquaspersa* with 99% sequence similarity. *Rhinocladiella* spp. are ascomycetes known as black yeasts that form gray to black colonies on the agar plate, the appearance of which is consistent with the colonial morphology and microscopic observation of MU-1. *Rinocladiella aquaspersa* is known to inhabit the general environment, but there have been no reports that any strains have shown BPH degradation ability, implying that MU-1 is a novel strain of *R. aquaspersa*. A more detailed phylogenic analysis should be carried out to ascertain whether MU-1 is in fact a novel strain of *R. aquaspersa*.

Construction of a Wastewater Treatment System Using EDC-Degrading Microbe

The strain S-3 degraded 1,000 ppm of NP, which is 10,000- to 1,000,000-fold higher than levels found in urban aquatic environments, almost completely within 10 days as described above. However, the utility of the strain in the biotreatment of NP-contaminated wastewater has yet to be determined. Therefore, the capacity of S-3 to degrade NP in wastewater to environmentally acceptable concentrations should be evaluated. We studied a potential of S-3 for bioremedial applications by constructing and evaluating a wastewater treatment system using the S-3 colonized carriers [41].

First, NP-degrading activity of the S-3 colonized carriers was evaluated in the flask-scale experiment. The strain was colonized on porous polypropylene carriers (Figure 6a) by incubating the carriers in S-3 culture at 25 °C. After 24 hrs of incubation, the carriers were recovered with a strainer, washed gently with 0.85% NaCl, and were added to YNB or wastewater containing NP at 200 ppb (μg/l) or 1 ppm as the S-3 colonized carriers, followed by cultivation at 25 °C, 120 rpm. NP remaining in each culture (respectively 0, 3, 6, 9 days old) was analyzed by a HPLC. The carriers degraded NP efficiently in the NP-YNB medium, and degraded NP almost completely within several days (Figure 6b). Interestingly, the carriers degraded trace concentrations of NP; in fact, NP concentration in the medium was 3-4 orders of magnitude lower than the concentration of carbon sources generally added to microbiological media. Next, the degrading activity in actual industrial wastewater was assessed. Surprisingly, the S-3 colonized carriers also degraded both 200 ppb and 1 ppm of NP in the wastewater without the addition of nutrients or adjustment of pH (Figure 6b). We also performed a control experiment using carriers not colonized by S-3, in which NP degradation was not observed, indicating that the disappearance of NP was not due to adsorption of NP onto the carriers.

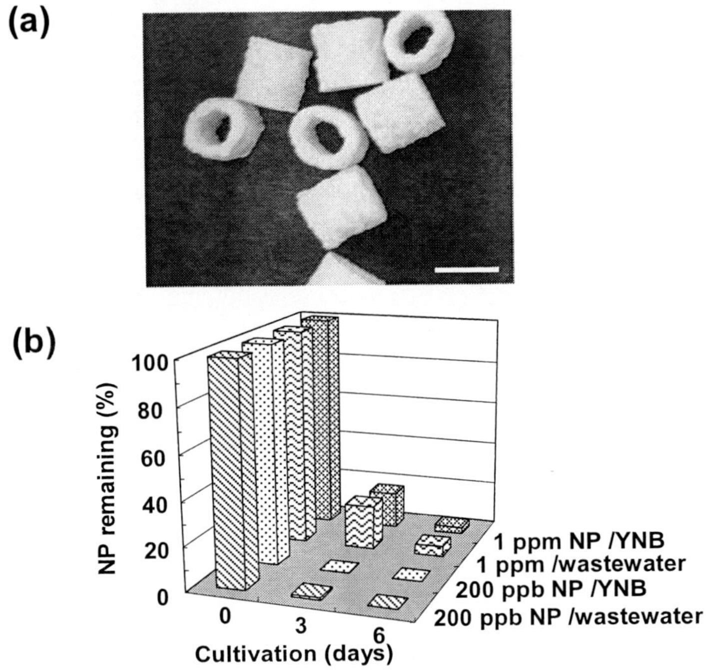

Figure 6. NP degradation by the S-3 colonized carriers. (a) Porous polypropylene carriers used in this study. Bar, 10 mm. (b) Results of the flask scale experiments using the S-3 colonized carriers.

A 3-liter benchtop wastewater-treating bioreactor using the S-3 colonized carriers was constructed (Figure 7a). Three-liter column was filled with NP-containing wastewater and 2-liter bed volume of the S-3 colonized carriers, and the reactor was operated continuously for 1 month. The NP concentration in the treated wastewater was monitored by HPLC. During the 1-month running period, approximately 90% of the 200 ppb NP was removed, without addition of nutrients, pH adjustment, or temperature control of the wastewater (Figure 7b).

We also tested for NP removal using carriers not colonized with strain S-3. During 1 month of operation, 130–160 ppb of NP leaked constantly into the treated wastewater. Therefore the major contributor in successful NP removal is the S-3 colonized carriers, although adsorption of NP onto carriers occurs and seems to help NP biodegradation by S-3 colonizing on the carriers. We also colonized S-3 on hydrophilic polyvinyl alcohol carriers and compared their degrading activity in the bioreactor with that of the polypropylene carriers.

The S-3 colonized polyvinyl alcohol carriers did not degrade NP as efficient as the polypropylene carriers. Since most EDCs, including NP, are hydrophobic, temporal adsorption of NP to a surface of the polypropylene carriers may explain the efficiency of its trapping and degrading capacity. Therefore, carriers made from hydrophobic polymers such as polypropylene are recommended.

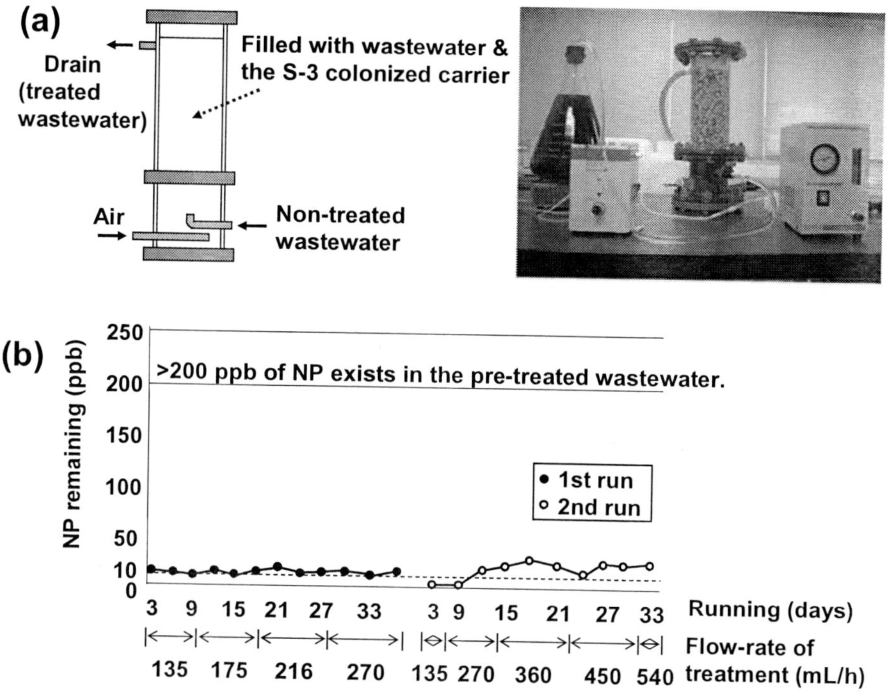

Figure 7. Wastewater treatment using the benchtop wastewater-treating bioreactor. (a) the 3-liter bioreactor constructed in this study. (b) NP removal from wastewater by the reactor.

A 300-liter wastewater-treating bioreactor with the S-3 colonized carriers was constructed and evaluated its NP-removing capacity for NP-containing sewage. Figure 8a shows an outline of the reactor, while Figures 8b, 8c, 8d show its photographs. The bioreactor equipped three air spargers and a waterwheel to supply air and to stir the S-3 colonized carriers in the reactor, respectively.

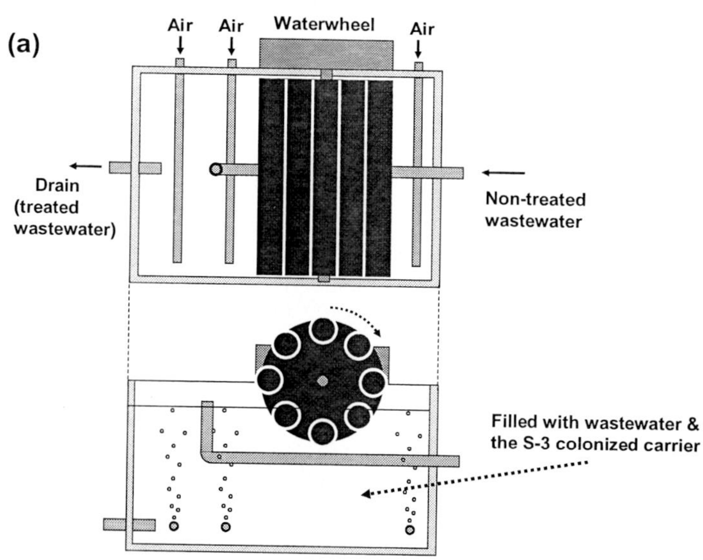

Figure 8. (Continued).

Figure 8. The 300-liter wastewater-treating bioreactor. (a) Outline of the bioreactor. (b) The whole view of the bioreactor. (c) A waterwheel equipped in the bioreactor. (d) The S-3 colonized carriers in the bioreactor.

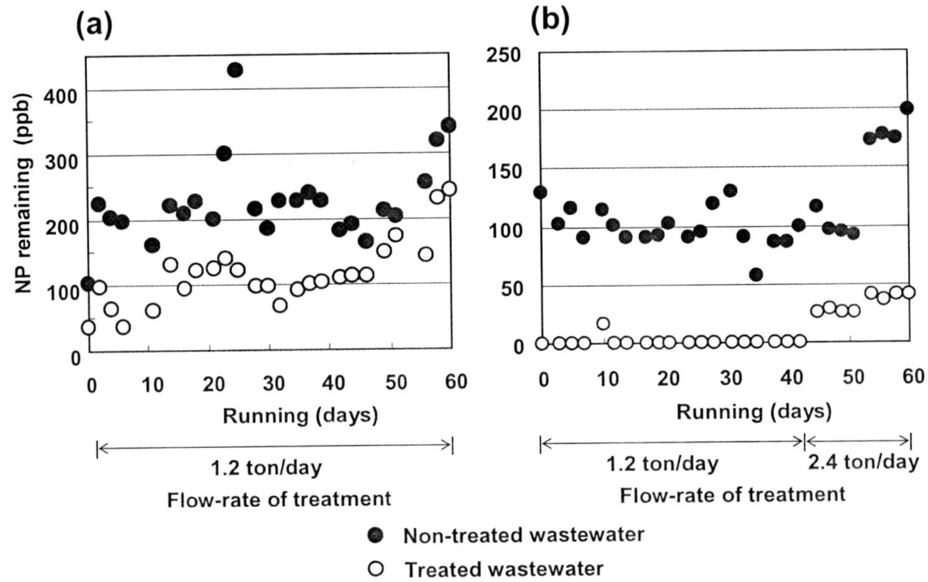

Figure 9. NP removal from sewage by the 300-liter bioreactor (a) and the 600-liter system in which two 300-liter bioreactor were serially connected (b).

Both 100 liter of nutrient broth and 200 liter bed volume of the porous polypropylene carriers were boiled in 300-liter drum for 20 min, poured in the reactor, cooled to room temperature. Then 3 liter of S-3 preculture was inoculated to the reactor and cultivated at room temperature. After 24 hrs cultivation, supernatant of the culture was drained completely and NP-containing sewage was irrigated for treatment. During the 2-month running period, the reactor removed approximately 40-80% of NP at a stable flow rate of 1.2 ton/day, and 40-140 ppb of NP still remains in treated wastewater (0-40 days in Figure 9a). Moreover, NP-removing capacity declined in the end of the running period (40-60 days in Figure 9a). Since it was reported that concentrations of more than 10 ppb NP appear to cause feminization in riverine fish [42], a 600-liter system was next constructed by connecting two 300-liter reactors serially, and evaluated its NP-removing capacity (Figure 9b). As a result of the improvement, NP was degraded more efficiently and the NP concentration in treated wastewater was reduced to less than 10 ppb at a stable flow rate of 1.2 ton/day (0-40 days in Figure 9b). When the flow rate was increased to 2.4 ton/day, however, NP concentration in treated wastewater was 30-50 ppb (40-60 days in Figure 9b), suggesting that a further improvement of the system is required for application to a large scale sewage treatment.

CONCLUSION

In this study, EDC-degrading microbes, a NP-degrading bacterium, an E2-degrading bacterium, and a BPH-degrading yeast, were isolated from STPs and characterized. The wastewater-treating bioreactor using NP-degrading bacterium worked well for removing NP in actual wastewater even at the concentrations found in the environment. Since addition of nutrients, pH adjustment, or temperature control of wastewater was not required during running, a wastewater treatment using EDC-degrading microbes can contribute to preserve aquatic environments. Since the microbes were isolated from STPs in this study, they appear to inhabit in an activated sludge. However they are not a dominant species, because EDCs are generally not biodegraded efficiently in STPs. Van Limbergen et al pointed out that a certain cell density of EDC-degrading microbes in a wastewater treatment system is the requisite to accomplish efficient EDC removal from wastewater [43]. Colonization of EDC-degrading microbes on carriers seems useful to keep a high cell density of EDC-degrading microbes as demonstrated in this study.

While a microbial community of activated sludge is recently analyzed using culture-independent methods, enrichment and isolation works, classical and simple culture-dependent methods, still play an important part to study biochemical and physiological characters and examine an application potential of microbes. The isolates S-3 and ARI-1 were the genera *Sphingomonas* and *Novosphingobium*, respectively. Recently, study of the genus *Sphingomonas* and *Novosphingobium* have become one of the most competitive research fields, because various pollutant-degrading strains have been reported. For example, *S. paucimobilis* and *S. yanoikuyae* are known to assimilate naphthalene and biphenyl [44-48]. *Sphingomonas chlorophenolica*, *Sphingomonas subarctica* and *Sphingomonas* sp. RA2 are able to assimilate chlorinated phenols [49-53]. *Sphingomonas herbicidovorans* has been found to utilize herbicides as carbon sources [54-56]. *Sphingomonas xenophaga* and *Sphingomonas* sp. HV3 have been reported to assimilate several aromatic and chloroaromatic compounds [57-59]. As for *Novosphingobium*, *N. aromaticivorans*, *N. stygium* and *N. subterraneum* were reported to degrade benzoate, cresol, naphthalene or xylene [60] and *N.* subarcticum can degrade chlorophenol [51]. Moreover, it was recently reported that both genera were found as ubiquitous microbes in activated sludge [61-63]. Considering the fact that more than 99% of microbes in the activated sludge have not been isolated yet, an employment of suitable culture conditions (e.g. medium compositions, pH, temperature, aeration, inocula) can bring a finding of novel pollutant-degrading microbes in an activated sludge.

References

[1] Cooke, W.B., and Pipes, W.O. (1970) The occurrence of fungi in activated sludge. *Mycopathologia* 40, 249-270.

[2] Sudo, R., and Aiba, S. (1984) Role and function of protozoa in the biological treatment of polluted waters. *Adv. Biochem. Eng/Biotechnol* 29, 117-141.

[3] Chu, C.P., and Lee, D.J. (2004) Multiscale structures of biological flocs. *Chem. Eng. Sci.* 59, 1875-1883.

[4] Snaidr, J., Amman, R.,Huber, I., Ludwig, W. and Schleifer, H. (1997) Phylogenetic analysis and in situ identification of bacteria in activated sludge. *Appl. Environ. Microbiol.* 63, 2884–2896.

[5] Onuki, M., Satoh, H., Mino, T. and Matsuo, T. (2000) Application of molecular methods to microbial community analysis of activated sludge. *Water Sci. Technol.* 42, 17–22.

[6] Wagner, M., and Loy, A. (2002) Bacterial community composition and function in sewage treatment systems. *Curr. Opin. Biotechnol.* 13, 218–227.

[7] Hiraishi, A. (1988) Respiratory quinine profile as tools for identifying bacterial populations in activated sludge. *J. Gen. Appl. Microbiol.* 34, 39-56.

[8] Hiraishi, A., Ueda, Y., and Ishihara, J. (1998) Quinone profiling of bacterial communities in natural and synthetic sewage activated sludge for enhanced phosphate removal. *Appl. Environ. Microbiol.* 64, 992–998.

[9] Amann, R.I., Ludwig, W., Schleifer, K.H. (1995) Phylogenetic identification and in situ detection of individual microbial cells without cultivation. *Microbiol. Rev.* 59, 143-169.

[10] Janssen, P.H., Yates, P.S., Grinton, B.E., Taylor, P.M., Sait, M. (2002) Improved culturability of soil bacteria and isolation in pure culture of novel members of the divisions *Acidobacteria*, *Actinobacteria*, *Proteobacteria*, and *Verrucomicrobia*. *Appl. Environ. Microbiol.* 68, 2391-2396.

[11] Madigan, M.T., Martinko, J.M., Parker, J. (2003) Methods in microbial ecology. In: Madigan, M.T. (ed.) *Brock Biology of Microorganisms*. Pearson Education, New Jersey. pp. 618-619.

[12] Harries, J.E., Sheahan, D.A., Jobling, S., Matthiessen, P., Neall, P., Routledge, E.J., Rycroft, R., Sumpter, J.P., and Tylor, T. (1996) A survey of estrogenic activity in United Kingdom island waters. *Environ. Toxicol Chem.* 15, 1993-2002.

[13] Harries, J. E., Sheahan, D.A., Jobling, S., Matthiessen, P., Neall, P., Sumpter, J.P., Tylor, T., and Zaman, N. (1997) Estrogenic activity in five United Kingdom rivers detected by measurement of vitellogenesis in caged male trout. *Environ. Toxicol. Chem.* 16, 534-542.

[14] Jobling, S., M. Noran, C. R. Tyler, G. Brighty, and J. P. Sumpter. 1998. Widespread sexual disruption in wild fish. *Environ. Sci. Technol.* 32:2498–2506.

[15] Sole, M., Lopez de Alda, M.J., Castillo, M., Porte, C., Ladegaard-Pedersen, K., and Barcelo, D. (2000) Estrogenicity determination in sewage treatment plans and surface waters from the Catalonian area (NE Spain). *Environ. Sci. Technol.* 34, 5076-5083.

[16] Prakasam T.B.S., and Dondero, N.C. (1970) Aerobic heterotrophic bacterial populations of sewage and activated sludge IV. Adaptation of activated sludge to utilization of aromatic compounds. *Appl. Microbiol.* 19, 663-670.

[17] Watanabe, K., Teramoto, M., Futamata, H., and Harayama, S. (1998) Molecular Detection, Isolation, and Physiological Characterization of Functionally Dominant Phenol-Degrading Bacteria in Activated Sludge. *Appl. Environ. Microbiol.* 64, 4396-4402.

[18] Fujii, K., Urano, N., Kimura, S., Nomura, Y., and Karube, I. (2000) Microbial degradation of nonylphenol in some aquatic environments. *Fisheries Sci* 66, 44-48.

[19] Fujii, K., Urano, N., Ushio, H., Satomi, M., Iida, H., Ushio-Sata, and N., Kimura, S. (2000) Profile of a nonylphenol-degrading microflora and its potential for bioremedial applications. *J. Biochem.* 128, 909-916.

[20] Fujii, K., Urano, N., Ushio, H., Satomi, M., and Kimura, S. (2001) *Sphingomonas cloacae* sp. nov., a nonylphenol-degrading bacterium isolated from wastewater of a sewage-treatment plant in Tokyo. *Int. J. Syst. Evol. Microbiol.* 51, 603-610.

[21] Shore, L.S., Gurevits, M., and Shemesh, M. (1993) Estrogens as an environmental pollutant. *Bull. Environ. Contam. Toxicol.* 51, 361-366.

[22] Folmar, L.C., Denslow, N.D., Rao, V., Chow, M., Crain, D.A., Enbrom, J., Marcino, J., and Guillette, L.J. (1996) Vitellogenin induction and reduced serum testosterone concentrations in feral male carp (*Cyprinus carpio*) captured near major metropolitan sewage treatment plant. *Environ. Health Perspect* 104, 1096-1101.

[23] Routledge, E.J., Sheahan, D., Desbrow, C., Brighty, G.C., Waldock, M., and Sumpter, J.P. (1998) Identification of estrogenic chemicals in STW effluent.2. In vivo responses in trout and roach. *Environ. Sci. Technol.* 32:1559–1565.

[24] Metcalfe, C.D., Metcalfe, T.L., Kiparissis, Y., Koenig, B.G., Khan, C., Hughes, R.J., Croley, T.R., March, R.E., and Potter, T. (2001) Estrogenic potency of chemicals detected in sewage treatment plant effluents as determined by in vivo assays with Japanese Medaka (*Oryzias latipes*). *Environ. Toxicol. Chem.* 20, 297-308.

[25] Williams, R. J., Jürgens, M.D. and Johnson, A.C. (1999) Initial predictions of the concentrations and distribution of 17□-oestradiol, oestrone and ethinyl oestradiol in 3 English rivers. *Water Res.* 33, 1663-1671.

[26] Baronti, C., Curini, R., D'Ascenzo, G., Di Corcia, A., Gentili, A., and Sampeli, R. (2000) Monitoring natural and synthetic estrogens at activated sludge sewage treatment plants and in a receiving river water. *Environ. Sci. Technol* 34, 5059-5066.

[27] Ishii, Y., Okita, S., Torigai, M., and Yun, S. (2000) Determination of estrogens in environmental water samples by LC/MS/MS. *Bunseki Kagaku* 49, 753-758. (*In Japanese*)

[28] Matsui, S., Takigami, H., Matsuda, T., Taniguchi, N., Adachi, J., Kawami, H., and Shimizu, Y. (2000) Estrogen and estrogen mimics contamination in water and the role of sewage treatment. *Water Sci. Technol.* 42, 173-179.

[29] Pentreath, R.J. (1997) The identification and assessment of oestrogenic substances in sewage treatment works effluents, p. 39. Environment Agency, London, United Kingdom.R. J.

[30] Routledge, E.J., Waldock, M., and Sumpter, J.P. (1999) Response to comment on "Identification of estrogenic chemicals in STW effluent. 1. Chemical fractionation and in vitro biological screening." *Environ. Sci. Technol.* 33, 371.

[31] Routledge, E.J., Sheahan, D., Desbrow, C., Brighty, G.C., Waldock, M., and Sumpter, J.P. (1998) Identification of Estrogenic Chemicals in STW Effluent. 2. In Vivo Responses in Trout and Roach. *Environ. Sci. Technol.*, 32, 1559 -1565.

[32] Fujii, K., Kikuchi, S., Satomi, M., Ushio-Sata, N., and Morita, N. (2002) Degradation of 17□-estradiol by a gram-negative bacterium isolated from activated sludge in a sewage treatment plant in Tokyo, Japan. *Appl. Environ. Microbiol.* 68, 2057-2060.

[33] Fujii, K., Satomi, M., Morita, N., Motomura, T., Tanaka, T., and Kikuchi, S. (2003) *Novosphingobium tardaugens* sp. nov., an oestradiol-degrading bacterium isolated from activated sludge of a sewage treatment plant in Tokyo. *Int. J. Syst. Evol. Microbiol.* 53, 47-52.

[34] Benfenatia, E., Di Toro, N., Fanelli, R., Lualdi, Z.G., Tridico, R., Stella, G., Buscaini, P. and Stimilli, L. (1992) Characterization of organic and inorganic pollutants in the Adige river (Italy). *Chemosphere* 25, 1665–1674.

[35] Paxeus, N. and Schroder, H.F. (1996) Screening for non-regulated organic compounds in municipal wastewater in Goteborg, Sweden. *Water Sci. Technol.* 33, 9–15.

[36] The Japanese Ministry of Environment (1999) The environmental disrupter chemicals in aquatic environments of Japan (*In Japanese*). *J. Environ. Chem.* 9, 160–208.

[37] Nakagawa, Y., Suzuki, T., and Tayama, K. (2000) Metabolism and toxicity of benzophenone in isolated rat hepatocytes and estrogenic activity of its metabolites in MCF-7 cells. *Toxicol* 156, 27–36.

[38] Nakagawa, Y. and Tayama, K. (2001) Estrogenic potency of benzophenone and its metabolites in juvenile female rats. *Arch. Toxicol.* 75, 74-79.

[39] Nakagawa, Y. and Tayama, K. (2002) Benzophenone-induced estrogenic potency in ovariectomized rats. *Arch. Toxicol.* 76, 727-731.

[40] Fujii, K., and Kikuchi, S. (2005) Degradation of benzophenone, a potential xenoestrogen, by a yeast isolated from the activated sludge of a sewage treatment plant in Hokkaido. *World J. Microbiol. Biotechnol.* 21, 1311-1315.

[41] Fujii, K., Yamamoto, R., Tanaka, T., Hirakawa, T., and Kikuchi, S. (2003) Potential of a new biotreatment: *Sphingomonas cloacae* S-3^T degrades nonylphenol in industrial wastewater. *J. Ind. Microbiol. Biotechnol.* 30, 531-535.

[42] Jobling, S., Sheahan, D., Osborne, J.A., Matthiessen, P., and Sumpter, J.P. (1996) Inhibition of testicular growth in rainbow trout (*Oncorhynchus mykiss*) exposed to estrogenic alkylphenolic chemicals. *Environ. Toxicol. Chem.* 15, 194-202.

[43] Van Limbergen, H., Top, E.M., and Verstraete, W. (1998) Bioaugmentation in activated sludge: current features and future perspectives. *Appl. Microbiol. Biotechnol.* 50, 16-23.

[44] Gibson, D. T., Roberts, R. L., Wells, M. C. and Kobal, V. M. (1973) Oxidation of biphenyl by a *Beijerinckia species*. *Biochem. Biophys. Res. Commun.* 50, 211-219.

[45] Furukawa, K., Simon, J. R., and Chakrabarty, A. M. (1983) Common induction and regulation of biphenyl, xylene}toluene, and salicylate catabolism in *Pseudomonas paucimobilis*. *J. Bacteriol.* 154, 1356-1362.

[46] Yabuuchi, E., Yano, I., Oyaizu, H., Hashimoto, Y., Ezaki, T. and Yamamoto, H. (1990) Proposals of *Sphingomonas paucimobilis* gen. nov. and comb. nov., *Sphingomonas parapaucimobilis* sp. nov., *Sphingomonas yanoikuyae* sp. nov., *Sphingomonas adhaesiva* sp. nov., *Sphingomonas capsulata* comb. nov., and two genospecies of the genus *Sphingomonas*. *Microbiol Immunol* 34, 99-119.

[47] Kuhm, A. E., Stolz, A. and Knackmuss, H. J. (1991) Metabolism of naphthalene by the biphenyl-degrading bacterium *Pseudomonas paucimobilis* Q1. Biodegradation 2, 115-120.

[48] Khan, A. A., Wang, R. F., Cao, W. W., Franklin, W. and Cerniglia, C. E. (1996) Reclassification of polycyclic aromatic hydrocarbon-metabolizing bacterium, *Beijerinckia* sp. strain B1, as *Sphingomnas yanoikuyae* by fatty acid analysis, protein pattern analysis, DNA-DNA hybridization, and 16S ribosomal DNA sequencing. *Int. J. Syst. Bacteriol.* 46, 466-469.

[49] Radehaus, P. M. and Schmidt, S. K. (1992) Characterization of a novel *Pseudomonas* sp. that mineralizes high concentrations of pentachlorophenol. *Appl. Environ. Microbiol.* 58, 2879-2885.

[50] Karlson, U., Rojo, F., van Elsas, J. D. and Moore, E. (1995) Genetic and serological evidence for the recognition of four pentachlorophenol-degrading bacterial strains as a species of the genus *Sphingomonas. Syst. Appl. Microbiol.* 18, 539-548.

[51] Nohynek, L. J., Nurmiaho-Lassila, E. L., Suhonen, E. L., Busse, H.-J., Mohammadi, M., Hantula, J., Rainey, F. and Salkinoja-Salonen, M. S. (1996) Description of chlorophenol-degrading Pseudomonas sp. strains KF1T, KF3, and NKF1 as a new species of the genus *Sphingomonas*, *Sphingomonas subarctica* sp. nov. *Int J Syst Bacteriol* 46, 1042-1055.

[52] McCarthy, D. L., Claude, A. A. and Copley, S. D. (1997) In vivo levels of chlorinated hydroquinones in a pentachlorophenol-degrading bacterium. *Appl. Environ. Microbiol.* 63, 1883-1888.

[53] Ohtsubo, Y., Miyauchi, K., Kanda, K., Hatta, T., Kiyohara, H., Senda, T., Nagata, Y., Mitsui, Y. and Takagi,M. (1999) PcpA, which is involved in the degradation of pentachlorophenol in *Sphingomonas chlorophenolica* ATCC39723, is a novel type of ring cleavage dioxygenase. *FEBS Lett* 459, 395-398.

[54] Horvath, M., Ditzelmuller, G., Loidl, M. and Streichsbier, F. (1990) Isolation and characterization of a 2-(2,4-dichlorophenoxy)propionic acid-degrading soil bacterium. *Appl. Microbiol. Biotechnol.* 33, 213-216.

[55] Zipper, C., Nickel, K., Angst, W. and Kohler, H. P. E. (1996) Complete microbial degradation of both enantiomers of chiral herbicide mecoprop [(RS)-2-(4-chloro-2-methylphenoxy)propionic acid] in an enantioselective manner by *Sphingomonas herbicidovorans* sp. nov. *Appl. Environ. Microbiol.* 62, 4318-4322.

[56] Zipper, C., Bunk, M., Zehnder, A. J. B. and Kohler, H. P. E. (1998) Enantioselective uptake and degradation of the chiral herbicide dichlorprop [(RS)-2-(2,4-dichlorophenoxy)propanoic acid] by *Sphingomonas herbicidovorans* MH. *J. Bacteriol.* 180, 3368-3374.

[57] Kilpi, S., Backstrom, V. and Korhola, M. (1980) Degradation of 2-methyl-4-chlorophenoxyacetic acid (MCPA), 2,4-dichlorophenoxyacetic acid (2,4-D), benzoic acid and salicylic acid by *Pseudomonas* sp. HV3. *FEMS Microbiol. Lett* 8, 177-182.

[58] Yrjala, K., Suomalainen, S., Suhonen, E. L., Kilpi, S., Paulin, L. and Romantschuk, M. (1998) Characterization and reclassification of an aromatic- and chloroaromatic-degrading *Pseudomonas* sp., strain HV3, as *Sphingomonas* sp. HV3. *Int. J. Syst. Bacteriol* 48, 1057-1062.

[59] Stolz, A., Schmidt-Maag, C., Denner, E. B. M., Busse, H.-J., Egli, T. and Kampfer, P. (2000) Description of *Sphingomonas xenophaga* sp. nov. for strains BN6T and N,N which degrade xenobiotic aromatic compounds. *Int. J. Syst. Evol. Microbiol.* 50, 35-41.

[60] Balkwill, D. L., Drake, G. R., Reeves, R. H. and 7 other authors (1997) Taxonomic study of aromatic-degrading bacteria from deep terrestrial-subsurface sediments and

description of *Sphingomonas aromaticivorans* sp. nov., *Sphingomonas subterranea* sp nov., and *Sphingomonas stygia* sp. nov. *Int J Syst Bacteriol* 47, 191–201.

[61] Snaidr, J., Amann, R., Huber, I., Ludwig, W., and Schleifer, K. (1997) Phylogenetic analysis and in situ identification of bacteria in activated sludge. *Appl. Environ. Microbiol.* 63, 2884-2896.

[62] Kämpfer, P., Erhart, R., Beimfohr, C., Böhringer, J., Wagner, M., Amann, R. (1996) Characterization of bacterial communities from activated sludge: Culture-dependent numerical identification versus in situ identification using group- and genus-specific rRNA-targeted oligonucleotide probes. *Micobial. Ecol.* 32, 101-121.

[63] Neef, A., Witzenberger, R. and Kämpfer, R. (1999) Detection of sphingomonads and in situ identification in activated sludge using 16S rRNA-targeted oligonucleotide probes. *J. Ind. Microbiol. Biotechnol.* 23, 261-267.

In: Sewage Treatment: Uses, Processes and Impact
Editor: Anna Stephens and Mark Fuller
ISBN 978-1-60692-959-9

Chapter 6

THERMAL CONVERSION OF SEWAGE SLUDGE BY PYROLYSIS

A. Moran and M. E. Sanchez*
Natural Resources Institute. Chemical Engineering Group.
University of Leon, Avda. de Portugal, 41, 24071

ABSTRACT

First of all a thermogravimetric (TG) and mass spectrometric (MS) study was carried out of the process of devolatilization of sewage sludge. This involved an analysis of the heating profiles in an inert atmosphere for this residue. The study of mass spectrometry shows major emissions of energy-producing gases although fewer pollutants are emitted. It contents of volatile matter represents a valuable resource, a possible alternative is the conversion into fuels by means of pyrolysis that yields gas, bio-oil and biochar. The composition and heating value of the gas makes it suitable for use as a fuel. The bio-oil may be evaluated as an environmentally-friendly green biofuel. The biochar product is carbon-rich and a potential solid biofuel. Other ways it might be used as a C and N source in soil amendment. The process of thermolysis of sewage sludge in a pilot plant is also described and the results are compared with the others obtained under laboratory conditions (semi-pilot scale). The aim of this study was to get the energetic valorisation of the products; gas and carbonized solid, no liquid fraction was obtained due to the specific characteristics of the plant. The combustion gases were subjected to a cleaning process. Finally an energy production system based on the Rankine cycle is proposed.

INTRODUCTION

Sewage sludge, a residue generated in wastewater treatment plants during the primary, secondary and the tertiary treatment, is one of the environmental problems in industrialized countries [1]. The increase in the production of this inevitable by-product of the treatment of

* Tel. +34 987291841, Fax +34 987 291839; E-mail: amoran@unileon.es

wastewater is causing more and more concern owing to the risks to health and the environment. The quantity and nature of sludge generated relates to character of the raw wastewater and processing units employed. Production, estimated at around 40-50g of dry matter per person per day, is increasing as more new plants are set up [2]. In recent years, methods formerly used for the disposal of sewage sludge, including landfill, ocean dumping and disposal on agricultural land, have become much less acceptable. There is an urgent need for alternative solutions to the sludge disposal problem [3]. Therefore, if an overall solution to the problem of wastewater is sought, this must entail rational sludge management, so that treatment does not mean simply converting the waste from one form to another [4]. Obviously, in order to treat and dispose of the sludge that is produce in a wastewater plant effectively, it is crucial to know the characteristics of the sludge that will be processed [5]: the control of pH levels, alkalinity and organic acid content is an important parameter in the process of anaerobic digestion. Additionally, the content of heavy metals, pesticides and hydrocarbons needs also to be determined when sewage sludge is to be incinerated or land filled. Lastly, the energy (thermal) content of sludge is significant when thermal processes (combustion, gasification and pyrolysis) are considered. Over the last decade, new treatments that result in less sludge being generated have appeared, such as breaking up certain bacteria and preventing the reproduction of others [6], the destruction of cell walls in secondary sludge, which improves the later processing of the sludge [7] and also mechanical disruption, thermal destruction and pre-treatment with ultrasound [8], although use of these has barely begun in the larger plants. Sludge has traditionally been dealt with in three different ways: application to the soil to recycle nutrients and organic matter, energy production and dumping. New possibilities have also arisen which are as yet only small-scale but which will be feasible in the future, including the production of adsorbents (active carbons) [9], the manufacture of bricks and co-digestion with the organic fraction of municipal solid wastes, although this last case is not really a means of use but a treatment.

The European Union's target to reduce final waste disposal by 20% compared with 2000 by 2010 and by 50% by 2050 [10]. To do this, it has drawn up a strategy setting the following priorities: prevention of waste; waste recovery through, reuse, recycling and energy recovery; improved treatment conditions; regulation of transport.

In this chapter, the main focus is on energy potential, with consideration of the possibilities that the thermochemical processes offer, including combustion, gasification and pyrolysis, specifically the last of the three, because their high volatile-matter content makes these sludges potentially energy-rich. Although currently the growing demand for energy is largely satisfied by fossil fuels, the limited reserves of these fuels along with the environmental concern about green-house gases have led to the search for new energy sources. Sewage sludge and other organic solid wastes such as biomass are examples of renewable energy resources, which offer a sustainable and renewable energy system [11, 12]. These technologies include: combustion, gasification and pyrolysis, offering an alternative trend to the sewage sludge disposal. They remove the organic part of the sludge, leaving only the ash component for final disposal. The main goal of thermal processing is the utilization of the stored energy in sludge and the minimization of environmental impacts. The principal problem concerning thermal processes includes the following [13]: excessive energy necessary to reach high temperatures, high capital costs, need for extensive air pollution equipment.

Combustion and co-combustion are perhaps the most established technologies, being the first one even more predominant. Multiple hearth and fluidized bed furnaces are the most popular and the latter is becoming widely applied [1]. The differences between the two types of furnaces is that multiple hearth furnaces usually burn mechanically dewatered (wet) sludge, while fluidized bed furnaces can burn both wet and semi-dried sludge, with dried matter content in the range of 41-65 wt%. Based on the composition of sewage sludge, combustion may be seen as potential source of various pollutants and care must be taken during its disposal. The following sources of pollutants are important for public health [14]: release of heavy metals; the handling of solid residues (e.g. bed and filter ash); emissions of dioxins and furans, NO_x, N_2O, SO_2, HCl, HF and C_xH_y.

Gasification technology can be applied to convert sewage sludge into usable energy and to reduce the environmental problem. Gasification is the thermal process during which carbonaceous content of sewage sludge is converted to combustible gas and ash in a net reducing atmosphere. The optimum targets of sewage sludge gasification are the production of clean-combustible gas at high efficiency. During the gasification process, the sludge undergoes a complex physical and chemical change, starting with the drying or removal of water contained as moisture. The dried sewage sludge is then pyrolysed or thermally decomposed. In the final step, the pyrolysis products, condensable and non-condensable vapours and char undergo gasification, where they are concurrently oxidized and then reduced to permanent gases at the reduction zone. This results to almost total conversion of all or the organic carbon in sludge to combustible gas and mineral residue [15]. The quality product gas from sewage sludge was estimated in $4 MJm^{-3}$ [16]. The actual process is constituted of several repeated cycles of mixing char residue, from previous gasification, with moist sludge, drying and gasification. The totally process is energetically self-sustaining.

Compared to incineration, gasification, due the fact that it is a net chemically reductive process, can prevent problems from occurring, including the need for supplemental fuel, emissions of sulphur oxides, nitrogen oxides, heavy metal and fly ash and the potential production chlorinated dibenzodioxins and dibenzofurans [17].

Pyrolysis is a thermal process through which, organic substances are decomposed in an oxygen-free atmosphere, at temperatures varying in the range of 300 and 900°C. The reactions that take place are endothermic thermal cracking and condensation reactions. According to Khiari et al. [13] pyrolysis process can be thoroughly studied using TG and MS. They distinguished the process into three stages:

1) Vaporization of volatile materials.
2) Primary decomposition of non-volatile components, resulting to solid char. Apart from the char, tar and gases are also produced.
3) The char may undergo secondary higher pyrolysis. In this stage many o the hydrocarbons and aromatic compounds are found in the final volatile phase.

The pyrolysis process involves the heating of sewage sludge in an inert atmosphere with the consequently release of organic matter and its potential recycling. The major fractions formed are the following:

- A non-condensable gaseous fraction which contains mainly hydrogen, methane, carbon monoxide, carbon dioxide and several other gases in smaller concentrations.

- A liquid fraction consists of a tar and/or a bio-oil is one of the desirable products of pyrolysis. They are usually mixed with condensed water. Their composition is important to evaluate the potential for direct combustion of the bio-oils or for their further processing and utilisation as bio-diesel or chemicals.
- A solid fraction that is carbon-rich and a potential solid biofuel. Other ways it might be used as a C and N source in soil amendment. The long persistence of biochar in soil makes it a prime candidate for the mitigation of climate change as a potential sink for atmospheric carbon dioxide. All organic matter added to soil improves soil functions, not the least the retention of several nutrients that are essential to plant growth. Biochar is more effective in retaining most nutrients and keeping them available to plants and more persistent in soil than other organic matter for example compost or manures.

This technique appears to be less pollutant than conventional methods (incineration, conmbustion), as it concentrates the heavy metals in a solid carbonaceous residue, so its leaching is not that crucial as that in ashes from incineration [18].

In recent years, developments in pyrolysis as new methods of organic waste treatment have led to the building of many pyrolysis plants all over the world. Pyrolysis (thermolysis) is a thermal technique catalogued in Directive 2000/76/CE as an incineration process [19]. It may be considered a special kind of pyrolysis, where no liquid fraction is produced, just gases and a carbonized solid residue. It involves thermal decomposition in the absence of oxygen, taking place in a temperature range from 300 to 900°C.The reactors normally used are rotary ovens [20] or fluidized bed reactors [21]. Rotary oven processes are characterized by high residence times (>20 minutes), while fluidized beds suppose a few seconds, 90 at most.

The solid residue can be used in combustion or co-combustion processes to produce energy. Active carbons may also be obtained by physical activation in air or CO_2, or by chemical activation of the carbon waste with, for example, H_3PO_4, $ZnCl_2$ or H_2SO_4 [9]. This would mean that the waste could be recycled for use in air pollution control or in the elimination of smells [22]. The other option, dumping, has the advantage that the heavy metals present would be fixed in the carbon matrix, with the subsequent reduction in leaching [23], whereas the gas could be used as a fuel to produce energy [24]. Apart from their use as fuels, the oils are a source of aromatic compounds such as benzene, toluene and xylene (BTX), with a higher market price than the oils themselves [25]. The high heating value of these products means that their main use is to be found in combustion to produce energy.

The possible joint management of sewage sludge is studied in this chapter, some experimental works related to the possible use of sludge as an energy resource in a process of pyrolysis was carried out, with trials run in a pilot plant, with the purpose to comment the results comparing pyrolysis of this waste at a laboratory level (semi-pilot scale) and at a pilot scale.

Two main type of pyrolysis are used: fast pyrolysis (Figure 1) where three products are obtained (bio-oil, gas and char) and slow pyrolysis were the liquids are cracked and only gas and char is obtained (Figure 2). In this exposition are shown the results of: fast pyrolysis in the second set of experiments called "Set-up in the Laboratory" and the results of slow pyrolysis in the third set of experiments that are carry out at "pilot plant". Yields of these two scales are compared, along with the composition of the resulting gas and solid fractions.

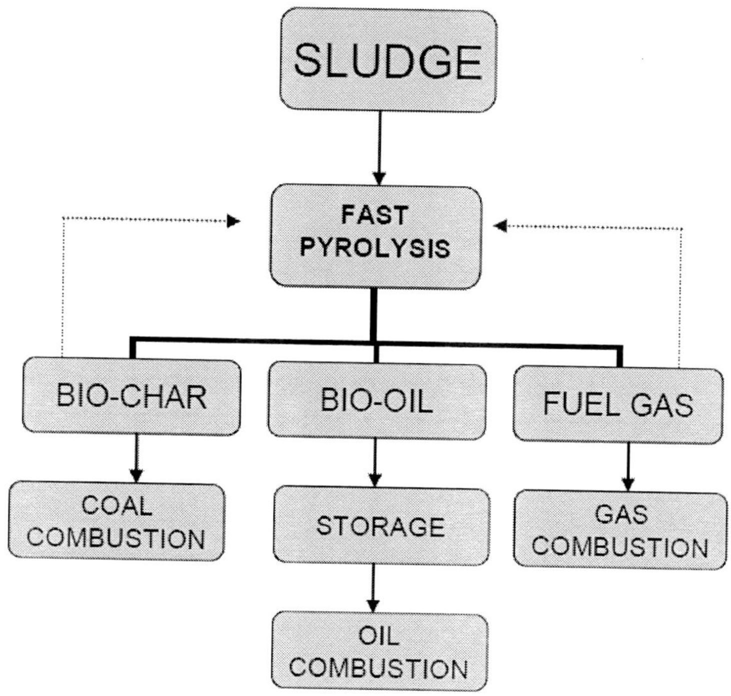

Figure 1. Fast pyrolysis scheme.

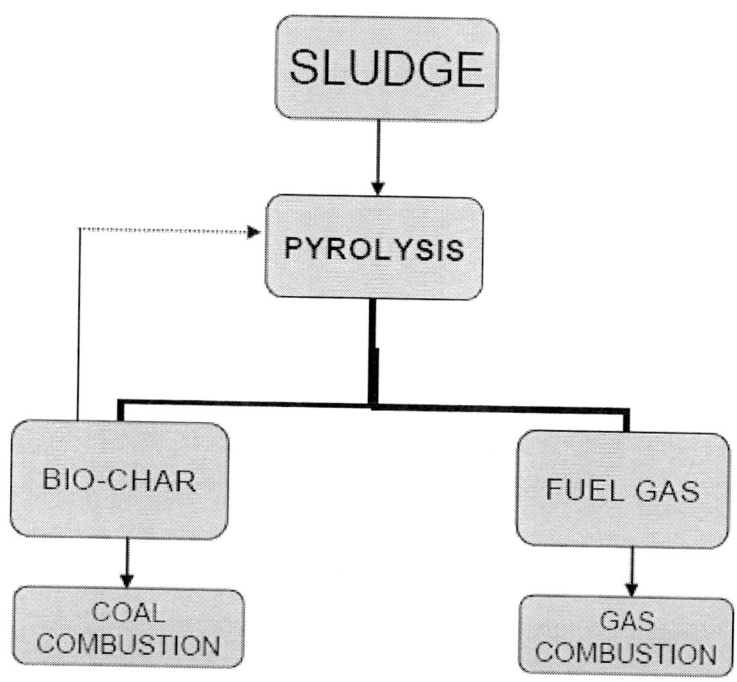

Figure 2. Slow pyrolysis scheme.

The products of the process, gas and carbonized solids, were characterized with a view to their being used as fuels. Thermogravimetric analyses of the carbonized fraction are also carried out in order to study its reactivity. They were burnt both separately and together, with monitoring of gas emissions and the incorporation of a gas cleaning treatment. Finally, emissions from the combustion of carbonized solids and gas yields achieved from the pilot scale pyrolysis process are subjected to a gas clean-up process using sodium bicarbonate as the neutralizing agent.

At last, an energy production scheme is proposed.

2. Experimental Details

2.1. Material

A representative sample of anaerobically stabilised sewage sludge from an urban wastewater treatment plant from León (Spain) was used in this work. It goes through a dehydration and thermal drying in the wastewater treatment plant of origin. The raw material was analysed, taken a representative sample, to determine the main properties affecting emissions. Moisture was determined in a Selecta THREOVEN with forced ventilation. Volatile matter and ash were analysed with a Thermolyne 48000 muffle furnace with automatic temperature control. Spanish standards UNE 32002, 32019 and 32004 were followed for the determination of moisture, volatile matter and ash, respectively. An elemental analysis was also performed to determine the amounts of the main chemical elements present in the sample: carbon, hydrogen, nitrogen, sulphur, chlorine and oxygen, the last by subtraction. For the first three elements, a LECO CHN-600 apparatus following ASTM standard D-5373 was used. Total sulphur was determined with a LECO SC-132 according to ASTM standard D-4239. The Eschka method was used to determine chlorine content according to ASTM D-2361. Calorific value was determined with a LECO AC-300 using the adiabatic method in accordance with the UNE 32006 standard.

2.2. Thermogravimetric and Mass Spectrometry Analysis

2.2.1. Thermogravimetric Analysis

Thermogravimetric analysis was carried out using a TA Instruments SDT2960, which is able to provide a continuous measurement of simple mass as a function of time or of temperature and also give a DTG signal (rate of mass loss) when used with the appropriate software. Thermogravimetric runs were performed for the residue studied; sewage sludge. The initial sample mass was 5 mg, this being distributed uniformly over the base of the sample holder so as to facilitate the transfer of heat and of material. The atmosphere used was helium with a flow rate of 100 mL/min. Prior to the start of heating, a flow of helium was passed through the oven for 10 minutes in order to remove any remaining air. Once this purge period was completed, heating from room temperature up to a final temperature of 800 °C commenced, at a heating rate of 50 °C/min. The experiments were repeated three times in order to evaluate their reproducibility [26].

2.2.2. Mass Spectrometry Analysis

Gases emitted during the thermal conversion process were monitored with a Balzers ThermoStar GSD 300 mass spectrometry apparatus. Analysis was carried out with the mass spectrometer directly linked to the gas off-take of the thermal analysis equipment. The connection between the thermobalance and the mass spectrometer was by means of a capillary kept at a temperature of 200 °C. In this way, the data obtained from the mass spectrometer corresponded to the emissions released by the samples undergoing devolatilization in the thermobalance. Several authors, for instance Hatton [27], have studied the optimal linkage between TA and MS. Such a linkage allows the relative concentrations of gas emitted to be determined under near-equilibrium conditions, although it lengthens the response time for the MS signals. For this reason, thermal and spectrometric analyses in this study were not in precise synchrony [28].

The evolution profiles for m/z ratios of 2, 15, 18, 26, 28, 39, 42, 43, 44, 55, 64 and 91 were monitored. Table 1 shows the assignments of each of these signals to different compounds. These assignments are in accordance with published details in the literature and take into account the ions of the compounds expected from pyrolysis [29]. These gas emissions are of importance from two points of view, their energy value and the environmental damage they may cause [1].

As N_2 has the same molecular mass as CO, helium was used instead of nitrogen for the inert atmosphere, making CO determination possible by mass spectrometry. No appreciable differences were detected in TG profiles [30].

Table 1. Assignments for each signal from the mass spectrometer

m/z	Ion	Mainly due to:
2	H_2^+	Hydrogen
15	CH_3^+	Methane
18	H_2O^+	Water
26	CN^+, $C_2H_2^+$	HCN, ethylene, aromatics
28	CO^+	Carbon monoxide
39	$C_3H_3^+$	Propylene, aromatics
42	CH_2CO^+, $C_3H_6^+$	Aldehyde, propane
43	CH_3CO^+, $C_3H_7^+$	Carboxylic acid, propane
44	CO_2^+	Carbon dioxide
55	$C_4H_7^+$	Butane
64	SO_2^+	Sulphur dioxide
91	$C_7H_7^+$	Aromatics

2.3. Experimental Set-up in the Laboratory

The experimental design used in the laboratory pyrolysis tests is shown in Figure 3. Pyrolysis was carried out in a reactor consisting of a quartz tube 40 cm long with a diameter of 7 cm placed in an electrically heated horizontal oven [31]. A 30 g sample was pyrolized. Helium was fed into the front of the reactor, and, before heating began, it was fed through at a flow rate of 200 mL/min for 60 minutes to remove the air from the system.

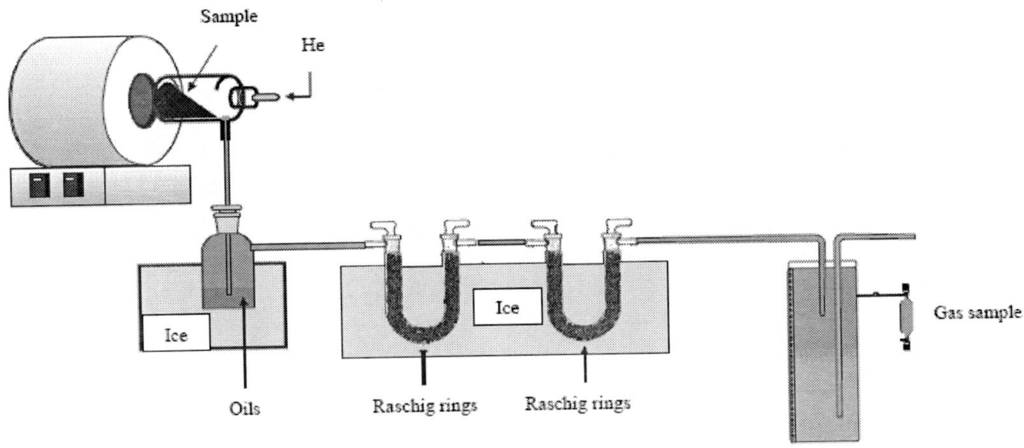

Figure 3. Laboratory Experimental Set-up

Pyrolysis conditions were a maximum heating ramp of 30ºC/min, starting from room temperature to a final temperature of 550ºC and at a He flow rate of 100 mL/min. When the reactor reached 550ºC, this temperature was maintained for approximately 1 minute and then the process was stopped. As for the heating ramp, the maximum allowed by the oven was used, and no tests were made with other gradients, as the bibliographical references showed that within the range 5-30˚C/min no variation in the heating gradient significantly affected product formation [32]. The final temperature was chosen from a previous thermogravimetric study, as the heating gradient was within the range studied [33]. It was observed that the range of temperatures over which devolatilization of the residue occurs was quite large, running from 200ºC to 550ºC, with devolatilization in thermobalance being almost complete. The gases produced in the pyrolysis process were carried by the He through the lower part of the reactor, where they passed through a number of "traps", where the oils (condensable gas fraction) were collected. The traps comprised a balloon and two U-shaped glass tubes packed in ice and filled with Raschig rings. Finally, the gases were fed into a methyl methacrylate container with water inside, which was displaced, and a sample taken for gas analysis. The gas yield was checked by directly measurement of the volume of gas produced in the pyrolysis process. The liquid products collected in the traps were oils with an aqueous fraction. Dichloromethane was added to separate the aqueous and the organic fractions. The aqueous fraction recovered in the tubes was then separated from the organic fraction by decantation and weighed. The organic fraction yield was calculated by difference.

2.3.1. Characterization of Gas

The gas fraction was analysed by GC with an HP-5890 series II with a flame ionization detector (FID) and a thermal conductivity detector (TCD). Different pyrolysis experiments established that the relative values of gas composition do not deviate more than 5%. To analyse hydrocarbons (C_xH_y), an HP-AL/S semi-capillary column with a length of 50 m and an internal diameter of 0.53 mm was used, with He as carrier gas and the FID detector was used. H_2, O_2, N_2, CH_4 and CO were analysed in a 5 Å molecular exclusion packed column of 1/8" outer diameter, 6 feet long, with a mesh of 60/80 and He as carrier gas and the TCD detector was used. CO_2 was analysed in a Chromosob 102 packed column specific for this substance with an outer diameter of 1/8" and a length of 6 feet, with a mesh of 80/100 and He as carrier gas, the TCD also being used. For the identification and quantification of the gaseous species, three SUPELCO gas reference standards were used.

2.3.2. Characterization of Bio-Oil

Elemental analysis of the bio-oil was performed according to ASTM D 5373. The heating value was determined according to ASTM D 3286-91a. These analyses were repeated at least three times for good accuracy, the results differing by ± 0.1 wt %.

The bio-oil was analysed by GC-MS, a Hewlett-Packard HP 6890 gas chromatograph coupled to an HP 5973 mass spectrometer quadrupole detector. Different pyrolysis experiments established that the relative values of bio-oil composition do not deviate more than 5%. A 60m X 0.25mm capillary column coated with a 0.25 μm thick film of 5% phenylmethylpolysiloxane (HP-5) was used and helium was employed as a carrier gas at a constant flow of 0.9 mL/min. The initial oven temperature of 40 °C was held for 5 min and then programmed from 40 to 300 °C at 5 °C/min with an isotherm held for 30 min. The split ratio was 20:1 and the injection was carried out at 300 °C. The ion source and transfer line temperatures were 230 and 325 °C, respectively. Data were collected in the full-scan mode between m/z 25-533 and a solvent delay of 6 min was used. The chromatographic peaks were identified with the help of the NIST mass spectral data library.

2.3.3. Characterization of biochar

In the characterization of the solid materials, biochar, a proximate analysis was run as we have explained before to characterize the original sludge sample. Moisture, volatile matter and ash were analysed according regulantions and fixed carbon was determined by difference. An elemental analysis was run, total sulphur and Heating value were also determined.

These analyses were repeated at least three times for good accuracy, the results differing by ± 0.1 wt %. The Kjeldahl technique was used to determine total nitrogen, and for assimilable fhosphorus the Olsen Watanabe method was followed.

The carbonized solid fraction was subjected to combustion at a programmed temperature and thermogravimetric analysis in an STD 2960 thermobalance. The initial sample mass was 5 mg, this being distributed uniformly over the base of the sample holder so as to facilitate the transfer of heat and of material. The atmosphere used was air with a flow rate of 100 mL min^{-1}. The heating rate was the 50°C min^{-1} from room temperature up to a final temperature of 800°C. TG runs were repeated at least three times, the average relative error being ± 5%.

2.4. Description of the Pilot Plant

Figure 4a is a picture and Figure 4b is a diagram of the pilot plant, which basically includes four phases:

1. Thermolysis Reactor. The waste to be thermolysed was fed into the reactor semi-continuously, through a hopper and a system of two valves designed to ensure conditions completely free of oxygen for each of the partial charges making up the total charge (about 15kg). The electrically heated reactor reached temperatures of between 500 and 600°C, which brought about the thermal decomposition of the wastes in an inert atmosphere (nitrogen) as it was carried through the reactor by an endless screw.

Figure 4a. RMD pilot plant.

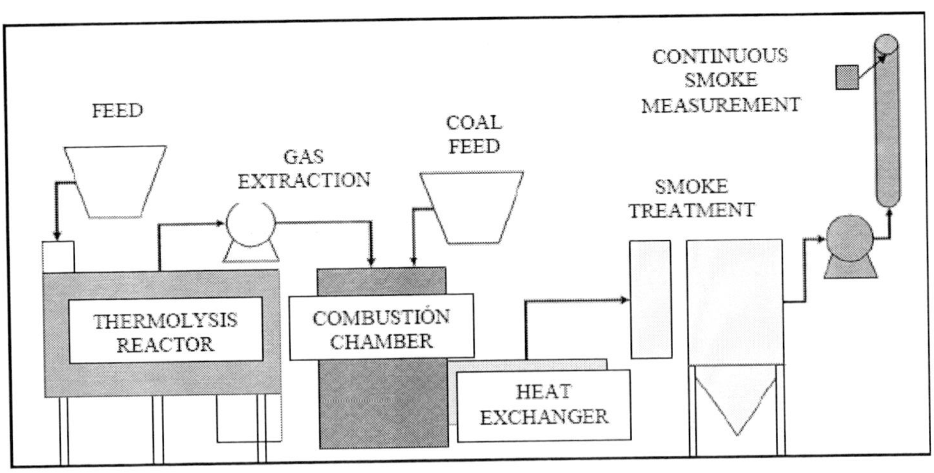

Figure 4b. Diagram of the pilot plant.

The gases produced by the thermolysis were extracted by a vacuum pump and fed into a combustion chamber, which brought about a slight depression in the reactor. Residence time was approximately 20 minutes. When the process was over, the carbonized solids were collected for subsequent discontinuous feeding into the combustion chamber. No liquid products were obtained as the tube taking the gases to the combustion chamber was kept at the same temperature at which they left the reactor, so that they did not condense into pyrolysis oils.

Pyrolysis presents certain advantages; it gives rise to gases and carbonized solid with a high energetic value which could be used as potential fuels. Pyrolysis is carried out at lower temperatures than in the case of incineration, as the process is carried out in the absence of air no dioxins are produced. Wastes direct combustion brings environmental problems due to gaseous pollutant emissions like nitric and sulphur oxides, unburnt organic compounds, dioxins and heavy metals. The gases and carbonized solid are valorised as traditional fuels. The energy obtained in the combustion is recuperated to produce steam and/or electricity.

2. *Combustion Chamber*. Here the gases and carbonized solids obtained by thermolysis were burnt. The gas consisted principally of hydrocarbons with a high calorific value, and hydrogen. The solid fraction was also a fuel of high heating value. The entry of gas was continuous, while the solids entered the chamber in batches. 4% oxygen excess respect stoichiometric is used. To ensure the stability of the flame, propane was used as a support fuel (1Nm^3/h propane). The temperature reached is over 850°C and the residence time over 2 s. The ashes obtained are collected for use in the manufacture of cement.

3. *Heat Exchanger*. A heat exchanger of concentric tubes was used to cool the combustion gases with a stream of air. The gases entering the exchanger flowed through the inner tube at a temperature of over 850°C, while cooling air flowed through the outer tube in the opposite direction, bringing the temperature of the gas down to below 200°C. The heat generated was used for heating the facility.

4. *Gas Cleaning Treatment*. Once cooled, the combustion gases were subjected to cleaning with sodium bicarbonate as a neutralizing agent, fed in through a hopper with an endless screw. Sodium bicarbonate dispersal in the combustion gases subjected to cleaning produced the adsorption of gaseous pollutants. To remove particles, the gases with the sodium bicarbonate dispersal were passed through filter bags.

2.4.1. Materials Used and Trials Performed

The sludge was subjected to a process of heat drying at the treatment plant and supplied in granulated form, suitable for use in the thermolysis plant. Batches of approximately 15 kg were fed into the reactor as partial charges every 2 minutes 40 seconds, approximately. Residence time of the material in the reactor was 20 minutes (while is carried through the reactor by an endless screw).

Two cases were studied: firstly, while thermolysis is effected, the gas produced was simultaneously burnt in the combustion chamber, and secondly, the gas was burnt together with the carbonized solid residue from the first case.

2.4.2. Sample Collection and Techniques of Analysis

As it has been mentioned, the process gave two products: gas and carbonized solid. The gas sample was taken from the tube taking it from the reactor to the combustion chamber by means of a vacuum pump to compensate for the slight depression in the tube. The system used involves two bubblers in an ice bath to cool the gases and attempt to collect a liquid fraction, if it exists, and finally a glass ampoule to collect the gas sample. In none of the trials it was possible to reduce the temperature enough to obtain a sample of oils, as the characteristics of the system implied that the tubes were kept at the same temperature as the reactor (550°C), which, with the long residence time (20 minutes), favoured the production of non-condensable gases. The solid was collected in a container at the bottom of the reactor and, once cooled, was weighed to calculate the yield of the thermolysis process. The percentage of the gas fraction was calculated by difference.

The gas fraction was analysed chromatographically with an HP-5890 series II with a flame ionization detector (FID) and a thermal conductivity detector (TCD).

In the characterization of the solid materials, the sludge and carbonized product, an immediate and an elemental analysis were run. Calorific value was also determined.

The carbonized solid fraction was subjected to combustion at a programmed temperature and thermogravimetric analysis in an STD 2960 thermobalance.

2.4.3. Emission Monitoring

In order to check effectiveness of the gas cleaning treatment, a study was undertaken of the emissions from the combustion of the thermolysis gas and of this together with the solid fraction, before and after gas cleaning treatment. Emissions (CO, TOC, SO_2, NO_x, HCl, and HF) were monitored by means of an infra-red interferometer with Fourier transformation based on Michelson's interferometer in accordance with Directive 2000/76/CE [19].

3. Results and Discussion

3.1. Material

Table 2 summarizes the main chemical characteristics of the sludge. It is carbon and oxygen rich feedstock, containing hydrogen and nitrogen at a low ratio and trace amounts of sulphur. The volatile matter content was 59.2 wt.%, making it a valuable source of organic compounds when it is pyrolysed, while the amount of ashes was 32.4 %. The moisture content was around 6.8 %. And the lower heating value was 15.3 MJ/kg.

Table 2. Chemical analyses (%) and heating values of original sewage sludge

	Sludge
Moisture	6.8
Volatile Matter[a]	59.2
Ash[a]	32.4
Fixed Carbon [a,c]	8.4
Carbon[b]	55.3
Hydrogen[b]	7.8
Nitrogen[b]	9.8
Sulphur[b]	1.4
Oxygen[b,c]	25.7
L.H.V. (MJ Kg^{-1})	15.3

[a] results expressed as a percentage of dry matter.

[b] results expressed as a percentage of dry matter free of ashes.

[c] calculated by difference.

3.2. Thermogravimetric and Mass Spectrometry Analysis

3.2.1. Thermogravimetric Analyse

Figure 5 shows the TG and DTG curves for pyrolysis of sewage sludge, which relate to pyrolysis at a heating rate of 50 °C per minute. On the basis of the mass loss curve for the process of pyrolysis it may be observed that there is a tendency for a percentage of residues of around 40% to remain by 800 °C. Devolatilization should be practically complete by then under the conditions in which the procedure is performed. There is an initial loss of mass, between 90 and 200 °C, corresponds to the driving out of moisture from the sample. It is between 200 °C and 450 °C that devolatilization chiefly takes place. After 450 °C mass is lost more slowly up to 800 °C, the end temperature of the process, due to an organic material more stable and difficult to degrade. As may be observed, the range of temperatures over which devolatilization of the residues of an organic nature occurs is quite large, running from 200 °C up to 550 °C for sewage sludge, probably as a consequence of the heterogeneity of the organic matter present in such wastes.

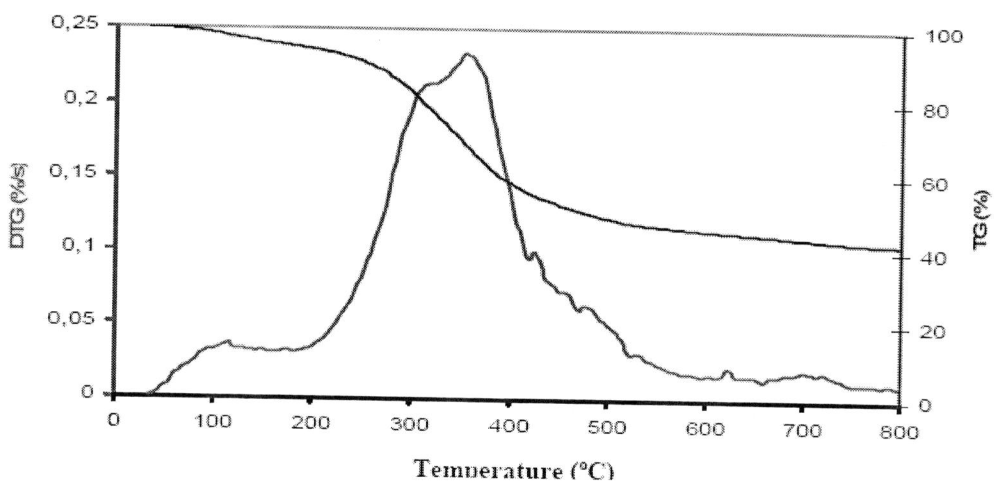

Figure 5. TG and DTG profiles of the sludge pyrolysis process.

3.2.2. Mass Spectrometry Analyse

In the light of results reported in the literature [29], the following mass to charge ratios were chosen for study: 2, 15, 18, 26, 28, 39, 42, 43, 44, 55, 64 and 91. Figures 6-9 show the results for each mass to charge (m/z) relationship.

Data were normalized so that each compound (ion) detected in MS had its own response factor and so that the intensities of the same compound for different samples could be compared [34]. In the case of intensities with different m/z signals, however, only the shape and the characteristic temperatures of the peak can be compared [28].

In general, a single peak is to be seen in the various emission profiles studied. For sewage sludge this peak had shoulders, leading to the conclusion that the process of decomposition takes place over more than one stage. The temperature ranges at which the emission profiles lie run from 300 °C up to 600 °C for hydrocarbons (m/z =26, 39, 42, 43, 55), with a maximum at around 500 °C. Emission of methane (m/z =15) begins at some point between 400 °C and 450 °C and increases up to 550 °C, whereupon it tends to decrease.

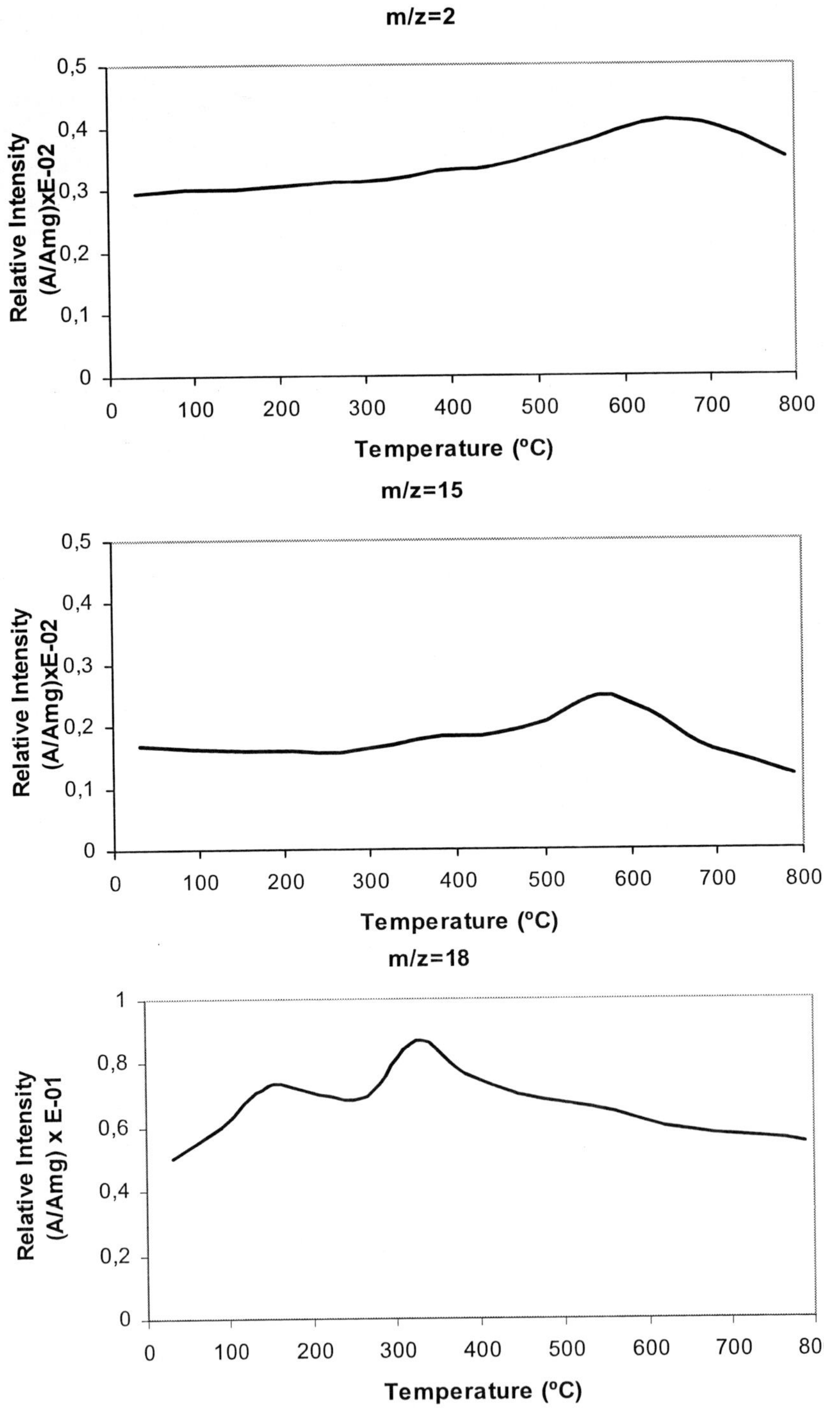

Figure 6. Evolution profiles of the sewage sludge pyrolysis products at 50°C/min, m/z = 2, 15 and 18.

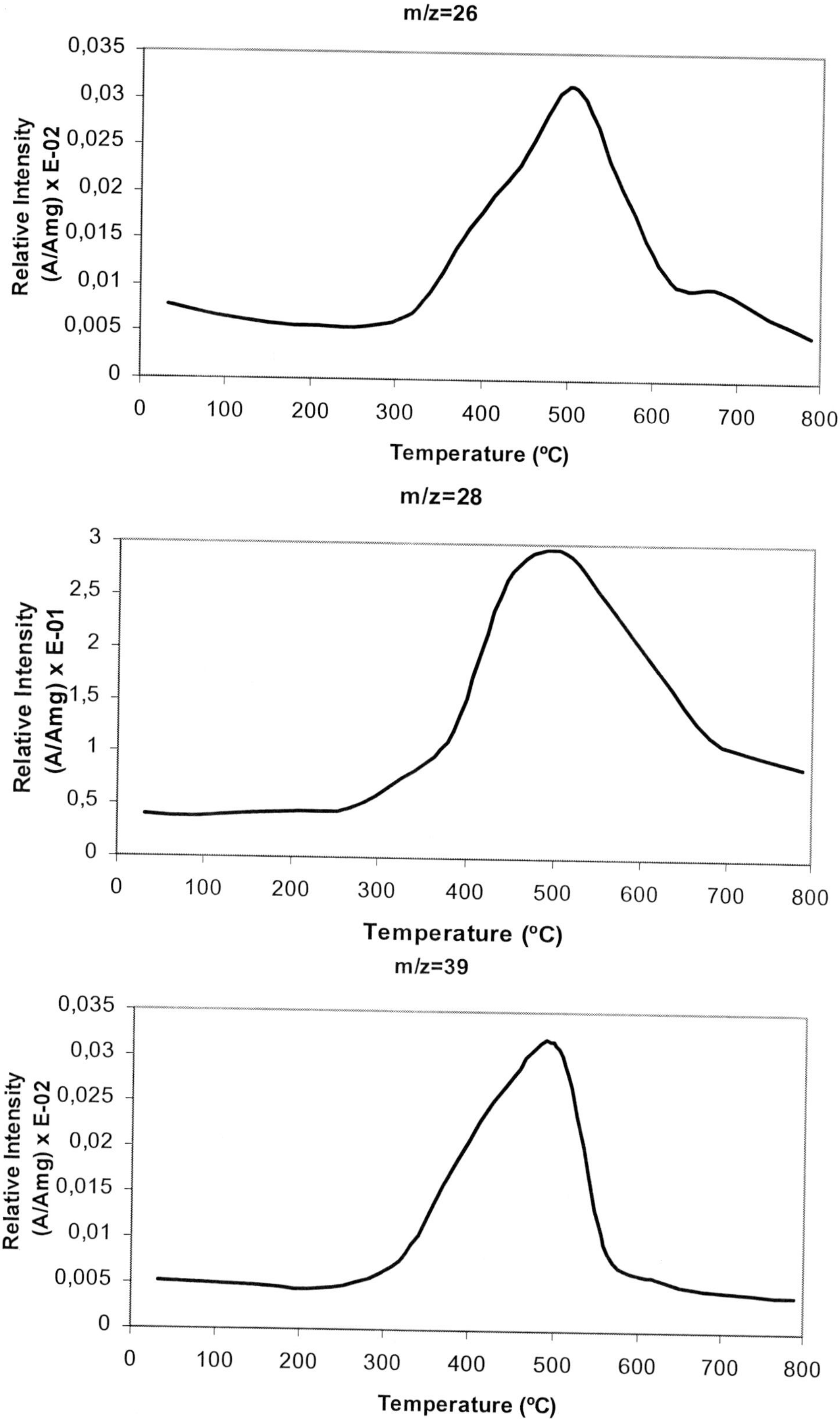

Figure 7. Evolution profiles of the sewage sludge pyrolysis products at 50°C/min, m/z = 26, 28 and 39.

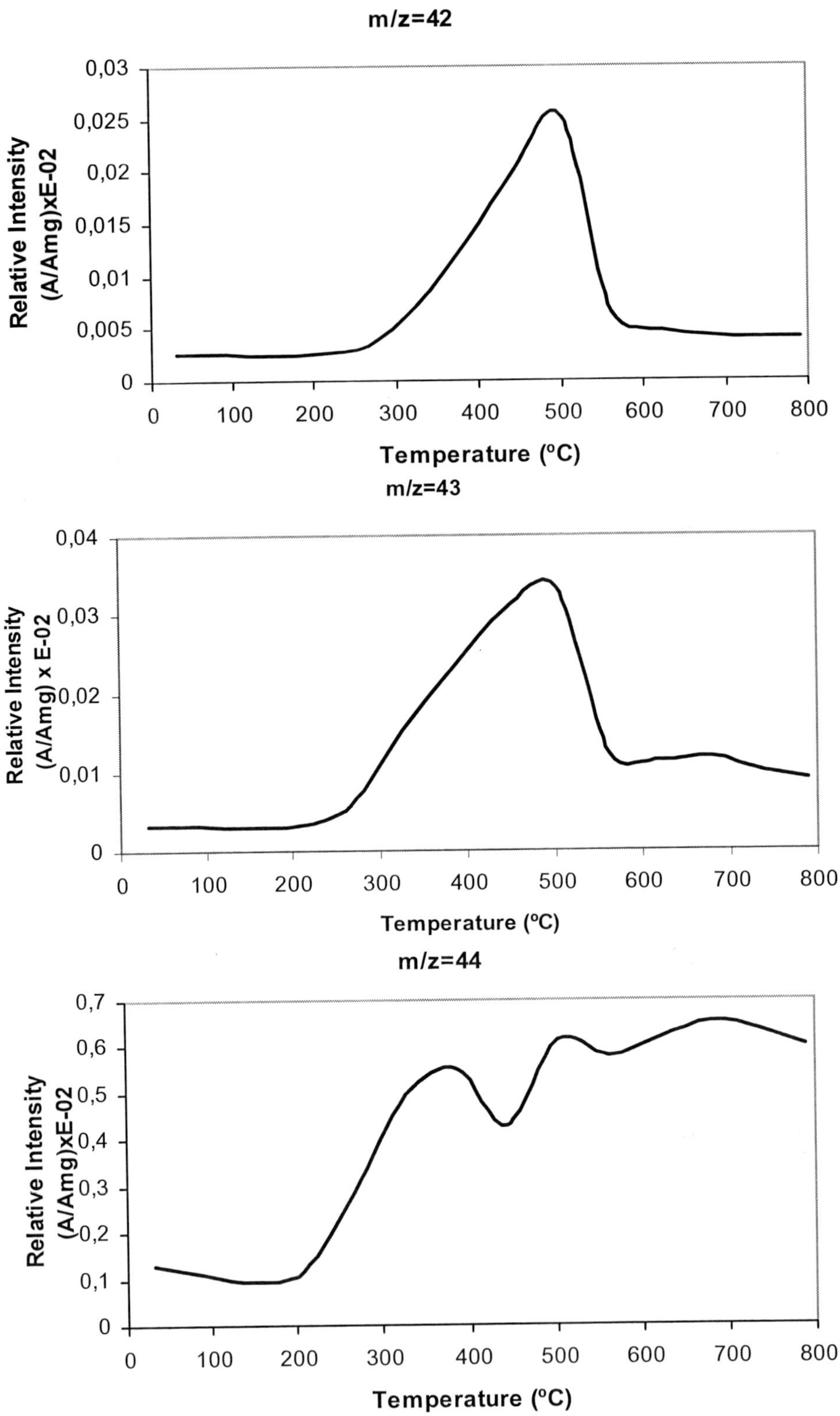

Figure 8. Evolution profiles of the sewage sludge pyrolysis products at 50ºC/min, m/z = 42, 43 and 44.

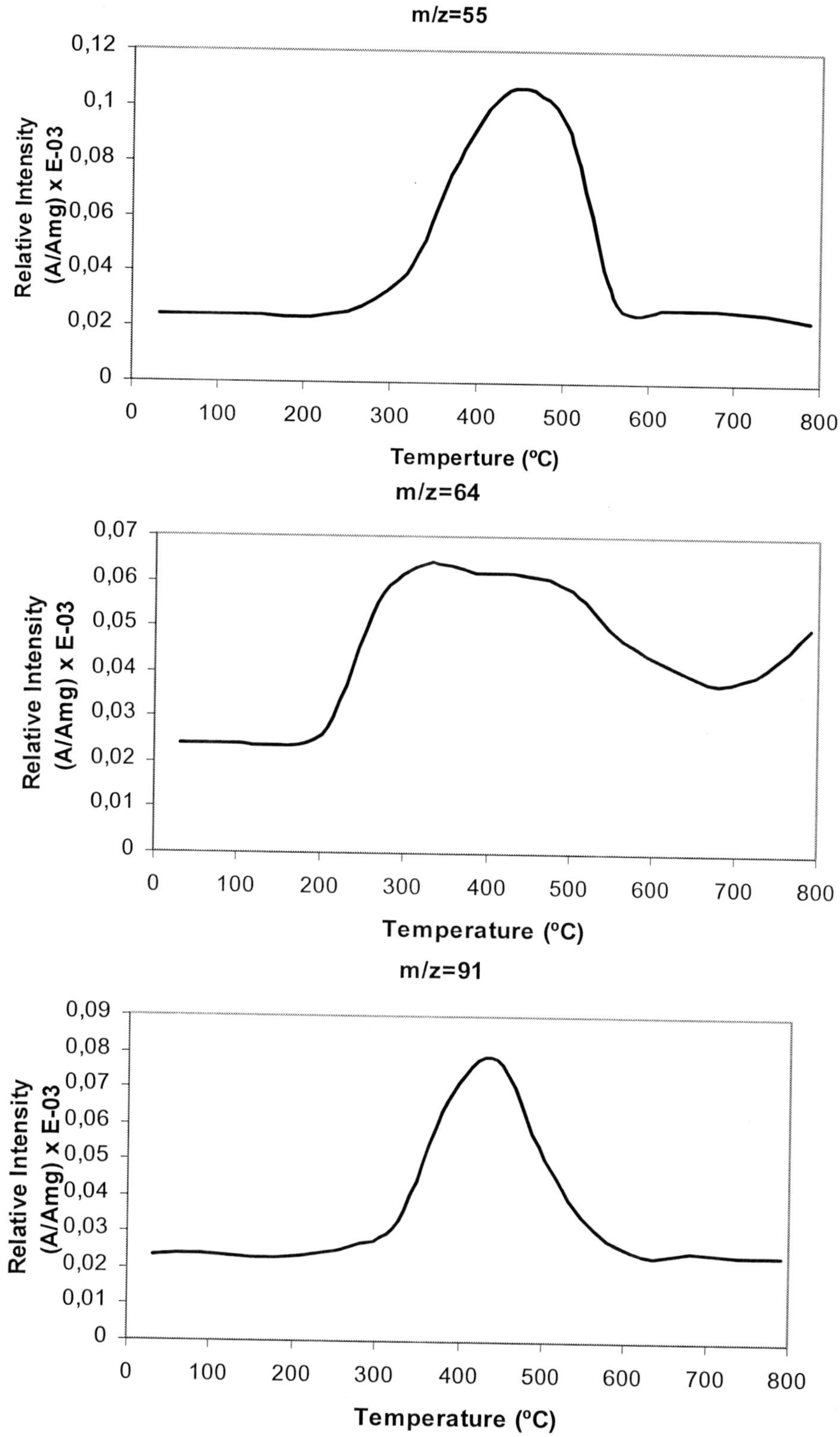

Figure 9. Evolution profiles of the sewage sludge pyrolysis products at 50°C/min, m/z = 55, 64 and 91.

Methane (CH_4) is a gas of importance for its energy value, because of its high calorific value, 12.99 MJNm^{-3} [35]. The higher the volatile matter content, the greater the likelihood of groups producing CH_4 during pyrolysis processes [28]. A similar pattern is shown by the profile for hydrogen (m/z = 2). Hydrogen peaks were detected in the final stages of thermal decomposition, this being due to the fact that, as the temperature rises, a fragmentation of hydrocarbons takes place, showing up as an increase in hydrogen and methane. The hydrogen derived from this is only part of the hydrogen present in the sample. Oxygen (O_2) content plays a major part in the release of hydrogen, due to the formation of water (H_2O). Hydrogen is also a gas of considerable value from the point of view of energy, thanks to its high calorific value, 41.22 MJ Nm^{-3} [35].

During thermal decomposition two stages may be observed in the evolution of water (m/z = 18). Up to 200 °C a peak is observed, corresponding to the evaporation of the water contained in the sample. Development of H_2O above 300 °C is due to the breaking down of groups that contain oxygen, principally hydroxide (OH) groups, giving rise to the formation of pyrolysis water over a wide range of temperatures.

The maximum emission of carbon monoxide (CO) is observed at higher temperatures depending on the oxygen content of the waste. Carbon monoxide is significant for energy production because of its high calorific value, 12.87 MJ Nm^{-3} [35]. This mass to charge ratio, m/z = 28, does not just relate to CO, but also to volatilized nitrogen which has crossed over to a gaseous state and to ethylene.

The emission profiles for carbon dioxide (m/z = 44) go up to higher temperatures as a consequence of the break-down of thermically stable structures.

Emissions of sulphur dioxide (SO_2) were detected from 200 to 700°C, sulphur in its reduced form of hydrogen sulphur (H_2S) was not detected.

Emissions of aromatic compounds were also detected at m/z = 91, a signal corresponding to toluene.

3.3. Experimental Set-up in the Laboratory

3.3.1. Pyrolysis Product Yields

The bio-fuels and biochar yields (% weight) under the process conditions fixed in the reactor were: 27% gas, 28% bio-oil and 45% bio-char.

3.3.2. Characterization of the Gases

The main components of the gas samples are H_2, CO, CO_2 and CH_4 (Table 2). In the pyrolysis process the heavier hydrocarbons are cracked, giving rise to lighter ones, methane and hydrogen. Those of greatest interest are the light hydrocarbons, as their heating values are higher than the other gas components [36]. The CO and CO_2 contents of the gaseous fraction are indicators of the oxygen present in the sample [37]. This oxygen derives from the pyrolytic decomposition of partially oxygenated organic compounds (cellulose, lignin, lipids and carbohydrates).The high H_2 content of the pyrolysis gas may be due to the cracking of hydrocarbons, processes favoured by high temperatures, apart from the contribution of the H in the elemental composition of the sludge itself.

The net heating value that could be theoretically be recovered through stoichiometric combustion of the evolved pyrolytic volatiles at 550°C is calculated accounting for the amounts of measured combustible volatiles at 550°C and using the corresponding heat of combustion for each volatile. The heating value obtained (24.4 MJ m^{-3}) is higher than those recorded in studies by other authors also performed in electric ovens, where the heating value was within the range 13-14 MJ m^{-3} [24], the different conditions under which the process is carried out must be considered, which imply a change in the composition of the gases obtained and therefore in the heating value.

3.3.3. Characterization of the Bio-Oil

Bio-oil is one of the desirable products of pyrolysis. They are usually mixed with condensed water. Bio-oil is a carbon rich oxy-fuel containing a small proportion of nitrogen and trace amounts of sulphur (Table 3). Bio-oil can be an alternative for heating fuel. The low sulphur content of bio-oil is a very important advantage [2].

Table 3. Composition (% vol) and heating value (MJ m^{-3}) of the gases

	(% vol)
Methane	23.38
Ethane	4.11
Ethylene	4.50
Propane	1.43
Propylene	2.69
i-butane	0.12
n-Butane	0.16
Acetylene	<0.01
t-2-Butene	0.13
1-Butene	0.31
Isobutylene	0.42
c-2-Butene	0.10
Isopentane	0.03
n-Pentane	0.04
1,3-Butadiene	0.31
C5 Rest	0.24
C6	0.34
H_2	25.76
N_2	6.83
CO	14.03
CO_2	14.15
Others	0.90
L.H.V. (MJ m^{-3})	24.40

Figure 10 shows a GC-MS chromatogram of the pyrolysis oils obtained from the pyrolysis of sewage sludge at 550°C. These are complex mixtures of organic compounds: n-alkanes and 1-alkenes (C_{11}-C_{31}); monoaromatic hydrocarbons (benzene, toluene, styrene, phenol, 1H-indene and their alkyl derivatives), aromatic compounds bearing nitrogen and oxygen such as pyridine, alkylpyridines, 1H-pyrrole, 1H-indole, isoquinoline, methylquinolines, 9H-carbazole and benzofurane; aliphatic and aromatic nitriles; carboxylic acids (RCOOH), where R represents long aliphatic chains with 14, 15, 16 and 18 carbon atoms; long chain aliphatic amides and steroids (cholestene, cholestadiene or formyl cholestene), and polycyclic aromatic hydrocarbons (PAHs) were also detected in these bio-oils.

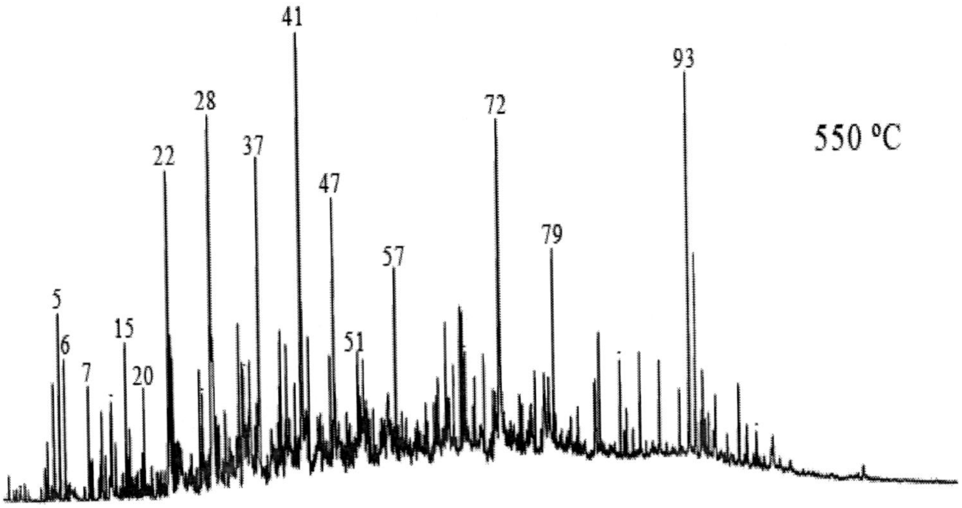

Figure 10. GC-MS chromatogram of the pyrolysis oils.

Table 4. Elemental analysis (%) and heating value (MJ/kg) of bio-oil

C[(a)]	H[(a)]	N[(a)]	S[(a)]	Cl[(a)]	O[(a),(b)]	L.H.V.
59.4	7.5	6.2	0.71	<0.02	17.8	29.01

[(a)] results expressed as a percentage of dry matter, [(b)] calculated by difference.

Table 5 shows the concentration (ppm) of the main compounds present in the pyrolysis oils. The major compounds are carboxylic acids, phenol, p-cresol and 1H-indole. The concentration of some monoaromatic hydrocarbons also shows significant values. Significant quantities of PAH were also found.

It is necessary to improve the quality of this product by modification of the pyrolysis using some catalytic process or using some post treatments. Bio-oil may be used as a raw material in a fractioning process for obtaining fuels. By its distillation, gasoline, diesel fuel or alternative fractional products to fuel-oils are obtained. These fractions may either be used directly or mixed with other conventional fuels. By the application of various processes (such as Fischer-Tropsch synthesis, cracking, hydrogenation, etc.) fuel characteristics may be improved and oil under these circumstances may either be used directly or its fractions may be evaluated as an alternative gasoline, diesel fuel and fuel-oil [38].

Table 5. Concentration (ppm) of the main compounds in the pyrolysis oils

Pk nº	Compound	RT (min)	ppm
2	Benzene	7,89	2706
4	Pyridine	10,50	6467
5	1H-Pyrrole	10,97	9731
6	Toluene	11,37	11333
14	1,4-dimethylbenzene	15,36	2601
15	Styrene	16,21	5994
16	1,2-dimethylbenzene	16,30	1522
18	2-Ethylpyridine	16,69	539
21	3-ethylpyridine	18,71	588
22	Phenol	19,62	10919
23	Benzonitrile	19,77	3539
26	1H-Indene	21,93	5235
27	o-Cresol	22,15	2165
28	p-Cresol	22,87	17589
29	1-Undecene	23,22	1012
36	1-Dodecene	26,36	1240
37	Naphthalene	26,58	5481
39	Isoquinoline	28,21	1550
40	1-Tridecene	29,28	445
41	1H-Indole	29,79	10271
42	2-Methylnaphthalene	29,84	1943
43	1-Methylnaphthalene	30,35	1328
45	1-Tetradecene	32,01	917
51	Acenaphthylene	34,15	1187
52	1-Pentadecene	34,58	864
53	n-Pentadecane	34,76	431
57	1-Hexadecene	37,00	809
58	n-Hexadecane	37,16	310
59	9H-Fluorene	37,44	489
60	1-Heptadecene	39,29	249
61	n-Heptadecane	39,45	206
62	Tetradecanoic acid	40,89	15873
63	1-Octadecene	41,47	372
64	Phenanthrene	42,01	1607
72	Hexadecanoic acid	45,04	66686
76	Fluoranthene	47,73	902
77	9-Octadecenoic acid	48,44	12483
78	Pyrene	48,77	1071
80	n-Docosane	49,28	145
82	n-Tricosane	51,00	106
85	n-Tetracosane	52,63	407
86	n-Pentacosane	54,22	203
87	Benzo[a]anthracene	54,52	263
88	Chrysene	54,71	487
90	n-Hexacosane	55,74	219
91	n-Heptacosane	57,22	173
92	n-Octacosane	58,74	122
99	Benzo[a]pyrene	61,28	313
101	n-Triacontane	62,33	-
103	n-Untriacontane	64,56	226

It is necessary to improve the quality of this product by modification of the pyrolysis using some catalytic process or using some post treatments. Bio-oil may be used as a raw material in a fractioning process for obtaining fuels. By its distillation, gasoline, diesel fuel or alternative fractional products to fuel-oils are obtained. These fractions may either be used directly or mixed with other conventional fuels. By the application of various processes (such as Fischer-Tropsch synthesis, cracking, hydrogenation, etc.) fuel characteristics may be improved and oil under these circumstances may either be used directly or its fractions may be evaluated as an alternative gasoline, diesel fuel and fuel-oil [38].

3.3.4. Characterization of the Bio-Char

Table 6 shows the chemical characterization of the biochar. As a consequence of pyrolysis, the carbon content of the carbonized residue increased with regard to the original sludge, along with a deoxygenation as a consequence of the loss of functional groups during the process. There was also a noticeable decrease in the hydrogen content, probably due to the great proportion of hydrogen compounds in the volatile matter. There was a concentration of sulphur and chlorine in the solid fraction, which caused emissions of sulphur dioxide and chlorine, compounds during combustion, being necessary a gas cleaning treatment of the combustion gases.

Table 6. Chemical analyses (%) and heating values of carbonized sludge

	Biochar
Moisture	2.3
Volatile Matter[a]	8.1
Ash[a]	70.2
Fixed Carbon [a,c]	8
Carbon[b]	88.0
Hydrogen[b]	1.8
Nitrogen[b]	8.0
Sulphur[b]	1.8
Chlorine[b]	0.3
Oxygen[b,c]	<0.1
L.H.V. (MJ Kg^{-1})	9.42

[a] results expressed as a percentage of dry matter.
[b] results expressed as a percentage of dry matter free of ashes.
[c] calculated by difference.

Reactivity of the carbonized solid in air was studied by means of thermogravimetric analysis. The evolution of the mass loss rate curve (DTG) against temperature is shown in Figure 11. A single mass loss peak is to be observed over a wide temperature range (300-600°C), owing to reactive material remaining in the carbonized solid. At low temperatures, it is attributed to a volatile material that had remained in the carbonized solid being driven off and to the presence of structures more reactive in air, capable of reacting at low temperatures. At high temperatures, the reaction of more condensed structures occurs. The heating value of the pyrolysed residue was quite low in comparison with that of other fuels [24], lower even than that of the sludge before pyrolysis, the high ash content (70%) might explain the low heating value of the char. Co-combustion of this pyrolysed residue and coal may provide an attractive option for the disposal and utilization of a renewable waste resource such as sewage sludges in an economic and environmentally safe manner [39].

Other ways it may be employed include use as a source of C and N. During the conversion of sewage sludge to biochar the original carbon is retained in the biochar, which offers a significant opportunity for creating a carbon sink.

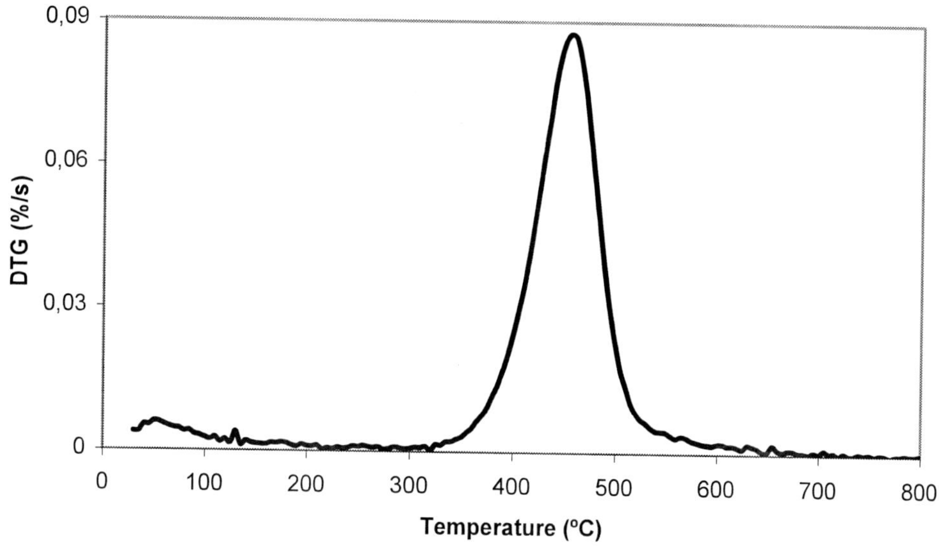

Figure 11. DTG profile of the programmed-temperature combustion of the biochar.

The long persistence of biochar in soil makes it a prime candidate for the mitigation of climate change as a potential sink for atmospheric carbon dioxide [40]. The success of effective reduction of greenhouse gases depends on the associated net emission reductions through biochar sequestration [41]. All organic matter added to soil significantly improves soil functions, not the least the retention of several nutrients that are essential to plant growth. The biochar from sludge contains 0.86% N, 0.91% P_2O_5 and 1.52% K_2O. The biochar persistence in soil is also important related with any other form of organic matter applied to the land. It is undisputed that biochar is much more persistent in soil than any other form of organic matter that is commonly applied to the land [42].

What is special about biochar is that it is much more effective in retaining most nutrients and keeping them available to plants than other organic matter for example compost or manures [43]. Therefore, all associated benefits with respect to nutrient retention and soil fertility are longer lasting than with alternative management.

3.4. Pilot Plant

3.4.1. Product Yield

The solid fraction was weight directly and the gas fraction calculated by difference as the pilot plant produced no liquid fraction. The yields obtained were 38% by weight of carbonized solid and 62% of gas. A large proportion of gases was produced under the operation conditions of temperature and heat gradient. When the residue was fed into the thermolysis reactor, it was already at a temperature of around 550°C and the residence time was 20 minutes, which allowed for rapid thermal decomposition. Temperature choice was based on a prior thermogravimetry study of the sewage sludge (Figure 5) and was the plant's planned operation temperature and that of the total thermal decomposition of the residue.

The production of char was observed to be higher in the laboratory (45%) than in the pilot plant process, about 38% of the original material; it is a logic result due the residence time at pilot plant is greater. Comparing showed a higher percentage of gases at the plant (62%), as was to be expected, for at the laboratory level (27%), this value was broken down into condensable (liquid fraction, 36%) and non-condensable (gas fraction, 40%) compounds.

Through mass balances, it has been calculated C, H, and O contents with an error of 8%, 2% and 0.5%, respectively.

3.4.2. Characterization of the Gas Fraction

The same conditions used in the laboratory for the collection of the condensable fraction were reproduced in the pilot plant. However, the amount of condensable fraction collected in the pilot plant was insignificant due to the operation conditions that avoid the formation of condensables [44].

Table 7 shows the composition of the gases in the pilot plant trial. Since the analysis of N, S and Cl is more reliable in the carbonized fraction, N_2 was determined through a material balance, also S and Cl were determined through a material balance, to determine other concentrations (mainly H_2S and HCl) and the remaining gases were normalized as percentages.

Table 7. Composition and heating value of the gases at the pilot plant (G_{pp})

	G_{pp} (% vol)
Methane	17.42
Ethane	3.61
Ethylene	2.54
Propane	1.43
Propylene	1.85
Isobutane	0.12
n-Butane	0.18
Acetylene	0
t-2-Butene	0.12
1-Butene	0.35
Isobutylene	0.26
c-2-Butene	0.09
Isopentane	0.03
n-Pentano	0.01
1,3-Butadiene	0.27
C5 Rest	0.41
C6	0.61
H_2	21.68
N_2	3.95
CO	14.58
CO_2	29.75
Others	0.75
L.H.V. (MJ m^{-3})	20.07

The main components of the gas samples are H_2, CO, CO_2 and CH_4. The pilot plant sample had a lower percentage of light hydrocarbons than the laboratory one. Sampling may also have an influence, although in principle it was to be expected that the operating conditions at the pilot plant, 550°C, would favour the cracking of gases to increase the proportion of light hydrocarbons. The laboratory sample had a higher percentage which is not only due to experimental errors because all the compounds present a similar tendency.

H_2O was no analysed, gas compositions are on dry basis. Gases were no analysed online. At the pilot plant H_2O came out as vapour because we have no condensates.

At laboratory level, a thermogravimetric and mass spectrometric study was carried out of the process of devolatilization of the organic fraction of municipal solid wastes. Development of H_2O above 300°C could be due to the breaking down of groups that contain oxygen, principally hydroxide (OH) groups, giving rise to the formation of pyrolysis water over a wide range of temperatures.

The water could be also consumed during pyrolysis; this could be attributed to the following reactions:

$$CO+H_2O \leftrightarrow CO_2+H_2 \quad \Delta H_{298K} = -41.5 \text{ kJmol}^{-1}$$

$$C+2H_2O \leftrightarrow CO_2+H_2 \quad \Delta H_{298K} = 88000 \text{ kJmol}^{-1}$$

Calorific power is estimated taking into account the densities of each of the gas species considered, the combustion enthalpies and specific heats. The value obtained is 20 MJ/m^3, higher than those recorded in studies by other authors [24].

3.4.3. Characterization of the Solid Fraction

The possible use of the solid fraction as a fuel required its characterization by chemical analysis and combustibility study based on thermal analysis.

Table 8 shows the chemical characterization of the carbonized solid fraction obtained at the pilot plant. There were no significant differences with the one obtained at the laboratory. Despite the different operating conditions, mainly in the heating gradient, the two processes may be said to be similar.

The process of devolatization is slower at higher temperatures, a tendency also found in the carbonization of biomass and coal, as other authors have previously pointed out [43], there was still a volatile content in the carbonized fraction (13%), indicating that thermolysis was not complete. The calorific value of the pyrolysed residue was low, 9.4 MJ/Kg, appropriate for energetic valorisation by co-combustion.

Reactivity of the carbonized solid obtained at the pilot plant in air was studied by means of thermogravimetric analysis. The evolution of the mass loss rate curves (DTG) against temperature is shown in Figure 12. A single mass loss peak is to be observed over a wide temperature range (300-550°C), owing to reactive material remaining in the carbonized solid. In this interval is possible to distinguished different origins of mass loss; firstly it is attributed to a volatile material that had remained in the carbonized solid being driven off and to the presence of structures more reactive in air, capable of reacting at low temperatures. At high temperatures, the reaction of more condensed structures occurs. The initial temperature for the combustion process is slightly lower for the carbonized products from the pilot plant, owing to the higher volatile matter content of these solid samples.

Table 8. Chemical analysis (%) of the carbonized residue obtained at pilot plant (S_{PP})

	S_{PP}
Moisture	2.3
Volatile Matter[a]	12.7
Ash[a]	67.5
Fixed Carbon [a,c]	17.5
Carbon[b]	81.4
Hydrogen[b]	2.1
Nitrogen[b]	9.3
Sulphur[b]	1.8
Chlorine[b]	0.4
Oxygen[b,c]	5.1
H.H.V. (MJ Kg^{-1})	9.5
L.H.V. (MJ Kg^{-1})	9.4

[a] results expressed as a percentage of dry matter.
[b] results expressed as a percentage of dry matter free of ashes.
[c] calculated by difference.

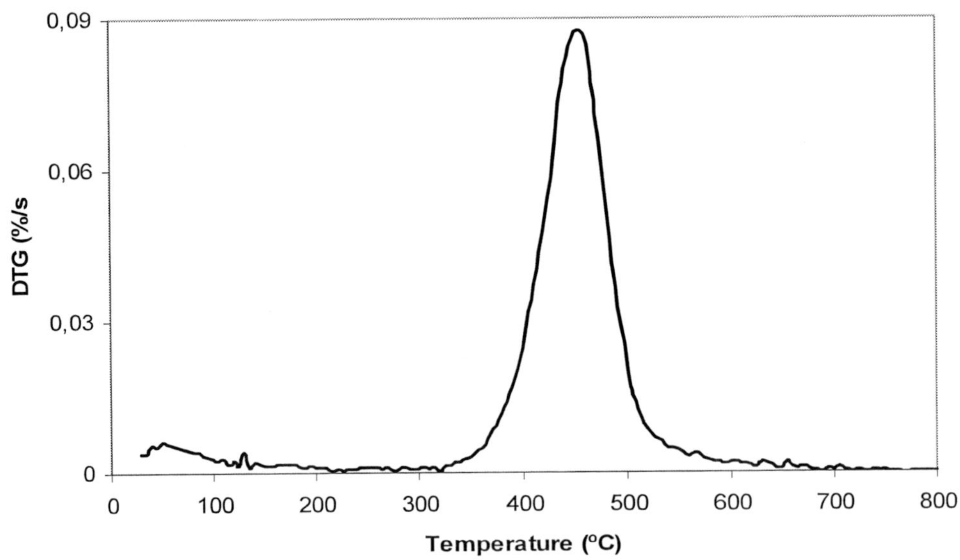

Figure 12. DTG profile of the programmed-temperature combustion of the carbonized fraction of the sludge collected at the pilot plant.

3.4.4. Gas Cleaning Treatment of the Combustion Gases

Figure 13 show the characteristic compositions for some of the combustion products of thermolysis gas alone and with carbonized solid co-combustion before and after the neutralization treatment.

The variations registered in the concentrations of carbon monoxide (CO) and total organic carbon (TOC) were not great, and the incorporation of char (co-combustion) implied a rise in emissions as the carbon content of the fuel was increased, while a reduction was noticed once the cleaning treatment was applied. The values were below legal limits, 50

mg/m^3 for carbon monoxide but slight higher than 10 mg/m^3 of legal limit for total organic carbon.

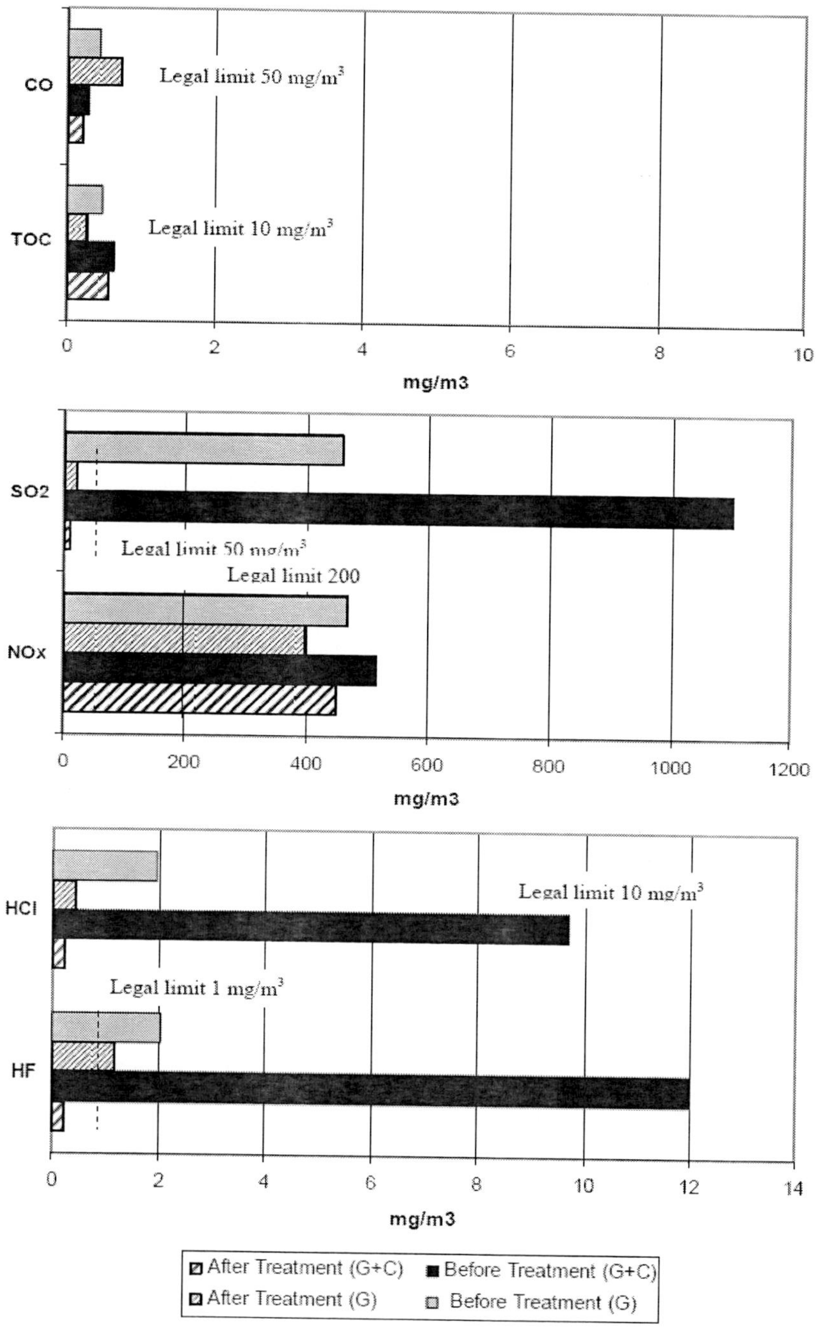

Figure 13. CO, COT, SO2, NOx, HCl and HF emissions associated with the combustion of thermolysis gas alone and together with the carbonized solid fraction of the sludge before and after cleansing treatment. The limits shown are those set out in Directive 2000/76/CE.

Sulphur dioxide emissions reached 450 mg/m^3 when the gases were not treated, far above the European legal limit of 50 mg/m^3. This value could more than duplicate when solid residue was present in the combustion chamber due to its high sulphur concentration, as stated previously, and its discontinuous input (the solid enter the chamber in batches) may imply a greater amount of solid residue than there would be if it were continuous. Gas cleaning treatment brought the values down to acceptable levels in both cases. Desulphurization of these gases may even be greater because of the ash in the carbonized fraction.

NO_x levels measured before treatment were above the 200 mg/m^3 set out in Directive 2000/76/CE. These compounds were less affected by the burning of the carbonized fraction along with the gas. In this case, treatment proved less effective in terms of the reduction percentage.

Regarding hydrogen chloride and hydrogen fluoride emissions, the first one was greatly affected by the inclusion of the carbonized residue, they were reduced when bicarbonate was added, especially those of hydrogen chloride, which dropped remarkably, having already been below the legal limit. Hydrogen fluoride reduction was also sufficient to comply with legal limits. As for sulphur dioxide, combustion together with the carbonized residue increased its emissions but also favoured the cleaning process, final emissions being reduced more that they would have been otherwise.

In general, the graphs show that the combustion of the gaseous and solid fractions together implied an increase in emissions. The values for SO_2, HCl and HF were much higher, while those for COT and NO_x were slightly higher. Nevertheless, the cleaning process was effective, and emissions were below the legal limits. The effectiveness of the treatment was reflected in a major reduction in SO_2, HCl and HF emissions. Fly ashes from the carbonized fraction increased retention of HF and SO_2.

3.4.5. Energy Yields

The energy production scheme proposed is shown in Figure 14. The process begins with the preparation of the material to be treated at the plant. Then the thermolysis of the material gives two products, gas and a solid residue, which are assessed from their combustion in a boiler generating steam. The energy of the steam will be used, by means of a combination of turbine and alternator, to generate electricity, the residual heat of the system being used for the drier of raw materials and heating water. In turn, the products generated by the boiler, combustion gases and ash, will be treated appropriately.

Table 9 shows the energy yields with the data obtained from the trials at the pilot plant. The thermal power (MJ/h) obtained from the joint combustion of the gas and the carbonized fraction is calculated along with that of the direct combustion of the raw material. A fixed consumption of energy has been allowed for the heating of the reactor, although this will depend on the material treated and the particular devolatization reactions occurring. On the other hand, yield data of both the boiler and the cycle have been obtained for similar combustion processes.

The total thermal power is obviously lower than that of the combustion of the residue. It will be observed that the differences between one value and the other in all cases should be similar. Any discrepancy is due to possible errors of analysis or sample taking.

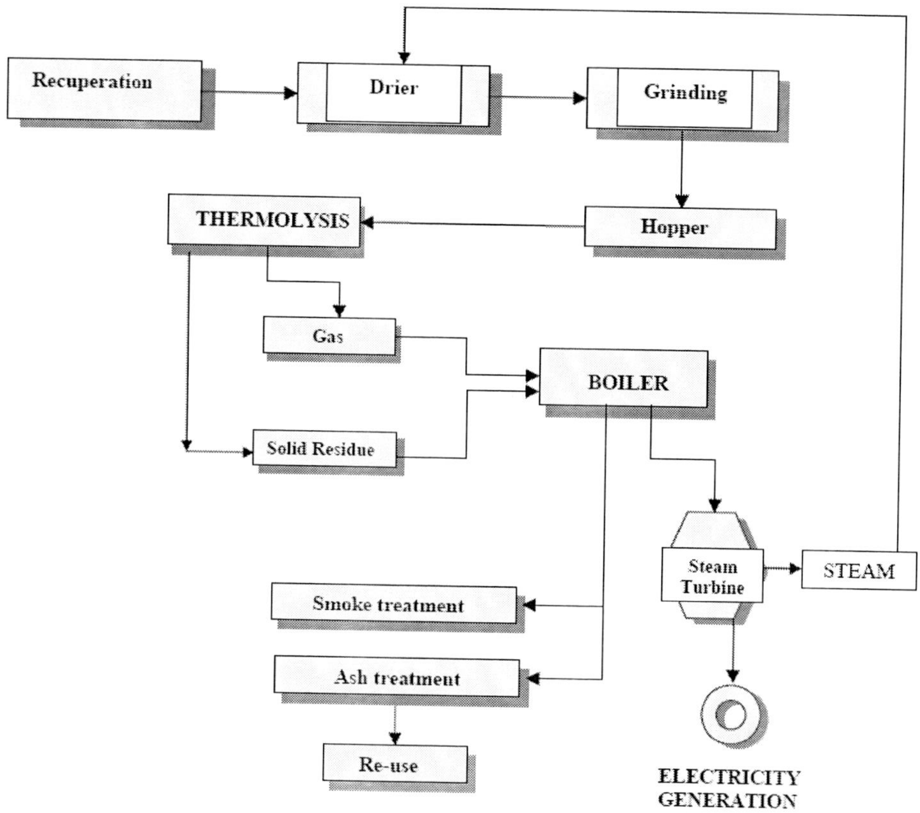

Figure 14. Energy production system.

Table 9. Energy yield of the pilot plant trial

	Sludge
BASE (kg/h)	1000
Thermolysis gas (kg/h)	620
LHV gas (MJ/Nm^3)	20.07
Carbonized residue (kg/h)	380
LHV carbonized residue (MJ/kg)	9.36
Total thermal power (MJ/h)	14192
Gas density (kg/m^3)	1.17
Boiler yield (%)	84
Net thermal power (MJ/h)	11921
Yield of the cycle (%)	25
Electricity produced (kW.h)	828
LHV o.f.MSW (MJ/kg)	15.32
Thermal power of the combustion of the o.f.MSW (MJ/h)	15320
Cost of pyrolysis (kW.h)	306
Net electricity production (kW.h)	522
Net electrical yield of the plant (%)	13.2
Net thermal yield of the plant (%)	20

CONCLUSIONS

Pyrolysis of Sewage Sludge Was Investigated by Thermal and Spectrometry Analysis

The range of temperatures over which devolatilization of sewage sludge occurs is quite large (200-500 °C), probably as a consequence of the heterogeneity of the organic matter. The rate with which the loss of mass takes place (DTG), together with the percentage of carbonized product from the process, is in accordance with what was to be expected on the basis of immediate analysis of sewage sludge. Nonetheless, the percentage of carbonized material in this organic waste indicates that devolatilization is not total under the operating conditions used. Nevertheless, it is possible to state that devolatilization is practically total by 550-600°C.

A study of the mass spectrometry results reveals major emissions of energy-producing gases during pyrolysis, although fewer pollutants are emitted. The products of the pyrolysis process are principally hydrocarbons, hydrogen, carbon monoxide and carbon dioxide.

From the Study Carried Out at the Experimental Set-Up in the Laboratory

The composition of the pyrolysis gas makes it suitable for use as a fuel, given the concentration of light hydrocarbons and hydrogen, and its heating value.

In this work it has been shown that sewage sludge pyrolysis oils contain significant quantities of potentially high value hydrocarbons such as monoaromatic hydrocarbons and phenolic compounds. The oils were also found to contain substancial concentrations of PAHs, an important constituent from an environmental point of view since some of these compounds have been shown to be carcinogenic and/or mutagenic.

Bio-oil, with its very low sulphur content, may be considered from the point of view of environmental as a potential alternative to fuel-oil. After some improving applications it can also be used as an alternative for heating fuel. By other side, new catalyst formulations and processing conditions for efficient conversion of bio-oil to value-added fuel and chemical products will be investigated.

Biochar may be evaluated as an amendment to improve soil fertility and nutrient retention, but on this point it would be necessary to undertake a more complete study. Combustion at a programmed temperature of the biochar revealed its good combustibility, but its applicability as a fuel will depend on the use of techniques for the treatment of combustion gases, as it concentrates much of the sulphur of the original waste.

The Study at Pilot Plant Led to a Positive Assessment of the Possible Use of the Process Products as Fuel

The composition of the thermolysis gas makes it suitable given the concentration of light hydrocarbons and hydrogen, and its heating value.

The calorific power of the pyrolysed residue is appropriate for energetic valorisation by co-combustion, but its applicability as a fuel will depend on the use of techniques for the treatment of combustion gases, as it concentrates much of the sulphur and chlorine of the original waste.

Neutralization treatment proves effective in the cleaning of combustion gases; both from the combustion of thermolysis gas alone and mixed with the carbonized fraction. The effectiveness of the treatment is reflected in a major reduction in sulphur dioxide, hydrogen chloride and hydrogen fluoride emissions. Fly ashes from the carbonized fraction increased retention of hydrogen fluoride and sulphur dioxide on the filter.

The low yields obtained regarding the production of electrical energy by means ofW a steam cycle show the need to include a cogeneration process.

REFERENCES

[1] Werther, J., Ogada, T. (1999). Sewage sludge combustion. *Prog. Energy Combust. Sci. 25*, 55-116.

[2] Morán A, Gómez X, Aller A. (2004). Management of sludge from waste water treatment plant. Current approaches and perspectives. *Tecnología del Agua, 245*, 34-38.

[3] Domínguez, A., Menéndez, J.A., Inguanzo, M., Pis, J.J. (2006). Production of bio-fuels by high temperature pyrolysis of sewage sludge using convencional and microwave heating. *Bioresource Technol. 97*, 1185-1193.

[4] Matthews, P.J. (1996). European Progress on the Political and Economic Concerns of Sludge Disposal-Is the future Biosolids or Ash? *Proceedings EWPCA-NVA Conference. Future of Water Quality Management in Europe.* Aquatech, Amsterdam. September.

[5] Metcalf, E. Wastewater engineering-treatment, disposal and reuse. 3rd Ed. New York, USA: McGraw Hill. 1991.

[6] Mene R, Lebrun T. (2002). CIWEM/Aqua Enviro *7th European Biosolids Organic Residuals Conference.*

[7] Hogan F. xx. (2003). *Water and Wastewater Internacional, 18*, 35.

[8] Stephenson R. (2002). *Water 21*, 20.

[9] Otero, M., Rozada, F., Calvo, L.F., García, A.I., Morán, A., (2003). Elimination of organic water pollutants using adsorbents obtained from sewage sludge. *Dyes and Pigments. 57*, 55-65.

[10] Lundin, M., Olofsson, M., Pettersson, G., Zetterlund, H. (2004). Environmental and economic assessment of sewage sludge handling options. *Resource Conservat Recycl. 41*, 255-278.

[11] Chumm H.L., Overend, R.P. (2001). Biomass and renewable fuels. *Fuel Process. Technol. 71*, 187-195.

[12] McKendry, P., (2002). Part 2. Energy production from biomass. Conversion Technologies. *Bioresource Tecnol. 83*, 47-54.

[13] Khiari, B., Marias, F., Zagrouba, F., Vaxelaire, J. (2004). Analytical study of the pyrolysis process in a wastewater treatment pilot station. *Desalination, 167*, 39-47.

[14] Fytili, D., Zabaniotou, A. (2008). Utilization of sewage sludge in EU application of old and new methods. A review. *Renewable and Sustainable Energy Reviews. 12*, 116-140.

[15] McAuley, B., Kunkel,J., Manahan, S.E. (2001). A new process for drying and gasification of sewage sludge. *Water Eng Manage*. 18-23.

[16] Hamilton, C.J. (2000). Gasification as an innovative method of sewage sludge disposal. *Water Environ Manage. 14*, 89-93.

[17] Marrero, T., McAuley, B., Sutterlin, W., Morris, S., Manahan, S. (2003). Fate of heavy metals and radioactive metals in gasification of sewage sludge. *Waste Manage. 24*, 193-198.

[18] Menéndez, J.A., Inguanzo, M., Pis, J.J. (2002). Microwave-induced pyrolysis of sewage sludge. *Water Res. 36*, 3261.6264.

[19] Directive 2000/76/CE of the European Parliament and Council of the 4th December 2000 concerning the incineration of wastes. Official Gazette of the European Communities L 332/91 of the 28.12.2000.

[20] Fink, J.K. (1999). Pyrolysis and combustion of polymer wastes in combination with metallurgical processes and the cement industry. *J. Anal. Appl. Pyrol. 51* , 239-252.

[21] W. Kaminsky, Kim, J-S. (1999). Pyrolysis of mixed plastic into aromatics. *J. Anal. Appl. Pyrol. 51* 127-134.

[22] Inguanzo, M.; Menéndez, J.A.; Fuente, E.; Pis, J.J., 2001. Reactivity of pyrolyzed sewage sludge in air and CO_2. *J. Anal. Appl. Pyrol. 58-59*, 943-954.

[23] Caballero, J.A.; Font, R.; Marcilla, A. ; Conesa, J.A., 1997. Characterization of sewage sludges by primary and secondary pyrolysis. *Journal of Analytical and Applied Pyrolysis. 40-41*, 433-450.

[24] Inguanzo M., Domínguez A., Menéndez J. A., Blanco C.G., Pis J.J. (2002). On the pyrolysis of sewage sludge: the influence of pyrolysis conditions on solid, liquid and gas fractions. *J. Anal. Appl. Pyrol., 63*, 209-222.

[25] Domínguez, A.; Menéndez, J.A.; Inguanzo, M.; Bernard, P.L.; Pis, J.J., 2003. Gas chromatographic-mass spectrometric study of the oil fractions produced by microwave-assisted pyrolysis of different sewage sludges. *Journal of Chromatography A. 1012*, 193-206.

[26] Sánchez, M.E., (2004). Gestión de residuos orgánicos y plásticos. Aprovechamiento energético mediante pirólisis. Tésis Doctoral. Departamento de Física, Química y Expresión Gráfica. (Organic and plastic residues management. Energetic valorization by means of pyrolysis.)Universidad de León.

[27] Hatton, P.J., Southward, J. (2003). Optimisation of the connection between TA-MS systems together with improved data interpretation for TA-MS applications. *J. Therm. Anal. Calorim., 72*, 83-92.

[28] Arenillas, A., Rubiera, F., Pis, J.J. (1999). Simultaneous thermogravimetric-mass spectromectric study on the pyrolysis behaviour of different rank coals. *J. Anal. Appl. Pyrol., 50*, 31-46.

[29] Conesa, J.A., Marcilla, A. Moral, Moreno-Caselles R., Pérez-Espinosa, J., (1998). Evolution of gases in the primary pyrolysis of different sewage sludges. *Themochim. Acta., 313*, 63-73.

[30] Calvo, L.F., Otero, M., Jenkins, B.M., García, A.I., Moran, A. (2004). Heating process characteristics and kinetics of sewage sludge in diferent atmospheres. *Thermochim. Acta., 409*, 127-135.

[31] Díez, C., Sánchez, M.E., Haxaire, P., Martínez, O., Morán, A. (2005). Pyrolysis of tyres: A comparison of the results from a fixed-bed laboratory reactor and a pilot plant (rotatory reactor). *J. Anal. Appl. Pyrol., 74*: 254–258.
[32] Helt, J.E., Agrawal, R.K., Soltes, J., Milne, T.A., (1998). Pyrolysis Oils from Biomass; Producing, Analysing and Upgrading. American Chemical Society, Washington.
[33] Sánchez, M.E., Martínez, O., Gómez, X., Morán. A. (2007) Pyrolisis of mixtures of sewage sludge and manure: a comparison of the results obtained in the laboratory (semi-pilot) and pilot plant. *Waste Manage. 27*, 1328-1334.
[34] Font, R., Fullana, A., Conesa, J.A., Llavador, F (2001). Análisis of the pyrolysis and combustión of different sewage sludges by TG. *J. Anal. Appl. Pyrol., 58-59*, 927-941.
[35] Calvo, L.F., Sánchez, M.E., Morán, A., García, A.I. (2004). TG-MS as a technique for a better monitoring of the pyrolysis, gasification and combustión of two kinas of sewage sludge. *J. Therm. Anal. Calorim. 78*, 587-598.
[36] Menéndez, J.A., Domínguez, A., Inguanzo, M., Pis, J.J., 2004. Microwave pyrolysis of sewage sludge: analysis of the gas fraction. *J. Anal. Appl. Pyrol.71*, 657-667.
[37] Fullana, A, Conesa, J.A., Font, R., Martín-Gullón, I. (2003). Pyrolysis of sewage sludge: nitrogenated compounds and pre-treatment effects. *J. Anal. Appl. Pyrol., 68-69*, 561-575.
[38] Özçimen, D., Karaosmanoglu, F. (2004). Production and characterization of bio-oil and biochar from rapeseed cake. *Renewable Energy, 29*, 779-787.
[39] Otero, M., Díez, C., Calvo, L.F., García, A.I., Morán. (2002). Analysis of the co-combustion of sewage slugge and coal by TG-MS. *Biomass and Bioenergy, 22*, 319-329.
[40] Rausa R., Mascolo G., Bassetti A. (1999). Thermal treatment of sediments as a function of temperature and reacting atmosphere. *J. Anal. Appl. Pyrol. 52*, 115-135.
[41] Rondon, M., J.A. Ramírez, J. Lehmann. (2005). Charcoal additions reduce net emissions of greenhouse gases to the atmosphere. In: *Proceedings of the 3rd USDA Symposium on Greenhouse Gases and Carbon Sequestration*, Baltimore, USDA, March, 21-24, p. 208.
[42] Lehmann, J., Gaunt, J. Rondon, M.. Biochar sequestration in terrestrial ecosystems-a review. Mitigation and Adaptation Strategies for Global Change 2006; 11: 403-427.
[43] Laird, D. (2007). The charcoal vision: a recipe for simultaneously producing bioenergy, permanently sequestering carbon, and enhancing soil and quarter quality. *Forth USDA Greenhouse Gas Conference. Positioning Agriculture and Forestry to meet the Challenges of Climate Change*. February 6-8, Baltimore, MD.
[44] Sánchez, M.E., Cuetos, M.J., Martínez, O., Morán, A. (2007). Pilot scale thermolysis of municipal solid waste. Combustibility of the products of the process and gas cleaning treatment of the combustion gases. *J. Anal. Appl. Pyrol. 78*, 125-132.
[45] Lu, G.Q., Low, J.C.F., Liu, C.Y., Lua, A.C. (1995). Surface area development of sewage sludge during pyrolysis. *Fuel 74 (3),* 344-348.

In: Sewage Treatment: Uses, Processes and Impact
Editor: Anna Stephens and Mark Fuller
ISBN 978-1-60692-959-9

Chapter 7

SEWAGE SLUDGE DISPOSAL – LAND APPLICATION - ENVIRONMENTAL PROBLEMS – AN OVERVIEW

Wasim Aktar*

Pesticide Residue Laboratory, Department of Agricultural Chemicals,
Bidhan Chandra Krishi Viswavidyalaya,
Mohanpur-741252, Nadia, West Bengal, India

1. INTRODUCTION

Most wastewater treatment processes produce a sludge which has to be disposed of. Conventional secondary sewage treatment plants typically generate a primary sludge in the primary sedimentation stage of treatment and a secondary, biological, sludge in final sedimentation after the biological process. The characteristics of the secondary sludge vary with the type of biological process and, often, it is mixed with primary sludge before treatment and disposal. Approximately one half of the costs of operating secondary sewage treatment plants in Europe can be associated with sludge treatment and disposal. Land application of raw or treated sewage sludge can reduce significantly the sludge disposal cost component of sewage treatment as well as providing a large part of the nitrogen and phosphorus requirements of many crops. Very rarely do urban sewerage systems transport only domestic sewage to treatment plants; industrial effluents and storm-water runoff from roads and other paved areas are frequently discharged into sewers. Thus sewage sludge will contain, in addition to organic waste material, traces of many pollutants used in our modern society. Some of these substances can be phytotoxic and some toxic to humans and/or animals so it is necessary to control the concentrations in the soil of potentially toxic elements (PTE) and their rate of application to the soil. The risk to health of chemicals in sewage sludge applied to land has been reviewed by Dean and Suess [1].

Sewage sludge also contains pathogenic bacteria, viruses and protozoa along with other parasitic helminths which can give rise to potential hazards to the health of humans, animals and plants. A WHO (1981) Report on the risk to health of microbes in sewage sludge applied

* Correspondence to: wasim04101981@yahoo.co.in

to land identified salmonellae and *Taenia* as giving rise to greatest concern. The numbers of pathogenic and parasitic organisms in sludge can be significantly reduced before application to the land by appropriate sludge treatment and the potential health risk is further reduced by the effects of climate, soil-microorganisms and time after the sludge is applied to the soil. Nevertheless, in the case of certain crops, limitations on planting, grazing and harvesting are necessary.

Apart from those components of concern, sewage sludge also contains useful concentrations of nitrogen, phosphorus and organic matter. The availability of the phosphorus content in the year of application is about 50% and is independent of any prior sludge treatment. Nitrogen availability is more dependent on sludge treatment, untreated liquid sludge and dewatered treated sludge releasing nitrogen slowly with the benefits to crops being realised over a relatively long period. Liquid anaerobically-digested sludge has high ammonia-nitrogen content which is readily available to plants and can be of particular benefit to grassland. The organic matter in sludge can improve the water retaining capacity and structure of some soils, especially when applied in the form of dewatered sludge cake.

2. What Is Sludge?

Residuals, biosolids, septage, sewage, wastewater byproduct, compost: there are many names for sludge and sludge products. The term "sludge" is used as most people understand it: the sometimes solid, sometimes liquid material generated by wastewater treatment plants and used as fertilizer on fields, in gravel pits, and on forestry lots throughout the state. Sludge may classified as "Class A" if it has been treated to reduce germs to background levels (levels normally found in soils) and "Class B" if it has been treated so that germs are reduced by an estimated 90%.

3. Composition of Sewage Sludge

The nature of the sewage sludge depends on the waste water treatment process and on the source of the sewage. In general it contains both toxic and non-toxic organic wastes. Of the two, non-toxic compounds are most prevalent comprising all materials of plant and animal origin, including proteins, amino acids, sugar and fats. Toxic organic compound comprises Poly-nuclear aromatic hydrocarbons (PAHs), alkyl phenols, polychlorinated biphenyls (PCBs) organo-chlorine pesticides, monocyclic aromatics, chloro-benzenes, aromatic and alkyl amines, polychlorinated dioxins, phenols etc. In addition to these organic waste material sewage sludge also contains traces of many pollutants like Copper, Zinc, Nickel, Cadmium, Lead, Arsenic, Chromium, Selenium etc. Some of these substances can be phytotoxic and some toxic to humans and / or animals, so it is necessary to control the concentrations in the soil of potentially toxic elements and their rate of application to the soil. Sewage sludge also contains pathogenic bacteria, viruses and protozoa along with other parasitic helminthes which can give rise to potential hazards to the health of humans, animals and plants. Apart from those components of concern sewage sludge also contains useful concentrations of N, P and organic matter. Each component of the sludge has its own environmental impact, which must be taken into account when choosing the disposal route.

4. Processing of Sludge

Increasing urbanization and Industrialisation have resulted in a dramatic increase in the volume of waste water produced around the world. The waste water treatment step concentrates the various pollutants (upto 90%) in the waste water into sludge, normally containing between 1% and 2% by weight dry solids. The waste water treatment commonly involves the following processes to process the sludge for the production of suitable end products for utilization or disposal:

Sludge processing methods

Process	Description
Sludge pasteurization	Minimum of 30 minutes at 70°C or minimum of 4 hours at 55°C (or appropriate intermediate conditions), followed in all cases by primary mesophilic anaerobic digestion.
Mesophilic anaerobic digestion	Mean retention period of at least 12 days primary digestion in temperature range 35°C ± 3°C or of atleast 20 days primary digestion in temperature range 25°C ± 3°C followed in each case by a secondary stage which provides a mean retention period of at least 14 days.
Thermophilic aerobic digestion	Mean retention period of at least 7 days digestion. All sludge to be subjected to a minimum of 55°C for a period of at least 4 hours.
Composting	The compost must be maintained at 40°C for at least 5 days and for 4 hours during this period at a minimum of 55°C within the body of the pile followed by a period of maturation adequate to ensure that the compost reaction process is substantially complete.
Lime stabilization of liquid sludge	Addition of lime to raise pH to greater than 12.0 and sufficient to ensure that the pH is not less than 12 for a minimum period of 2 hours. The sludges can then be used directly.
Liquid storage	Storage of untreated liquid sludge for a minimum period of 3 months.
Dewatering and storage	Conditioning of untreated sludge with lime or other coagulants followed by dewatering and storage of the cake for a minimum period of 3 months. If sludge has been subject to primary mesophilic anaerobic digestion storage to be for a minimum period of 14 days.

5. Agricultural Application

The application of sewage sludge as a " safe fertilizer " started in earnest after the 1988 ban on dumping sewage sludge into the ocean. When the Ocean Dumping Ban Act of 1988 went into effect, the municipalities and the Govts. left with a new problem – how to get rid of the tons of sludge they generate on a daily basis. The federal Environmental Protection Agency (EPA) stepped in with a plan to "solve" this problem by promoting sludge (sometimes called 'biosolids', a public relations term that is used interchangeably by EPA with the technical term "sewage sludge") as fertilizer to be spread on land – where people live, work and play. Though, the viscous, black cake adds free Organic Matter and Fertilizer to poor soils, making them productive and profitable, the main limitations arising from such factors are: pathogens, heavy metals, toxic organics. Therefore, the plan of EPA has allowed toxic chemicals into air, water, soil, crops and into us. So, to call this sludge " fertilizer" is tantamount to call a soup "food" which, though it contains some meat and vegetables, also contains a bit of lead, a little arsenic, and perhaps hundreds or even thousands of other toxic organic and inorganic materials whose impact ranges from carcinogenic to teratogenic (birth defect inducing). "Most people want a simple answer; is it good or is it bad. The answer is not that simple. It is not completely risk free, but it has benefits. Just like driving a car", Sanden said.

The benefits of sewage sludge on agricultural land:

- Valuable agricultural nutrients like Nitrogen, Phosphorus, Potassium and Sulphur can be returned to the land
- Soil organic matter levels have been increased to 12% – 15%
- Ground water and surface water quality are maintained
- Decrease bulk density and increase the non-capillary pore space
- Improve the aggregation of soil particles
- No significant health or nuisance problems occur

6. Problem of Sludge

Sludge contains measurable quantities of pollutants, such as heavy metals, dioxin, and other toxic chemicals. Sludge also contains pathogens--human germs, bacteria, viruses, and parasites. And sludge smells: sludge odor is more than just a nuisance; it is a public health threat, which has been linked to respiratory problems and death. The land application of sludge distributes pollutants from large towns and cities to rural areas, far from where they were originally produced. State and federal agencies of various countries regulate sludge spreading, but regulation of this waste is difficult and problematic. Many scientists agree that the current land application rules do not protect human health, agricultural productivity, or the environment. The lack of funding to provide proper regulatory oversight and the very nature of sewage allow for sludge spreading of an unknown quality to occur on our lands.

The problems with sludge include:

- Sludge contains heavy metals, toxic chemicals, and pathogens.
- The testing and regulation of sludge is inadequate and problematic.
- Sludge odors pose a public health threat and lower quality of life.

7. THE TROUBLE WITH SLUDGE

7.1. How Toxic Sludge Become Fertilizer

In traditional agricultural societies, human waste was often used to enrich the soil. The Industrial Revolution caused increased urbanization and the need for cities to develop primitive sewer systems to remove human waste. Pipes and gutters were built to dump sewage directly into our lakes, rivers, and oceans. As industry increased in World, factories began using these primitive sewer systems to get rid of their waste. This practice continued well into 20th century, when industry began widely using toxic chemicals. Using the local sewer system as a dumping ground for toxic waste was an easy solution to their disposal problems and was cheaper than treating their waste on site. Sewage loaded with toxic chemicals created major public health and environmental disasters throughout the World: rivers caught fire, public drinking water supplies became polluted, and waste washed up on our beaches. Public outcry from the growing number of disasters led to the passage of the federal Clean Water Act in 1972. This act set water quality standards nationally and provided money to communities to improve sewer systems and create wastewater treatment facilities. Unfortunately, instead of addressing the root of the problem by stopping industrial use and disposal of toxic chemicals, the act instead regulated the amount of pollution large industries could release into sewer systems.

By the late 1970s, extensive sewage systems had been built across the country. Wastewater treatment plants were built to separate solid waste from water, and, following natural and chemical treatment, release water back into the environment, clean of human waste. Unfortunately, they were not built to treat toxic chemical waste. While these sewage systems and wastewater treatment plants improved public health standards and water quality, they have an ironic flaw. The treatment process creates cleaner water but also creates a toxic byproduct: sludge. In fact, the Clean Water Act rightly defines sludge as a pollutant. Like all waste, sludge must be disposed of in some way. What to do with sludge has been a source of controversy for the past three decades in the World. Through the 1970s and 80s, the federal Environmental Protection Agency (EPA) strictly regulated the land spreading of sludge, effectively prohibiting much of the waste from being used on agricultural land. Wastewater treatment facilities could only dispose of sludge in one of three ways: by sending it to a landfill, by incinerating it, or by dumping it 100 miles offshore into the ocean [2].

Ocean dumping eventually created large under-sea dead areas. In response to public concern, Congress passed the Ocean Dumping Act, which banned ocean dumping of sludge in 1992 [3]. Sludge disposals was then largely limited to landfills and incineration that became expensive for wastewater treatment plants.

Municipal treatment facilities then pressured the EPA to relax its standards for the land spreading of sludge on agricultural fields. Following a number of draft rewrites of EPA regulations, corporate sludge marketing companies and municipal wastewater treatment facilities were successful in relaxing the limits of toxins in sludge for land spreading. What was once considered hazardous waste became a fertilizer? By classifying sludge as a fertilizer, it became exempted from several waste management regulations.

7.2. Marketing of Toxic Sludge

Municipal water treatment facilities depend upon corporate sludge brokers to dispose of their sludge. To dispose of it, these private corporations convince farmers and landowner across the country to spread sludge on their fields as a nutrient supplement for their crops. Sludge is marketed to landowners and consumers in two different ways. The first, and most obvious, is by offering them free sludge. By convincing individual property owners that sludge is of "agronomic benefit" to their land, sludge brokers are finding extremely cheap disposal sites for sludge that would otherwise have to be shipped to landfills or incinerators at a cost of approximately $70 a ton [4].

Companies then claim that everyone wins: treatment plants have a cheap disposal option for their sludge, which gives taxpayers a break, and landowners get free nutrients for their fields. As an accurate result, the sludge brokers walk away with the disposal fees from the treatment facility. The sludge brokers also escape from potential liability, which is now assumed by the farmer or property owner. The second way sludge is marketed is by composting or palletizing it. Then it can be sold or given away as compost or fertilizer. Since the weakening of sludge regulations in the late 1980s, citizens cross the World have been fighting to keep sludge from being spread on fields and farmland in their communities. Activists fighting sludge are up against formidable opponents. Water treatment facilities and sludge brokers have formed powerful trade groups, such as the New England Biosolids and Residuals Association (NEBRA). NEBRA, in turn, is part of an even larger and more powerful group: the National Biosolids Partnership, which is a coalition of groups such as the EPA and Water Environment Federation, whose primary responsibility is to change "public perception" about sludge spreading.

7.3. Toxic Secrets of Sludge

Land applied sludge is required laws to have toxic levels below certain limits and it is treated with lime to reduce pathogen levels. However, no sludge in World is completely free of toxic chemicals or pathogens. In fact, after it is treated, Class B sludge still contains a significant amount of pathogens [5].

7.4. Toxic in Sludge

A. Heavy Metals

All sludge in world contains heavy metals like arsenic, cadmium, chromium, copper, lead, mercury, molybdenum, nickel, selenium, and zinc[6]. These metals are persistent—that is, they do not break down in the environment and therefore build up over time. As the Cornell Cooperative Extension states, "most heavy metals remain in the soil for long periods of time, ranging from several decades to many centuries." The heavy metals in land spread sludge therefore become permanent additions to the total quantity in the soil. Even extremely small amounts of heavy metals in sludge, therefore, are dangerous [7]. High levels of arsenic in food or water can be fatal. Cadmium, chromium, nickel, and selenium have been linked to cancer. Cadmium has also been linked to kidney problems, miscarriages, and stillbirths. Copper, nickel, and zinc are known to cause growth problems in crops. Children exposed to lead can develop behavioral and learning problems. Mercury exposure at key moments in fetal development can cause learning disabilities and neurological disorders. Molybdenum bioaccumulates in grass eating livestock; ingested in excess, it can cause anemia, diarrhea, and growth problems [8].

These metals can be taken up by the plants that are grown on sludge and re-enter the human food chain via livestock feed. These metals can also leach into groundwater. Highly acidic soils, like those found in Maine, can exacerbate heavy metal leaching [9].

B. Pathogens: Bacteria, Viruses, and Parasites

Sludge, by its very nature, contains human pathogens: germs such as bacteria, viruses, and parasites. Whereas exposure to heavy metals can cause problems over time, exposure to these germs is more acute and can cause health problems almost immediately. Because of the extremely large numbers of pathogens that exist in the world, it is impossible to test sludge for all types of pathogens.

Some common pathogens in sludge include the bacteria E-coli and Salmonella, the virus Hepatitis A, and parasitic worms. Pathogens can cause intestinal problems, other serious illnesses, and death. Land spread sludge can be treated to nearly eliminate pathogens. By composting sludge, for example, pathogen levels can be reduced significantly.

Unfortunately, federal and state laws allow “Class B” sludge, which has not been treated to the strictest pathogen reduction methods, to be spread. In other words, sludge with live pathogens is being spread throughout the state. Unfortunately for the residents and workers of Northern New England, wet and overcast climates encourage pathogen growth. Researchers have found that pathogens can survive in sludge for weeks, months, or even years after reduction treatment processes.

Humans can be exposed to sludge pathogens in a number of ways. We might consume vegetables that have pathogens on them. Children might accidentally gain access to a sludge field and become exposed to the germs. Pathogens can also be spread by pets or wildlife, such as deer, that walk through a sludge field.

C. Dioxin: “The Darth Vader of Chemicals”

Dioxin is the unwanted byproduct of chemical processes involving chlorine. According to the EPA, sludge spreading is the largest land distributor of dioxin nationally [10] Dioxin is a known carcinogen and has been linked to reproductive problems, genetic damage, and endometriosis. Scientific evidence suggests there is no safe exposure level to dioxin [11].

As one well-known dioxin expert called it, dioxin is “the Darth Vader of chemicals,” because you can't see or taste it, but it is deadly. The source of dioxin contamination in sludge is not known. It might be discharged into the sewer system by unknown industrial or residential sources. Dairy cattle grazing on sludged land may ingest dioxin and the chemical will then enter humans via milk and meat.

7.5. What We Don't Know Can Hurt Us

The federal Environmental Protection Agency estimates that there are 70,000 synthetic (not naturally occurring) chemicals. Yet, only 2% of these chemicals have been fully tested. In fact, even the most basic toxicity testing results cannot be found in the public record for nearly 75% of the most widely used of these chemicals. The ways in which these chemicals affect human health and the ways in which they interact with one another in the environment (their "synergistic effects") are not always known. Despite this, industry only needs to report the discharge of 1% of these chemicals into the waterways and sewers. Although industries and households release thousands of chemicals, World sludge is only regularly tested for few heavy metals and occasionally tested for dioxin and toxic pesticides.

8. SOURCE OF TOXIC CHEMICALS

Sludge contains heavy metals and other pollutants because industry and households use and release far too many toxic chemicals. The sources of contaminates in sludge are many, depending upon the specific water treatment facility and the community that it serves. Sources of contamination include industrial releases, small business discharges, hospital releases, household waste, leachates from landfills and Superfund sites, including nuclear waste dumps, and municipal water and sewer systems as a whole [12].

Everything that is discharged into a sewer that leads to a water treatment plant could potentially become part of the sludge that the facility produces. If a worker at an industrial facility accidentally dumps toxic chemicals down the drain instead of disposing of it properly, those chemicals could end up in the sludge. Likewise, if a home gardener rinses out a bottle containing toxic pesticides in the sink, those toxic pesticides could find their way to the sludge.

8.1. Industrial Hazards

As discussed earlier, many chemicals used by industry have not been properly tested and are not regulated or reported. Additionally, even at the safest facilities, accidents happen and toxic chemicals can be released into the waste stream. World requires wastewater treatment plants to work with large industries on reducing and monitoring their waste discharge.

This "pretreatment process" is required of companies that discharge a large amount of waste into the sewer system or use a large amount of chemicals that could affect the operation of the sewer system. Unfortunately, once companies release heavy metals, or other toxins, into the sewer system, there is no process to remove these chemicals from the sludge. In addition, every industry in the country can discharge 33 pounds of hazardous waste *every month* into wastewater treatment plants, without penalty or reporting [13].

8.2. Small Business Hazards

Many small businesses are not regulated for their toxic releases. Nor are they included in the pretreatment processes. While auto garages, dentist offices, photo developers, dry cleaners, and other small businesses may not individually release a large amount of toxic chemicals, taken as a whole their contribution to chemicals in sludge could be dangerous.

8.3. Hospital Hazards

All hospitals are required to dispose of toxic chemicals and biohazards in a state approved manner. Nevertheless, accidents do happen: from a broken mercury hermometer to additional human pathogens being washed down the drain, hospitals can contaminate sludge.

8.4. Contamination from Municipal Water and Sewer Systems

Many towns and cities have water and sewer systems made with lead and copper pipes. Lead, copper, and other metals often leach into the waste stream and contaminate sludge. Contamination of sludge can also occur if a town's reservoir is polluted with pesticides and other chemicals for which testing are not required.

8.5. Household Hazards

From pesticides (including flea shampoos), to heavy duty cleaning agents and hair coloring products, toxic chemical containing products abound. Any of these chemicals dumped down the drain could end up being spread on a farm field or in a forest.

9. Sludge Regulation

It is nearly impossible to know the exact levels of toxic materials in each batch of sludge because what is released into the waste stream varies day to day. While sewage waste is treated at wastewater facilities for several days, not every batch of sludge is tested before it leaves the plant. It is more due to economics than to concerns for health protection, that sludge generators do not test the waste more frequently. For example, waste is often only tested for dioxin twice a year because of the cost of the test. A worker may accidentally spill pesticides into a sink or storm drain, or someone might illegally dump other toxic chemicals down the drain, and no matter how strict regulations are in the law books, testing could miss these sudden increases in contaminants. Regulations and testing cannot guarantee sludge safety until toxic chemicals are removed from industrial household use.

10. Sludge Consequences

"Temporary odors are a necessary inconvenience in the practice of agriculture." [14]. Sludge smells similar to manure and that the smell will dissipate "within several days." Despite industry propaganda, studies have shown that sludge odors are more than just a nuisance; they are a public health threat. Harmful gases, called organic amines, can develop from chemical reactions that occur in sludge. These gases are released when the pH of sludge is raised above 10, such as when lime is added. Studies suggest that sludge odor can cause health problems in humans as far as 1600 feet from a site [15]. A study performed by a former EPA sludge regulator linked sludge odors to "severe irritation to mucous membranes followed by respiratory infections" in residents living near a sludge site. Irritation of the eyes, throat and skin make infection from pathogens in sludge more likely. The study was conducted following the death of a New Hampshire man suffering from respiratory distress in the vicinity of a sludge site [16]. Residents near sludge sites have not been the only victims of sludge odor. Symptoms associated with organic amine poisoning frequently occur among waste treatment plant workers and drivers who haul sludge.

10.1. Deaths Associated with Sludges

At least two deaths have been associated with sludge spreading. In October 1994, an eleven-year-old boy, named Tony Behun, went dirt bike riding near his home in Osceola Mills, Pennsylvania. Unknowingly, the boy rode through a field covered in Class B sludge. He came home covered in dirt and grime. Two days later, he developed a sore throat, headache, and a boil on his left arm. Brenda Robertson, his mother, took him to the doctor, who prescribed flu antibiotics.

The next day, Tony had trouble breathing. He died after being flown by helicopter to a hospital in Pittsburgh. The final diagnosis was that Tony had died from a bacterial infection. How her son contracted the infection remained a mystery to Brenda Robertson until five years later when she read about an investigation into her son's death by the Pennsylvania Department of Environmental Protection. Without consulting Brenda, the state published a report concluding that Tony died of a bee sting and that Class B Sludge *was not* spread on property that he went riding on.

Another sludge related death occurred in Greenland, New Hampshire. In late October of 1995, the Marshall family had their otherwise quiet lives tragically disrupted. Sludge was being dumped on a field in their rural neighborhood.

This was just the beginning of the residents' problems. On Halloween, Joanne Marshall rushed home from work to take her little girl trick-or-treating. When she arrived home and jumped out of her car, she was "greeted by such a stench, it took her breath away. "[17]. The Marshalls and their neighbors began suffering from nausea, vomiting, stomach cramps, migraine headaches, flu-like symptoms, slowed reflexes and respiratory problems.

10.2. Environmental Assessment and Some Remedy

Recycling sewage sludge to agricultural land to gain benefit from the essential plant nutrients and organic matter it contains, would seem a reasonable and rational method of managing a material which would otherwise need disposing of by some other non-beneficial route. But sludge also contains inorganic, organic and biological contaminants and so careful, management is required to avoid the potential environmental problems. The problems are listed in following Table. Large application of sewage sludge can decrease the soil pH. This can be avoided, if the soil pH is increased by application of lime, or if sludge application rates are limited in some way.

The no. of bacteria of different genera in sludge varies. In general, a total coliform count of 10 to 10 can be found per gram of dry wt., while fecal coliform bacteria generally represent 10 to 10 per gram of dry wt. The pathogens should be reduced to levels that are unlikely to cause a threat to public health and the environment under specified use conditions processes to significantly reduce pathogens, such as digestion, drying, heating and high pH or their equivalent are the most commonly used one.

For the removal of OCs from sludge mainly two approaches ar there – physico chemical or microbiological which involves either high temperature oxidation (incineration) or reductive dechlorination (pyrolysis in an atmosphere of hydrogen). To achieve allow level of risk, presticide concentrations in the combined soil and sludge mixture must be less than 1.25 mg/kg dry wt.

Environmental impact risk and benefit assessment for sewage sludge recycling to agricultural land (B= beneficial effect, L=Low risk, P=Possible risk, NA=Not applicable)

Environ-mental parameter	PTEs	Organic contaminants	Pathogens	Nitrogen	Phosphorus	Organic matter
Human health	L	P	L	B	B	B
Crop yields	L	L	L	B	B	B
Animal health	L	L	L	B	B	B
Ground water quality	L	L	L	P	L	L
Surface water quality	L	L	L	P	P	B
Air quality	L	L	L	P	NA	NA
Soil fertility	P	L	L	B	B	B
Natural ecosystem	P	P	L	P	P	B

11. Sludge Regulation

Sludge, by its very nature, is difficult to regulate. Depending upon what chemicals are being released into various sewer systems minute to minute, the toxicity of the state's sludge could vary day-to-day, minute-to-minute. Regulations of sludge do not adequately protect public health and the environment.

11.1. Regulations Problems

- Have weak pollution standards;
- Allow for the spreading of sludge containing live pathogens;
- Discourage municipalities from being precautionary and public health oriented by not allowing them to make stricter standards than the state's; and
- Marginalize citizens' voices in the process as the sludge industry has greater access to state regulators than the average citizen.

11.2. Heavy Metals Standards (in ppm)

Heavy Metal	Denmark	Sweden	Finland	Germany	Netherlands	Norway	European Union
Arsenic	25	N/A	N/A	N/A	0.15	N/A	N/A
Cadmium	0.8	2.0	1.5	5 or 10*	1.25	2.5	20
Chromium	100	100	N/A	900	75	100	N/A
Copper	1000	600	N/A	800	75	1000	1000
Lead	120	100	100	900	100	80	750
Mercury	0.8	2.5	1	8	0.75	3	16
Nickel	30	50	100	200	30	50	.300
Zinc	4000	800	1500	2500	300	800	2500

*Source Harrison, et al. 1999 [7].

11.3. Sludge vs. Natural soil

Heavy Metal	Average Sludge (ppm)	Natural Soil (ppm)	Times Higher than Natural Soil
Arsenic	5.6	7.4	1.3
Cadmium	2.4	0.37	6.4
Copper	388.0	23.3	16.6
Chromium	33.3	30	1.1
Lead	61.5	17	3.6
Mercury	1.2	0.003	400
Molybdenum	7.5	0.79	9.4
Nickel	22.8	18	1.2
Selenium	2.6	0.45	5.7
Zinc	468.5	68.5	6.8

11.4. Standard Values for Organic Compounds

Compounds	Concentration in sludge
PAHs	1-10 mg./Kg.
Compounds	Concentration in sludge
Alkyl phenols	100 – 3000 mg./Kg.
PCBs	1 - 20 mg./Kg.
Poly chlorinated dibenzo-p-dioxins	Very low < μg /Kg.
OC pesticides	< low mg./Kg.
Monocyclic aromatics	<1 – 10 mg./Kg.
Chloro benzenes	<0.1 – 50 mg./Kg.
Aromatic and alkyl amines	0 – 1mg./Kg.
Phenols	0 – 5mg./Kg.

12. The Sludge Solution

If spreading sludge in our communities is dangerous, where should it go? What are we supposed to do with this waste? The real question is, how can we eliminate the spreading of toxic pollutants on our land and how can we eliminate these contaminants from our wastewater treatment plant so that human waste becomes a truly useful and safe commodity? Because sludge contains toxic chemicals and other pollutants, the best solution to our sludge problem is reducing these contaminants at their source. By dramatically reducing the use and disposal of industrial and household toxic chemicals we can greatly cut the chemical levels in sludge. Until the long-term goal of eliminating the use and disposal of toxic chemicals is achieved, the state should:

- Ban the use of sludge that contains industrial discharges.
- Require the strictest level of pathogen reduction.
- Broaden and strengthen sludge testing and toxic limits.
- Allow municipalities to enact ordinances that are more stringent than the state's regulations through the town meeting or a town-wide vote process.
- Provide for the long-term pH maintenance and metal monitoring of sludge sites.

In addition to statewide protections, municipalities should also enforce their own protections through strong ordinances controlling sludge. It is, after all, local communities that are most threatened by sludge spreading.

13. Disposal of Sludges

Sludge disposal is a worldwide problem and a wide variety of disposal routes have been adopted as directed by local conditions. The final resting place of the sludge must be either on the land, in the air or in the water. Disposal of sludge to the air largly employs high temperature incineration or pyrolysis. Although, this reduction is sufficient to "stabilise" the sludge, a large volume remains for disposal. Disposal of sewage sludge to the ocean in now banned because of its perceived environmental effects. The major sludge disposal methods employed by the waste water treatment plants are alienation or selling lagooning, used for municipal gardens, used for instant lawn cultivation, land application. The remaining of the sludge is either stockpiled or land filled [18].

Disposal and Application of Sludge's Should Involve the Following

1. The application must contain a summary of the types of crops to be grown on the proposed site, the method of sludge application, and an anticipated spreading schedule. The application must also include a representative soil nutrient analysis for the site.
2. The sludge must provide "agronomic benefit" to the crops grown on this soil--meaning the generator must show that the site has a need for the nutrients provided for by the sludge. Farms utilizing sludge are required to have a licensed nutrient management specialist develop a whole farm nutrient management plan. This plan is the basis for the above determination that additional nutrients are needed on the farm [19].
3. The application must show that "the water of the state will be protected." In practice, state regulators assume that the waters of the state will be protected as long as certain setbacks and spreading requirements are provided for in the application.
4. To this end, sludge cannot be spread when soil is frozen, snow covered, and water logged. Sludge cannot be spread on land that favors the growth of water loving plants such as wetlands, swamps and others.
5. The soil of a proposed sludge site must have a six-inch soil cap and a minimum depth to bedrock of 10 inches for perennial crops (such as hay) and 20 inches for row crops (such as corn).
6. For Class B sludge, spreading may not occur within 25 feet of on-site waterways, including gullies, ravines, and swales. Sludge sites may not be located within 75 feet of a river, perennial stream, or great pond.
7. The application must include a statement as to whether or not the site is located on or next to a protected natural resource, a sensitive area, and/or a direct watershed to waters.
8. The generator must demonstrate that the sludge spreading activity will meet traffic standards for the site. This standard is assumed to be met if the sludge spreading activity will result in 16 or less vehicle trips a day [20].
9. The application must include a site-specific odor control plan to prevent nuisance odors at neighboring properties. It assumes that odor, air quality, and nuisance standards will be met at the site if the site is 300 feet from occupied buildings, if there is a site specific odor control plan [21]
10. The application must prove that the sludge is "non-hazardous". To prove this, the application must include an analysis of the heavy metal levels in the sludge. If the generator's sludge contains heavy metal concentrations above screening concentrations then the application must include a sampling and monitoring plan as well as demonstrate that the maximum heavy metal soil concentration will not be exceeded.

11. The application must also include an analysis of the dioxin level in the sludge. If a generator's sludge contains 27 parts per trillion of dioxin, then the application must include a statement signed by the generator, the landowner, and the operator acknowledging the dioxin in the sludge to be spread.

The statement must also include an agreement to the following conditions:

- The site will be tested for dioxin within 3 months of the last sludge spreading.
- If the soil on the site contains 27 parts per trillion of dioxin, then livestock intended for human consumption may not be pastured on site, crops for human consumption may not be grown on the site, and the deed to the site must record this information.

12. The application must also include a sampling plan: how often and in what manner the sludge will be tested for heavy metals and other toxins [22].
13. Sludge will be spread at a minimum of 15 inches above groundwater surfaces. Food crops grown on the site with harvested parts that touch the soil will not be harvested for 14 months after the last sludge spreading.
14. If the sludge remains on the land for four months or more before being incorporated into the soil, food crops that grow below the soil cannot be harvested for at least 20 months after the last sludge spreading.
15. Food crops, feed crops, and fiber crops grown on the site but do not have harvested parts that might touch the sludge cannot be harvested for at least 30 days after the last sludge spreading.
16. Domestic animals are not allowed to graze on the land for at least 30 days after the last sludge spreading.
17. Turf grown on the site cannot be harvested for one year after the last sludge spreading. The application must contain site maps, including: a topographical map; a sketch of the site; a tax map; soils map (from U.S. Department of Agriculture); sand and gravel aquifer map; and a flood zone map.
18. The site sketch should include all the set backs and buffers that will be incorporated, as well as the location of onsite and abutting roads, wells, and buildings. The topographical maps are used to determine slopes at the site. The soils, sand and gravel aquifer, and flood zone maps are used to determine if the site is suitable, in a regulatory sense, for sludge spreading activities [23].

14. Conclusion and Recommendation

14.1. Policy Recommendation

- Prohibit sludge that contains industrial discharges from being land applied. The best way to ensure that our rural land is protected from industrial contamination is to ban the use of sludge that contains these toxins.

- Require land spread sludge to undergo the strictest pathogen reduction method available. Sludge with viruses, bacteria, and parasites above background levels should not be land applied.
- Broaden and strengthen sludge testing parameters. Sludge needs to be tested more frequently for more contaminants. In order to best protect public health and the environment, allowable pollutant levels should be guided not only by toxicology but also by natural background levels as well.
- Allow municipalities to enact ordinances that are more stringent than the state's through a town meeting or town-wide vote. The people who are most affected by sludge sites are local residents. It is important that these residents have a voice when it comes to decisions that affect their community.
- Provide for long-term maintenance of sludge sites. Sludge generators should be responsible for testing the pH of all sludge application sites, whether active or closed, and cover the costs of lime (or other amendments) to maintain safe soil pH. All large volume sludge activities should be recorded on deeds so that future potential buyers are aware of past use of the property.

15.2. Recommendation for Municipalities

- Sludge is an especially important issue for municipalities to oversee: it is local residents that have the most to lose from the threat of sludge.
- In municipalities that are home to a wastewater treatment facility, local residents, town officials, and directors of the facility can work together to implement the above statewide recommendations at the local level.
- All towns have the authority to ban the use of sludge, or sludge materials (such as compost) on municipal property.
- Municipalities can also pass strict ordinances controlling sludge application. Although the state preempts local control on setting strict standards, there are several ways towns can discourage sludge spreading.

15.3. What Concern Citizen Can Do?

Citizens can protect themselves and their community from the dangers of sludge by being proactively engaged in sludge reform. Depending upon the needs of the community, citizens can reform sludge rules through engaging town officials, local and statewide public health and environmental groups.

References

[1] Dean and Suess (*1995)*. Toxic Sludge Is Good For You!, Center for Media and Democracy. Published by Common Courage Press, Monroe, ME. p. 101-107.

[2] www.vpirg.org , On the Ground, The Spreading of Toxic Sludge in Vermont, Vermont Public Interest Research Group,VPIRG, 64 Main St., Montpelier, VT 05602. (802) 223-5221.

[3] vpirg@vpirg.org. 7-9.

[4] www.vpirg.org , Conversations with DEP Officials and Staff of Portland Water District.

[5] vpirg@vpirg.org On the Ground, The Spreading of Toxic Sludge in Vermont, Vermont Public Interest Research Group, VPIRG, 64 Main St., Montpelier, VT 0560. (802) 223-5221. 35-36.

[6] State of Maine Solid Waste Management Regulations Chapters 419, 400, 405, and Appendix A of Chapter 418, as well as repealed *Chapter* 567, Department of Environmental Protection, Bureau of Solid Waste Management, 17 State House Station, Augusta, ME 04333-0017. Chapter 419, 17.

[7] Harrison, Ellen Z. et al, (1999) The Case for Caution, Recommendations for Land Application of Sewage Sludge and an Appraisal of the US EPA's Part 503 Sludge *Rules*, Cornell Waste Management Institute, Center for the Environment, Cornell University, Ithaca, NY 14853. February.

[8] www.vpirg.org ,vpirg@vpirg.org On the Ground, The Spreading of Toxic Sludge in Vermont, Vermont Public Interest Research Group, VPIRG, 64 Main St., Montpelier, VT 05602. (802) 223-5221. 12-14.

[9] www.essential.org/cchw America's Choice Children's Health or Corporate Profit, Center for Health, Environment, and Justice, PO Box 6806, Falls Church, VA 22040 703.237.2249, 546

[10] Gibbs, Lois Marie et al. (1995) Dying from Dioxin: A Citizen's Guide to Reclaiming Our Health and Rebuilding Democracy. South End Press, Boston.. p. 25.

[11] vpirg@vpirg.org 10-11.

[12] Scott, Laura, et al. (1998) The Sludging of New Hampshire. Answers for Local City and Town Officials in New Hampshire. New Hampshire Sierra Club. "Land Application of Wastewater Biosolids in Maine." Maine Wastewater Control Association brochure.

[13] Lewis, David L., et al. Enhanced Susceptibility to Infection From Exposure to Gases Emitted by Sewage Sludge: *A Case Study*, Departments of Marine Sciences, Biological and Agricultural Engineering, and Medical Microbiology, University of Georgia, Athens, GA 30602, BIOSET, Inc, 13700 Veterans Memorial, Ste. 385, Houston, TX, 77014. (conclusions)

[14] Tuohy, John, (2000) "State probe wrongly followed path of bike ride to a bee sting," *USA Today*, July 13,. 20. Statement of Joanne Marshall.

[15] www.essential.org/cchw. "A Comparison of Heavy Metals in Sewage Sludge, Soil, and Applicable Regulatory.

[16] cchw@essential.org Standards," 10/10/00 fact sheet from Maine Department of Environmental Protection. 47.

[17] State of Maine Solid Waste Management Regulations Chapters *419, 400, 405, and* Appendix A of Chapter 418, as well as repealed Chapter 567, Department of Environmental Protection, Bureau of Solid Waste Management, 17 State House Station, Augusta, ME 04333-0017. Chapter 2, 21-22.

[18] Standards," 10/10/00 fact sheet from Maine Department of Environmental Protection.

[19] State of Maine Solid Waste Management Regulations Chapters *419, 400, 405, and Appendix A of Chapter 418,* as well as repealed *Chapter* 567, Department of Environmental Protection, Bureau of Solid Waste Management, 17 State House Station, Augusta, ME 04333-0017. Chapter 419, 7-10.

[20] State of Maine Solid Waste Management Regulations Chapters *419, 400, 405, and Appendix A of Chapter 418,* as well as repealed *Chapter* 567, Department of Environmental Protection, Bureau of Solid Waste Management, 17 State House Station, Augusta, ME 04333-0017. Chapter 419, 26.

[21] State of Maine Solid Waste Management Regulations Chapters 419, 400, 405, and Appendix A of Chapter 418, as well as repealed *Chapter* 567, Department of Environmental Protection, Bureau of Solid Waste Management, 17 State House Station, Augusta, ME 04333-0017. Chapter 419, 7-10 and 26.

[22] State of Maine Solid Waste Management Regulations Chapters 419, 400, 405, and Appendix A of Chapter 418, as well as repealed *Chapter* 567, Department of Environmental Protection, Bureau of Solid Waste Management, 17 State House Station, Augusta, ME 04333-0017. Chapter 400, 28.

[23] State of Maine Solid Waste Management Regulations Chapters 419, 400, 405, and Appendix A of Chapter 418, as well as repealed *Chapter* 567, Department of Environmental Protection, Bureau of Solid Waste Management, 17 State House Station, Augusta, ME 04333-0017. Chapter 419, 26-27.

In: Sewage Treatment: Uses, Processes and Impact
Editor: Anna Stephens and Mark Fuller

ISBN 978-1-60692-959-9

Chapter 8

WASTEWATER SEWAGE SLUDGE DISPOSAL BY PYROLYSIS AND VITRIFICATION

Marco Ischia[1], Maurizio Grigiante[2], Marco Baratieri[2] and Roberto Dal Maschio[1]

[1]Department of Materials Engineering and Industrial Technologies, University of Trento - Via Mesiano 77, 38100, Italy

[2]Department of Civil and Environmental Engineering, University of Trento - Via Mesiano 77, 38100, Italy

ABSTRACT

The feasibility to dispose wastewater sewage sludge through thermal degradation methods, was studied in consideration of the recent European environmental directives, which limited the use of this solid waste as fertilizer. In particular, the pyrolysis of sewage sludge was investigated, in order to obtain a gas phase, rich in hydrocarbons, for a fuel gas use. The characterization of this produced gas phase, and the determination of its heat of combustion, are fundamental data for the application of a pyrolysis disposal method. For this reason, thermogravimetric coupled mass spectrometric and thermogravimetric coupled gas chromatographic and mass spectrometric analyses were carried out for the obtainment of an accurate characterization of the sewage sludge pyrolysis process. Tests on a laboratory-scale reactor were furthermore carried out with charges of 0.3 kg of sewage sludge. Analogously to the ashes of an incinerator plant, the pyrolysis solid residue is a waste with high content of hazardous heavy metal ions, and in consequence it must be stabilized. For this aim, vitrification is considered the best method which bears the immobilization of the heavy metal ions into a vitreous matrix. This vitrified material doesn't need landfill collocation, but it can be used as inert additive in ceramics and concretes field. On the basis of these considerations, the disposal of wastewater sewage sludge, through a pyrolysis-vitrification process, was studied, considering this two steps process as an integration of the depuration systems present in the wastewater treatment plant.

In order to apply this disposal treatment for sewage sludge produced by the wastewater treatment plant, the resolution of the mass and energy balances was carried out by applying a thermochemical equilibrium model.

Keywords: pyrolysis study, thermogravimetric-mass spectrometric-gas chromatographic analyses, sewage sludge disposal, vitrification.

INTRODUCTION

European urban wastewater treatment directives limited pollutions due to insufficient treatment of waste waters, by suggesting secondary treatments. The Directives introduce general rules, regulations, and/or specific authorizations for the discharge of waste waters, also arising from industrial sectors, and finally regulate the disposal of sludge from urban wastewater treatment plants. The directives encourage the recycling of sludge arising from wastewater treatment, whenever this produces environmental benefits.

Disposal routes shall minimize the adverse effects on the environment. Moreover, the disposal of sludge by dumping from ships, by discharge from pipelines, or by other means is phased out.

The rules prevent the disposal of wastewater sewage sludge as fertilizer, because its considerable content of heavy metals, and the presence of bacteria and viruses.

The landfill collocation was the option chosen by many member countries, but this implies the increase of the areas devoted to landfill use, subtracting soils from agricultural or other uses. From many years, the Society has realized that the landfill collocation is not a solution for the waste problem, but instead the expression of the will to delay to others the problem of the solid wastes disposal.

Moreover, taking into account that the average production of dehydrated sewage sludge is around 0.2-0.3 kg per inhabitant per day [1], and that the chemical composition of sewage sludge presents high content of Carbon, a very high amount of energy is wasted when the sludge is collocated in landfill.

This consideration has prompted the use of biomasses and sludge as fuels. When the direct utilization of these waste is prohibitive for many factors (ignition, low calorific value, emission or burning problems), a co-combustion with coal was suggested. However, although incineration appears a useful method, the cost for the scrubbing of gaseous by-product up to acceptable limits cannot be disregarded.

For these reasons, pyrolysis and gasification was investigated as viable options. Pyrolysis process involves the heating of sludge and biomasses in inert atmosphere, leading to the production of carbon, tar and gas. The pyrolysis products can be used as fuel, or they can be convert into valuable chemicals [2].

In past, pyrolysis tests of sewage sludge was carried out on fluidized bed reactors, but the researchers reached the conclusion that the process presents high investment and running costs, and it was not interesting economically, because the pyrolysis products could be used only as fuel [1].

However, at present, the rapid increase in discarded sewage sludge legislation, beside the continuous increase of the fuels price, give to pyrolysis of sewage sludge renewed interest.

Generally, wastewater sewage sludge presents a carbon content of 20-40 %wt, and a calorific content which ranges from 9 to 20 MJ/kg [3].

INSTRUMENTAL TECHNIQUES IN THE CHARACTERIZATION OF THE PYROLYSIS PROCESS OF WASTEWATER SEWAGE SLUDGE

The accurate characterization of the sewage sludge pyrolysis process is a fundamental operation for its industrial scaling. Data concerning the amount of the mass loss during the process, the features of the produced gaseous and liquid compounds, the energy required by the thermal decomposition process, are indispensable to develop any hypothesis.

For this aim, thermogravimetric analysis is a versatile method to obtain more information such as the thermal stability of the sludge, the intensity of its mass loss and the thermal decomposition temperature range. Moreover, it is possible to observe the behaviour of the solid waste during its thermal treatment, at different atmosphere (inert or oxidizing conditions), or at different heating rate [4].

Thermal analysis is also applied to study the effect of the presence of more species, during a thermal treatment. Among this topic, co-pyrolysis or co-combustion of sewage sludge/coal and biomasses/coal mixtures, or blends of waste, were studied by TG [1, 5, 6].

Thermogravimetric analysis allows to obtain information about the kinetic of a solid thermal degradation process [7, 8]. Thus, the use of many mathematical models available, can furnishes an adequate kinetic description of the process, in terms of Arrhenius parameters (function of conversion, activation energy, and the pre-exponential factor).

The potentialities of the thermogravimetric analysis considerably increase when an evolved gas detection system is connected to the thermobalance. Generally, for simultaneous thermogravimetric-evolved gas analysis, mass spectrometry and Fourier transformation infrared spectroscopy are used. These techniques allow the identification of the compounds which are released from the solid sample during its thermal treatment. Moreover, they can furnish the real time monitoring of the process, giving a qualitative and quantitative analysis of the evolved species. Mass spectrometry is preferred for its higher sensitivity.

Usually, standard TG-MS instrumentation, although differently assembled, operates in electron impact ionization mode, and utilizes quadrupole mass spectrometer which are able to collect specific m/z ion signals [2, 9-11]. These signals are used to detect the evolution of the released compounds, and this operative approach, labelled "Selected Ion Monitoring", requires a preventive attribution of detectable m/z signals to pre-definite compounds. Indeed, some of these attributions could be inaccurate, because the evolution of more species with similar m/z ion pattern could take place contemporaneously. In this case, the evolution of the considered m/z signal is given by the contribution of all the contemporaneously evolved compounds which present this signal in their mass spectrum.

Typical examples are: the attribution of m/z=15 signal to CH_4 release, even if the CH_3^+ ion is commonly formed in the ionic fragmentation process of each hydrocarbon compound containing a methyl group. Another examples is the use of m/z=28 to monitor the release of CO, when this fragmentation ion can be due by the evolution of CO_2, unsaturated hydrocarbons (C_2H_4), and N_2.

A continuous Scan monitoring of the produced fragmentation ions may be preferable for a better analysis, although a small number of TG-MS experiments were carried out with the mass detector that works in "Scan mode", because the major complexity of the instrumentation.

Moreover, in case of contemporaneous evolution of more chemical species with similar ion fragmentation pattern, only the gas chromatographic separation of the gas mixture can give a correct identification of the evolved compounds, with a semi-quantitative determination.

A successive improvement in the pyrolysis process characterization, concerns the running of experiments on lab-scale reactors. Many pyrolysis experiments are reported in literature, with continuous [12-14] or batch reactors, also using different heating methods [15-19].

Regarding the chemical analysis of species evolved in gas phase during the pyrolysis of sludge samples in reactor experiments, generally, released compounds were first condensed in cold traps, and subsequently analyzed by using gas chromatographs equipped with thermo-conductivity or flame-ionization detectors, or MS instrumentation.

Sewage Sludge Pyrolysis Tests on a Home Assembled TG-MS/TG-GC-MS Instrumental Plant/Apparatus

For the study of wastewater sewage sludge pyrolysis process, a home-assembled instrumental plant was used.

A themogravimetric balance was connected by a mass spectrometer or by a gas chromatograph equipped with a mass spectrometer as detector, by two types of interface, realized by modifying the gas inlet system of the thermobalance.

In TG-MS configuration, gas species, released from the solid sample during thermal analysis, were drawn into an alumina tube fixed inside the furnace of the thermobalance close to the sample crucible and then connected to a capillary silica column heated at 300°C. The gases were then directly sucked and introduced into the ionization chamber of the mass spectrometer.

This instrumental interface allows:

i. the graph of the total ion current (TIC), obtained from the contribution of all ions present in each recorded mass spectra, plotted *vs.* time (or pyrolysis temperature), whose trend allows to detect any chemical species released from the sample;
ii. the plot of a selected m/z ion current (IC), which could monitor the evolution of a single specific molecular species (during the whole pyrolysis), if this ion arises from the fragmentation pattern of that chemical species only;
iii. the integrations of the m/z ion currents *vs.* time, calculated by the mass spectrometer software (in automatic or manual mode), yielding integrate values related to a quantitative information of those detected ion species.

In fact, each chemical compound (subsequently indicated with a 'j') can be represented by an appropriate ion among the ones formed from its ionization and fragmentation process occurring during mass spectrometric analysis.

Any fragment ion will be indicated as '$m/z(i)$'. Unfortunately, some of these ions can arise from the fragmentation pathway of different molecules and this fact must be taken into account. Generally, species observed in lower quantity or in traces are neglected in the following procedure.

To approximately quantify the amount of a released species, the signal of its representative ion has been first integrated during the whole pyrolysis process extrapolating this m/z ion current trend from the TG-MS analysis. Integrated values will be indicated as '$Im/z(i)$'.

Moreover, the relative intensity of this ion was calculated by considering all the peaks heights present in the mass spectrum of that compound:

$$r(i)\mathrm{j}=h(i)\mathrm{j}/\Sigma h(i)\mathrm{j}. \quad (1)$$

The total amount of that released species, indicated with '$A(j)$', has been then estimated by dividing the integrated m/z signal by the relative intensity of its representative ion:

$$A(j)=Im/z(i)/r(i)\mathrm{j}, \quad (2)$$

labelling it as integrated amount.

In turn from this last value, the contribution of each ion present in the mass spectrum of that j-species can be obtained, as integrated ion current

$$[Im/z(i)\mathrm{j}] \text{ where } Im/z(i)\mathrm{j}=A(j)r(i)\mathrm{j}. \quad (3)$$

Finally, the molar ratio, or percentage, of each chemical species present in the evolved gas mixture can be calculated by means of the integrated amount of all compounds:

$$\%\mathrm{mol}(j)=100A(j)/\Sigma A(j). \quad (4)$$

This procedure affords a gas phase semi-quantitative analysis by considering the mass spectrometer as a gas chromatographic detector presenting the same sensibility in revealing any different chemical compounds [20-23].

When the contemporaneous evolution of more species takes place, the revealed mass spectrum in TG-MS configuration is the sum of the contribution of all evolved species. In this case, the second instrumental configuration TG-GC-MS allows an accurate analysis, through the chromatographic separation of the contemporaneously evolved compounds and their identification by the successive mass spectrometric analysis.

This second instrumental interface is realized by connecting the capillary silica column, heated at 300°C, to the inlet port of a micro-sample valve. In this case, an external vacuum pump continuously sucked a fraction of the gas flow from the thermobalance through the valve, whereas, an appropriate gas chromatographic capillary column was connected to the out-port of the valve. Gas sample injections were made for temperatures corresponding to the most important mass losses during TG analyses.

This second instrumental configuration allows:

(i) to separate different compounds released during the same mass loss;
(ii) to identify released chemical species on the basis of their different retention times and mass spectra.

TG-GC-MS analysis allows to collect the relative intensities of all ions, $m/z(i)$, detected in the mass spectrum (recorded in the same operative conditions) of each chemical species, j, when these compounds were eluted as pure substances in the gas chromatographic measurements. From this global picture a correct choice of the suitable ions for the integration runs can be done.

In the pyrolysis tests carried out, a LabSys Setaram thermobalance was used. This TG/DTA thermobalance operates in the 20–1000°C range, under 100 $cm^3 \cdot min^{-1}$ He flow, with a heating rates of 5, 10, 15, and 20 $°C \cdot min^{-1}$ chosen for the conduction of kinetic runs. Samples were analysed by using a 0.1 cm^3 alumina crucible and α-Al_2O_3 as reference.

A VG-QMD-1000 Carlo Erba Instruments quadrupole mass spectrometer was used as detector of the evolved gas species. Electron impact mass spectra (70eV) were continuously registered and stored with frequency of 1 $scan \cdot s^{-1}$ ranging from 2 to 500 amu.

In TG-GC-MS measurements, gas chromatographic (GC) analyses were carried out on an HRGC Carlo Erba Instruments chromatograph, equipped with a GR8 Bimatic thermostatted micro-valve for gas sampling, and the VG-QMD-1000 Carlo Erba Instruments quadrupole mass spectrometer as detector.

Chromatographic elutions were performed by using OV1 Mega (25 m, 0.32 mm) and poraPLOT Q Chrompack (25m, 0.32 mm) capillary columns with a temperature program of 30°C for 5 min, followed by 10$°C \cdot min^{-1}$ heating rate up to 200°C, held for 15 min. Helium was used as carrier gas with 15 kPa inlet pressure.

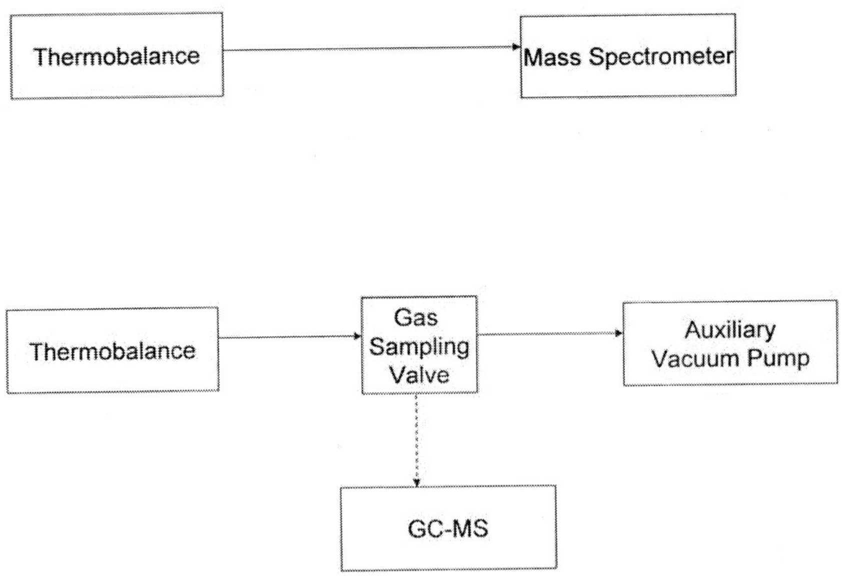

Figure 1. Thermogravimetric coupled mass spectrometric and thermogravimetric coupled gas chromatographic and mass spectrometric instrumental configurations (TG-MS / TG-GC-MS). In the first case, gas species evolved during the thermal treatment of the solid sample are continuously aspirated by the vacuum of the mass spectrometer. This configuration allows the real time monitoring of the pyrolysis process. In the second case, a gas sampling valve, equipped with an auxiliary vacuum pump, allows the sampling of the gas phase evolved from the solid sample in correspondence of a thermogravimetric event. The collected gas phase is then analyzed by gas chromatographic-mass spectrometric analysis.

Experimental Gasification Pilot Plant

Experimental activities concerning gasification/pyrolysis processes requires suitable experimental equipments. The proposed one consists of a bench-scale apparatus specifically designed for this experimental work. Core of the plant is an indirectly heated batch reactor operating at variable pressure.

The reactor, reported on the following Figure 2, is inserted from above in the laboratory cylindrical furnace, open on the top side, where all the control and acquisition devices are connected. The vessel, half filled of feeding material (biomass), can be continuously purged with nitrogen to sweep the product gases from the reaction zone, either to avoid high concentration of the products in the effluent gas or to reduce the extent of secondary reactions. The temperature is monitored by a set of thermocouples placed in the reactor. In order to assess the yields of the syngas and its composition as a function of the parameters able to influence the process principal reactions, the characterization of the effluent gas has been performed by means of gas-chromatographic analysis on gas samples.

The reactor consists of a cylindrical-shaped Incoloy vessel having internal diameter of 142mm, height of 400mm, thickness of 4mm and internal net volume of 6.3 L. Two blind flanges (210mm diameter and 20mm thickness), made of the same material, seal the tubular reactor on the top and bottom sides. The top flange is equipped with 8 pipes (diameter 10mm x length 300mm), used both for inserting the measure and control devices and for the gas inlets and outlets; on the same flange there is also a central flue gas discharge duct, provided with a ball valve. The total reactor height, including flanges and ancillary pipes, is equal to 746mm.

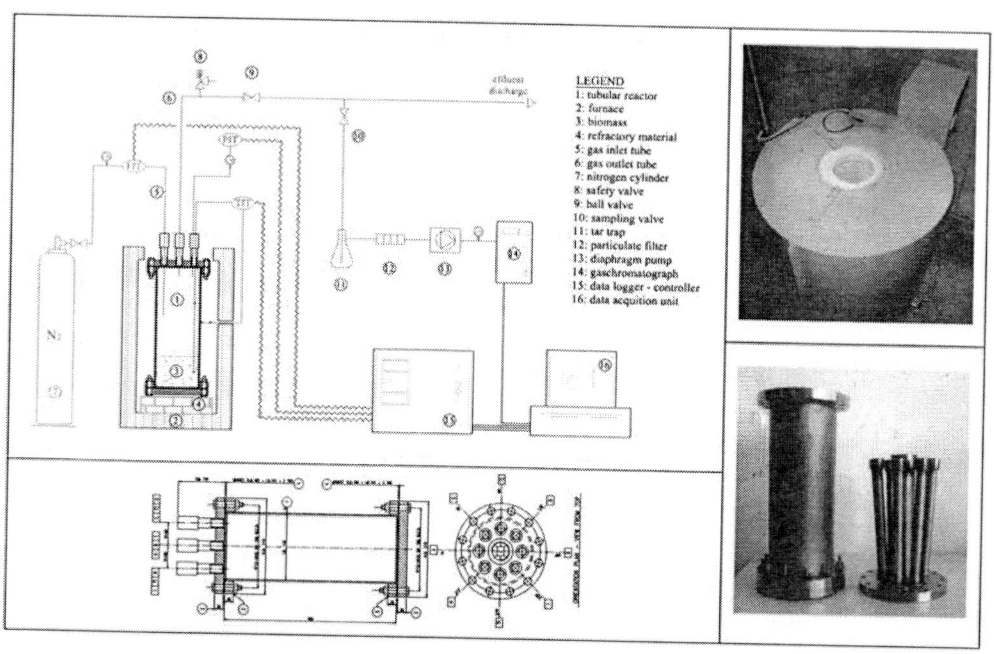

Figure 2. Gasification equipment: overview (top left); pit furnace (top right); reactor (bottom right); reactor design data (bottom left).

The reactor air-tightness is assured by 12 bolts and nuts, fastened on each flange; moreover, union sleeves with threads have been welded on the top of all the inlet/outlet tubes, in order to connect the necessary measure probes.

The reactor has been manufactured using Incoloy 800® (30%Ni, 21%Cr, 46%Fe, other metals in traces), an alloy suitable for operating at high temperatures and pressures. This is a material widely used for construction of equipment requiring high corrosion resistance, heat resistance, strength, and stability for service up to 800°C. At high temperatures it offers resistance to oxidation, carburization and sulphidation along with rupture and creep strength. The reactor design parameters are a maximum temperature of 800°C at 3barg pressure and 700°C at 7barg. same flange there is also a central flue gas discharge duct, provided with a ball valve.

The reactor is inserted in a pit furnace with a circular cross-section of 100cm (external diameter) and 98cm height; the net internal volume is about 51 L with 33cm diameter and 60cm height.

The furnace is open at the top side, with a 21cm circular hole. There is an insulation layer (33.5cm thick) that keeps the external shell at low temperature (lower than 60°C); moreover, the internal chamber is lined by a coat of refractory material. The reactor can be handled by means of an electric cable hoist.

The control panel is installed on the right side of the furnace and includes the master switch, the heating switch, the hoist switch, two temperature controllers, an ammeter, a voltmeter and a temperature safety alarm. The heating is provided by several electric resistances (maximum power 10.5kW at 230V, 50Hz), up to a maximum work temperature of 1000°C. The power supplied to the resistances is regulated by a PID temperature controller (Fuji PX3 series) controlling both the maximum temperature (i.e. the desired temperature) and the heating rate 6 (temperature ramp).

After selecting the heating rate and the set point (it is possible to select up to 8 different set points and as many ramp for the same experimental run), the furnace begins the operation cycle using a constant power, proportional to the heating rate value, until the temperature is close to the set point, and then it reaches the set value (and keeps it stable) by means of a pulse width modulation (PWM) control strategy. Pulse-width modulation control works by switching the power supplied to the furnace electric resistors (i.e. the heating system) with a variable duty cycle. A second safety temperature controller is installed on the furnace panel, cutting the power to the resistors in case the temperature exceeds the safety temperature set point (usually selected 20°C above the first controller set point).

The temperature is measured using two K-type thermocouples insulated by a refractory sheath and inserted into the furnace through the insulation layer at 26cm from the top side of the furnace.

The data acquisition unit consists of an Agilent34970 multimeter (DMM 6½ accuracy) with a 20-channel plug-in module connected to a PC by means of a RS-232 port.

For the temperature measurements inside the reactor, K-type thermocouples (Chromel® - Alumel®) have been used, because they are well suited for corrosive and oxidizing environments and have a wide operating temperature range, between -200°C and 1370°C (accuracy range 1.1°C to 2.2°C); the DMM internal reference junction has been used since its accuracy (0.5 °C) is of the same order of magnitude of the K-type thermocouples. To measure the temperature and its variation during the experimental runs, four thermocouples have been used placed to measure: the furnace heating rate, the wall heating rate, the feedstock thermal behaviour, the syngas temperature. Two different kinds of thermocouple junctions have been used: 'bare wire' for the 'furnace' and sheathed for the 'wall', 'bed' and 'internal' measurements.

A 90cm long tube, connected to the central reactor outlet (located on the top flange), acts as discharge duct. The discharge duct is open to the atmosphere and the tar compounds tend to condense on the chimney cold surfaces, thus a small container (for the condensate) has been located at the effluent discharge point. The gas sampling line consists of a 1/4' stainless steel tube connected to the discharge duct and equipped with a sampling valve and a diaphragm pump (N 820, KNF series, max flow rate 20 l/min) to let the desired syngas amount to flow towards the gas analysis unit. Between the gas sampling valve and the gas analyzer a tar trap and a particulate filter have been installed; in particular a filter flask collects the liquid coming from a condensation coil immersed in a cooling water tank, while the remaining gases are piped to the analyzing unit. When nitrogen has been used as the inert gas carrier, in order to continuously purge the vessel, a flowmeter (MKS Instruments series, operating range 2 - 20000 Scm3) has been installed between the gas cylinder and the reactor inlet.

The analyzing unit is an Agilent micro gas-chromatograph (3000 micro-GC series, repeatability less than 0.5% RSD at constant pressure and temperature) well suited for refinery and reformer gas analysis. It consists of two complementary chromatographic channels, each consisting of an injection device, a separation column and a detector, that analyze a portion of sample in parallel.

In the performed experiments Argon has been used as inert gas carrier, detecting 9 different gaseous compounds: H_2, O_2, N_2, CH_4, CO, CO_2, C_2H_4, C_2H_6 and C_2H_2.

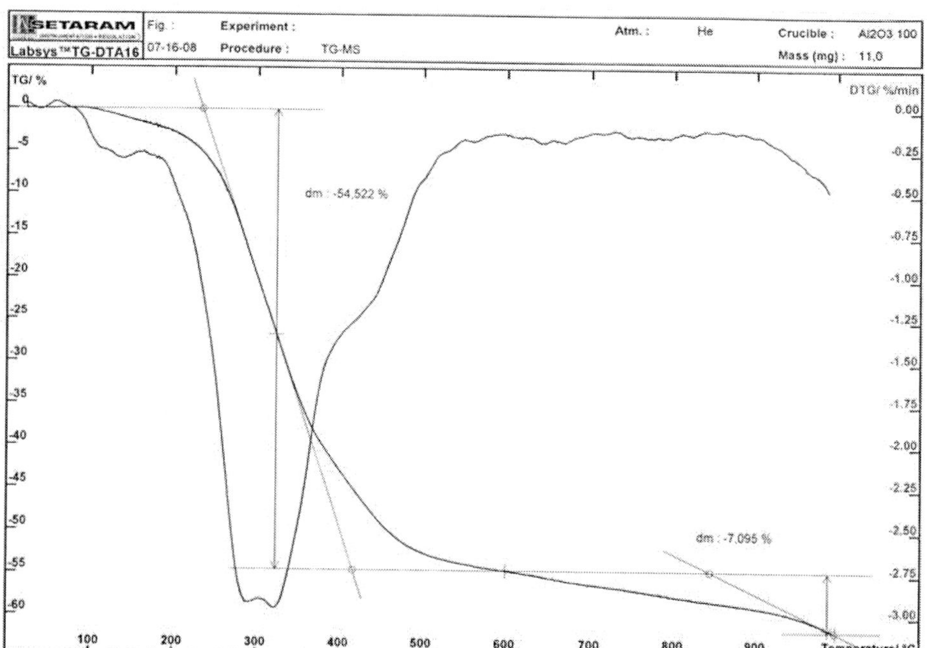

Figure 3. Thermal analysis of the dried sewage sludge. TG and DTG curves recorded in the TG-MS measurement.

Pyrolysis Results

Pyrolysis Tests on Thermobalance

The analyzed wastewater sewage sludge sample was furnished by the urban wastewater plant of Trento (population about 100.000). The sludge was collected in the sewage plant, after the drying step. Elemental analysis reveals C 34 %, H 4.2%, N 5.2%, and S 0.5%.

Some hazardous metals, present in the dried sludge was detected by ICP-OES Spectroscopy; their concentration values, as mg/kg, are: As 10, Ba 350, Cd <1, Cr 29, Cu 295, Mn 234, Ni 20, Pb 53, and Zn 805.

Thermogravimetric coupled mass spectrometric analysis on sewage sludge sample presents a continuous mass loss up to 1000°C. Two thermogravimetric intervals are evidenced. The first and most intense one, stretches from 80 to 600°C with intensity of -54.5%, and it is due to the thermal decomposition of the sludge organic content.

DTG curve shows a first weakly signal at 130°C, followed by an intense band which is constituted by two overlapped peaks, centred at 285 and 320°C, with a pronounced shoulder at 430°C. The mass loss in the second interval, from 600 to 1000°C, is generally produced by the thermal decomposition of carbonates that are present in the sludge, and by the elimination of carbon residues, that takes place at high temperature. In this interval thermogravimetric curve presents a mass loss of -7.1%, while DTG trend reveals any particular event.

The total mass loss is 61.6%, very close to other sewage sludge sample, characterized in the same manner [3]. The mass spectrometric analysis of the evolved gas phase (TG-MS) is characterized by a total ion current curve which presents more signals. The first peak at 140°C is produced by the release of small amount of moisture from the samples. At this first band follow two intense overlapped peaks, centred at 330 and 460°C. Mass spectra, at the most significant TIC peaks (Figure 4), indicate main evolution of CO_2 and H_2O at 330°C, while at 460°C the simultaneous release of different compounds is observed.

At temperature above 600°C, TIC curve presents two small bands, centred at 660 and 980°C, respectively. Mass spectra recorded at 660°C shows the release of CO_2 and CO. Evolution of carbon monoxide continues at higher temperature, due to elimination of carbon residues. The gas phase evolved in correspondence of the most significant TIC peaks was sampled and analyzed by gas chromatographic-mass spectrometric analysis (TG-GC-MS).

At 330°C the analyses reveal the evolution of CO_2 and H_2O, beside very small amount of acetaldehyde, acetone, and acetonitrile, while at 460°C the evolution of several compounds is showed. Table 1 summarizes the hydrocarbon composition of the gas phase released at this temperature. As observed by other authors [12] the pyrolytic decomposition of partially oxygenated organic compounds (lipids, carbohydrates, cellulose, lignin) is the cause of the evolution of CO_2 and CO at temperature below 600°C, and the release of the small amount of acetaldehyde and acetone. Cyano compounds may derive from proteins, nucleic acids and dead micro-organisms present in sewage sludge; they could be formed by dehydrogenation of amino groups [12]. TG-MS analysis allows to follow the evolution of a single chemical species, by selecting an appropriate m/z signal. Thus, water evolved from the samples during the thermal analysis can be easily monitored by its m/z molecular ion, m/z 18, also aromatic hydrocarbons such as styrene (m/z 104) and toluene (m/z 91) can be monitored during the thermal analysis.

Table 1. Chemical composition of the gas phase evolved at 460°C in the pyrolysis of the urban plant sewage sludge. Percent values are calculated from peak areas of gas chromatographic elutions (TG-GC-MS analyses)

Evolved chemical species	Gas phase composition/mol%
CH_4	2.5
CO	1.7
CO_2	17.6
C_2H_4	0.9
C_2H_6	4.1
C_3H_6	7.4
C_3H_8	4.8
H_2O	14.5
HCN	1.6
C_4H_8	9.8
C_4H_{10}	4.5
CH_3CN	4.0
C_5H_{10}	5.0
C_5H_{12}	3.1
Toluene	8.1
C_8H_{16}	2.3
C_8H_{18}	1.7
Styrene	4.2
C_9H_{18}	1.1
C_9H_{20}	1.1

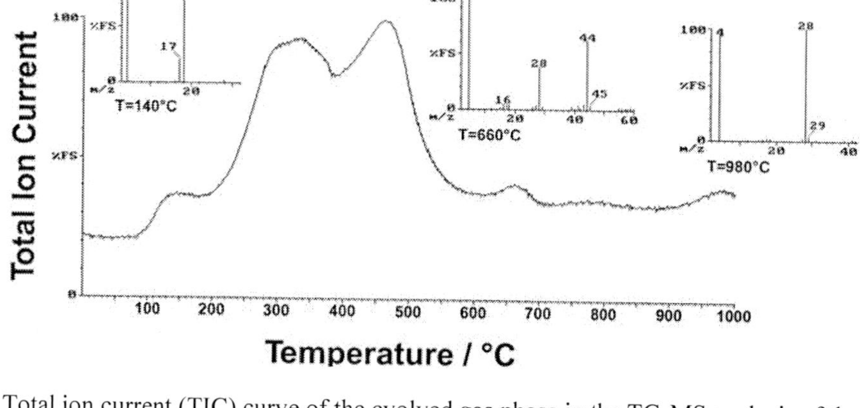

Figure 4. Total ion current (TIC) curve of the evolved gas phase in the TG-MS analysis of the dried sewage sludge. In the insets mass spectra recorded corresponding to selected pyrolysis temperatures.

A more complicated situation takes place when a considered m/z ion is common for more species. In this case it must be taken into account the amount of these compounds, determined by TG-GC-MS analysis, and the relative intensity of the considered m/z ion into the fragmentation pattern of the species.

This is the case of CO_2, monitored by *m/z* 44 ion. Although this fragmentation ion is common for other evolved species, its relative intensity into the fragmentation pattern of these compounds is low. Moreover amounts of these released compounds are also very low, if compared with the evolution of CO_2, so that it may be concluded that the evolution of *m/z* 44 ion current can be reasonably used for the monitoring of the CO_2 release.

More difficult is the monitoring of CO, because at the fragmentation ion m/z 28 corresponds also the evolution of N_2, ethylene and other hydrocarbons which mass spectrum presents this signal [3]. However, at high temperature (above 650°C) the m/z 28 ion current can reasonably attributed to the evolution of CO.

Finally, m/z 15 ion current, that represents the CH_3^+ ion, allows to monitor the evolution of hydrocarbons; another approximate way consists in the subtraction from the Total ion current curve (TIC) of the contribute of m/z ions that describe the release of H_2O, CO_2 and CO. On the basis of these assumption, water evolution (m/z 18) takes place in the 90-550°C interval; its release is characterized by two peaks centred at 145°C and 290°C. The second peak more intense presents a shoulder at 370°C and a tail up to 550°C (Figure 5).

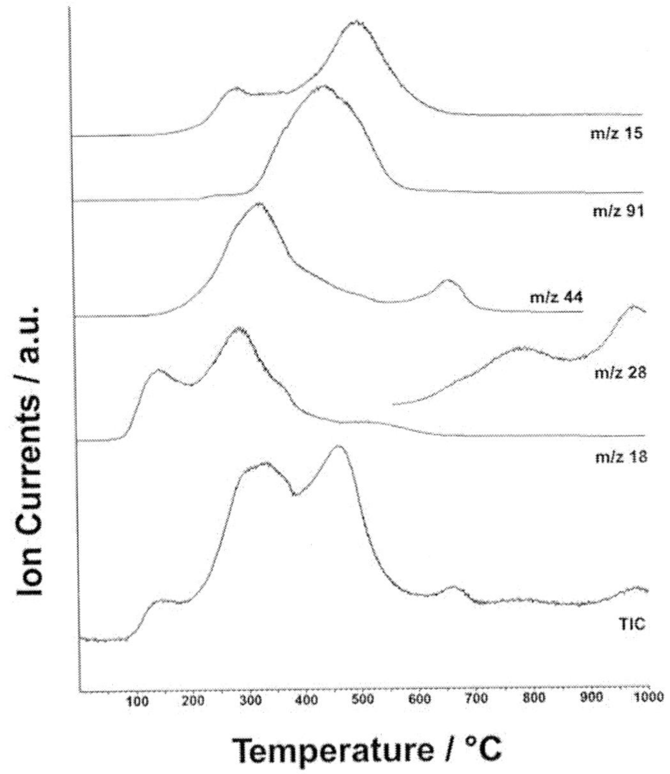

Figure 5. Ion currents of selected m/z values, recalled from TG-MS analysis, used to monitor the release of H2O (m/z 18), CO (m/z 28), CO2 (m/z 44), Toluene (m/z 91) and Hydrocarbons (m/z 15).

The evolution of toluene is showed by a large band in the 350-550°C interval, centred at 445°C. CO_2 is released from the sludge in two events; the first and more intense takes place in the 200-500°C interval, and regards the evolution of carbon dioxide from the solid organic component of the sludge. In fact, m/z 44 ion current presents in the considered interval an intense peak, centred at 330°C. This peak is followed by a second small one at 660°C, due to evolution of CO_2 from the sludge inorganic counterpart [6, 7].

CO release takes place at temperature above 600°C, from carbonaceous residue present in the pyrolyzed solid; m/z 28 presents a first peak at 790°C, followed by a second more intense one at 970°C.

Finally, m/z 15 allows to monitor the evolution of hydrocarbons; its ion current trend presents a first small band at 290°C, followed by an intense peak in the 400-600°C interval, with maximum at 505°C.

By comparing the integrations of the selected *m/z* ion currents, the chemical composition of the gas phase released at 460°C, and the elemental analysis of dried and pyrolyzed sludge samples, it can be obtain a semi-quantitative evaluation of the pyrolysis process [3].

Water represents the 22.2 % of the evolved gas phase, while CO_2 and CO are 20.9 and 5.4%, respectively, and the remaining 51.5% of the evolved gas phase is constituted by the produced hydrocarbons.

Data are similar to the one of sludge samples previously analyzed [3] and confirm the potentialities of the pyrolysis process as viable option for the disposal of wastewater sewage sludge.

Kinetic Experiments

An estimation of the activation energy of the sewage sludge pyrolysis process was carried out by thermogravimetric approach, although it must be taken into account that sewage sludge samples can present very different properties, which are produced by the different origin and the different treatment at which the sludge undergone in the wastewater treatment plant.

The kinetic of a reaction is a function of the temperature and the reaction model; considering α the extent of a reaction, t the time, $k(T)$ the temperature-dependent constant and $f(\alpha)$ a function which describes the reaction model, the rate of the reaction can be described by this formula:

$$\frac{d\alpha}{dt} = k(T) f(\alpha) \quad (5)$$

The Arrhenius equation describes the temperature dependence of the rate constant:

$$k(T) = Ae^{-E/RT}. \quad (6)$$

The kinetic of a solid thermal decomposition can be studied by isothermal and non-isothermal methods. The main disadvantages of the first method are excessively longs times of measure, beside the possibility that the sample can decompose before the starting of the isothermal step.

In non-isothermal methods, the kinetic of the reaction is studied as a function of the constant heating rate ($\beta = dT/dt$=constant).

So, the previously introduced formula can be rewritten into a non-isothermal rate expressions which describes the reaction rates as a function of temperature at a constant β:

$$\frac{d\alpha}{dT} = \frac{1}{\beta} A e^{-E/RT} f(\alpha) \tag{7}$$

Its integration, up to conversion α, gives:

$$\int_0^{\alpha} \frac{d\alpha}{f(\alpha)} = g(\alpha) = \frac{A}{\beta} \int_{T_0}^{T} e^{-E/RT} dT \tag{8}$$

The so called "kinetic triplet" includes the value of the three parameters which describe the kinetic of a process: the Activation Energy, the Pre-exponential Factor, and the reaction model (E, A, and $f(\alpha)$).

The analysis of a solid-state kinetic can be carried out by many methods, which can be subdivided in model-fitting and isoconversional (free model) methods.

In the model-fitting method, $f(\alpha)$ is determined by fitting various reaction models to experimental data. Then, the other two parameters can be evaluated by the chosen form of $f(\alpha)$. With this approach E and A, and $f(\alpha)$ cannot be independently determined. Moreover, it must be taken into account that generally solid states reactions are multiple-step processes characterized by serial and parallel reactions that influence the activation energy that value can change during the course of the reaction.

Model-free isoconversional methods are preferable, because they allow the estimation of the activation energy as a function of α, without pre-fixing the reaction model. The main assumption is that the reaction rate for a constant interval of conversion, α, depends only by the temperature. Constant E values are founded in the case of single-state process, while in a multi-step one, the activation energy varies with α and this fact is due to the variation in the relative contributions of the single step to the overall reaction rate.

In the isoconversional methods, a series of experiments must to be conducted at different heating rates. Flynn, Wall and Ozawa method [24, 25] is one of the most popular approach used to calculate the Arrhenius parameters, used as standard method by the American Society for Testing and Materials [26].

The integration of the formula:

$$\int_0^{\alpha} \frac{d\alpha}{f(\alpha)} = g(\alpha) = \frac{A}{\beta} \int_{T_0}^{T} e^{-E/RT} dT \tag{9}$$

gives this expression:

$$\ln g(\alpha) = \ln (AE / R) - \ln \beta + \ln p(x) \tag{10}$$

where $p(x) = \frac{\exp - x}{x} - \int_0^{\alpha} \frac{\exp - x}{x}$, and x = E / RT (11)

For the resolution of p(x), Doyle [27] proposed two approximation:

The first is that Log p(x) = -2,315 -0,4567x, for 60 ≥ x ≥ 20 (12)

(In p(x) = -5,3305 -1,052x in natural logarithm base); (13)

and $p(x) = e^{-x} / x^2$, for 50 ≥ x ≥ 20. (14)

By using the first approximation, it has been obtained the Ozawa-Flynn-Wall method.

In the Ozawa-Flynn-Wall method, ln p(x) = -5,3305 -1,052x, (15)

and by using 10-base logarithms:

Log g(α) = Log (AE / R) - Log β -2,315 -0,4567x (16)

or: Log g(α) = Log (AE / R) - Log β -2,315 -0,4567 E /RT (17)

and the final expression of the formula is:

$$Log(\beta) = Log\left[\frac{AE}{g(\alpha)R}\right] - 2.315 - 0.4567\frac{E}{RT} \quad (18)$$

By plotting Log β versus 1/T at several conversion degree α, a set of lines is obtained, from these lines the activation energy is calculated by the above formula. Five different percentages of conversion were considered for the kinetic evaluation of the pyrolysis process (Figure 6), with regard to the first thermogravimetric event. Table 2 reports, at the selected degree of conversion, the respective Activation Energy value, and the goodness of the linear fitting.

Table 2. Activation Energy at different degrees of conversion α, and correlation coefficients R for the thermal decomposition in inert atmosphere of the analyzed wastewater sewage sludge

Conversion α	R	Activation Energy ($E^{\neq}$, kcal/mol)
0.10	0.9798	31.4
0.15	0.9732	33.8
0.20	0.9605	34.0
0.25	0.9541	33.9
0.30	0.9383	32.8

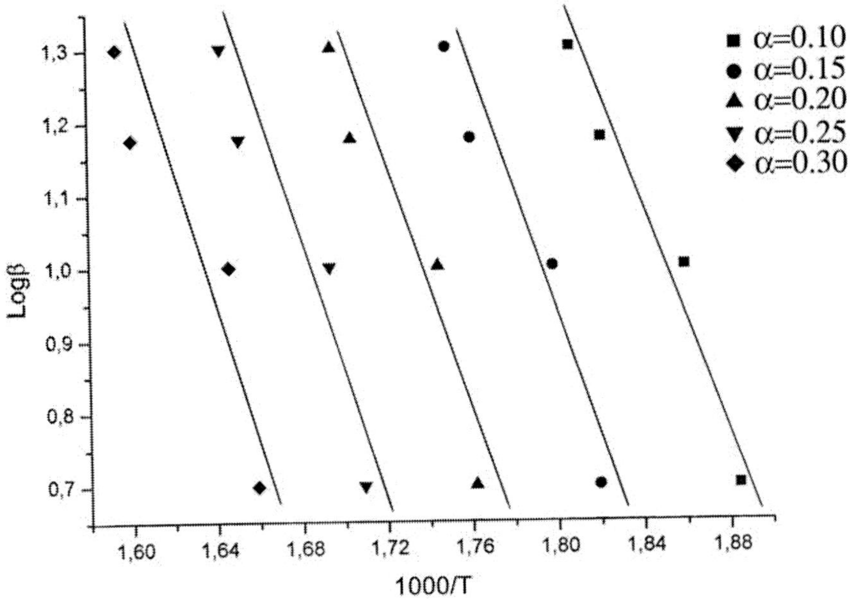

Figure 6. Arrhenius plot of heating rate and temperature at constant conversion data.

The value, determined for the Activation energy, are close to that found by Calvo et al. for two samples of urban sewage sludge, stabilized by anaerobic digestion and dehydrated by centrifugation [28].

Pyrolysis Tests on the Pilot-Scale Reactor

The described batch-plant has been utilized to investigate the sludge pyrolysis/gasification process in order to extend the results formerly obtained on the laboratory-scale investigation. The designed reactor enables to process a more consistent amount of sludge (about 6 dm^3 of net volume available) at nearly the same conditions imposed for the laboratory-scale investigation.

In particular the evolution of the guide compounds CO, CO_2, H_2, CH_4, C_2H_2 and C_2H_6 has been investigated by processing, on the cited pilot plan, a charge of 0.3 kg of sewage sludge. A selection of the results is presented on the following Figure 7 which reports the concentration of the guide compounds vs. time. The reported temperature trend refers to the syngas temperature detected inside the reactor above the initial sludge bed, at a level of 130mm from the reactor top. It can be reasonable assumed as an average temperature value characterizing the thermal sludge conversion.

In all the performed experiments, the biomass thermal conversion process starts with a significant production of CO_2, then, gradually, there is an increase of CH_4, CO and H_2 concentrations. This early CO_2 peak has been found also by other authors [29], reporting that for the thermal conversion of wooden biomass (chips or sawdust) carbon dioxide precedes carbon monoxide formation.

The effective reaction range occurs between 1500 and 7000s run time and is characterized by the typical pyrolysis and gasification processes; specifically, the gasification reactions (gas shift and reforming) use the steam coming from vaporization of the biomass water content. In particular it interesting to observe a significant production of CH_4 and H_2 corresponding to a decrease of CO and CO_2 molar concentrations (around 6000s)

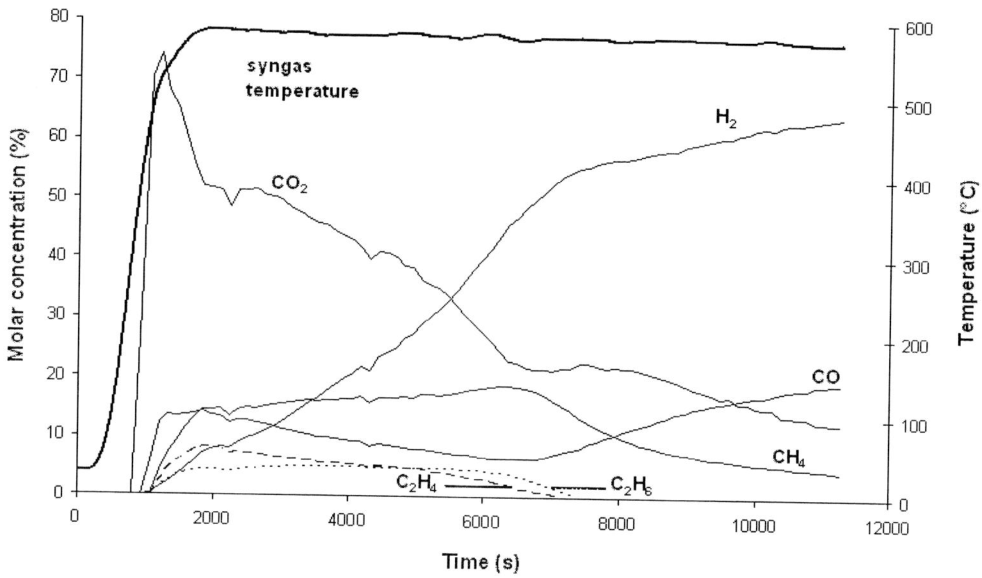

Figure 7. Pilot plant pyrolysis results: syngas molar composition versus time.

The gaseous effluents mainly consist of the above mentioned gases with low concentrations of C_2H_4 and C_2H_6. These small hydrocarbon molecules found in the analyzed mixture are probably the results of the cracking action of the pyrolysis process that tends to split large molecules (tar) into smaller ones.

Vitrification of the Pyrolysis Solid Residue

Pyrolysis process carried out on sewage sludge represents a plausible and viable option as an ecological way for sludge disposal, along with the exploitation of produced gas as an alternative energy source. Although this process considerably reduces the sludge mass, it is not a complete solution for the disposal of the solid waste. Analogously to the ashes of incineration plants, the solid residue from the pyrolysis process must be stabilized.

For this purpose, vitrification is a reliable method [30]. Vitrification considerably reduces the environmental impact of the heavy metals, which are immobilized into a stable vitreous-matrix. Moreover, possible organic-chlorinated compounds are totally destroyed by the high temperatures (1300-1450°C) of the process. The vitreous matrix presents high resistance to leaching, and high durability. Moreover, it doesn't need landfill disposal, and it may be used as sintering additive in the ceramics industry, or as roadbed, with considerable savings on raw materials. Since 30 years, vitrification is considered the best available technology for the immobilization of radioactive wastes.

Moreover, fly ashes from thermoelectric power plants, or from incinerators, absestos, muds of different origin, metallurgical slags can be vitrified, giving a excellent leaching resistant material [31]. The costs of the treatment, due to the high consumption of energy, during the melting, can be reduced by the transformation of the obtained glass into a marketable product.

Among the heating methods adopted in the glass industry for the melting of the kiln charge, oxy-combustion has been revealed in recent years a valuable option. Compared with traditional processes in air or with electric furnaces, the oxy-combustion process allows the drastic decrease of the off-gas volume (by over 70% [31-35]) and of the particulate content, a flame temperature higher than the one of a traditional combustion process, and a higher efficiency. Otherwise, the energetic consumption sensitively affects the cost of production, which is justified only if the final product presents higher quality, for remunerative applications. The oxy-combustion kiln is similar to the one used in glass industry, without the presence of regenerative heat exchangers.

Generally, the chemical composition of the pyrolysis solid residue (40-50 % of the sludge mass) presents a high percentage of SiO_2, which allows a good vitrification of the solid residue. Moreover, the presence of a substantial amount of CaO, makes the resultant vitreous matrix resistant against leaching tests [36].

The melted glass can be discharged from the kiln by gravity casting in water (glass frit), or through two cylindrical roller rotating in the same direction of the casting. In the glass frit, a pellet size material is obtained, useful for road bed or inert additive for mortars and plasters. In the second case, the vitreous matrix appears as solid scales with high mechanical resistance, and it is used as inert additive in concretes production.

Vitrification tests on pyrolysis solid residue samples were carried out on a lab-scale oxy-fuel fired kiln. Leaching tests on the obtained glass show heavy metals concentration values which are lower than the limit values, determined by IRSA and US-EPA rules [37]. In most cases, these values are lower even of an order of magnitude. Table 3 reports the chemical composition of the pyrolysis residue, and the detected values of concentration of heavy metals in leaching tests, compared with their limits.

Table 3. Inorganic oxides in the pyrolysis residue of the 600°C thermally treated wastewater sewage sludge samples, with the detected values of heavy metals concentration in leaching tests, compared with their limits

Chemical composition of the sewage sludge pyrolysis residue		Heavy metals	Detected value / $mg \cdot L^{-1}$	Limit value / $mg \cdot L^{-1}$
Species	% wt.	As	< 0.001	0.5
SiO_2	59	Cd	< 0.001	0.02
Al_2O_3	6	Cr	< 0.001	2
$CaCO_3$	15	Cu	0.09	0.1
$MgCO_3$	3	Mn	0.006	2
Fe_2O_3	16	Ni	0.007	2
Na_2O	0.4	Pb	< 0.001	0.2
K_2O	0.2	Zn	0.01	0.5
Li_2O	0.1			
Metals	0.3			

A Pyrolysis-Vitrification Process for the Disposal of Wastewater Sewage Sludge

In order to verify the feasibility to dispose wastewater sewage sludge through a pyrolysis-vitrification process, an energy balance analysis of both single units was carried out. Both processes were scaled on the basis of the sludge flow produced in the wastewater treatment plant of Trento, where the analyzed sample was collected.

After anaerobic digestion, the sludge (flow rate 350 m^3/day) presents a solid concentration of 25 kg/m^3, and undergoes belt-press filtration, where solid concentration is increased to 200 kg/m^3, resulting in a sludge outlet flow rate of 44 m^3/day.

At present, this sludge is addressed, with sludges arising from other close wastewater treatment plants, to a drying plant, with capability of 70 m^3/day. The drying process bears a sewage sludge with a water content lower than 10% wt., and requires an energetic consumption of 800 kcal/kg H_2O evaporated. The sludge going out from the drying step with a temperature of 90-100°C.

The dried sludge, about 5000 ton/year, is then addressed to landfill collocation.

Considering the treatment of this sludge by the proposed pyrolysis and vitrification processes, the following hypothesis was taken into account for the scaling of the plants:

- 315 weekday per year, and a consequent sludge flow rate, addressed to pyrolysis, of 660 kg/h;
- a water content of 10% wt in the dried sludge;
- a working temperature of pyrolysis reactor of 600°C;
- a heating of the dried sludge (C_p=0.3 kcal/kg°C) from 100 to 600°C;
- a heat required by the pyrolysis reaction of the 113 kcal/kg, as deduced by a series of DSC analyses carried out on sludge samples;
- a nominal calorific value of the released gas mixture of 7850 kcal/kg, calculated taking into account both the chemical composition of the gas phase (determined by TG-MS and TG-GC-MS measurements), and the $\Delta H°$ values of the combustion reactions of evolved compounds.

On the basis of the above hypothesis, the pyrolysis reactor requires: 99000 kcal/h for the heating of the sludge up to the working temperature, 52800 kcal/h for the evaporation of the present water, about 67800 kcal/h for the thermal decomposition of the sludge. Taking into account a heat transfer efficiency of 30%, the process requires 732000 kcal/h. Moreover, the heat loss with the outgoing solid residue is quantified about 52250 kcal/h, so the total endothermicity of the pyrolysis process is 784250 kcal/h.

The combustion of the hydrocarbons ($\Delta H°$ = 7850 kcal/kg) produced by the pyrolysis reaction (51.5% of the produced gas phase, which is the 54.5% of the sludge mass), generates a heat of about 1454170 kcal/h, which satisfy the endothermicity of the pyrolysis step.

An oxy-fuel fired furnace of 3.3 m^2 bottom, with a productivity of 8 ton/day for 315 weekday per year, was considered for the vitrification disposal of the pyrolysis solid residue, about 300 kg/h.

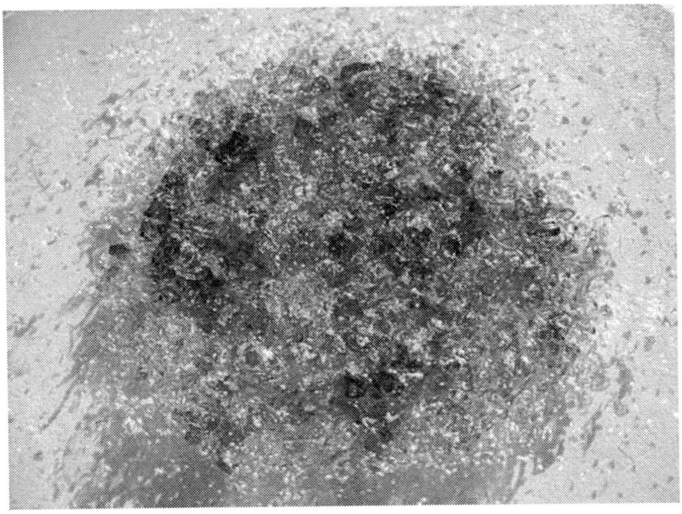

Figure 8. The wastewater sewage sludge pyrolysis residue, vitrified in a lab-scale oxy-fuel fired kiln.

The melting temperature of the furnaces was fixed at 1450°C while the heat required was estimated 1921 kcal/kg; with these operative conditions, the heat requires by the kiln is 640333 kcal/h.

It must be taken into account that above 600°C, the sludge presents a 7.1% mass loss, due to the release of CO_2 and CO. The first species is produced by thermal decomposition of carbonates, while the second one by elimination of carbonaceous residues. The quantitative analysis of the gas phase evolved in this temperature interval reveals that CO_2 constitutes the 23.8 %wt of the released gas, while CO the remaining 76.2%wt. This fact indicates that the sludge mass loss above 600°C (7.1%) is constitutes by CO_2 1.7%, and CO 5.4%. On the other hand, 2.3 %wt of graphitized carbon is present in the pyrolysis residue.

The combustion of this carbon, which heat of combustion was estimated 5000 kcal/kg, takes place into the oxy-fired furnace, and produces 75900 kcal/h.

The comprehensive heat balance of both pyrolysis-vitrification steps is: + 784250 kcal/h required by the pyrolysis, - 1454170 kcal/h produced by the combustion of the pyrolysis fuel gas, + 640333 kcal/h for the melting of the pyrolysis residue, - 75900 kcal/h produced by the oxycombustion of the carbonaceous residue, giving an exothermicity of -105480 kcal/h that confirms the energetic feasibility of the proposed processes.

Another approach is to consider the pyrolysis and the vitrification as two steps of the same plant. In this case, the heating of the pyrolysis reactor may be carried out with the oxy-combustion off-gas heat. For this aim, it must be consider an oxi-fired furnace at least of 30 t/day. The pyrolysis solid fraction corresponds to a little percentage of the melted raw materials, but with this plant other waste can be disposed, such as incinerator ashes, polluted soils, industrial sludge with high content of inorganic compounds, asbestos ecc., because the required composition for the obtainment of a vitreous material with satisfying properties, is: SiO_2 ~50%, Al_2O_3 10÷20%, and CaO 10÷12% [36].

In this case, the combustion of the fuel gas produced by pyrolysis, may be available for other neighbouring devices of wastewater treatment plant, firstly the drying of the sludge after the belt-press filtration.

A Thermodynamic Equilibrium Model for the Sludge Pyrolysis Process

Although the accurate characterization of the pyrolysis conversion products requires complex kinetic models in order to account for all pyrolysis reactions taking place, the present analysis is based on a simpler thermodynamic approach that is capable to give only a rough estimation of the syngas composition (as normally it is not possible to achieve complete thermodynamic equilibrium inside a pyrolysis reactor).

Nevertheless, the predictions of equilibrium models are not far-away from the experimental results both for gasification of organic matter [38-42] and for gasification of coal [43, 44]. The thermodynamic equilibrium model described in the following paragraphs is a two-phase (gas - solid) one, allowing to predict not only the syngas composition but also the char fraction.

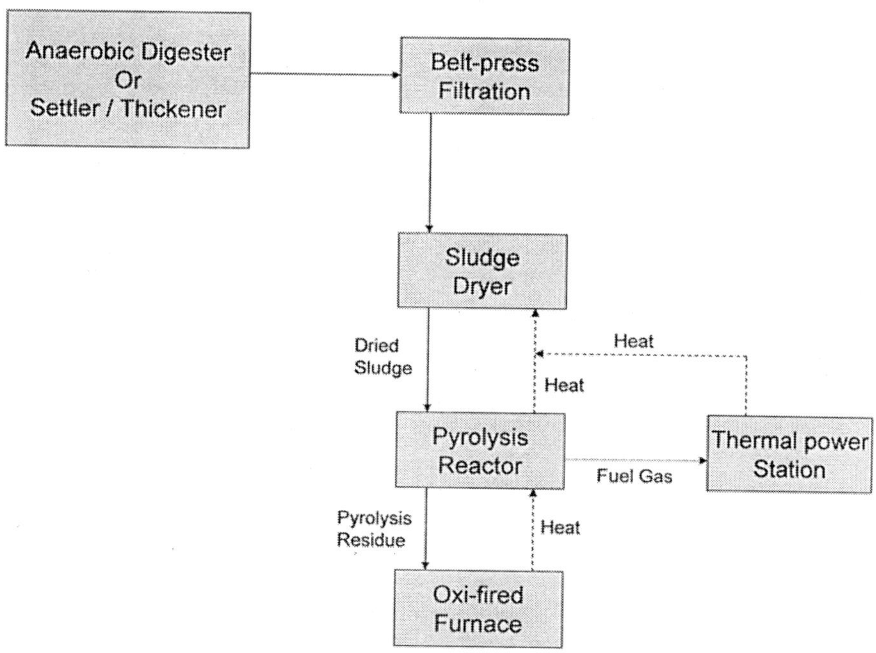

Figure 9. Flow diagram of the possible application of the pyrolysis / oxy-combustion process as a step unit of the wastewater treatment plant.

Equilibrium Theory

The composition of a mixture at equilibrium can be estimated using different methods. The approach used here, called the Gibbs energy minimization method, consists in evaluating the concentrations of the species present that minimize the total Gibbs energy of the products, in accordance with the constraints imposed by the principle of conservation of mass and of the stoichiometry (elements conservation). The chief advantage of this method is that it does not require to select a number of "representative" chemical reactions allowing the formation of (equilibrium) products; it is nevertheless necessary to establish the chemical species expected in the product mixture. This method is based on the fact that the Gibbs energy reaches a minimum value at thermodynamic equilibrium. In fact a for a closed chemical system with an arbitrary number of species present in several phases at uniform temperature and pressure (not necessarily constant) evolving from a non-equilibrium initial state to a closer-to-equilibrium final state, we can write the following expression.

$$dU^t + PdV^t - TdS^t \leq 0 \tag{19}$$

The expression above is valid for any closed system of uniform T, P and it could be applied to every change of the system through an irreversible path (inequality) or a reversible path (equality). Moreover, for a process restricted to constant T and P , equation (19) becomes

$$d\left(G^t\right)_{T,P} \leq 0 \tag{20}$$

where G is the total Gibbs Energy ($G = U + PV - TS$). All irreversible process (occurring at constant T and P) evolve in the direction that causes a decrease of the Gibbs energy. The equilibrium state of a closed system is the one for which the total Gibbs energy reaches a minimum with respect to all possible changes at the given T and P. The Gibbs energy minimization method, then, consists in writing an expression for (G^t) as a function of the number of moles of the species present in the several phases and then finding the set of values for the mole number that minimizes this function, subject to the constraints of mass conservation and stoichiometry (i.e. elements conservation). Moreover, the expression for the Gibbs function can be extended for open systems, in which the number of moles of the species can vary because of mass exchange with the surroundings. In this case it is necessary to introduce a chemical potential μ_i, that is a function of the n_i moles of the different compounds.

$$dG^t = -SdT + VdP + \sum_{i=1}^{S} \mu_i dn_i \tag{21}$$

Model Structure

A code written in Matlab environment has been developed using the Cantera software library (a collection of object-oriented software tools for problems involving chemical kinetics, thermodynamics, and transport processes; [45]). The solver implemented in Cantera is a version of the Villars-Cruise-Smith (VCS) algorithm (a well suited method to handle multiphase problems), that finds the composition minimizing the total Gibbs free energy of an ideal mixture [46]. The NASA [47] and the GRI-MECH [48] databases have been used to evaluate the thermodynamic properties of the chemical species considered in the model.

There are two major procedures for solving the minimization problem; one procedure is the non-stoichiometric formulation in which the closed-system constraint, represented by the element-abundance equations, is treated by means of Lagrange multipliers.

$$F = G' + \sum_k \lambda_k \left(\sum_{i=1}^{N} n_i a_{ik} - A_k \right) \tag{22}$$

where:

λ_k Lagrange multiplier for the *k- th* element
A_k total number of atoms of the *k- th* element *(k = 1, ..., M)*
a_{ik} number of atoms of the *k- th* element found in the molecule of the *i-th* specie

The minimization procedure applied to the F function, implies the computation of its partial derivatives with respect to the n_i mole number and gives the equilibrium condition:

$$\mu_i + \sum_k \lambda_k a_{ik} = 0 \tag{23}$$

This procedure may be the more efficient one when there are a relatively large number of independent stoichiometric equations. The computer implementations of this algorithm for ideal systems are the RAND, NASA, and Brinkley-NASA-RAND (BNR) algorithms.

A second procedure is the stoichiometric formulation, in which the closed-system constraint is treated by means of the linearly independent stoichiometric equations so as to result in an unconstrained minimization problem.

$$\vec{n} = \vec{n}^0 + \sum_{j=1}^{R} \nu_j \xi_j \tag{24}$$

where:

$\vec{n}^0$ initial composition vector
ν_j stoichiometric coefficients vector
ξ_j extents of the reaction

The minimization procedure applied to the G function, implies the computation of its partial derivatives with respect to the ξ_j extents of the reactions (25) and gives the equilibrium condition (26).

$$\left(\frac{\partial G}{\partial \xi_j} \right)_{T,P,\xi_{k \neq j}} = \sum_{i=1}^{N} \left(\frac{\partial G}{\partial n_i} \right)_{T,P,n_{k \neq j}} \left(\frac{\partial n_i}{\partial \xi_j} \right)_{\xi_{k \neq j}} \tag{25}$$

$$\sum_i \nu_{ij} \mu_i = 0 \tag{26}$$

The VCS algorithm utilized this procedure, that may be the more efficient when there are only a few independent stoichiometric equations. Stoichiometric algorithms search among a set of states that all satisfy the element constraints for the one state that satisfies the conditions of chemical equilibrium.

The VCS algorithm implemented in Cantera calculates the equilibrium composition for specified temperature and pressure - TP. If any other property pair other than temperature or pressure is specified (for example it is possible to solve the problem for constant enthalpy and pressure - HP, or for constant entropy and pressure - SP), then an outer iteration loop is used to adjust T and P so that the specified property values are obtained. The algorithm iteration continues until the difference of the chemical potential between two computational steps is less than a fixed value (relative error tolerance = 10^{-9} RT, where R = 8.314 J mol^{-1} K^{-1} represents the universal gas constant, and T is the equilibrium temperature), that can be set by the user. The maximum number of steps to take while solving the equilibrium problem for specified T and P is also fixed equal to 1000.

In order to estimate the yield both of gaseous products and solid carbon, a multiphase formulation of the model has been used. To this purpose, the products have been assumed to consist of a two-phase mixture (solid-gas), i.e. an ideal gas mixture consisting of several gaseous species and a solid phase component represented by the graphite allotropic form of carbon. The description of the product mixture then needs a further coordinate, to account for the mass fraction of each phase of carbon. The model considers 63 chemical species, 60 for the gaseous phase and 3 for the solid one (shown in Table 4).

These chosen compounds are combinations of C, H, O, N and S that are the elements typically constituting the biomass. Moreover the tar fraction has been taken into account, selecting some Polycyclic Aromatics Hydrocarbons – PAH (C_6H_6, $C_{10}H_8$, $C_{12}H_{10}$); although, as expected, the simulation results have shown that their computed equilibrium concentrations are negligible.

Table 4. Chemical species considered by the equilibrium model

Phase	Group	Compounds
Gas phase	Inorganic carbon compounds	C(g) CO CO_2
	Hydrogen compounds	H H_2 O O_2 OH H_2O HO_2 H_2O_2 HCO
	Nitrogen compounds	N N_2 NH NH_2 NH_3 NNH NO NO_2 N_2O HNO CN HCN H_2CN HCNN HCNO HOCN HNCO NCO
	Sulphur compounds	S SO_2 SO_3 H_2S COS CS_2
	Hydrocarbons	CH CH_2 CH_3 CH_4 C_2H C_2H_2 C_2H_3 C_2H_4 C_2H_5 C_2H_6 C_3H_7 C_3H_8 C_6H_6 $C_{10}H_8$ $C_{12}H_{10}$
	Other organic compounds	CH_2O CH_2OH CH_3O CH_3OH HCCO CH_2CO HCCOH CH_2CHO CH_3CHO
Solid phase	Carbon	C(s)
	Ash	CaO SiO_2

The model results show that only a small number of gaseous species is present in the syngas in a significant fraction: hydrogen (H_2), steam (H_2O), methane (CH_4), carbon monoxide (CO), carbon dioxide (CO_2), and nitrogen (N_2); and this is in accordance with experimental data.The feedstock standard enthalpy of formation is calculated from its Lower Heating Value (*LHV*) i.e. the exploitable energy obtained from the complete combustion, where the reaction products are only CO_2, H_2O (vapour), N_2, SO_2 and *ash*.

Once defined the reactants composition (i.e. the X biomass average molecular formula, based on its ultimate analysis), the moisture content and the ash fraction (considered inert) as follows

$$X = C_a H_b O_c N_d S_e + jH_2O(l) + Ash \tag{27}$$

the stoichiometric combustion reaction, used to define the *LHV* (i.e. using O_2 as comburent) is:

$$X + \left(a + \frac{1}{4}b - \frac{1}{2}c + e \right) O_2 \rightarrow$$
$$aCO_2 + \left(\frac{1}{2}b + j \right) H_2O(g) + \frac{d}{2} N_2 + eSO_2 + Ash + LHV \tag{28}$$

The lower heating value *LHV* (or the higher heating value *HHV*) of the feedstock is usually measured experimentally (by means of a calorimetric bomb), but its standard enthalpy of formation at 25°C (ΔH_f^0) is often unknown.

Nevertheless, it is possible to compute the biomass enthalpy of formation ΔH_f^0 value using the stoichiometric combustion equation (28) as follows

$$\Delta H_f^0(X) = \Delta H_f^0(C_a H_b O_c N_d S_e) + j\Delta H_f^0(H_2O,l) + \Delta H_f^0(Ash) =$$
$$a\Delta H_f^0(CO_2,g) + \left(\frac{1}{2}j + b \right) \Delta H_f^0(H_2O,g) + \frac{d}{2}\Delta H_f^0(N_2,g) + e\Delta H_f^0(SO_2,g) +$$
$$+ LHV - \left(a + \frac{1}{4}b - \frac{1}{2}c + e \right) \Delta H_f^0(O_2,g) \tag{29}$$

The enthalpy of the gasifying agent ΔH^T (air, oxygen or steam) fed to the reactor has been computed by means of the following expression:

$$\Delta H^T(gas.agent) = \Delta H_f^0(gas.agent) + R\int_{T_0}^{T} \frac{C_p}{R} dT \tag{30}$$

where the specific constant pressure heat C_p is a function of the process temperature and can be approximated with an appropriate function [47].

The total input enthalpy is the sum of the contribution of the biomass and the gasifying agent. The total output enthalpy from the system is calculated as the sum of the syngas enthalpy, the char enthalpy and the ash enthalpy. The global enthalpy variation along the conversion process represents the amount of energy that is released or has to be supplied.

Pyrolysis Products Characterization at Equilibrium

In Figure 10, 11 and 12 some results obtained by means of the thermodynamic simulation of the pyrolysis of sewage sludge are presented. Two different water content have been taken into account, 10% and 20%, considering both a high and a low drying process.

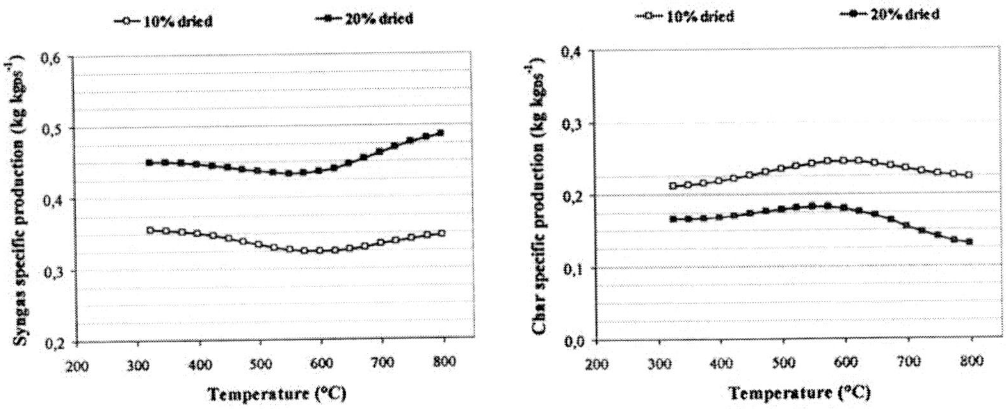

Figure 10. Pyrolysis products specific production (on a basis of 1kg of dried sludge): syngas (on the left) and char (on the right).

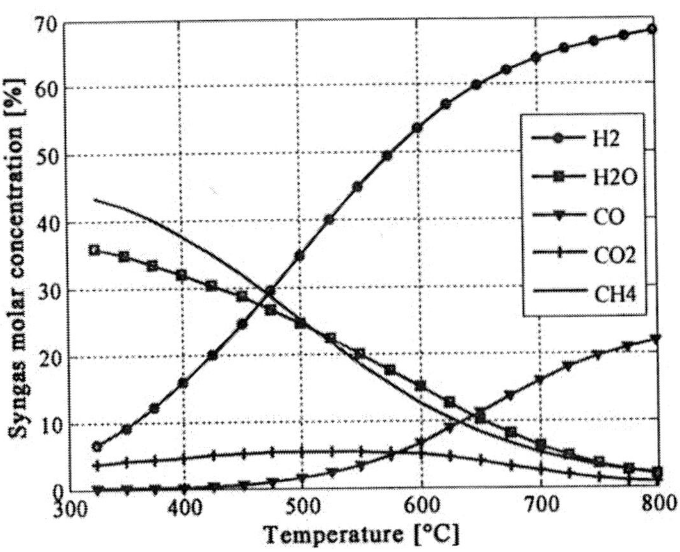

Figure 11. Molar concentration of the main chemical compounds in the produced synthesis gas (10% dried).

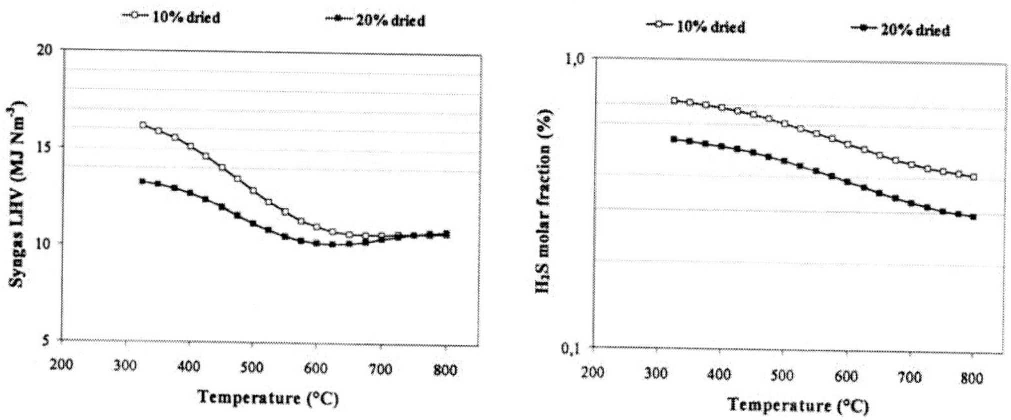

Figure 12. Syngas lower heating value (on the left) and H2S molar fraction (on the right, logarithmic diagram): all the values have been plotted as wet basis.

The pyrolysis products (syngas and char) specific production (on a basis of 1 kg of dried sludge) is presented in Figure 11. It possible to observe a quite constant behaviour of the production as a function of the process temperature with a slight minimum (or maximum) around 600°C for syngas (or char). It is interesting to observe that the moisture in the sludge acts as gasifying agent reducing the production of char and respectively increasing the syngas obtained. The molar concentration of the main chemical compounds in the produced synthesis gas is plotted in Figure 12 - as an example - for sewage sludge to a 10% value of moisture. The characterization of the syngas has been also carried out computing the lower heating value (LHV) as a function of the pyrolysis temperature (Figure 12, on the left) and providing an estimate of the H_2S presence in the syngas (Figure 12, on the right).

The syngas energy content is interesting with LHVs ranging between 10 and 16 MJ Nm^{-3}; as a whole, an increase of temperature causes a significant loss in the syngas LHV as the CH_4 production decreases. However, for particular applications, in order to maximize the H_2 production it seems important to adopt pyrolysis temperatures greater than 600°C.

Regarding the H_2S presence, it is important to notice that a low initial sulphur content in the sludge (0.5% mass) can lead to significant concentrations in the syngas (0.3÷0.7% volume). In fact, if the syngas utilization was different from direct combustion (e.g. in internal combustion engines), the plant should require a proper clean-up stage for the sulphides abatement.

Process Simulation

By means of the developed thermodynamic equilibrium model it is possible to perform process simulations involving an entire plant. The balance of plant can be carried out supplementing the chemical equilibrium analysis for the reacting stages with an evaluation of the mass and energy fluxes arising from the other plant components. The layout of the proposed process, shown in Figure 13, consists of two main stages: the first thermal unit, a pyrolysis oven, represents the main step of the process which is followed by a second thermal unit, a vitrification oxy-fuel furnace, introduced to stabilize the ashes residual fraction discharged by the pyrolysis unit, because of its high concentration of hazardous heavy metals.

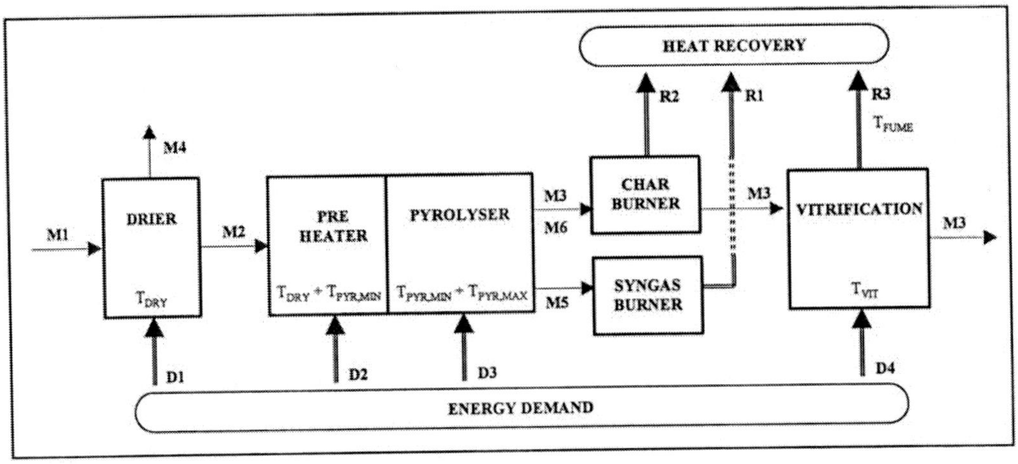

Figure 13. Considered plant layout: plant units, process temperatures, mass and energy fluxes.

The simulation of the conversion process requires the knowledge of the thermal capacity of the matter subjected to the thermochemical treatment. Therefore, the values of the specific heat have been computed as a function of the temperature by means of experimental-derived formulas [49]. In Figure 13 and in Table 5 the capital letter *M* refers to mass flux, *R* to recovered energy and *D* to energy demand.

Table 5. Mass and energy fluxes arising from the plant stages

Mass and Energy Fluxes description		
M1	(kg h^{-1})	High moisture content sludge, either digested or undigested, into the thermal dryer
M2	(kg h^{-1})	Dried sludge (DS) discharged from the dryer and fed into the pyrolysis system
M3	(kg h^{-1})	Residual ash discharged from the pyrolysis process
M4	(kg h^{-1})	Evaporated water from the drier
M5	(kg h^{-1})	Synthesis gas from the pyrolyser
M6	(kg h^{-1})	Residual char discharged from the pyrolysis process
D1	(kWh kg_{DS}^{-1})	Drier energy demand (water evaporation and wet sludge pre-heating)
D2	(kWh kg_{DS}^{-1})	Pre-heater energy demand (sludge heating)
D3	(kWh kg_{DS}^{-1})	Pyrolysis stage energy demand
D4	(kWh kg_{DS}^{-1})	Vitrification stage energy demand
R1	(kWh kg_{DS}^{-1})	Energy recovered form the syngas combustion (fumes, steam generation)
R2	(kWh kg_{DS}^{-1})	Energy recovered form the char combustion (fumes, steam generation)
R3	(kWh kg_{DS}^{-1})	Energy recovered form the vitrification stage combustion (fumes)

Table 5 describes the energy and mass fluxes arising from the different plant units. The main components of the plant comprehend the first thermal unit, i.e. the pyrolysis oven, followed by a second thermal unit, i.e. the vitrification oxy-fuel furnace, which stabilizes the char and ashes residual fraction discharged by the pyrolysis unit. The high moisture content sludge (*M1*), either digested or undigested, is introduced into a thermal dryer. The two common types of dryers are direct contact, which uses heated air as the drying medium, or indirect contact, which uses steam or heated oil as the drying medium.

The discharge dried sludge (*M2*) presents in real plant a moisture content varying from 10 to 20% which makes then available a quality bio-solids fuel. The proposed scenarios make reference to these two moisture content limits for both the two type of sludge. The evaporated water is removed (*M4*). The dried bio-solids are fed into the pre-heater and the pyrolysis conversion stage, while the residual ash (*M3*) and char (*M6*) are discharged. The synthesis gas (or pyrolysis-gas) (*M5*) and the char (*M6*) are oxidized in dedicated burning stages and the released thermal energy is recovered to support the energy balance of the process.

The thermal energy obtained from the syngas is indicated as *R1* and the one arising form the char combustion stages as *R2*. The energy for the evaporation of the moisture contained in the wet sludge and its preheating from ambient temperature is named as (*D1*). The energy need for the pre-heating up to the pyrolysis temperature is (*D2*) and for the pyrolysis stage is (*D3*). All the energy fluxes have been evaluated on a basis of $1 kgh^{-1}$ of dried sludge, indicated in the following tables as kg_{DS}. The energy balance, presented in the following section of this work, provides some information on the sustainability of the whole process, considering also the vitrification stage energy demand (*D4*) and the corresponding energy flux recovery (*R3*).

The following Table 6 shows the values assumed for the main parameters of the process. Two different solutions have been chosen for the pyrolysis process corresponding to the maximum reached temperatures ($T_{PYR, MAX}$) at 600 or 800°C respectively. The temperatures at which the pyrolysis process begins ($T_{PYR, MIN}$) has been chosen at 330°C according to the TG-MS analysis.

Table 6. Process parameters

Process Temperatures		
T_{DRY}	(°C)	105
$T_{PYR, MIN}$	(°C)	330
$T_{PYR, MAX}$	(°C)	600 - 800
T_{VIT}	(°C)	1500
T_{FUME}	(°C)	1170

The simulation of the considered plant has been carried out by taking into account the whole balance of plant described by the following Eqs. 31, 32 and supplementing the chemical equilibrium analysis - applied to the reacting stages - with an evaluation of the energy fluxes arising from the other components. In Table 7 the final values of the mass and energy fluxes obtained from the assumed hypothesis have been reported for each of the process units

$$Mass\ Balance := M1 - (M3 + M4 + M5 + M6) \quad (31)$$

$$Energy\ Balance := R1 + R2 + R3 - (D1 + D2 + D3 + D4) \quad (32)$$

The sustainability of the process is given by a positive result of the *Energy Balance* reported on Table 7. For the recovery stages a 50% and 75% efficiencies have been assumed for the char and syngas burners respectively, based on the heating values of the considered fuel. The recovered heat from the vitrification exhausts has been computed between the fumes and the drying temperatures. The *ASH need* values reported in Table 7 represent the amount of ashes that the vitrification oxy-fuel furnace would require in order to supply the heat necessary to sustain the pyrolysis stage (including pre-heating).

Table 7. Mass and energy balances of the process

$T_{PYR,\ MAX}$	600		800	
moisture content (%mass)	10	20	10	20
M1	4.500	4.000	4.500	4.000
M2	1.000	1.000	1.000	1.000
M3	0.432	0.384	0.432	0.384
M4	3.500	3.000	3.500	3.000
M5	0.339	0.441	0.338	0.450
M6	0.229	0.175	0.230	0.166
Mass Balance	*0.000*	*0.000*	*0.000*	*0.000*
D1	2.565	2.200	2.565	2.200
D2	0.178	0.192	0.178	0.192
D3	0.404	0.492	0.565	0.690
D4	0.826	0.734	0.826	0.734
R1	1.438	1.491	1.506	1.658
R2	1.042	0.798	1.048	0.756
R3	0.099	0.088	0.099	0.088
Energy Balance	*-1.395*	*-1.241*	*-1.482*	*-1.315*
ASH need (kg h^{-1})	2.118	2.616	2.825	3.481

The drying energy demand (*D1*) almost balances the summary of the syngas (*R1*) and char (*R2*) recovered energy (considering the mean values among the four proposed scenarios). This means that the energy demand for the preheating and pyrolysis unit (*D2* and *D3*) must be externally supplied. If the vitrification stage is added, considering that the recovery of this stage (*R3*) is very limited, the vitrification energy demand (*D4*) can be considered as a net energy input of the process. Then the total energy demand (*D1*+ *D2*+ *D3*+ *D4*) amounts to

3,884 kWh kg_{DS}^{-1} while the total recovered energy (R1+R2+R3) amounts to 2,527 kWh kg_{DS}^{-1} providing then the 65% of the total energy demand. This value is approximately equal to the energy percentage required by the drying unit itself (61%). The reported values correspond to the mean of the four values reported on table 2.4 that corresponds moreover to the four hypothesis selected in this analysis.

It clearly emerges that the sewage sludge thermal treatment requires a supply of energy as the recovery amount meets only the 65% of the total energy demand. This is due to the relevant amount of energy required in the drying stage to reduce the moisture content from 80% to the assumed 20-10% content.

However, considering that the advantage of these processes have to be evaluated also in terms of the stabilization of the solid residue, the proposed analysis looks promising in improving both modelling and experimental investigation to promote alternative solutions to sludge thermal treatment.

CONCLUSION

The pyrolysis of the sewage sludge sample was carried out under inert atmosphere and it was studied by TG-MS and TG-GC-MS analyses. A detailed identification of gaseous species, evolved during the thermal treatment up to 1000°C, was obtained. Obtained data show that the pyrolysis process carried out on wastewater sewage sludge bears a remarkable reduction of its mass, yielding an evolved gas mixture which can be exploited as fuel gas in neighbouring power plants.

The feasibility of the process has been tested on a bench-scale apparatus specifically designed for this experimental work. The main part of the plant is a batch reactor operating at variable pressure and inserted inside a laboratory cylindrical furnace providing the required heat power. In order to evaluate the syngas composition, the evolution Vs. time of compounds as CO, CO_2, H_2, CH_4 C_2H_2 and C_2H_6 has been investigated by processing several charge of sewage sludge. The effective reaction range, occurring between 1500 and 7000s run time, is characterized by the typical pyrolysis and gasification processes and a significant production of CH_4 and H_2 corresponding to a decrease of CO and CO_2 molar concentrations is observed.

The stabilization of the solid residue, with or without the presence of other waste, by vitrification in a oxy-fuel fired furnace, allows to obtain an inert vitreous matrix. This vitreous material shows excellent resistance against leaching of heavy metal ions, and it is suitable for many uses in more fields.

The energy balance of the overall process indicates that the heat required by the oxy-combustion furnace and by pyrolysis reactor is satisfied by the heat developed by the combustion of the fuel gas obtained from the pyrolysis and the one produced by the combustion of carbonaceous residues into the oxy-combustion furnace.

The application of a thermodynamic equilibrium model to the thermal conversion of sewage sludge has been presented in this chapter. Even if the thermodynamic equilibrium is a rough approximation of the reality, its application to thermochemical conversion processes (pyrolysis and gasification) can often give a useful insight, because these processes are usually quite slow and then, given the long residence time, the product are not far from equilibrium. Thus, this simple approach can be used to assess the energy balance and to design sludge treatment plants.

The innovative plant for the complete thermal treatment of sewage sludge proposed in this work couples a pyrolysis process unit followed by a vitrification oxy-fuel furnace. This last process is required to stabilize the ashes coming from the pyrolysis unit due to their high concentration of hazardous heavy metals. The present study investigates specifically the mass and energy balances in order to evaluate the energy feasibility and the chemical species of the syngas generated of this specific thermal process. The obtained results show that the sewage sludge thermal treatment requires a remarkable supply of energy corresponding to the 35 % of the total energy demand.

In terms of the stabilization of the solid residue the advantage of these processes are significant. The proposed analysis looks promising in improving both modelling and experimental investigation to promote innovative solutions for future sludge thermal treatment.

REFERENCES

[1] Font, R.; Fullana, A.; Conesa, J.A.; Llavador, F. *J. Anal. Appl. Pyrolysis* 2001, vol 58-59, 927-941.

[2] Conesa, J.A.; Marcilla, A.; Moral, R.; Moreno-Caselles, J.; Perez-Espinosa, A. *Thermochim. Acta* 1998, vol 313, 63-73.

[3] Ischia, M.; Perazzolli, C.; Dal Maschio R.; Campostrini, R. *J. Therm. Anal. Cal.* 2007, vol 87 n 2, 567-574.

[4] Haines, P.J. *Principles of Thermal Analysis and Calorimetry*; RSC Paperbacks, Cambridge, UK, 2002.

[5] Otero, M.; Calvo, L.F.; Estrada, B.; García, A.I.; Morán, A. *Thermochim. Acta* 2002, vol 389, 121-132.

[6] Biagini, E.; Lippi, F.; Petarca, L.; Tognotti, L. *Fuel* 2002, vol 81, 1041-1050.

[7] Scott, S.A.; Dennis, J.S.; Davidson, J.F.; Hayhurst, A.N. *Fuel* 2006, vol 85, 1248-1253.

[8] Otero, M.; Gómez, X.; García, A.I.; Morán, A. *J. Therm. Anal. Cal.* 2008, vol 93 n 2, 619-626.

[9] Arenillas, A.; Rubiera, F.; Pis, J.J. *J. Anal. Appl. Pyrolysis* 1999, vol 50, 31-46.

[10] Otero, M.; Díez, C.; Calvo, L.F.; García A.I.; Morán, A. *Biomass and Bioenergy* 2002, vol 22, 319-329.

[11] Calvo, L.F.; Sánchez, M.E.; Morán, A.; García, A.I. *J. Therm. Anal. Cal.* 2004, vol 78, 587-598.

[12] Fullana, A.; Conesa, J.A.; Font, R.; Martín-Gullón, I. *J. Anal. Appl. Pyrolysis* 2003, vol 68-69, 561-575.

[13] Shen, L.; Zhang, D. *Fuel* 2003, vol 82, 465-472.

[14] Shen, L.; Zhang, D. *Fuel* 2005, vol 84, 809-815.

[15] Barrio, M.E.; Llibera, J.L.; Comellas, L.; Broto-Puig, F. *J. Chromatography* A 1996, vol 719, 131-139.

[16] Inguanzo, M.; Dominguez, A.; Menéndez, J.A.; Blanco, C.G.; Pis, J.J. *J. Anal. Appl. Pyrolysis* 2002, vol 63, 209-222.

[17] Dominguez, A.; Menéndez, J.A.; Inguanzo, M.; Bernad, P.L.; Pis, J.J. *J. Chromatography* A 2003, vol 1012, 193-206.
[18] Menéndez, J.A.; Dominguez, A.; Inguanzo, M.; Pis, J.J. *J. Anal. Appl. Pyrolysis* 2004, vol 71, 657-667.
[19] Dominguez, A.; Menéndez, J.A.; Inguanzo, M.; Pis, J.J. *Fuel Processing Tec*hnology 2005, vol 86, 1007-1020.
[20] Campostrini, R.; Ischia, M.; Palmisano, L. *J. Therm. Anal. Cal.* 2003, vol 71, 997-1009.
[21] Campostrini, R.; Ischia, M.; Palmisano, L. *J. Therm. Anal. Cal.* 2003, vol 71, 1011-1021.
[22] Campostrini, R.; Ischia, M.; Palmisano, L. *J. Therm. Anal. Cal.* 2004, vol 75, 13-24.
[23] Campostrini, R.; Ischia, M.; Palmisano, L. *J. Therm. Anal. Cal.* 2004, vol 75, 25-34.
[24] Ozawa, T. Bull. *Chem. Soc. Jpn.* 1965, vol 38, 1881.
[25] Flynn, J.H.; Wall, L.A. *Polym. Lett.* 1966, vol 4, 323.
[26] ASTM International E 1641-04. Standard Test Method for Decomposition Kinetics by *Thermogravimetry*, 100 Barr Harbor Drive, PO Box C700, West Conshohocken, PA 19428-2959 US.
[27] Doyle, C.D. *J. Appl. Polm. Sci.* 1961, vol 5, 285.
[28] Calvo, L.F.; Otero, M.; Jenkins, B.M.; Garcia, A.I.; Moran, A. *Thermochimica Acta* 2004, vol 409, 127-135.
[29] [29] Simmons, W.W.; Ragland, K.W., *Combustion Science and Technology* 1986, vol 46, 1-15.
[30] Rigo, H.G.; Chandler, A.J.; Lanier, W.S. In *Solid Waste Management: Thermal Treatment and Waste to Energy Technologies*, Washington, DC, April 18-21, 1995.
[31] Colombo, P.; Brusatin, G.; Bernardo, E.; Scarinci, G. *Current Opinion in Solid State and Materials Science* 2003, vol 7, 225-239.
[32] U.S. Environmental Protection Agency. Handbook on vitrification technologies for treatment of hazardous and radioactive waste. Office of Research and Development, Washington DC. Report EPA/625/R-92/002, May 1992.
[33] Wakamura, Y.; Nakazato, K. In: *National Waste Processing Conference Proceedings ASME* 1994, 91-6.
[34] Hollander, H. *Solid Waste Technol.* 1995, vol 9, 31-40.
[35] Richards, R.S.; Plodinec, M.J. In: *Proceedings of the XVIII International Congress on Glass* (San Francisco, CA, July 5-8, 1998). Paper N. A7-I (CD ROM). The American Ceramic Society, Westerville, Ohio.
[36] Falcone, R.; Hreglich, S.; Profilo, B.; Vallotto, M. *Rivista della Stazione Sperimentale del Vetro* 2003, vol 4, 25-29.
[37] Pisciella, P.; Crisucci, S.; Karamanov, A.; Pelino, M. *Waste Management* 2001, vol 21, 1-9.
[38] Kinoshita, C.M.; Wang, Y.; Takahashi, P.K. *Energy Sources* 1991, vol 13, 361–368.
[39] Wang, Y.; Kinoshita, C.M. *Solar Energy* 1992, vol 49, 153–158.
[40] Schuster, G.; Löffler, G.; Weigl, K.; Hofbauer, H. *Bioresource Technology* 2001, vol 77, 71–79.
[41] Altafini, C.R.; Wander, P.R.; Barreto, R.M. *Energy Conversion and Management* 2003, vol 44, 2763–2777.
[42] Baratieri, M.; Baggio, P.; Fiori, L.; Grigiante, M. *Bioresource Technology* 2008, vol 99, 7063–7073.

[43] Altafini, C.R.; Mirandola, A. In: *Proceedings of Flowers'97*, Florence, Italy, (1997) 419–427.

[44] Li, X.; Grace, J.R.; Watkinson, A.P.; Lim, C.J.; Ergüdenler, A. *Fuel* 2001, vol 80, 195–207.

[45] Goodwin, D. Cantera: object oriented software for reacting flows, California Institute for Technology (2005) <http://www.cantera.org> (accessed July, 2007).

[46] Smith, W.R.; Missen, R.W. *Chemical Reaction Equilibrium Analysis: Theory and Algorithm.* New York, Wiley Interscience,1982.

[47] McBride, B.J.; Gordon, S.; Reno, M.A. *Coefficients for Calculating Thermodynamic and Transport Properties of Individual Species*, NASA Report TM-4513,1993.

[48] Smith, G.P. et al, GRI-Mech thermodynamic values. www.me.berkeley.edu/gri_mech, 2000.

[49] Arlabosse, P.; Chavez, S.; Prevot, C. *Brazilian Journal of Chemical Engineering* 2005, vol 22, 227–232.

In: Sewage Treatment: Uses, Processes and Impact
Editor: Anna Stephens and Mark Fuller
ISBN: 978-1-60692-959-9

Chapter 9

BEHAVIOR OF PHARMACEUTICALS AND THEIR DEGRADATION PRODUCTS DURING WASTEWATER TREATMENT

Anastasia D. Nikolaou[1,a], Despo Fatta Kassinos[b], Maria Kostopoulou-Karadanelli[a] and Sureyya Meric[c]

[a] University of the Aegean, Faculty of Environment, Department of Marine Sciences, Mytilene, Greece
[b]Department of Civil and Environmental Engineering, School of Engineering, University of Cyprus, Nicosia, Cyprus
[c]Department of Civil Engineering, Engineering Faculty, University of Salerno, Fisciano (SA), Italy

ABSTRACT

During the latest years there has been a growing level of concern related to the adverse effects of chemical compounds on the environment. Pharmaceuticals is one of the categories of pollutants that during the last 20 years have drawn the attention of the scientific community as they are ubiquiteously present in wastewater effluents at trace levels, and several of them may have adverse effects on aquatic organisms and human health. Their transformation pathways during wastewater treatment are the subject of a growing current research, while many gaps still exist in the knowledge of the degree into which they are decomposed during wastewater treatment, as well as their degradation products. Analytical techniques for the determination of pharmaceuticals are being optimized and applied in order to provide answers regarding their behaviour during water treatment and their degradation products. Current research and obtained results on this subject are summarized and discussed in ththis chapter.

[1] Tel: + 302251036848, Fax: + 302251036809, E-mail: nnikol@aegean.gr

1. Introduction

During the recent years, a large number of chemicals released into the environment is continuously contributing to the degradation of water quality and posing risks for human health. Pharmaceuticals consist a major category of environmental pollutants (Purdom et al., 1994; Hartmann et al., 1998), being investigated and detected in many environmental matrices worldwide (Sacher et al., 2001; Jux et al., 2002; Weigel et al., 2004; Bound and Voulvoulis, 2006; Moldovan, 2006; Verenitch et al., 2006; Radjenovic et al., 2007) while the formation, transport, fate and toxicity of their degradation products in the environment are largely unknown subjects requiring focused research in order to estimate the actual health risks deriving from their presence in the environment.

Pharmaceuticals are increasingly being used, and as a result they are finally released into sewage, wastewater treatment plants and surface water receivers. Recent research results reported in the literature reveal their occurrence in trace concentrations in sewage treatment plant effluents, surface water, seawater and groundwater in many countries (Möder et al., 2000; Ollers et al., 2001; Huggett et al., 2003; Rodriguez et al., 2003; Koutsouba et al., 2003; Andreozzi et al., 2003; Carballa et al., 2004; Lee et al., 2005; Kosjek et al., 2005; Lin and Chen, 2005). Diclofenac, clofibric acid, acetaminophen, ibuprofen, aspirin, carbamazepine, artorvastatin, gemfibrozil, fluoxetine and 17β-ethynylestradiol are among the pharmaceutical compounds most commonly used and most frequently detected in water and wastewater samples (Halling-Sorensen et al., 1998; Daughton and Ternes, 1999; Hirsch et al., 1999; Metcalfe et al., 2003; Boyd et al., 2003; Oaks et al., 2004; Hernando et al., 2006).

Pharmaceuticals can be subject to various reactions in the human body and in the acqueous matrices such as water and wastewater, forming metabolites and degradation products. Two common examples are ibuprofen, forming the metabolites hydroxy- and carboxy-ibuprofen, and clofibrate, forming clofibric acid (Weigel et al., 2004). The metabolites of pharmaceuticals can be further transformed in the wastewater treatment plants, forming degradation products via hydrolysis and/or photolysis (Lishman et al., 2006; Winkler et al., 2001; Boree et al., 2003; Tixier et al., 2003). The toxicity of these degradation by-products in some cases can be higher than the parent compounds (Andreozzi et al., 2003; Doll and Frimmel, 2003), therefore the occurrence and behavior of these compounds in the environment could raise more health concerns that the occurrence of the initial pharmaceuticals, and needs to be further investigated. Figure 1 presents selected degradation products of pharmaceuticals that have been recently investigated and reported in the relevant literature (Gomez et al., 2006; Perez and Barcelo, 2007; Kosjek et al., 2007; Zuehlke et al., 2007; Diaz-Cruz et al., 2008; Focazio et al., 2008; Feldmann et al., 2008; Zuccato et al., 2008).

The determination of pharmaceuticals in complex environmental matrices such as wastewater in trace concentration levels still constitutes a challenge for analytical scientists, although currently advanced gas and liquid chromatographic techniques in combination with modern extraction, derivatization and cleanup methods have provided the opportunity of quantification of many pharmaceutical compounds down to ng/L levels. In parallel, the laboratory efforts to investigate the metabolites of pharmaceuticals in environmental matrices are also in progress (Ternes, 2001; Martin and Thomas, 2003; Petrovic et al., 2003; Gomez et al., 2007; Perez and Barcelo, 2007; Kosjek et al., 2007; Zuehlke et al., 2007; Diaz-Cruz et al.,

2008; Focazio et al., 2008; Feldmann et al., 2008; Zuccato et al., 2008). Gas chromatography – mass spectrometry (GC-MS and GC-MS/MS) as well as liquid chromatography – mass spectrometry (LC-MS and LC-MS/MS) are very useful techniques utilized to determine metabolites and transformation products of pharmaceuticals as well as to obtain detailed information on their molecular structure.

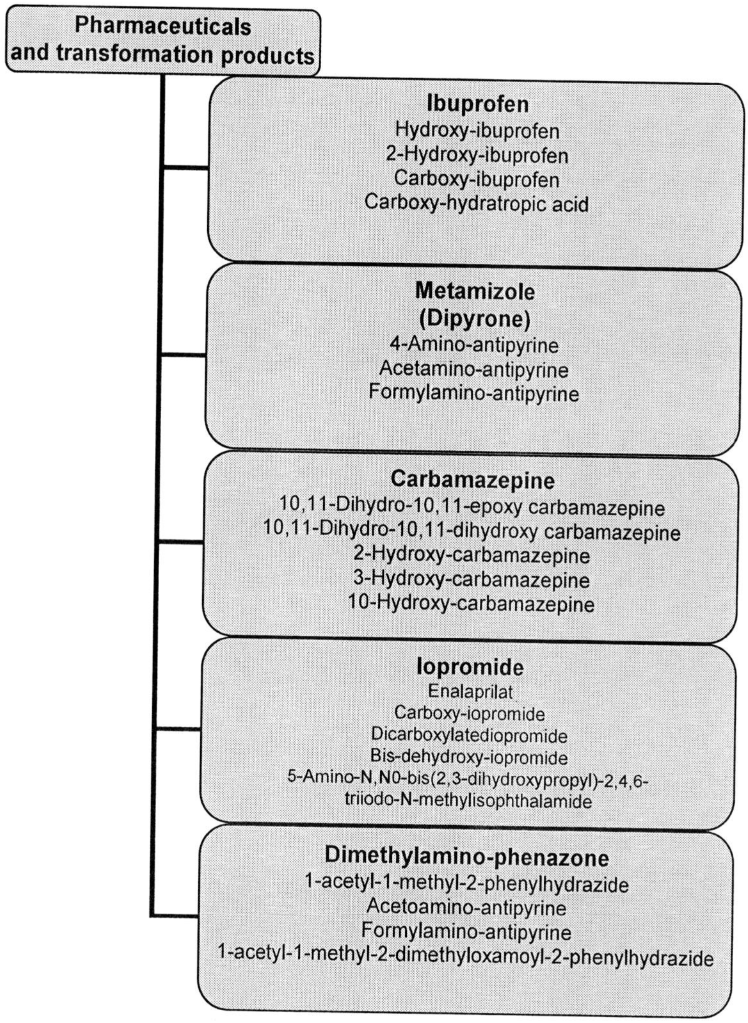

Figure 1. Selected degradation products of pharmaceuticals recently investigated

The subject of the present review paper is the current research and obtained results on the behavior of pharmaceuticals and degradation products in wastewater. The up-to-date knowledge of the transformations and fate of these compounds are discussed, along with the recent developments of the analytical techniques utilized in order to provide more insight on their occurrence and effects in the environment.

2. Transformations and Fate of Pharmaceuticals and Their Degradation Products in Wastewater

Pharmaceuticals and their metabolites enter the environment mainly through excretion and disposal via wastewater. Municipal and hospital wastewaters, wastewater from industries and landfill leachates, as well as disposal of unused medicines, are the most important sources of human pharmaceuticals in the aquatic environment. Pharmaceuticals are not completely removed in wastewater treatment plants, therefore their residues or products can finally reach the surface waters (Giger et al., 2003; Rizzo et al., 2005; Fent et al., 2006; Moldovan, 2006; Pavlogeorgatos et al., 2006; Mok et al., 2006).

Table 1 presents concentrations of selected pharmaceuticals that have been detected in effluents of sewage treatment plants (Nikolaou et al., 2007).

Pharmaceuticals are excreted by the organisms partly in their parent form and partly as metabolites. The procedures of biodegradation result in modifications of the chemical structures of the active molecules. In most cases, the metabolism of pharmaceuticals is incomplete. The metabolites can be divided into two categories: (a) phase I metabolites, resulting from the modification of the active compound by hydrolysis, oxidation, reduction, alkylation and dealkylation, and (b) phase II metabolites which are formed by phase I metabolites via glucuronation or sulfatation to enhance excretion. As a result, the excretions may contain the parent compound, glucuronide and/or sulfate conjugates, a major metabolite, or a mixture of many metabolites. The pharmaceutical molecules are also affected by the environmental conditions, especially pH, and can be neutral, cationic, anionic or zwitterionic (Nikolaou et al., 2007; Fatta Kassinos et al., 2008; Kostopoulou and Nikolaou, 2008).

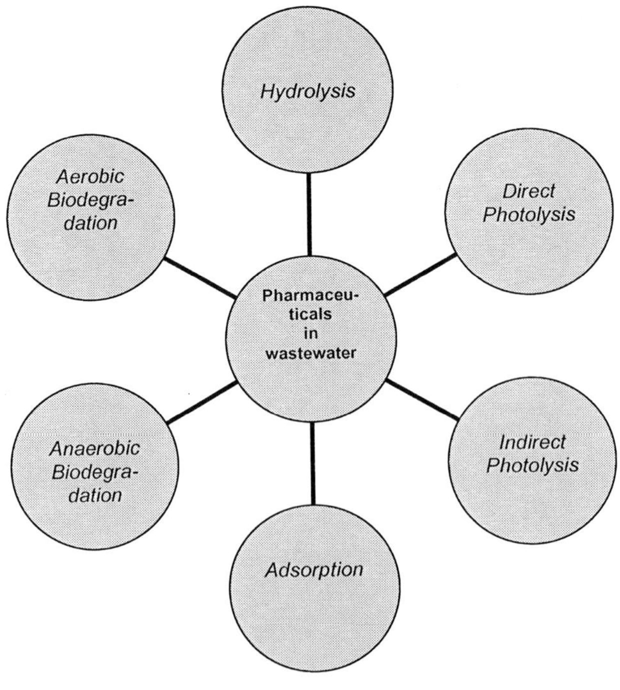

Figure 2. Transformation processes of pharmaceuticals in wastewater

Further transformations of pharmaceuticals and metabolites occur in the wastewater treatment plants as well as in surface water receivers, as schematically presented in Figure 2. Biotic and/or abiotic processes such as hydrolysis and photolysis lead to the formation of degradation products of pharmaceuticals. These degradation by-products raise concerns, because their adverse effects on the environment can be similar to or even higher than the parent compounds (Andreozzi et al., 2003; Doll and Frimmel, 2003). Parameters affecting the transformation reactions of pharmaceuticals in the wastewater treatment plant are the composition of the sewage, weather conditions and the design and operation of the treatment process (Roberts and Thomas, 2006). Pharmaceuticals and metabolites in wastewater can be mineralized to carbon dioxide and water, adsorbed on the suspended solids, or released in the effluent in the form of the initial compound or as degradation products (Lishman et al., 2006).

Table 1. Concentrations of selected pharmaceuticals in effluents of sewage treatment plants (Adapted from Nikolaou et al., 2007).

Compound name	*Concentration (μg/l)*	*Reference*
Benzafibrate	*0.02-0.07*	*Miao et al., 2002*
	<0.25-4.6	*Ternes, 1998*
Clofibric acid	*0.044*	*Roberts and Thomas, 2006*
	<LOD-0.51	*Heberer et al., 2001*
	0.25-4.55	*Stan & Heberer, 1997*
	<LOD-2	*Garrison et al., 1976*
Diclofenac	*0.289*	*Roberts and Thomas, 2006*
	0.03-0.07	*Miao et al., 2002*
	0.01-0.56	*Heberer et al., 2001*
Ibuprofen	*2.972*	*Roberts and Thomas, 2006*
	0.03	*Lin et al., 2005*
	<0.005-0.01	*Miao et al., 2002*
	<0.05-1.43	*Gans et al., 2002*
	0.002-0.081	*Buser et al., 1999*
	0.22 ± 0.15	*Khan, 2002*
Naproxen	*0.017*	*Lin et al., 2005*
	0.02-0.30	*Miao et al., 2002*
	0.35 ± 0.12	*Khan, 2002*
	<LOD-0.12	*Heberer et al., 2001*
	>0.05-0.52	*Ternes, 1998*
Metoprolol	*>0.025-2.2*	*Hirsch et al, 2003; Ternes, 1998*
Carbamazepine	*0.042*	*Lin et al., 2005*
	0.28-1.11	*Gans et al., 2002*
	>0.05-6.3	*Ternes, 1998*
Triclosan	*0.01-0.02*	*Boyd et al., 2003*
	0.2-2.7	*McAvoy et al., 2002*

Direct and indirect photolysis can be a significant pathway for abiotic transformation of pharmaceuticals in aqueous matrices. Direct photolysis is caused by direct absorption of solar light, while indirect photolysis involves natural photosensitizers, e.g. nitrate and humic acids. Under solar irradiation, these photosensitizers form strong oxidant species such as hydroxyl radicals and singlet oxygen. Photolysis of many pharmaceuticals, such as sulphamethoxazole, clofibric acid, oflaxocin, propanolol, prednisone, carbamazepine, iomeprol and furosemide, has been demonstrated by laboratory experiments (Andreozzi et al., 2003; Doll and Frimmel, 2003; DellaGreca et al., 2004; Isidori et al., 2006).

The abiotic photodegradation of the pharmaceuticals carbamazepine, diclofenac, clofibric acid, ofloxacin, sulfamethoxazole and propranolol was evaluated during experiments of exposure to solar irradiation at 40° N latitude during spring and summer. Carbamazepine and clofibric acid had half-life times of the order of 100 days at the highest latitudes (50° N) in winter, whereas under the same conditions sulphamethoxazole, diclofenac, ofloxacin and propanolol underwent fast degradation with halving time ($t_{1/2}$) respectively of 2.4, 5.0, 10.6 and 16.8 days. The presence of nitrate ions in aqueous solutions resulted in a reduction of $t_{1/2}$, with only exception the case of propanolol. When present, humic acids acted as inner filters towards carbamazepine and diclofenac, and as photosensitizers towards sulphamethoxazole, clofibric acid, oflaxocin and propanolol (Andreozzi et al., 2003). In another study (DellaGreca et al., 2004), 60% of prednisone exposed to solar simulator for 8 h, was found to be transformed into seven photoproducts, some of which were stable and more toxic than the parent drug. The photodegradation of clofibric acid, iomeprol and carbamazepine was also investigated (Doll and Frimmel, 2003), in batch experiments with simulated UV-sunlight, and their photodegradation showed a pseudo-first-order kinetics. Carbamazepine absorbed most of the incoming radiation in relation to the other tested pharmaceuticals, but had the lowest degradation rate constants. During the irradiation of carbamazepine, degradation products were found. Clofibric acid absorbed only little of the incoming radiation but it degraded over the intermediates 4-chloropenol, hydroquinone, p-benzoquinone and phenol. New intermediate products were also reported to be formed during the degradation of iomeprol (Doll and Frimmel, 2003). Furosemide has also been reported to be subject of photodegradation, forming intermediate products (Isidori et al., 2006).

The adsorption mechanism is important for some pharmaceuticals, such as basic pharmaceuticals and zwitterions, e.g. fluoroquinolone antibiotics, which tend to adsorb on sludge to a significant extent (Golet et al., 2002). However, acidic pharmaceuticals such as acetylsalicylic acid, ibuprofen, fenoprofen, ketoprofen, naproxen, diclofenac and indomethacin having pKa values ranging from 4.9 to 4.1, as well as clofibric acid, bezafibrate (pKa 3.6) and gemfibrozil, occur as ions at neutral pH and therefore are not easily adsorbed on the sludge and remain in the aqueous phase. Limited adsorption may occur at lower pH (Buser et al., 1998; Fent et al., 2006).

Aerobic or anaerobic biodegradation is a very important removal process of pharmaceuticals in the dissolved phase. Biodegradation increases with increasing hydraulic retention time and with increasing age of the sludge in the activated sludge treatment. For example, diclofenac was shown to be significantly biodegraded only when the sludge retention time was at least 8 days. However, carbamazepine is less than 10% biodegraded. In some cases, conjugates of pharmaceuticals can be cleaved in sewage treatment plants, resulting in the release of active parent compound, as in the case of estradiol and the steroid hormone in the contraceptive pill, 17α-ethinylestradiol (Fent et al., 2006). Recent studies have shown the partial removal for some pharmaceuticals, such as carbamazepine, atenolol, metoprolol, trimethoprim or diclofenac (removal <10% for most of them and 10-39% for diclofenac) (Hernando et al., 2006). Some compounds may be released in a modified form, either hydrolysed or conjugated, such as ibuprofen which is removed in more than 90%, and transformed in hydroxyl and carboxy transformation products.

3. Analytical Techniques for The Determination of Pharmaceuticals and Their Degradation Products in Wastewater

The determination of pharmaceuticals in environmental matrices where they usually occur at very low concentration levels is a difficult task. Analytical techniques have become available only recently, and they are still under development and optimization in order to achieve accurate determination of pharmaceuticals and metabolites in complex aqueous matrices, such as wastewater. Another important challenge of modern environmental analytical chemistry is to identify and quantify a continuously increasing number of new transformation products of pharmaceuticals in water and wastewater. During the last years there have been remarkable developments of increasing chromatographic resolution, detection sensitivity and specificity, as well as more effective techniques for the extraction of pharmaceuticals and transformation products from wastewater samples (Fatta et al., 2007).

The most common analytical techniques currently applied for the determination of pharmaceuticals in environmental samples are gas chromatography-mass spectrometry (GC-MS and GC-MS/MS) and liquid chromatography-mass spectrometry (LC-MS and LC-MS/MS). The major advantages of these tehniques is that they can provide compound confirmation and detailed information on the structure of the compounds analyzed, and that they can perform separation and detection of compounds having the same molecular mass but different product ions, even if they co-elute. Low detection limits can be achieved even in complex matrices such as wastewater, when these techniques are combined with advanced extraction and purification sample preparation protocols, and further information on new compounds, metabolites or transformation products of pharmaceuticals can be obtained via their mass spectra. (Verenitch et al., 2006; Fatta et al., 2007).

During application of GC-MS and GC-MS/MS analysis, derivatization is necessary for the analysis of polar pharmaceuticals. Various derivatization agents are being used for this purpose, depending on the compounds investigated (Table 2). The use of diazomethane was very common previously, however due to its high toxicity, other alternatives are being applied recently, such as acid anhydrides, benzyl halides and alkylchloroformates. During the derivatization step, losses of analytes or incomplete derivatization reaction can occur, resulting in a decrease of accuracy of the method in some cases, and it is also a time-consuming step (usually requiring at least 1h). This is one of the main reasons that LC-MS or LC-MS/MS analysis have become very popular for pharmaceuticals analysis, as with these techniques derivatization is not needed. However analytical problems are still not eliminated, especially during analysis of complex wastewater samples containing many different organic compounds that can interfere during analysis. Addition of a clean-up step prior to the analysis of the final extract can be very helpful to minimize such problems (Fatta et al., 2007; Mutavdzic Pavlovic et al., 2007; Kostopoulou and Nikolaou, 2008).

Table 2. Selected GC-MS and LC-MS analytical methods applied during the last ten years for the determination of pharmaceuticals in aqueous samples (adapted from Kostopoulou and Nikolaou (2008))

Pharmaceuticals studied	Sample preparation	Column and temperature program used	LOD (ng/l)	Reference
GC-MS methods				
Acetylsalicylic acid *Clofibric acid* *Fenofibric acid* *Bezafibrate* *Gemfibrozil* *Ibuprofen* *Diclofenac* *Ketoprofen* *Fenoprofen* *Indometacine* *Naproxen* *Meclofenamic acid* *Tolfenamic acid*	*SPE with RP-C18 Lichrolut, derivatization with N-Methyl-N-(trimethylsilyl)-trifluoroacetamide (MSTFA), elution with methanol*	*HTI-5 (50 °C for 2 min, 16 °C/min to 180 °C, 4°C/min to 290 °C, held for 7 min)*	*50-250 for sewage 10-25 for surface water 1-25 for drinking water*	*Stumpf et al., 1999*
Ibuprofen *Paracetamol* *Phenazone* *Carbamazepine*	*SPME with fibre coating polyacrylate, Carbowax-DVB*	*HP5-MS (70 °C for 2 min, 10 °C/min to 250 °C, 5 min, increased to 280 °C and held for 10 min)*	*200-50000*	*Möder et al., 2000*
Ibuprofen *Paracetamol* *Phenazone* *Carbamazepine*	*SPE with LiChrolut-EN and LiChrolut-RP-C, derivatization with BSTFA, elution with acetone, methanol*	*HP5-MS (70 °C for 2 min, 10 °C/min to 250 °C, 5 min, increased to 280 °C and held for 10 min)*	-	*Möder et al., 2000*
Carbamazepine *Ibuprofen* *Diclofenac* *Ketoprofen* *Naproxen* *Clofibric acid*	*SPE with Oasis HLB, derivatization with diazomethane for acidic compounds, elution with ethyl acetate, acetone*	*RTX-5MS (90 °C for 1 min, 15 °C/min to 150 °C, for 15 min, 5 °C/min to 200 °C, for 5 min, 15 °C/min to 290 °C, held for 6 min)*	*1-10*	*Ollers et al., 2001*
Diclofenac *Ibuprofen* *Ketoprofen* *Indomethacine* *Naproxen* *Fenoprofen* *Clofibric acid* *Bezafibrate* *Gemfibrozil* *Etofibrate* *Fenofibrate* *Fenofibric acid* *Carbamazepine* *Pentoxifylline* *Diazepam*	*SPE with RP-C18, derivatization with pentafluorobenzyl bromide, elution with acetone*	*DB-35 (65°C for 2 min, 30°C/min to 180°C, 5°C/min to 300°C, held for 12 min)*	*3.3-9.6*	*Sacher et al., 2001*

Pharmaceuticals studied	Sample preparation	Column and temperature program used	LOD (ng/l)	Reference
Gemfibrocil *Clofibric acid* *Clofibrat* *Etofibrat* *Etofyllinclofibrat* *Diclofenac* *Ibuprofen* *(hydroxy- and carboxy-ibuprofen)* *Ketoprofen* *Indomethacin* *Fenoprofen* *Sarkosin-N-(phenylsulfonyl) (SPS)*	*SPE with LiChrolut-EN, "on column" reaction with trimethylsulfonium hydroxide (TMSH), trimethylanilinium hydroxide (TMAH) and "pre column" with diazomethane, elution with methanol and ethyl acetate*	*HP Ultra-2 (50 °C for 1 min, 40 °C/min to 170 °C, 1 min, 4 °C/min to 250 °C, 15.6 min 40 °C/min to 280 °C, 4.3 min)*	*3-12*	*Jux et al., 2002*
Clofibric acid *Diclofenac* *Fenofibrate* *Fenoprofen* *Flurbiprofen* *Gemfibrozil* *Ibuprofen* *Ketoprofen* *Naproxen*	*SPE with RP C18, derivatization with diazomethane, elution with methanol, dichloromethane*	*HP-5 (80 °C for 8 min, 5 °C/min to 280°C 3 °C/min to 300°C, held for 5 min)*	-	*Andreozzi et al., 2003*
Metoprolol *Nadolol* *Propranolol*	*SPE with Empore SDB-XC extraction disks, derivatization with N-Methyl-N-(trimethylsilyl)-trifluoroacetamide (MSTFA), elution with methanol*	*DB5-MS (160 °C for 1 min, 10 °C/min to 290 °C, for 10 min)*	*1-10*	*Huggett et al., 2003*
Phenazon *Clofibric acid Propyphenazone* *Ibuprofen* *Diclofenac*	*SPE with C18, derivatization with pentafluorobenzyl bromide (PFBBr) (triethylamine as catalyst), elution with methanol*	*HP-5MS (100 °C for 1 min 30 °C/min to 150 °C for 1 min, 3 °C/min to 205 °C 10 °C/min to 260 °C for 23 min)*	*0.6-20*	*Koutsouba et al., 2003*
Ibuprofen *Naproxen* *Ketoprofen* *Tolfenamic acid Diclofenac*	*SPE with Oasis HLB, derivatization with* N-*Methyl*-N *(*tert. *butyldimethylsilyl) trifluoroacetamide (MTBSTFA), elution with ethyl acetate*	*BP5 (50 °C for 1 min, 10 °C/min to 180 °C, for 7 min, 10 °C/min to 230 °C, for 25 min, 20 °C/min to 250 °C, for 5 min)*	*20-50 **	*Rodriguez et al., 2003*
Carbamazepine *Diazepam*	*SPE with Oasis HLB, elution with ethyl acetate*	*BP5 (50 °C for 1 min, 10 °C/min to 180 °C, for 7 min, 10 °C/min to 230 °C, for 25 min, 20 °C/min to 250 °C, for 5 min)*	*18.9-22.2*	*Carballa et al., 2004*

Table 2. (continued)

Pharmaceuticals studied	Sample preparation	Column and temperature program used	LOD (ng/l)	Reference
Diclofenac *Ibuprofen* *Naproxen* *Roxithromycin* *Sulfamethoxazole* *Iopromide*	*SPE with Oasis HLB, derivatization with* N-*methyl*-N-*(*tert.-*butyldimethylsilyl) trifluoroacetamide (MTBSTFA), elution with ethyl acetate*	*BP5 (50 °C for 1 min, 10 °C/min to 180 °C, for 7 min, 10 °C/min to 230 °C, for 25 min, 20 °C/min to 250 °C, for 5 min)*	*0.5-16.7*	*Carballa et al., 2004*
Ibuprofen *Hydroxy-Ibuprofen* *Carboxy-Ibuprofen* *Clofibric acid* *Diclofenac* *Triclosan*	*SPE with Oasis HLB, derivatization with methyl chloromethanoate, elution with methanol*	*HP-5MS 353 K (80 °C) for 2 min, at 7 K/min to 533 K (260 °C) for 10 min*	*0.05-0.38**	*Weigel et al., 2004*
Ibuprofen *Naproxen* *Ketoprofen* *Diclofenac*	*SPE with Strata X, derivatization with MSTFA (N-methyl-N-(trimethylsilyl) trifluoroacetamide), elution with methanol*	*HP-5- MS (2 min at 100 °C, 4 °C/min to 180 °C, 10 °C/min to 230 °C, held for 20 min, 20 °C/min to 270 °C held for 7 min)*	*2-6*	*Kosjek et al., 2005*
Ibuprofen *Salicylic acid* *Gemfibrozil* *Naproxen* *Ketoprofen* *Diclofenac* *Indomethacin*	*SPE with Oasis MAXSPE, derivatization with pentafluoropropionic acid anhydride (PFPA)* N-t-*butyldimethylsilyl-* N-*methyltrifluoroacetamide (MTBSTFA), elution with methanol*	*Restek Rtx-5Sil MS (70 °C for 1 min, 30 °C/min to 190 °C, 5 °C/min to 260 °C, 15 °C/min to 300 °C, held for 5 min)*	*10*	*Lee et al., 2005*
Ibuprofen *Naproxen* *Ketoprofen* *Diclofenac* *Carbamazepine* *Clofibric acid*	*SPE with Oasis HLB, large-volume on-line derivatization with tetrabutylammonium salts, elution with methanol, acetone-ethyl acetate, acetone-methanol*	*DB5-MS (70 °C for 4 min, 10 °C/min to 300 °C, held for 4 min)*	*1-8*	*Lin et al., 2005*
Clofibric acid *Ibuprofen* *Ketoprofen* *Mefenamic acid Diclofenac*	*SPE with ENVI-18 reverse phase packed tube, derivatization with pentafluorobenzyl bromide (triethylamine as catalyst), elution with methanol*	*RTX-5 (80 °C for 1 min, 30 °C/min to 150 °C, 3.5 °C/ min to 280 °C, held for 30 min)*	*5-15*	*Tauxe-Würsch et al., 2005*
Paracetamol *Ibuprofen* *Salbutamol* *Mefenamic acid Propranolol*	*SPE with LiChrolut ENV+, derivatization with MSTFA (N-methyl-N-trimethylsilyl-trifloroacetamide) elution with methanol*	*DB5MS (50 °C for 7.5 min, 30 °C/min to 270 °C, held for 10 min)*	*2-4**	*Bound and Voulvoulis, 2006*

Pharmaceuticals studied	Sample preparation	Column and temperature program used	LOD (ng/l)	Reference
Ibuprofen *Acetaminophen* *Dipyrone* *Diclofenac* *Carbamazepine* *Codeine* *Caffeine* *Metabolites:* *1,7-dimethylxanthine,* *carbamazepine 10,11-epoxide*	*SPE with Oasis HLB, elution with ethyl acetate*	*ZB-5MS (105 °C for 1 min, 17 °C/min to 200 °C for 1 min, 2 °C/min to 220 °C for 2 min, 5 °C/min to 290 °C for 1 min)*	*1-100*	*Gomez et al, 2006*
N,N-*bis(3,3-dimethyl-2-oxetanyl)-3,3-dimethyl-2-oxetanamine* *Ibuprofen* *Aspirin* *Galaxolide* *Tonalide* *Cyclophosphamide* *Triclosan* *1,1'-Sulfonylbis 4-chloro-benzene* *4-Formylaminoantipyrine* *4-Acetylaminoantipyrine* *Carbamazepine* *Codeine* *Pentoxyfylline* *Diazepam*	*SPE with Oasis, elution with acetonitrile and dichloromethane*	*HP5MS (90 °C for 1 min, 10 °C/min to 120 °C, 3.5 °C/min to 200 °C, 5 °C/min to 315 °C, held for 11 min)*	*30**	*Moldovan, 2006*
Acetylsalicylic acid Ibuprofen *Gemfibrozil* *Fenoprofen* *Naproxen* *Ketoprofen* *Diclofenac*	*SPE with LC-18 , derivatization with BF3/MeOH, elution with methanol*	*CP-SIL 8CB-MS (50 °C for 0.75 min, 20 °C/min to 120 °C, 2 °C/min to 200 °C, 9 °C/min to 290 °C, held for 10 min)*	*0.1-1*	*Verenitch et al., 2006*
Clofibric acid *Ibuprofen* *Salicylic acid* *Gemfibrozil* *Naproxen* *Ketoprofen* *Diclofenac*	*SPE with Oasis HLB, derivatization with MSBSTFA, elution with acetone*	*HP-5MS (100 ◦C for 1 min, 20 ◦C/min to 280 ◦C, 280 ◦C for 10 min)*	*0.005 to 1 in spring water, 0.5 to 100 in wastewater.*	*Gibson et al., 2007*
Ibuprofen *Flufenamic acid* *Naproxen* *Mefenamic acid* *Tolfenamic acid* *Meclofenamic acid*	*SPME with PTFE fibre coating, derivatization with dimethyl sulfate and addition of tetrabutylammonium hydrogen sulfate as ion pairing agent*	*HP-5MS (50 ◦C for 3 min, 30 ◦C/min to 250 ◦C)*	*0.3-2.9*	*Araujo et al, 2008*
Carbamazepine *Ibuprofen* *Diclofenac* *Sulfamethoxazole*	*SPE with Oasis HLB, derivatization with MSTBSTFA, elution with methanol*	*HP-5MS 50 °C (1 min), 10 °C/min to 180 °C (7 min), 10 °C/min to 230 °C (25 min), 20 °C/min to 250 °C (5 min).*	*5.0-20.0*	*Kostopoulou et al, 2008*

Table 2. (continued)

Pharmaceuticals studied	Sample preparation	Column and temperature program used	LOD (ng/l)	Reference
Aspirin *Ibuprofen* *Ketoprofen* *Naproxen* *Paracetamol* *Gemfibrozil* *Salbutanol* *Clenbuterol* *Terbutalin* *Diclofenac* *Diazepam* *Caffeine* *Carbamazepine* *Amitryptiline* *Imipramine* *Doxepine* *Nordiazepam*	*SPE with Oasis MCX, derivatization with MSTFA, elution with ethyl acetate/acetone*	*HP-5MS (70 ∘C for 2 min, 10 ∘C/min to 250 ∘C for 5 min)*	*0.1-2.6 for surface water, 3.2-28.6 for wastewater*	*Togola and Budzinski, 2008*
LC-MS methods				
Phenazone *Dimethylaminophenazone* *Propyphenazone* *Metoprolol* *Propranolol* *Atenolol* *Bisoprolol* *Sotalol* *Pindolol* *Betaxolol* *Salbutamol* *Clenbuterol* *Terbutaline* *Ifosfamide* *Cyclophosphamide* *Simvastatin*	*SPE with PPL Bond-Elut, elution with methanol*	*Nucleosil 120-3-C18 (Solvent A: 20 mM ammonium acetate in Milli-Q water, pH 6.8 Solvent B: 20 mM ammonium acetate in acetonitrile-methanol, 2:1 v/v)*	*2.3-13*	*Sacher et al., (2001)*
Iopamidol *Iopromide* *Iomeprol* *Amidotrizoic acid*	*SPE with LiChrolut EN, elution with methanol, acetonitrile*	*Nucleosil 120-3-C18 (Solvent A: 2 mM ammonium formiate in Milli-Q water, pH 7.0 Solvent B: 2 mM ammonium formiate in acetonitrile-methanol, 2:1 v/v)*	*2.3-4.8*	*Sacher et al., (2001)*
Sulfamethoxazole *Sulfadiazine* *Sulfadimidine* *Sulfamerazine* *Ronidazole* *Metronidazole* *Furazolidone* *Trimethoprim* *Dapsone*	*SPE with Isolut ENV+, elution with acetonitrile, water, triethylamine*	*Nucleosil 120-3-C18 (Solvent A: 20 mM ammonium acetate in Milli-Q water, pH 6.8 Solvent B: 20 mM ammonium acetate in acetonitrile-methanol, 2:1 v/v)*	*1-3.2*	*Sacher et al., (2001)*

Pharmaceuticals studied	Sample preparation	Column and temperature program used	LOD (ng/l)	Reference
Chloramphenicol *Virginiamycin* *Oleandomycin* *Erythromycin* *Anhydro-erythromycin* *Roxithromycin* *Clarithromycin* *Spiramycin* *Tylosin*	*SPE with Isolut ENV+, elution with acetonitrile, water, triethylamine*	*Nucleosil 120-3-C18 (Solvent A: 20 mM ammonium acetate in Milli-Q water, pH 6.8 Solvent B: 20 mM ammonium acetate in acetonitrile-methanol, 2:1 v/v)*	*1.8-4.5*	*Sacher et al., (2001)*
Amoxicillin *Oxacillin* *Cloxacillin* *Dicloxacillin* *Nafcillin* *Penicillin G* *Penicillin V*	*SPE with Isolut ENV+, elution with acetonitrile, water, triethylamine*	*Nucleosil 120-3-C18(Solvent A: 2 mM ammonium formiate in Milli-Q water, pH 7.0 Solvent B: 2 mM ammonium formiate in acetonitrile-methanol, 2:1 v/v)*	*3.6-6.5*	*Sacher et al., (2001)*
Clofibric acid *Bezafibrate* *Gemfibrocil* *Fenofibrate* *Atenolol* *Sotalol* *Metoprolol* *Betaxolol*	*SPE Isolute C18, elution with methanol*	*RP-18 (Solvent A: acetonitrile Solvent B: HPLC-grade water)*	*0.017-1.25**	*Hernando et al., (2004)*
Clarithromycin *Erythromycin* *Spiramycin* *Lincomycin* *Ciprofloxacin* *Ofloxacin* *Amoxycillin* *Sulfamethoxazole* *Oleandomycin* *Tilmicosin* *Tylosin* *Oxytetracycline* *Cyclophosphamide* *Methotrexate* *Ibuprofen* *Salbutamol* *Atenolol* *Enalapril* *Diazepam* *Carbamazepine* *Furosemide* *Hydrochlorothiazide* *17α-Ethinylestradiol* *Omeprazole* *Ranitidine* *Bezafibrate* *Clofibric acid* *Demethyl diazepam* *17β-Estradiol* *Estrone*	*SPE Oasis MCX and Lichrolut EN, elution with methanol, ethyl acetate*	*Luna C8 (Positive ion mode: Solvent A: formic acid 0.1% in Milli-Q water, pH 2 Solvent B: acetonitrile)* *(Negative ion mode: Solvent A: triethylamine, pH 8, 0.05% in water Solvent B: acetonitrile)*	*0.1-5.2 **	*Castiglioni et al., (2005)*

Table 2. (continued)

Pharmaceuticals studied	Sample preparation	Column and temperature program used	LOD (ng/l)	Reference
Carbamazepine Mefenamic acid Indomethacine Ibuprofen Naproxen Diclofenac Ketorolac Acetaminophen Codeine Fluoxetine Paroxetine Atenolol Propranolol Trimethoprim Metronidazole Erythromycin Ranitidine	*SPE with Oasis HLB, elution with methanol*	*C8 ZB-5MS (Positive ion mode: Solvent A: acetonitrile Solvent B: formic acid 0.1% in MilliQ water) (Negative ion mode: Solvent A: acetonitrile Solvent B: MilliQ Water)*	*3.8-47*	*Gomez et al., (2006)*
Sulfaguanidine Sulfacetamide Sulfadiazine Sulfathiazine Sulfapyridine Sulfamerazine Sulfamethazine Sulfamethoxazole Sulfadimethoxine Sulfasalazine	*SPME with carbowax/ divinylbenzene (CW/DVB) fiber*	*C18 (Solvent A: 20mM aqueous ammonium acetate, 0.1% formic acid, pH 3 Solvent B: 20mM ammonium acetate in 2:1 CH_3CN:MeOH)*	*9000-55300*	*Balakrishnan et al., (2006)*
Sulfamethoxazole Acetyl-sulfamethoxazole Trimethoprim Erythromycin Paracetamol Ibuprofen Mefenamic acid Diclofenac Clofibric acid Propranolol Dextropropoxyphene Lofepramine Tamoxifen	*SPE with Strata X, elution with methanol*	*Luna C18 (mobile phase of water, methanol and 40mM ammonium acetate in water, adjusted to pH 5.5 by the addition of formic acid)*	*10-50*	*Hilton and Thomas (2003) Ashton et al. (2004) Roberts and Thomas (2006)*

* limit of quantitation (LOQ) is provided

Extraction techniques are a very important part of the analytical process. In order to increase the efficiency of the extraction, acidification of wastewater samples is necessary for the determination of pharmaceuticals containing acidic groups in their structure. Samples are acidified after collection by the addition of an acid to a pH= 2–3 (Renew and Huang, 2004). Filtration of the samples is also performed, in order to remove suspended solids which may reduce the extraction efficiency. In general, the samples are filtered through 1.7, 0.45 or 0.22 μm glass-fiber filters. The most common extraction methods increasingly being used for the analysis of pharmaceuticals and related compounds in wastewater are Solid-Phase Extraction (SPE) and Solid-Phase Micro-Extraction (SPME). These techniques have the advantages of being faster and more environmentally friendly compared to the previously used Soxhlet extraction, or liquid-liquid extraction methods, which were time-consuming and required use of large volumes of toxic solvents. Much research has been performed recently on the assessment of several types of SPE stationary phases (Rossi and Zhang, 2000; Weigel et al., 2004; Lee et al., 2005; Tauxe-Wuersch et al., 2005), and the most effective ones have proved to be ENV+, Oasis HLB, Oasis MAXSPE, Oasis MCX, Strata-X, Lichrolut C18, Lichrolut EN for the preconcentration as well as for the clean-up of pharmaceuticals in aqueous samples (Table 2). Regarding SPME procedures, the most commonly used fibre coatings are polyacrylate, poly(dimethylsiloxane–di-vinylbenzene) (PDMS-DVB), C18 and Carbowax-DVB (Möder et al., 2000).

4. Conclusions and Outlook

The occurrence of pharmaceuticals in the environment and their possible effects on living organisms is of major concern, as they are partly eliminated during conventional wastewater treatment, and partly transformed to degradation products that have not yet been investigated and with largely unknown toxicological properties. The development of more sensitive analytical methods for the determination of pharmaceuticals and degradation products in wastewater samples and in aquatic receivers is in a constant progress and optimization, and aims to clarify many existing gaps on this subject. In parallel, the execution of environmental risk assessment studies for pharmaceuticals and degradation products is necessary, with a further step being the investigation of mixtures of such compounds, in order to simulate better the real-world situation where pharmaceuticals and related compounds are present in the environment simultaneously, having combined adverse effects on living organisms.

References

Andreozzi R., Marotta R., Paxéus N. (2003) *Chemosphere* 50: 1319.

Andreozzi R., Raffaelea M., Nicklas P. (2003) *Chemosphere* 50: 1319.

Balakrishnan V.K., Terry K.A., Toito J. (2006) *J. Chromatogr.* A 1131: 1.

Boree A.N., Arnold W.A., McNeill K. (2003) *Aquat Sci* 65: 320–341

Bound P.J., Vourvoulis N. (2006) *Water Res.* 40: 2885.

Boyd G.R., Reemtsma H., Grimm D.A., Mitra S. (2003) *Sci Total Environ* 311: 135..

Buser H.R., Poiger T., Müller M.D. (1998) *Environ Sci Technol* 32: 3449.

Buser H.R., Poiger T., Müller M.D. (1999) *Environ Sci Technol* 33: 2529.

Carballa M., Omil F., Lema M.J., Liompart M., Garcia-Jares C., Rodrıguez I., Gomez M., Ternes T. (2004) *Water Res.* 38: 918.

Castiglioni S., Bagnati R., Calamari D., Fanelli R., Zuccato E. (2005) J. *Chromatogr. A* 1092: 206.

Daughton C.G., Ternes T.A. (1999) *Environ. Health Perspect.* 107: 907.

DellaGreca M., Fiorentino A., Isidori M., Lavorgna M., Previtera L., Rubino M., Temussi F. (2004) *Chemosphere 54:* 629.

Diaz-Cruz M.S., Garcia-Galan M.J., Barcelo D. (2008) *J. Chromatography A* 1193: 50.

Doll T.E., Frimmel F.H. (2003) *Chemosphere 52*: 1757.

Fatta D., Nikolaou A., Achilleos A., Meric S. (2007) *Trends in Anal. Chem.*, 26: 515.

Fatta Kassinos D, Meric S, Nikolaou A (2008) *Advances in analytical methods for the determination of pharmaceutical residues in waters and wastewaters,* Elsevier, book chapter, submitted for publication

Feldmann D.F., Zuehlke S., Heberer T. (2008) *Chemosphere 71*: 1754.

Fent K., Weston A.A., Caminada D (2006) *Aquatic Toxicol,* 76: 122.

Focazio M.J., Kolpin D.W., Barnes K.K., Furlon E.T.g, Meyer M.T., Zaugg S.D., Barber L.B., Thurman M.E. (2008) *Sci. Total Environ.* 402: 201.

Gans O., Sattelberge R., Scharf S. (2002) *Vom Wasser* 98: 165.

Garrison A.W., Pope J.D., Allen F.R. (1976) GC/MS analysis of organic compounds in domestic wastewaters. In: *Identification and Analysis of Organic Pollutants in Water* (Ed, Keith LH) Ann Arbor Science Publishers Inc, Ann Arbor, 517.

Giger W., Alder A.C., Golet E.M., Kohler H.P.E., McArdell C.S., Molnar E., Siegrist H., Suter M.J.F. (2003) *Chimia 57*: 485.

Golet E.M., Alder A.C., Giger W. (2002) *Environ Sci Technol* 36: 3645.

Gomez M.J., Martinez Bueno M.J., Lacorte S., Fernandez-Alba A.R., Aguera A. (2007) *Chemosphere* 66: 993.

Gomez M.J., Petrovic M., Fernandez-Alba A.R., Barcelo D. (2006*) J. Chromatogr. A* 1114: 224.

Halling-Sorensen B., Nors Nielsen S., Lanzky P.F., Ingerslev F., Holten Lutzhoft H.C., Jorgensen S.E. (1998) *Chemosphere* 36: 357.

Hartmann A., Alder A.C., Koller T., Widmer RM (1998) *Environ. Toxicol. Chem.* 17: 377.

Heberer T., Fuhrmann B., Schmidt-Baumler K., Tsipi D., Koutsouba V., Hiskia A. (2001) Occurrence of pharmaceutical residues in sewage, river, ground and drinking water in Greece and Berlin (Germany). In: Pharmaceuticals and Personal Care Products in the Environment: Scientific and Regulatory Issues (Eds, Daughton CG and Jones-Lepp TL) *American Chemical Society,* Washington, 70.

Hernando M.D., Petrovic M., Fernandez-Alba A.R., Barcelo D. (2004) *J. Chromatogr. A* 1046: 133.

Hernando M.D., Mezcua M., Fernández-Alba A.R., Barceló D. (2006) *Talanta* 69: 334.

Hilton M.J., Thomas K.V. (2003) *J. Chromatogr. A* 1015: 129.

Hirsch R., Ternes T., Haberer K., Kratz K.L. (1999*) Sci. Total Environ.* 225: 109.

Hirsch R., Ternes T.A., Haberer K., Kratz K.L. (1996) *Vom Wasser* 87: 263.

Huggett D.B., Khan I.A., Foran C.M., Schlenk D. (2003) *Environ. Poll.* 121: 199.

Isidori M., Nardelli A., Parrella A., Pascarella L., Previtera L. (2006) *Chemosphere* 63: 785.

Jux U., Baginski R., Hans - Gunter A., Kronke M., Seng P. (2002) *Inter. J. Hyg. Environ.l Health* 205: 393.

Khan SJ, Ongerth JE (2002) *Water Sci. Tech.* 46: 105-113

Kosjek T., Heath E., Krbavcic A. (2005) *Environ. Int.* 31: 679.

Kosjek T., Heath E., Petrovic M., Barcelo D. (2007) *Trends in Anal. Chem.* 26: 1076.

Kostopoulou M.N., Nikolaou A.D. (2008) Determination of pharmaceuticals and their metabolites in aqueous environmental matrices: analytical problems and needs during sample preparation, *Trends in Anal. Chem.*, accepted for publication.

Koutsouba V., Heberer Th., Fuhrmann B., Schmidt-Baumler K., Tsipi D., Hiskia A. (2003) *Chemosphere* 51: 69.

Lee H.B., Thomas E.P., Lewina S.M. (2005) *J Chromatogr. A* 1094: 122.

Lin W.C., Chen H.C., Ding W.H. (2005) *J. Chromatogr. A* 1065: 279.

Lishman L., Smyth S.A., Sarafin K. Kleywegt S., Toito J., Peart T., Lee B., Servos M, Beland M., Seto P. (2006) *Sci Total Environ* (In Press)

McAvoy D.C., Schatowitz B., Jacob M., Hauk A., Eckhoftf W. (2002) *Environ. Toxicol. Chem.* 21: 1323-1329

Metcalfe C.D., Miao X.S., Koenig B.G., Struger J. (2003) *Environ. Toxicol. Chem.* 22: 2881.

Miao X.S., Koenig B.G., Metcalfe C. (2002) *J Chromatogr A* 952: 139.

Möder M., Schrader S., Winkler M., Popp P. (2000), *J. Chromatogr. A* 873: 95.

Mok K.M., Wong H., Fan X.J. (2006) *Global Nest Journal* 7: 1-16

Moldovan Z. (2006) *Chemosphere 64*: 1808.

Mutavdzic Pavlovic D., Babic S., Horvat A.J.M., Kastelan-Macan M. (2007) *Trends in Anal. Chem.* 26: 1062.

Nikolaou A., Meric S., Fatta D. (2007) Occurrence patterns of pharmaceuticals in environmental matrices. *Analytical and Bioanalytical Chemistr* 387: 1225.

Oaks J.L., Gilbert M., Virani M.Z., Watson R.T., Meteyer C.U., Rideout B.A., Shivaprasad H.L., Ahmed S., Chaudhry M.J.I., Arshad M., Mahmood S., Ali A., Khan A.A. (2004) *Nature 427*: 630.

Ollers S.J., Singer H., Fassler Ph., Muller S.R. (2001) *J. Chromatogr.* A 911: 225.

Pavlogeorgatos G.D., Thomaidis N.S., Nikolaou A.D., Lekkas T.D. (2006) *Global Nest Journal* 8: 61.

Perez S., Barcelo D. (2007) *Trends in Anal. Chem.* 26: 494.

Petrovic M., Gonzalez S., Barcelò D. (2003) *Trends Anal. Chem.* 22: 685.

Purdom C.E., Hardiman P.A., Bye V.J., Eno N.C., Tyler C.R., Sumpter J.P. (1994) *J Chemical Ecology* 8: 275.

Radjenovic J., Petrovic M., Barcelo D. (2007) *Trends in Anal. Chem.* 26: 1132.

Renew J.E., Huang C.H. (2004) *J. Chromatogr. A* 1042: 113.

Rizzo L., Selcuk H., Nikolaou A., Belgiorno V., Bekbolet M., Meric S. (2005) *Global Nest Journal* 7: 95.

Roberts P.H., Thomas K.V. (2006) *Sci. Total Environ.* 356: 143.

Rodriguez I., Quintana J.B., Carpinteiro J., Carro A.M., Lorenzo R.A., Cela R. (2003) *J. Chromatogr. A,* 985: 265.

Rossi D.T., Zhang N. (2000) *J. Chromatogr.* A 885: 97.

Sacher F., Lange F., Brauch H., Blankenhorn I. (2001) *J Chromatogr.* A 938: 199.

Stan H.J., Heberer T. (1997) *Analysis* 25: 20.

Stumpf M., Ternes T., Wilken R., Rodrigues S., Baumann W. (1999) *Sci. Total Environ.* 225: 135.

Tauxe-Wuersch A., De Alencastro L.F., Grandjean D., Tarradellas J. (2005) *Water Res.* 39: 1761.

Ternes T.A. (2001) *Trends Anal. Chem.* 20: 419.

Ternes T.A. (1998) *Water Res* 32: 3245.

Tixier C., Singer H.P., Oellers S., Müller S.R. (2003) *Environ Sci Technol* 37: 1061.

Verenitch S.S., Lowe C.J., Mazumder A. (2006) *J. Chromatogr. A* 1116: 193.

Weigel S., Berger U., Jensen E., Kallenborn R., Thoresen H., Hühnerfuss H. (2004) *Chemosphere* 56: 583.

Weigel S., Kallenborn R., Hühnerfuss H. (2004) *J Chromatogr. A* 1023: 183.

Winkler M., Lawrence J.R., Neu T.R. (2001) *Water Res* 35: 3197.

Zuccato E., Castiglioni S., Bagnati R., Chiabrando C., Grassi P., Fanelli R. (2008) *Water Res.* 42: 961.

Zuehlke S., Duennbier U., Heberer T. (2007): *Chemosphere* 69: 1673.

In: Sewage Treatment: Uses, Processes and Impact
Editor: Anna Stephens and Mark Fuller

ISBN 978-1-60692-959-9

Chapter 10

INTENSIVE AEROBIC BIOCONVERSION OF SEWAGE SLUDGE AND FOOD WASTE INTO ORGANIC FERTILISER

O. Stabnikova, J. Y. Wang and V. Ivanov
School of Civil and Environmental Engineering,
Nanyang Technological University, Singapore

ABSTRACT

The addition of food waste to sewage sludge with low content of heavy metals improved C/N ratio in initial material for bioconversion, increased the potassium content, diluted the content of heavy metals, and reduced the viscosity of sewage sludge. The mixture of sewage sludge and food waste was pre-treated at 100°C for 45 minutes to inactivate the pathogens and parasites, disrupt cells in both food waste and sewage sludge, and partially hydrolyze polymers. Bioconversion of the pre-treated mixture of sewage sludge and vegetable food waste was performed in the rotating bioreactor with air supply and controlled temperature 60°C for 10 days with the addition of starter culture, artificial bulking agent, and pH buffering component. Thermophilic bacterial strain of *Bacillus thermoamylovorans* SW25 with high digestion activity have been selected and used as the starter culture to enhance bioconversion of sewage sludge and food waste. The best organic fertiliser was obtained when the ratio of the thermally pre-treated sewage sludge and food waste was 1:1 by total solids. The reductions of organic matter, 16S rRNA of methanogens, and coenzyme F_{420} fluorescence during 10 days of the bioconversion were 67%, 54%, and 87% of the initial values, respectively. The produced organic fertiliser was a powder with the moisture content of 5%. The content of volatile solids was decreased by 25% of total solids. The plant tests indicated that the product of bioconversion can be graded as good commercial fertiliser. Addition of 1.0% to 1.5% of this fertiliser to the subsoil increased the yield and growth of different plants by 113 to 164%. The product was confirmed as a stable and non-toxic matter for the plants by the phytotoxicity test. Although the intensive aerobic thermophilic process consumes more energy than that of the conventional composting process but it saves time and land area by small footprint bioconversion of the mixture of sewage sludge and food waste into a commercially-compatible fertiliser in 10 days without curing and maturation phases that are required in conventional composting.

INTRODUCTION

Sewage sludge is a general term used to describe the solids produced during the wastewater treatment process at a municipal sewage plant, including the sludge removed from an aerobic or anaerobic digester. Main component of the sewage sludge is usually dewatered sludge from anaerobic digester of activated sludge. Sewage sludge is disposed of by landfilling, incineration, in cement production, in agriculture, or by composting (Butt et al., 1998; Edge 1999; Schmidt 2001). The advantage for using sewage sludge in agriculture is recycling of nitrogen, phosphorus, potassium, calcium, microelements, and organic matter for plant growth by the enhancement of soil. Although applying sewage sludge on land is concerned because of high content of heavy metals, low concentration of potassium, high moisture content, presence of pathogens and chemical pollutants (Furhacker and Haberl 1995), there is a trend to increase the use of sludge in agriculture (Ito et al., 1998; Stenger 2000; Vesilind and Spinosa 2001). However, sewage sludge in majority of Asian countries is not utilized yet (Vesilind and Spinosa 2001). Particularly, in Singapore, around 50,000 tonnes of sewage sludge are disposed of in incineration plants (www.env.gov.sg).

The simplest way for sewage sludge utilization is its application as an organic fertiliser after thermal treatment, drying or addition of alkali to inactivate the pathogens (Hassebrauck and Ermel 1996; Larsen et al., 1991; Matthews 2001). It is known that composting of sewage sludge, an aerobic mesophilic/thermophilic decomposition of organic constituents to a relatively stable material, can enhance its quality as an organic fertiliser (Furhacker and Haberl 1995; Krogmann 2001; Rantala et al., 2000; Wei et al., 2001). Sewage sludge in Singapore has low heavy metal content because of the separation of industrial and municipal wastewaters and could be used for composting (Wang et al., 2003a; 2004).

In conventional composting, the material is aerobically or anaerobically treated, then "cured" or "maturated", and dried in piles or windrows for several months (Furhacker and Haberl 1995; Haug 1993; Matthews 2001). To minimize the process duration and the space required for the large-scale bioconversion, an intensive in-vessel bioconversion of sewage sludge can be used (Furhacker and Haberl 1995; Ponti et al., 1995; Sharma et al., 1997). Composting in-vessel systems ensure constant temperature control and proper air supply, require little space, and minimize odour problems. Optimal conditions for in vessel biotransformation/composting are as follows: moisture content of the composting material must be at least 65%; bulking agent must be added to improve the mass transfer of oxygen and carbon dioxide between air and material; C/N ratio of the produced organic fertiliser must be between 25:1 and 30:1; additional aeration can improve the process; temperature of 60°C must be kept at optimal thermophilic composting (Haug 1993). However, C/N ratio in sewage sludge is approximately 7:1. The texture of the composting material must allow effective oxygen transfer into the bulk of the compost, but this is difficult due to the viscous texture of the material.

An important part of the process occurs in the thermophilic phase, at temperatures from 60°C to 80°C (Haug 1993). Many bacterial strains belonging to the genera *Bacillus, Thermus, Hydrogenobacter* have been isolated from thermophilic composts (Beffa et al., 1996a, b; Fujio and Kume 1991). The numbers of bacterial cells in such composts have been quoted as being from 1×10^7 to 1×10^{10} cells per g of dry matter (Beffa et al., 1996a). The most effective composting of sewage sludge has been reported in reactors with continuous mixing and aeration of the material, as well as with pH and temperature control (Furhacker and Haberl 1995).

Sewage sludge is not easily utilized by microorganisms (Fang et al., 1998; 2001; Warman and Termeer 1996); also it is not a good raw material for composting because its moisture content is high (about 80%), the texture is very viscous (and hence prevent oxygen transfer), and C/N ratio is too low (7:1). Therefore, to improve bioconversion of sewage sludge into organic fertiliser chemical modification, osmotic, thermal, and mechanical pre-treatment of sewage sludge (Erdincler and Vesilind 2001; Rivard and Nagle 1995), addition of the co-composting materials (Wong et al., 1997; 2000); addition of bulking agent (Nakasaki *et al.*, 2001) and starter bacterial culture with high sludge digestion activity (Fujio and Kume 1991) were proposed.

The aim of this research was to develop a technology for fast bioconversion of sewage sludge into organic fertiliser. The process was performed as single thermophilic stage in a closed system with addition of selected microbial strains without curing and maturation stages. An additional approach, which was used in this research, was to mix sewage sludge with vegetable food waste to allow their simultaneous bioconversion. Due to this mixing, both the C/N ratio and potassium content in the raw material could be increased and the texture of the mixture will be less viscous than that of sewage sludge. Such intensive aerobic bioconversion system requires small space and minimizes odour problems. Examination of the fertiliser's properties included analysis of chemical content, stability, phytotoxicity, effect on plant growth, and a comparison of produced fertiliser with commercial organic fertilisers and raw sewage sludge.

RESULTS

Heavy Metals in Sewage Sludge Used for Bioconversion

High content of heavy metals in sewage sludge is the most common objection for its application in agriculture. Due to the proper planning of the industrial and municipal zones, the municipal wastewater and sewage sludge in Singapore have low content of heavy metals and could be converted for agriculture use. The content of Cu, Ni, Zn, Cd, Cr and Pb in sewage sludge from the wastewater treatment plant in Singapore did not exceed the acceptable limits for heavy metal content in biosolids accepted in either the USA (US EPA 1999) or the EU (Krogmann 2001) (Table 1).

Table 1. Comparison of heavy metals contents in sewage sludge, food wastes, US and EU metal limits for biosolids, mg/kg

Samples	Cu	Ni	Zn	Cd	Cr	Pb
Metal limits in USA	4300	420	7500	89	3000	840
Metal limits in EU	1000-1750	300-400	2500-4000	20-40	NR	750-1200
Sewage sludge	1268 ±80	226±30	1394±100	8±1	194±8.0	66±7
Food waste	92±34	42±14	928±57	5±3	31±21	7±3
Organic fertiliser	898±60	178±20	863±70	5±1	159±10	41±6

Selection of Starter Bacterial Culture

38 bacterial strains were isolated from sewage sludge treated aerobically at 60°C. The ability of the 38 isolated strains to digest sewage sludge was tested by their growth on a mixed agar sludge medium (SAM) at 60°C. The medium was prepared by mixing 10 g of dry sewage sludge and 18 g of Bactoagar (Difco Laboratories, USA) in one litre of distilled water. The pH was adjusted to 7.0, and the medium was autoclaved for 15 min at 121°C. Five strains, SW06, SW09, SW11, SW19, and SW25, producing the biggest colonies on SAM, were selected for further screening by cultivation in 100 ml of 1% (w/v) suspensions of dry sewage sludge in 250 ml flasks shaken at 150 rev/min and 60°C. Strain SW25 with highest growth rate and bioconversion activity was selected for further testing as the starter culture in the bioconversion of sewage sludge (Wang et al., 2003b). The strain was identified from partial sequences of 16S rRNA gene as *Bacillus thermoamylovorans* (GenBank accession numbers AY197333).The determination of the nearest phylogenetic neighbour sequences for 16S rDNA sequences of the strain SW25 by BLAST search program (Altschul et al., 1997), showed that that strain was the representative of one species close to both the uncultured bacterium pPD10 (identities = 99%) isolated from synthetic food waste hot compost (Dees and Ghiorse 2001), and also to the facultative anaerobic *Bacillus thermoamylovorans*, isolated by Combet-Blanc with co-authors (1995, 1999) (identities = 98%).

Strain *Bacillus thermoamylovorans* SW25 was facultative anaerobic, Gram-positive, rod-shaped bacteria and did not form microscopically detected spores on a tryptic soy agar (TSA) at a temperature of 60°C. Strain grew and formed spores on TSA after treatment of the cell suspension at 100°C for 45 min. The coefficient of thermal sensitivity (K) for the incubation at 100°C for 45 min was evaluated by the equation:

$$K = (1/t)\ ln(N_o/N) \qquad (1)$$

where N_o and N are the initial and final number of viable cells, respectively, and t is time. The coefficient of thermal sensitivity for strain SW25 was 5.0.

Optimal pH for the growth of strain *Bacillus thermoamylovorans* SW25 was 6.7-8.2 and optimal temperature was 60°C.

Bioconversion of Sewage Sludge into Fertiliser

Sewage sludge has the following disadvantages as a raw material for intensive bioconversion. Its organic matter is poorly biodegradable, the texture was too viscous for mixing and oxygen transfer, and a C/N ratio was 6.8, which is a low value for composting (Table 2).

Vegetable, uncooked food waste was collected from a mixed food stall university canteen. Vegetable food waste had a C/N ratio 18.3, high content of potassium, and the texture was not viscous. Therefore, mixing of sewage sludge with vegetable food waste improved both the chemical composition (C/N ratio, content of nitrogen, phosphorus, and potassium) and the texture of the raw material (Table 1). Before the bioconversion, sewage sludge was pre-treated at 100°C for 45 minutes to inactivate the pathogens and to improve the bioconversion (Wang et al., 2003a; 2004). Pre-treated sewage sludge was mixed with solid food waste in ratio 1:1. Due to the mixing of food waste and sewage sludge, the C/N ratio in the raw material for bioconversion was increased and the texture of the mixture was less viscous than that of sewage sludge.

Table 2. Characteristics of the raw materials, initial matter and final product of bioconversion

Characteristics	Raw materials		Bioconversion	
	sewage sludge	food waste	initial matter	final product
pH	7.9	7.1 – 7.3	7.3	7.0
TS, %	21.7±0.4	7.7±1.5	16.6±0.3	95.9±0.3
VS, % of TS	73.6±0.6	91.7±1.5	82.8±0.6	62.0±0.2
C/N ratio	6.8	18.3	11.2	7.7
Content of phosphorus,%	1.8±0.3	0.1±0.0	0.5±0.0	0.5±0.1
Content of potassium,%	0.2±0.0	5.0±0.3	2.1±0.3	2.1±0.1

Note: mean ± standard deviation for three independent measurements are shown.

Drop of pH from 7.3 – 7.5 to 5.3 – 6.1 was observed during the bioconversion process. To maintain a neutral pH, the buffering substance, such as chalk, was added to adjust pH. Chalk was added at the beginning of bioconversion in a quantity of 5% by total solids to ensure pH stability. To improve the aeration of the mixture, plastic rings with diameter 10 mm and width 8 mm were added as an artificial bulking agent. Mass ratio of the rings and total solids (TS) was 0.25.

Biomass of the bacterial strain *Bacillus thermoamylovorans* SW25 was used as the starter culture for the bioconversion. The addition ensured an initial content of bacterial cells in the range from 4×10^7 to 6×10^8 cells/g of TS. The bacterial biomass for the inoculation of the solid waste was grown on Tryptic Soy Broth (Difco, USA) in a shaker at 130 rpm for 24 hours at 60°C. Biomass was recovered by centrifugation at 4000 rpm for 5 min in an Eppendorf centrifuge 5810R, and added into the reactor for bioconversion. The concentration of the introduced cells typically changed, for example, from 1.4×10^7 after inoculation to 5.6×10^8 colony-forming units/g of dry matter at the end of the process. The starter culture was only applied in the beginning of the bioconversion process. Similar maxima in the range 10^8-10^9 cells/g of the composted matter, have been found during composting of anaerobic sewage sludge mixed with wood chips in a closed 300 m^3 bioreactor, using aeration, and temperature control from 65°C to 82°C (Beffa *et al.,* 1996a). It was the same as the highest content of bacterial cells in hot synthetic compost (Dees and Ghiorse 2001).

An application of starter culture is not current practice in large-scale composting. However, the starter culture may be useful for start-up of the process (Fang et al., 2001). Additionally, the use of pure starter culture has theoretical advantages, such as both control over the desirable processes and lower risk of accumulation of harmful microorganisms in the final product of biodegradation. It can, probably, prevent or diminish the growth of actinomycetes and fungi releasing allergenic spores in air. After the start-up period, the produced compost/organic fertiliser, containing essential bacteria, can be recycled as an inoculum to initiate the microbial degradation (Beffa *et al.,* 1996a). There may be a diverse microbial community in the organic waste treated under thermophilic conditions (Ghazifard et al., 2001). However, the strain of *Bacillus thermoamylovorans* SW25 introduced at the beginning of the described thermophilic process was the only dominating strain. The cells of this species dominated in the biodegraded sewage sludge and food waste because they are facultative anaerobic bacteria and can grow under both aerobic and anaerobic conditions in the viscous matter in the reactor.

Bioconversion of the mixture of sewage sludge and solid food waste was carried out in the laboratory reactor under agitation, air supply and temperature at 60°C (Figure 1).

a

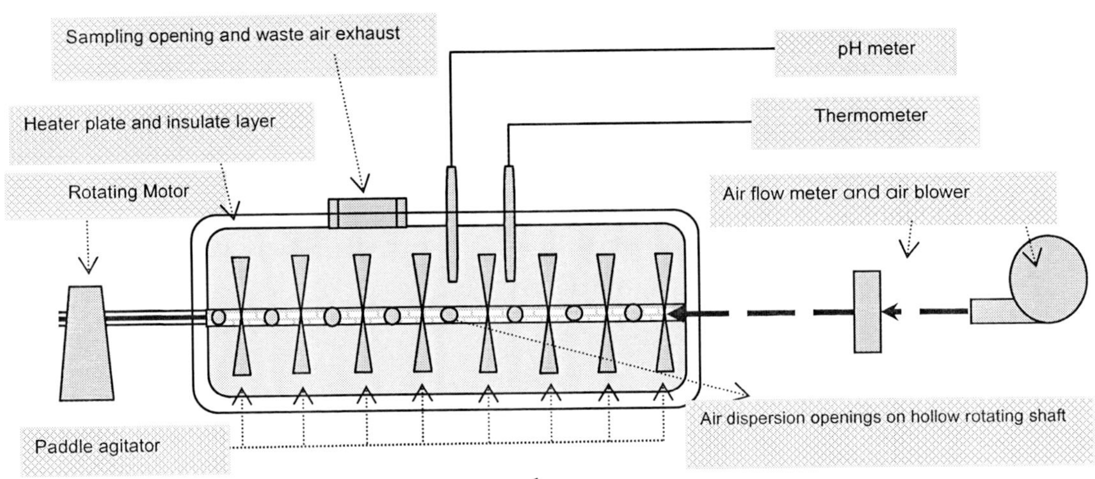

b

Figure 1. Photo (a) and schematic diagram (b) of the laboratory reactor for intensive aerobic bioconversion of sewage sludge.

The moisture content was maintained at 75-80%. The described process may be considered physically intermediate between conventional solid-phase composting with an optimum of moisture 55-65% and slurry-phase biotreatment of suspended solid wastes where the moisture content is usually greater than 85%. Water was added daily to the reactor, except during the last two days of the process. The high temperature and the effect of aeration ensured the drying of the organic fertiliser in the last two days. The organic fertiliser was a grey powder with a moisture content of 5%.

Degradation of Organic Matter and Methanogenic Biomass

The measurement of autofluorescence of coenzyme F_{420} (7,8-didemethyl-8-hydroxy-5-deazaflavin, electron transfer coenzyme) by fluorescence spectrometer can be used as a relative parameter of methanogenic biomass (Ashby et al., 2001; Gorris et al., 1994; Stabnikova *et al.,* 2005). The autofluorescence of coenzyme F_{420} dropped from 8050 relative units per 1 gram of dry matter to 804 relative units per 1 gram of dry matter during 10 days of treatment of the sewage sludge and food waste mixture, which demonstrated the biodegradation of methanogenic biomass within the matter. The content of 16S rRNA molecules of methanogenic *Archaea*, measured by specific binding (hybridization) of oligonucleotide probe Arc915, in the treated mixture also decreased during the biodegradation from 910 pmol per gram of dry matter to 421 pmol per gram of dry matter. This was correlated with the decrease of volatile solids (VS) content from 82.8 % from TS to 62.0 % from TS in the treated matter.

The percentage of organic matter biodegradation (α) was estimated by the equation:

$$\alpha = [1 - VS_2(1-VS_1)/VS_1(1-VS_2)]\times 100\ \% \qquad (2),$$

where VS_1 and VS_2 are the initial and final contents of volatile solids in the organic matter, respectively (Ivanov et al., 2004). This equation is based on the assumption that inorganic matter remained unchanged during the biodegradation of organic waste. The reduction of organic matter, the 16S rRNA of methanogens, and the coenzyme F_{420} fluorescence over the 10 d of treatment were 67%, 54%, and 87% of their initial values, respectively. Monitoring of volatile solids, autofluorescence of F_{420}, and binding of oligonucleotide probe Arc915, separately showed the biodegradation of total organic matter and its main component, which was the biomass of methanogens.

The degradation of F_{420} was 87% of their initial values. The degradation of VS (α) was 62% of initial quantity. This shows that degradation of anaerobic biomass was slightly faster than biodegradation of total organic matter (VS). In fact, methanogens as obligate anaerobic and O_2 intolerant bacteria cannot survive under aerobic conditions, where undergo lysis.

Characteristics of Final Product of Bioconversion

The heavy metal contents in the food waste were significantly lower than those in sewage sludge (Table 1). Therefore, the mixing of sewage sludge with food waste decreased the heavy metal contents in the final product of bioconversion below acceptable levels. The study of the contents of the main heavy metals in sewage sludge from Singapore municipal water reclamation plant showed that their contents were generally lower than the US and EU limits for the biosolids. Mixing of sewage sludge with food waste additionally diminished heavy metal contents in the final product (Table 1). Therefore, there is no limitation by the content of heavy metals to use the fertiliser from sewage sludge and food waste in agriculture.

The fertiliser stability index, evaluated according to the CO_2 evolution rate (United States Department of Agriculture and the United States Composting Council 2001), was 1.3 mg CO_2

/g organic matter/d (this value is graded as "very stable"). Meanwhile stability index of raw sewage sludge was 7.1 mg CO_2 /g organic matter/d (unstable).

The plant tests indicated that the produced fertiliser can be considered as good as a commercial organic fertiliser. Addition of 1.0% to 1.5% of fertiliser to the poor subsoil tested showed that the yield and growth of different plants increased by 113 to 164 %, respectively, depending on the types of plants tested. The results of tomato growth, for example, are showed in Figure 2.

Figure 2. Growth of tomato in soil with different contents of fertilizer.

The fertiliser produced by the bioconversion of sewage sludge and food waste was compared with that of raw sewage sludge, and commercial organic fertiliser in the germination and root elongation test and in the plant experiments. The fertiliser and raw sludge were added to the subsoil in dosage of 1% (weight of dry matter/weight of soil). Commercial organic fertiliser was added to the subsoil in quantity 0.7% (weight of dry matter/weight of soil) to equalize the content of nitrogen in the fertiliser and in commercial organic fertiliser. Best growth of the plants was obtained when fertiliser was applied. The parameters of the tomato growth in the subsoil with addition of the raw, commercial organic fertiliser and fertiliser produced from sewage sludge and food waste are shown in Table 3.

Table 3. The influence of raw sewage sludge, commercial organic fertiliser and fertiliser produced from sewage sludge and food waste on the tomato growth

Parameters	Growth on subsoil with		
	Raw sewage sludge	Commercial organic fertiliser	Produced organic fertiliser
Length of stem, cm	26.2±0.6	34.0±5.9	38.3±6.1
Length of root, cm	8.1±2.6	5.6±1.4	11.0±1.2
Dry weight of stem, g	0.56±0.07	0.65±0.35	0.91±0.34
Dry weight of root, g	0.12±0.08	0.11±0.10	0.20±0.10

The leaves had yellow spots when commercial organic fertiliser and raw sewage sludge were used, but there were no such spots when produced fertiliser was applied. The lengths of stems and roots of tomato grown in soil with produced fertiliser were bigger than those of tomato grown in soil with raw sewage sludge and commercial organic fertiliser by 146 and 136, 113 and 196%, respectively. The extract from the commercial organic fertiliser inhibited germination of the seeds but the germination rate in the extract of fertiliser was 100%.

Conclusion

The organic fertiliser can be produced from the mixture of the pre-treated sewage sludge and food vegetable waste, in the ratio 1:1 by total solids, in a closed reactor under controlled conditions of aeration, stirring, and pH at a temperature of 60°C. The additions were the starter bacterial culture *Bacillus thermoamylovorans* SW25, $CaCO_3$, and artificial bulking agent. Final product contained stable organic matter, nitrogen, phosphorus and potassium. Its application to the subsoil as organic fertiliser resulted in faster growth of agricultural plants. The chemical and agricultural properties of this fertiliser were significantly better than the properties of raw sewage sludge. The process seems to be expensive compared with composting since sewage sludge was thermally pre-treated, and bulking agent altogether with $CaCO_3$ need to be added to the bioreactor. However, this intensive and expensive method could be applied in a country such as Singapore where land is at a premium.

References

Altschul, S.F., Madden, T.L., Schäffer, A.A., Zhang, J., Zhang, Z., Miller, W. and Lipman, D.J. (1997). Gapped BLAST and PSI-BLAST: a new generation of protein database search programs. *Nucleic Acids Research, 25*, 3389-3402.

Ashby, K.D., Casey, T.A., Rasmussen, M.A. and Petrich, J.W. (2001). Steady-state and time-resolved spectroscopy of F_{420} extracted from methanogen cells and its utility as a marker for fecal contamination. *Journal of Agricultural and Food Chemistry, 49*, 1123-1127.

Beffa, T., Blanc, M., Lyon, P.-F., Vogt, G., Marchiani, M. and Fisher, J.L. (996a). Isolation of *Thermus* strains from hot compost (60 to 80°C). *Applied and Environmental Microbiology, 62*, 1723–1727.

Beffa, T., Blanc, M. and Aragno, M. (1996b). Obligately and facultatively autotrophic, sulfur- and hydrogen-oxidizing thermophilic bacteria isolated from hot composts. *Archives of Microbiology, 165*, 34–40.

Butt, E.P., Andrews, M.J., Guy, J.A., Morse, G.M. and Lester, J.N. (1998). Co-recycling of sludge and municipal solid waste: a cost-benefit analysis. *Environmental Technology, 19*, 475– 488.

Combet-Blanc, Y., Ollivier, B., Streicher, C., Patel, B.K., Dwivedi, P.P., Pot, B., Prensier, G. and Garcia, J.L. (1995). *Bacillus thermoamylovorans sp. nov.*, a moderately thermophilic and amylolytic bacterium. *International Journal of Systematic Bacteriology, 45*, 9-16.

Combet-Blanc, Y., Dieng, M.C. and Kergoat, P.Y. (1999). Effect of organic complex compounds on *Bacillus thermoamylovorans* growth and glucose fermentation. *Applied and Environmental Microbiology, 65*, 4582-4585.

Dees, P.M. and Ghiorse, W.C. (2001). Microbial diversity in hot synthetic compost as revealed by PCR-amplified rRNA sequences from cultivated isolates and extracted DNA. *FEMS Microbiology Ecology, 35*, 207-216.

Edge, D. (1999). Perspectives for nutrient removal from sewage and implications for sludge strategy. *Environmental Technology, 20*, 759–764.

Fang, M., Wong, J.W.C., Li, G.X. and Wong M.H. (1998). Changes in biological parameters during co-composting of sewage sludge and coal ash residues. *Bioresource Technology, 64*, 55-61.

Fang, M., Wong, M.H. and Wong, W.C. (2001). Digestion activity of thermophilic bacteria isolated from ash-amended sewage sludge compost. *Water, Air, and Soil Pollution, 126*, 1-12.

Fujio, Y. and Kume, S. (1991). Isolation and identification of thermophilic bacteria from sewage sludge compost. *Journal of Fermentation and Bioengineering, 72*, 334–337.

Furhacker, M. and Haberl, R. (1995). Composting of sewage sludge in a rotating vessel. *Water Science and Technology, 32*, 121-125.

Ghazifard, A., Kasra-Kermanshahi, R. and Far, Z.E. (2001). Identification of thermophilic and mesophilic bacteria and fungi in Esfahan (Iran) municipal solid waste compost. *Waste Management and Research, 19*, 257-261.

Gorris, L.G. and van der Drift, C. (1994). Cofactor contents of methanogenic bacteria reviewed. *Biofactors, 4*, 139-145.

Hassebrauck, M. and Ermel, G. (1996). Two examples of thermal drying of sewage sludge. *Water Science and Technology, 33*, 235–242.

Haug, R.T. (1993). *The Practical Handbook of Compost Engineering*. Boca Raton, FL: Lewis Publishers.

Ito, A., Umita, T., Aizawa, J. and Kitada, K. (1998). Effect of inoculation of iron oxidizing bacteria on elution of copper from anaerobically digested sewage sludge. *Water Science and Technology, 38*, 63-70.

Ivanov V.N., Wang J.Y., Stabnikova O.V., Tay S.T.L. and Tay J.H. (2004). Microbiological monitoring in the biodegradation of sewage sludge and food waste. *Journal of Applied Microbiology, 96*, 641–647.

Krogmann, U. (2001). Composting. In: Spinosa, L. and Vesilind, A. (Eds.), *Sludge into Biosolids: Processing, Disposal, Utilization* (pp. 259–277). London, UK: IWA Publishing.

Larsen, A.B., Funch, F.H. and Hamilton, H.A. (1991). The use of fermentation sludge as a fertiliser in agriculture. *Water Science and Technology, 24*, 33–42.

Matthews, P. (2001). Agricultural and other land uses. In: Spinosa, L. and Veslind, P.A. (Eds). *Sludge into Biosolids. Processing, Disposal, Utilization* (pp. 43–100). London, UK: IWA Publishing.

Nakasaki, K., Ohtaki, A. and Takano, H. (2001). Effect of bulking agent on the reduction of NH_3 emissions during thermophilic composting of night-soil sludge. *Waste Management and Research, 19*, 301-307.

Ponti, C., Sonnleitner, B. and Fiechter, A. (1995). Aerobic thermophilic treatment of sewage sludge at pilot plant scale. 1. Operating conditions. *Journal of Biotechnology, 38*, 173-182.

Rantala, P.-R., Vaajasaari, K., Juvonen, R., Schultz, E., Joutti, A. and Marela-Kurtto, R. (2000). Composting and utilization of forest industry wastewater sludges. *Water Science and Technology, 42*, 227-234.

Rivard, C.J. and Nagle, N. (1995). Pretreatment of microbial sludges. *Biotechnology Advances, 13*, 833-833.

Stabnikova, O., Ang, S.S., Liu, X.Y., Ivanov, V., Tay, J.H. and Wang, J.Y. (2005). The use of hybrid anaerobic solid-liquid (HASL) system for the treatment of lipid-containing food waste. *Journal of Chemistry Technology and Biotechnology, 80*, 455–461.

Schmidt, T. (2001). Sludge disposal management in the context of regional waste management (Zurich, CH). In: *Sewage Sludge disposal sustainable and/or reliable solutions*, OWAV/EWA-Workshop-Wien 2001. Sept. 2001, Technische Universitat Wien. Wiener Mitteilungen Band 171, pp. 157–166.

Sharma, V.K., Canditelly, M., Fotuna, F. and Cornacchia G. (1997). Processing of urban and agro-industrial residues by aerobic composting: review. *Energy Conversion Management, 38*, 453-478.

Stenger, A. (2000). Experimental valuation of food safety: application to sewage sludge. *Food Policy, 25*, 211–218.

Vesilind, P.A. and Spinosa, L. (2001) Sludge production and characterization. Production and regulations. In: Spinosa, L. and Vesilind, A. (Eds). *Sludge into Biosolids: Processing, Disposal, Utilization* (pp. 3 – 8). London, UK: IWA Publishing.

www.env.gov.sg Ministry of the Environment, Singapore.

Wang, J.Y., Stabnikova, O., Ivanov, V., Tay, S.T.L. and Tay, J.H. (2003a). Intensive aerobic bioconversion of sewage sludge and food waste into fertiliser. *Waste Management and Research, 21*, 405–415.

Wang, J.Y., Stabnikova, O., Tay, S.T.L., Ivanov, V. and Tay, J.H. (2003b). Intensive bioconversion of sewage sludge and food waste by *Bacillus thermoamylovorans. World Journal of Microbiology and Biotechnology, 19*, 427–432.

Wang, J.Y., Stabnikova, O., Tay, S.T.L., Ivanov, V. and Tay, J.H. (2004). Biotechnology of intensive aerobic conversion of sewage sludge and food waste into fertiliser. *Water Science and Technology, 49*, 147–154.

Warman, P.R. and Termeer, W.C. (1996). Composting and evaluation of racetrack manure, grass clippings and sewage sludge. *Bioresource Technology, 55*, 95-101.

Wei, Y.S., Fan, Y.B. and Wang, M.J. (2001). A cost analysis of sewage sludge composting for small and mid-scale municipal wastewater treatment plants. *Resources, Conservation and Recycling, 33*, 203– 216.

Wong, J.W.C. and Su, D.C. (1997). Growth of *Agropyron elongatum* in an artificial soil mix from coal fly ash and sewage sludge. *Bioresource Technology, 59*, 57-62.

Wong, J.W.C. and Fang, M. (2000). Effects of lime addition on sewage sludge composting process. *Water Research, 34*, 3691-3698.

United States Department of Agriculture and the United States Composting Council. (2001a) Carbon dioxide evolution rate. 05.08-B. In: Thompson, W.H. (Ed.), Leege, P.B., Milner, P.D. and Watson, M.E. (Co-eds.) *Test Methods for Examination of Composting and*

Compost. Jointly Published by the United States Department of Agriculture and the United States Composting Council. 05.08 - 11 – 05.08-12.

US EPA. United States Environmental Protection Agency. Standard for the use or disposal of sewage sludge, 40 CFR Part 503 sewage sludge rule. Federal Register, 1999.

In: Sewage Treatment: Uses, Processes and Impact
Editor: Anna Stephens and Mark Fuller
ISBN 978-1-60692-959-9

Chapter 11

ENVIRONMENTAL IMPACT OF SEWAGE WATER POLLUTION

Wasim Aktar
Pesticide Residue Laboratory, Department of Agricultural Chemicals,
Bidhan Chandra Krishi Viswavidyalaya,
Mohanpur-741252, Nadia, West Bengal, India

INTRODUCTION

India faces a turbulent water future. Unless water management practices are changed – and changed soon – India will face a severe water crisis within the next two decades and will have neither the cash to build new infrastructure nor the water needed by its growing economy and rising population. Water is one of the critical inputs for the sustenance of mankind. It is used both terrestrial and aquatic environment for various activities, balancing the ecological system of global environment. Water is the important natural source, which is abundant in nature and cover about 2/3ds of earth surface. However, only 1% of the water resource is available as fresh water (i.e., surface water-rivers, lakes, reams, and ground water) for human consumption and other activities. The major uses of water are for irrigation (30%), thermal power plants (50%), while other uses are domestic (7%) and industrial consumption (~12%) (A. K. De, 2002).The United Nation's report on "Water for People, Water for Life" (the first ever UN system wide evaluation on global water resources-2003) has put India a poor 120th for water quality among 122 nations covered. Only Belgium and Morocco are ranked worse than India. The quality indicator value was based on quality and quantity of fresh water (especially ground water), waste water treatment facilities, legalities like application of pollution regulations, India's quality indicator value stood at -3.1 while for based ranked country Finland it was 1.85.

The UN evaluation also ranked India 133 in a list of 180 countries for its poor water availability (1880m^3 per person per year). Kuwait was ranked the poorest on water availability. Against the National average target of 135 lpcd of water and 180 lpcd per capita in large cities, the per capita availability is low and ranges from 165 lpcd in a few larger town to about 50 lpcd in most smaller towns. The availability of water in urban slums is about 27 lpcd. Urbanisation has given rise to a number of environmental problems such as water supply, wastewater generation and its collection, treatment and disposal in urban areas. In most cases wastewater is let out untreated and it either percolates into the ground and in turn contaminates the groundwater or is discharged into the natural drainage system causing pollution in downstream areas. Sewage and not the industrial pollution accounts for more than 75 per cent of the surface water contamination in India. Due to negligence, groundwater is also increasingly getting contaminated. In India less than 50% of the urban population has access to sewage disposal system. Most of the existing collecting systems discharge directly to the receiving water without treatment. Garbage, domestic and otherwise, is directly dumped into water bodies or roadside, which can often be washed into streams and lakes. The municipalities disposes off their treated or partly treated or untreated wastewater into natural drains joining rivers or lakes or used on land for irrigation or fodder cultivation or into sea or combination of these. Toxic chemicals from sewage water transfer to plants and entire in the food chain and affect public health. Pathogens occurring in the sewage water directly affect the mammals causing severe diseases. About 60 per cent of urban deaths in India are due to lack of safe drinking water facilities. Further deaths due to water borne diseases are second only to malnutrition. It is estimated that around 80% of water consumed by a household is let of to the drains of sewers as wastewater.

There is substantial scope for segregated use of the water for further use for gardening, industrial cooling, street cleaning, vehicular washing, fire fighting, irrigation, yard cleaning, fountains, recreational lakes, etc. Though methods are available to improve the quality of recycled water to potable grade, the lack of social acceptance and prohibitive costs may prevent the adoption of these techniques. The importance of reuse and recycling of treated sewage and industrial effluents has been realized on account of two distinct advantages: reduction of pollution in the receiving water bodies and reduction in the requirement of fresh water for various uses. Reuse of municipal wastewater after necessary treatment to meet industrial water requirement is being practiced in India.

Thus, wastewater can be considered as both a resource and a problem. Wastewater and its nutrient content can be used extensively for irrigation and other ecosystem services. Its reuse can deliver positive benefits to the farming community, society, and municipalities. However, wastewater reuse also exacts negative externality effects on humans and ecological systems, which need to be identified and assessed. Before one can endorse wastewater irrigation as a means of increasing water supply for agriculture, a thorough analysis must be undertaken from an economic perspective as well. In this regard the comprehensive costs and benefits of such wastewater reuse should also be evaluated. Moreover, the economic effects of wastewater irrigation need to be evaluated not only from the social, economic, and ecological standpoint, but also from the sustainable development perspective.

Objectives

The objective is to provide a source of information of the characteristics of wastewater used for various purposes, and the reasoning behind the international guidelines presently used in regulating wastewater reuse for agriculture and others. Also to identify major impacts both positive and negative impacts of wastewater irrigation. Finally, to assess environmental valuation techniques for analyzing impacts of wastewater uses.

WASTEWATER CHARACTERISTICS

Sources of Wastewater

In general, municipal wastewater is made up of domestic wastewater, industrial wastewater, storm water, and by groundwater seepage entering the municipal sewage network.

1) Domestic wastewater consists of effluent discharges from households, institutions, and commercial buildings.
2) Industrial wastewater is the effluent discharged by manufacturing units and food processing plants.
3) Unlike in some developed cities where the systems are separate, there, the municipal sewage network also serves as the storm water sewer. Due to defects in the sewerage system, there is groundwater seepage as well, adding to the volume of sewage to be disposed.

Composition of Sewage Water

- Organic matter
- Nutrients (Nitrogen, Phosphorus, Potassium)
- Inorganic matter (dissolved minerals)
- Toxic chemicals (heavy metal and pesticides)
- Pathogens

Table 1. Major Constituents of Typical Domestic Wastewater

Constituent	Concentration (mg/l)		
	Strong	Medium	Weak
Total solids	1200	700	350
Dissolved solids (TDS)	850	500	250
Suspended solids	350	200	100
Nitrogen (as N)	85	40	20
Phosphorus (as P)	20	10	6
Chloride	100	50	30
Alkalinity (as CaCO3)	200	100	50
Grease	150	100	50
BOD5	300	200	100

Source: UN Department of Technical Cooperation for Development (1985).

QUALITY PARAMETERS OF IMPORTANCE

Parameters of Health Significance

Organic chemicals usually exist in municipal wastewaters at very low concentrations and ingestion over prolonged periods would be necessary to produce detrimental effects on human health. This is not likely to occur with agricultural/aquacultural use of wastewater, unless cross-connections with potable supplies occur or agricultural workers are not properly instructed, and can normally be ignored.

Table 2. Pollutants and contaminants in wastewater and their potential impacts

Pollutants/ Contaminants	Parameters	Impacts
Hydrogen ion concentration	pH	1. Possible adverse impact on plant growth due to acidity /alkalinity. 2. Impact sometimes beneficial to flora and fauna.
Suspended solids	Volatile compounds, settable, suspended and colloidal impurities	1. Development of sludge deposit.
Dissolved inorganic substances	TDS, EC, Na, Ca, Mg, Cl and B	2. Cause salinity and associated adverse impacts 3. Phytotoxicity 4. Affect permeability and soil structure
Plant food nutrients	N, P, K etc.	1. Excess N causes nitrogen injury, excessive vegetative growth, delayed growth season and maturity, causing economic loss of farmers. 2. Excessive of N and P cause excessive growth of undesirable aquatic life (eutropication) 3. Nitrogen leaching causes ground water pollution with adverse health and environmental impacts.
Heavy metals	Fe, Mn, Cu, Cd, Cr, Pb, Ni, Zn, Ag, Hg etc,	1. Accumulate in aquatic organisms 2. Accumulate in sewage water irrigates soils and transfer to the plants and entire in the food chain and affect public health. 3. Toxic to plants and animals. 4. May make sewage water unsuitable for irrigation.
Pesticide residues	Both parent molecules and metabolites	1. Ground and surface water contamination 2. Toxicity to mammals and aquatic organisms 3. residual organic compounds 4. Green-house effect.
Biodegradable organics	BOD,COD	1. Depletion of D.O. in surface water. 2. Development of septic conditions. 3. Unsuitable habitat and Environment. 4. Can inhibit pond-breeding amphibians. 5. Fish death. 6. Humus build up

Source: Asano et.al. (1985).

The principal health hazards associated with the chemical constituents of wastewaters, therefore, arise from the contamination of crops or groundwaters. Hillman (1988) has drawn attention to the particular concern attached to the cumulative poisons, principally heavy metals, and carcinogens, mainly organic chemicals. World Health Organization guidelines for drinking water quality (WHO 1984) include limit values for the organic and toxic substances given in the table 3 based on acceptable daily intakes (ADI).

These can be adopted directly for groundwater protection purposes but, in view of the possible accumulation of certain toxic elements in plants (for example, cadmium and selenium) the intake of toxic materials through eating the crops irrigated with contaminated wastewater must be carefully assessed.

Table 3. Organic and inorganic constituents of drinking water of health significance

Organic	Organic	Inorganic
Aldrin and dieldrin	1,1 Dichlorethylene	Arsenic
Benzene	Heptachlor and heptachlor epoxide	Cadmium
Benzo-a-pyrene	Hexachlorobenzene	Chromium
Carbon tetrachloride	Lindane	Cyanide
Chlordane	Methoxychlor	Fluoride
Chloroform	Pentachlorophenol	Lead
2,4 D	Tetrachlorethylene	Mercury
DDT	2, 4, 6 Trichloroethylene	Nitrate
1,2 Dichloroethane	Trichlorophenol	Selenium

Source: WHO (1984).

Sewage water contains pathogenic microorganisms like bacteria, viruses, fungi, algal etc., having the potential risks to causes diseases can causes immense harm to public health. The water borne diseases are typhoid, paratyphoid fevers, dysentery and cholera, polio and infectious hepatitis. The responsible organisms occur in the faces or urine or infected people. Where raw untreated sewage water is used to irrigate crops helminthic disease caused by Ascaris, and Trichuris spp. as occurred in West Germany. Melbourne, Australia and from Denmark (reported by Shuval et al. 1985) that cattle grazing on fields freshly irrigated with raw wastewater, or drinking from raw wastewater canals or ponds, can become heavily infected with the disease (cysticerosis).

In India sewage farm workers exposed to raw wastewater in areas where Ancylostoma (hookworm) and Ascaris (nematode) infections are endemic have significantly excess levels of infection with these two parasites compared with other agricultural workers in similar occupations.

From the health point of view important microbiological parameter are coliform , fecal coliform, fecal streptococci and clostridium perfringens. Finally, in respect of the health impact of use of wastewater in agriculture, Shuval et al. (1986) rank pathogenic agents in the order of priority shown in Table 4. They pointed out that negative health effects were only detected in association with the use of raw or poorly-settled wastewater, while inconclusive evidence suggested that appropriate wastewater treatment could provide a high level of health protection. high level of health protection.

Table 4. Relative health impact of pathogenic agents

High Risk	Helminths (*Ancylostoma, Ascaris, Trichuris* and *Taenia*)
Medium Risk	Enteric Bacteria *(Cholera vibrio, Salmonella typhosa, Shigella* etc.
Low Risk	Enteric viruses

Shuval *et al.* 1986.

Indicator Organisms

A) *Coliforms and Faecal Coliforms.* The Coliform group of bacteria comprises mainly species of the genera Citrobacter, Enterobacter, Escherichia and Klebsiella and includes Faecal Coliforms, of which Escherichia coli is the predominant species. They are not itself harmful but presesnce of coliform groups of bacteria indicate t he presence of pathogenic bacte4ria and fecal coliforms indicate fecal contamination and presence of enteric pathogens in surrounding water. Several coliforms are able to grow out side of the intestines , specially in hot climates. Hence their enumeration is unsuitable as a parameter. The fecal coliforms can grow at 44 degree C, so E.coli, is most s satisfactory indicator parameter in sewage water use.

B) *Faecal Streptococci.* Faecal Streptococci as an indicator in tropical conditions and especially to compare survival with that of Salmonellae.

C) *Clostridium perfringens.* This bacterium is an exclusively faecal spore-forming anaerobe normally used to detect intermittent or previous pollution of water, due to the prolonged survival of its spores. In sewage water studies it is useful as it may have survival characteristics similar to those of viruses or even helminth eggs.

Parameters of Agricultural Significance

Sewage water contains soluble salts that may accumulate in the root zone with possible harmful effect on soil health and crop yield. The quality of irrigation water is of particular importance in arid zones where extremes of temperature and low relative humidity result in high rates of evaporation, with consequent deposition of salt which tends to accumulate in the soil profile. The physical and mechanical properties of the soil, such as dispersion of particles, stability of aggregates, soil structure and permeability, are very sensitive to the type of exchangeable ions present in irrigation water. Thus, when effluent use is being planned, several factors related to soil properties must be taken into consideration.

Another aspect of agricultural concern is the effect of dissolved solids (TDS) in the irrigation water on the growth of plants. Dissolved salts increase the osmotic potential of soil water and an increase in osmotic pressure of the soil solution increases the amount of energy which plants must expend to take up water from the soil. As a result, respiration is increased and the growth and yield of most plants decline progressively as osmotic pressure increases. Important Agricultural Water Quality parameters include a number of specific properties of water that are relevant in relation to the yield and quality crops, maintenance of soil productivity and protection of the environment. These parameters mainly consist of certain physical and chemical characteristics of the water. The primary wastewater quality parameters of importance from an agricultural viewpoint are:

Table 5. Guidelines for interpretation of water quality for irrigation

Potential irrigation problem	Units	Degree of restriction on use		
		None	Slight to moderate	Severe
Salinity				
EC	dS/m	< 0.7	0.7 - 3.0	> 3.0
TDS	mg/l	< 450	450 - 2000	> 2000
Specific ion toxicity				
Sodium (Na)				
Surface irrigation	SAR	< 3	3 - 9	> 9
Chloride (Cl)				
Surface irrigation	me/I	< 4	4 - 10	> 10
Boron (B)	mg/l	< 0.7	0.7 - 3.0	> 3.0
Miscellaneous effects				
Nitrogen (NO_3-N)	mg/l	< 5	5 - 30	> 30
Bicarbonate (HCO_3)	me/I	< 1.5	1.5 - 8.5	> 8.5
pH	Normal range 6.5-8.0			

Source: FAO (1985).

A. pH

pH is an indicator of the acidity or basicity of water but is seldom a problem by itself. The normal pH range for irrigation water is from 6.5 to 8.4; pH values outside this range are a good warning that the water is abnormal in quality. Normally, pH is a routine measurement in irrigation water quality assessment.

B. Electrical Conductivity

Electrical conductivity is widely used to indicate the total ionized constituents of water. It is directly related to the sum of the cations (or anions). It should be noted that the electrical conductivity of solutions increases approximately 2 percent per °C increase in temperature. The symbol EC_w, is used to represent the electrical conductivity of irrigation water and the symbol EC_e is used to designate the electrical conductivity of the soil saturation extract. The unit of electrical conductivity is deciSiemen per metre (dS/m).

C. Total Salt Concentration

Total salt concentration (for all practical purposes, the total dissolved solids) is one of the most important agricultural water quality parameters. This is because the salinity of the soil water is related to, and often determined by, the salinity of the irrigation water. Accordingly, plant growth, crop yield and quality of produce are affected by the total dissolved salts in the irrigation water. Equally, the rate of accumulation of salts in the soil, or soil salinization, is also directly affected by the salinity of the irrigation water. Total salt concentration is expressed in milligrams per litre (mg/l) or parts per million (ppm).

D. Sodium Adsorption Ratio

Sodium is an unique cation because of its effect on soil. When present in the soil in exchangeable form, it causes adverse physico-chemical changes in the soil, particularly to soil structure. It has the ability to disperse soil, when present above a certain threshold value, relative to the concentration of total dissolved salts. Dispersion of soils results in reduced infiltration rates of water and air into the soil. When dried, dispersed soil forms crusts which are hard to till and interfere with germination and seedling emergence. Irrigation water could be a source of excess sodium in the soil solution and hence it should be evaluated for this hazard. The most reliable index of the sodium hazard of irrigation water is the sodium adsorption ration, SAR. The sodium adsorption ratio is defined by the formula and the ionic concentrations are expressed in me/l.

$$SAR = \frac{Na}{\sqrt{\frac{Ca + Mg}{2}}}$$

E. Toxic Ions

Irrigation water that contains certain ions at concentrations above threshold values can cause plant toxicity problems. The most common phytotoxic ions that may be present in municipal sewage and treated effluents in concentrations such as to cause toxicity are: boron (B), chloride (Cl) and sodium (Na). Hence, the concentration of these ions will have to be determined to assess the suitability of waste-water quality for use in agriculture.

F. Trace Elements and Heavy Metals

A number of elements are normally present in relatively low concentrations, usually less than a few mg/l, in conventional irrigation waters and are called trace elements. They are not normally included in routine analysis of regular irrigation water, but attention should be paid to them when using sewage effluents, particularly if contamination with industrial wastewater discharges is suspected. These include Aluminium (Al), Beryllium (Be), Cobalt (Co), Fluoride (F), Iron (Fe), Lithium (Li), Manganese (Mn), Molybdenum (Mo), Selenium (Se), Tin (Sn), Titanium (Ti), Tungsten (W) and Vanadium (V). Heavy metals are a special group of trace elements which have been shown to create definite health hazards when taken up by plants. Under this group are included, Arsenic (As), Cadmium (Cd), Chromium (Cr), Copper (Cu), Lead (Pb), Mercury (Hg) and Zinc (Zn). These are called heavy metals because in their metallic form, their densities are greater than 4g/cc. The threshold levels of trace elements for crop production are given in Table – 6.

Table 6. Threshold levels of trace elements for crop production

	Element	Recommended maximum concentration (mg/l)	Remarks
Al	(aluminium)	5.0	Can cause non-productivity in acid soils (pH < 5.5), but more alkaline soils at pH > 7.0 will precipitate the ion and eliminate any toxicity.
As	(arsenic)	0.10	Toxicity to plants varies widely, ranging from 12 mg/l for Sudan grass to less than 0.05 mg/l for rice.
Cd	(cadmium)	0.01	Toxic to beans, beets and turnips at concentrations as low as 0.1 mg/l in nutrient solutions. Conservative limits recommended due to its potential for accumulation in plants and soils to concentrations that may be harmful to humans.
Co	(cobalt)	0.05	Toxic to tomato plants at 0.1 mg/l in nutrient solution. Tends to be inactivated by neutral and alkaline soils.
Cr	(chromium)	0.10	Not generally recognized as an essential growth element. Conservative limits recommended due to lack of knowledge on its toxicity to plants.
Cu	(copper)	0.20	Toxic to a number of plants at 0.1 to 1.0 mg/l in nutrient solutions.
F	(fluoride)	1.0	Inactivated by neutral and alkaline soils.
Fe	(iron)	5.0	Not toxic to plants in aerated soils, but can contribute to soil acidification and loss of availability of essential phosphorus and molybdenum. Overhead sprinkling may result in unsightly deposits on plants, equipment and buildings.
Li	(lithium)	2.5	Tolerated by most crops up to 5 mg/l; mobile in soil. Toxic to citrus at low concentrations (<0.075 mg/l). Acts similarly to boron.
Mn	(manganese)	0.20	Toxic to a number of crops at a few-tenths to a few mg/l, but usually only in acid soils.
Mo	(molybdenum)	0.01	Not toxic to plants at normal concentrations in soil and water. Can be toxic to livestock if forage is grown in soils with high concentrations of available molybdenum.
Ni	(nickel)	0.20	Toxic to a number of plants at 0.5 mg/l to 1.0 mg/l; reduced toxicity at neutral or alkaline pH.
Pd	(lead)	5.0	Can inhibit plant cell growth at very high concentrations.
Zn	(zinc)	2.0	Toxic to many plants at widely varying concentrations; reduced toxicity at pH > 6.0 and in fine textured or organic soils.

Source: National Academy of Sciences (1972) and Pratt (1972).

Potential Impacts of Wastewater in Environment

This section provides the potential impacts of wastewater use in various substrates

1. Public Health and Other living organism
2. Crops
3. Social Resources
4. Ground Water resources
5. Property values
6. Ecological impacts
7. Social Impacts

1. Public Healthand Other Living Organisms

Use of untreated sewage water pose a high risk to human healthand other living organisms in all groups as it contain pathogenic microorganisms which have the potential to cause diseases.

2. Crops

Generally speaking, wastewater (treated and untreated) is extensively used in agriculture because it is a rich source of nutrients and provides all the moisture necessary for crop growth. Most crops give higher than potential yields with wastewater irrigation; reduce the need for chemical fertilizers, resulting in net cost savings to farmers.

3. Soil Resources

Impact from wastewater on agricultural soil, is mainly due to the presence of high nutrient contents (Nitrogen and Phosphorus), high total dissolved solids and other constituents such as heavy metals, which are added to the soil over time. Wastewater can also contain salts that may accumulate in the root zone with possible harmful impacts on soil health and crop yields. The leaching of these salts below the root zone may cause soil and groundwater pollution (Bond 1999). Prolonged use of saline and sodium rich wastewater is a potential hazard for soil as it may erode the soil structure and effect productivity. This may result in the land use becoming non-sustainable in the long run. Wastewater induced salinity may reduce crop productivity (Kijne et al. 1998). The net effect on growth may be a reduction in crop yields and potential loss of income to farmers. Wastewater irrigation may lead to transport and bio-accumulate heavy metals to soils, affecting soil flora and fauna. e.g., Cd and Cu, may be redistributed by soil fauna such as earthworms (Kruse and Barrett 1985). In general, heavy metal accumulation and translocation is more a concern in sewage sludge application than wastewater irrigation, because sludge formed during the treatment process consists of concentrations of most heavy metals. The impact of wastewater irrigation on soil may depend on a number of factors such as soil properties, plant characteristics and sources of wastewater.

4. Groundwater Resources

Wastewater application has the potential to affect the quality of groundwater resources in the long run through excess nutrients and salts found in wastewater leaching below the plant root zone. For instance the quality of groundwater would determine the magnitude of the impact from leaching of nitrates. Groundwater constitutes a major source of potable water for many developing country communities. Hence the potential of groundwater contamination needs to be evaluated before embarking on a major wastewater irrigation program. In addition to the accretion of salts and nitrates, under certain conditions, wastewater irrigation has the potential to translocate pathogenic bacteria and viruses to groundwater (NRC report 1996).

Farid et al. (1993), reported that the long-term use of wastewater for crop irrigation has interestingly led to an improvement in the salinity of the groundwater. This was offset by evidence of coliform contamination of groundwater which was also observed in Mexico (Downs et al. 1999, Gallegos et al. 1999). A companion study (Rashed et al. 1995), reveals that in the wastewater irrigated Gabar el Asfar region, concentrations of chloride, sulfate, TDS, and dissolved oxygen in groundwater is much higher than average concentrations in sewage effluents. The leaching and drainage of wastewater, applied for crop irrigation, to groundwater aquifer may serve as a source of groundwater recharge. In some regions, 50-70 percent of irrigation water may percolate to groundwater aquifer (Rashed et al. 1995).

5. Ecological Impacts

When drainage water from wastewater irrigation schemes drains particularly into small confined lakes and water bodies and surface water, and if phosphates in the orthophosphate form are present, the remains of nutrients may cause eutrophication (Smith et al. 1999). For example, overloading of organic material resulting in decreases in dissolved oxygen may lead to changes in the composition of aquatic life, such as fish deaths and reduced fishery. The eutrophication potential of wastewater irrigation can be assessed using biological indices or biomarkers, which in turn can be quantified in monetary units using appropriate economic valuation techniques.

6. Social Impacts

In the context of this analysis social impacts are the concerns/doubts expressed by the public about wastewater irrigation. These concerns can be classified as follows:

General concerns such as nuisance, poor environmental quality, poor hygiene, odor, noise, higher probability of accidents, etc.

Social concerns such as food safety, health and welfare, impaired quality of life, loss of property values, and sustainability of land use.

Natural resource concerns such as pollution of vital water resources, loss of fish, wildlife, exotic species, etc.

7. Economics of Wastewater Irrigation

To date, in relation to wastewater irrigation, economic analyses have been conducted with specific perspectives in mind viz that of a municipality optimizing treatment costs, or that of farmers or a regional entity maximizing income, or that of evaluating environmental impacts.

The researchers evaluated the effect of crop selection on cost and revenue streams and system efficiency by selecting three cropping patterns viz. reed canary grass, alfalfa, corn and forest plantations. Wastewater can also be used for producing rapidly growing pulpwood, such as eucalyptus, on public lands, along canal banks, roads and greenbelts etc. These plants can be harvested every 8 to 10 years to generate revenue, along with the added advantage of working as natural air conditioners and greenhouse gas sinks, for ameliorating the highly polluted urban environments.The main benefits from wastewater irrigation are effective water and nutrient recycling, higher crop yields, a diversified cropping pattern, and disposal cost savings. Segarra et al. (1996), suggested that alfalfa, wheat-corn, wheat-grain sorghum, and cotton are optimal crop combinations to maximize net revenue. It, therefore, implies that municipalities can benefit from cooperative arrangements with neighboring farmers for wastewater irrigation. A recent IWMI study (Scott et al. 2000), evaluated the economic value and risks associated with long-term use of urban wastewater for crop irrigation in Guanajuato, Mexico. The study was conducted to predict changes in water quality under various wastewater management scenarios. The study used an opportunity cost or replacement value approach to estimate dollar values for water and nutrient contents of wastewater. The findings suggest that wastewater is a valuable resource for the community and wastewater reuse for irrigation is an economical alternative to expensive treatment. However, the study recognizes that there could be negative health and environmental impacts of wastewater use, and that these impacts should be evaluated.

Waste Water Treatment Procedure Adopted in India

- Activated sludge process
- Trickling filter
- Oxidation pond and Waste stabilization pond

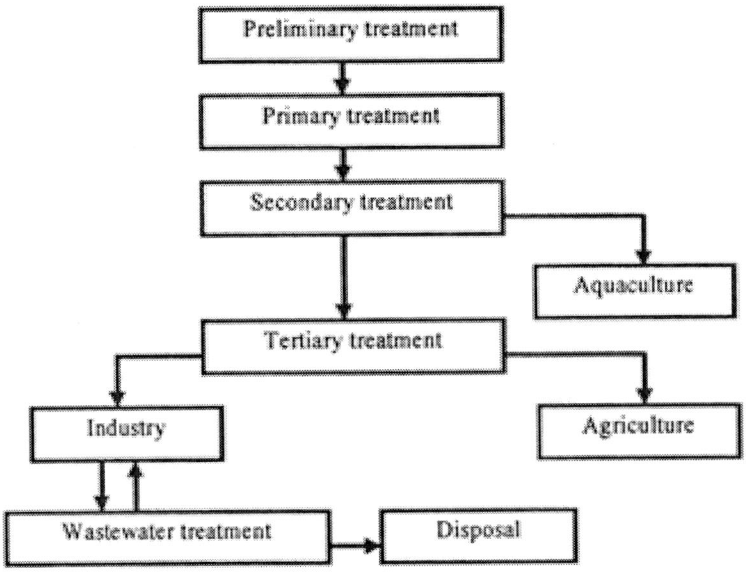

Schematics of Municipal Wastewater Reuse.

Status of Sewage and Sewage Treatment in India

The total wastewater generated by 23 metropolitan cities is 9,275 mld. Out of 9,275 mld of total wastewater generated, only 31% (2,923 mld) is treated before letting out and the rest i.e. 6,352 mld is disposed off untreated. Three cities have only primary treatment facilities and thirteen have primary and secondary treatment facilities. In India less than 50% of the urban population has access to sewage disposal system. Most of the existing collecting systems discharge directly to the receiving water without treatment. Garbage, domestic and otherwise, is directly dumped into water bodies or roadside, which can often be washed into streams and lakes. This vulnerable environment requires special attention and the solution of such complex and interdisciplinary problems call for an integrated water resources management approach.

The municipalities (governing bodies of metropolitan cities) disposes off their treated or partly treated or untreated wastewater into natural drains joining rivers or lakes or used on land for irrigation or fodder cultivation or into sea or combination of these. In four cities, it is disposed indirectly into the rivers/lakes, while in two cities it is disposed into sea/creek and the rest partly used for agriculture and partly disposed into rivers. It is found that in 12 metropolitan cities there is some level of organized sewage farming under the control of government or local body (CPCB, August 1997).

In India, till now very little emphasis has been laid on research on hydrology of urban areas. Taking into account that the trends of urban population concentration increase will continue in the future, a programme for encompassing all hydrological, ecological and socio-economic aspects of future urban planning and management needs to be taken up in right earnest. This would require improvement in the management of existing urban drainage systems, disseminate knowledge of integrated urban water management, identify the impact of urbanization on surface and ground water quality through point and nonpoint sources, to study impact of storm water (wastewater discharges) on ecosystem health of receiving water courses and to establish experimental urban catchments.

Water Quality Guidelines

From effect of sewage water several guidelines are produced to minimize the potential risk. WHO guidelines is used on the safe use of water for agriculture and aquaculture. The rationale behind the WHO guidelines was to develop criteria that would present the transmission of communicable diseases caused by microorganisms while optimizing resource conservation and recycling.

Recent evidence suggest that these guidelines are used only to crop consumers but not necessarily farmers, farm workers and their families, thereby meeting this guidelines debatable. In order to evaluate the financial feasibility of WHO and USEP a microbial health guidelines, Shuval et al. (1997), developed a risk assessment approach to conduct a comparative risk analysis. Most European countries, with the exception of Germany and France, have not established any guidelines for the use of wastewater for irrigation.

The EU guidelines, when formulated, propose to cover both agronomic aspects, of soil and groundwater protection, yield maximization, and the sanitary aspects, relating to public health protection.

CONCLUSION

Rapid urbanization places immense pressure on the world's fragile and dwindling fresh water resources and over-burdened sanitation systems, leading to environmental degradation. Thus, it is quiet justified and seems logistic to say that:

1. Wastewater (raw, diluted or treated) is a resource of increasing global importance.
2. Without proper management sewage water use poses high risks to human health and cause environmental degradation Thus scientists around the world refocus on conserving water, recycling of water and treatment of sewage water through sewage treatment plant.
3. With proper management, wastewater use contributes significantly to sustaining livelihoods, food security and the quality of the environment.
4. Sewage treatment cost studies shows that marginal cost are very high at higher levels of treatment at higher levels of treatment. However, these costs become justifiable in view of the value of the degree of water scarcity and public concern. Cost-effective and appropriate treatment suited to the end use of wastewater, supplemented by guidelines and their application.
5. Proposed guidelines should link heath, agriculture and environmental quality, which are implemented in a stepwise approach.
6. Reduction of toxic contaminants in sewage water is essential by improved management practices.
7. Where sewage water is insufficiently treated due to lack of treatment facilities there some steps should be taken, which are:
    a) Development and application of guidelines for untreated wastewater use that will safe livelihoods, public health and the environment.
    b) Application of appropriate irrigation, agricultural, post-harvest, and public health practices that limit risks to farming communities, vendors, and consumers.
    c) Education and awareness programs for all stakeholders, including the public at large, to disseminate these measures.
8. Therefore, we strongly urge policy-makers and authorities in the fields of water, agriculture, aquaculture, health, environment and urban planning, as well as donors and the private sector to:

> *"Safeguard and strengthen livelihoods and food security, mitigate health and environmental risks and conserve water resources by confronting the realities of wastewater use in agriculture through the adoption of appropriate policies and the commitment of financial resources for policy implementation".*

Parameters for Water Quality Characterization and Standards (Domestic Water Supply)

Parameters	USPH Standard	ISI Standard
Color, odour, state	Colorless, odorless, tasteless	-
Inorganic Chemicals		
pH	6.0-8.5	6.0-9.0
conductance	300mmho/cm	-
D.O	4.0-6.0 ppm	3.0
TDS	500	-
Suspended Solid	5.0	-
SO_4^{2-}	250	100
Cl^-	250	600
F^-	1.5	3.0
PO_4^{3-}	0.1	-
S^-	0.1mg/L	-
Ammonia	0.5	-
B	1.0	-
Ca^{2+}	100	-
Mg2+	30	-
As	0.05	0.2
Cd	0.01	-
Cr	0.05	0.05
Cu	1.0	-
Fe	Less than 0.3	-
Pb	Less than 0.05	0.01
Mn	Less than 0.05	-
Hg	0.001	-
Ag	0.05	-
U	5.0	-
Zn	5.5	-
Organics		
COD	4.0	-
Phenols	0.001	0.005
Pesticides(total)	0.005	-
Polycyclic aromatic hydrocarbons(PAH)	0.002ppm	-
Surfactants	200	-
Biological parameters		
Coliform cells/1000mL	100	Less than5000
Total bacteria count/100mL	$1x10^6$	

References

CPCB, August 1997, *Status of Water Supply and Wastewater Generation, Collection, Treatment and Disposal in Metrocities* (1994-94), Control of Urban Pollution Series; CUPS/42/1997-98.

Status of environment, India, *Ministry of Environment and Forests*, Govt. of India, 1993.

Report of the working Group on Water Related Ecological Matters for the 10th Five Year Plan (2002-2007), Government of India, Ministry of Water Resources, July 2001.

Ajmal, M.; and A. U. Khan. 1985. Effect of textile factory effluent on soil and crop plants. *Environmental Pollution.* Vol. 37, pp.131-148.

Al-Salem Saqer S. 1996. Environmental considerations for wastewater reuse in agriculture. *Water Science and Technology.* Vol. 33, no. 10-11, pp. 345-353.

Asano, T.; M. Maeda; and M. Takaki. 1996. Wastewater reclamation and reuse in Japan: overview and implementation examples. *Water Science and Technology.* Vol. 34, no. 11, pp. 219-226.

Barden, J. B.; R. S. Larson; and E. E. Herricks. 1991. Impact targets versus discharge standards in agriculrural pollution management. *American Journal of Agricultural Economics,* pp. 388-397.

Bole, J. B.; and R. G. Bell. 1978. Land application of municipal sewage wastewater: yield and chemical composition of forage crops. *Journal of Environmental Quality.* Vol. 7, pp. 222-226.

Carmichale, J. J.; and K. M. Strzepek. 2000. A multiple-organic-pollutant simulation/optimization model of industrial and municipal wastewater loading to a reverine environment. *Water resources research.* Vol. 36, no. 5, pp. 1325-1332.

Chattopadhyay, B. S. Datta; A. Chatterjee; and S. K. Mukhopadhyay. 2000. The environmental impact of waste chromium of tannery agglomerates in the east Calcutta wetland ecosystem. *Journal of the Society of Leather Technologists and Chemists* Mar-Apr. 84(2): 94-100.

Day, A. D.; and T. C. Tucker. 1977. Effects of treated wastewater on growth, fibre, protein and amino acid content of sorghum grains. *Journal of Environmental Quality.* Vol. 6, no. 3, pp. 325-327.

Day, A. D.; F. A. Taher; and F. R. H. Katterman. 1975. Influence of treated municipal wastewater on growth, fibre, acid soluble nucleotide, protein and amino acid content in wheat grain. *Journal of Environmental Quality.* Vol. 4, no. 2, pp. 167-169.

Musharrafieh, G. R.; R. C. Peralta; L. M. Dudley; and R. J. Hanks. 1995. Optimizing irrigation management for pollution control and sustainable crop yield. *Water Resources Research.* Vol. 31, no. 4, pp. 1077-1086.

Nakamura, M.; and E. D. Brill, Jr. 1979. Generating and evaluation of alternative plans for regional wastewater systems: An imputed value method. *Water Resources Research.* Vol. 15, no. 4, pp. 750-756.

National Research Council (NRC). 1996. *Use of reclaimed water and sludge in food crop production.* National Academy Press, Washington, D.C.

Scott, C. A.; J. A. Zarazua; and G. Levine. 2000. *Urban-Wastewater Reuse for Crop Production in the Water-Short Guanajuato River Basin, Mexico.* IWMI Research Report No. 41. International Water Management Institute, Colombo: Sri Lanka.

Singh, K. K.; and L. C. Mishra. 1987. Effect of fertilizer factory effluent on soil and crop productivity. *Water, Air and Soil Pollution.* Vol. 33, pp. 309-320. -2) pp. 69-95.

Smith, G. H. S.; A. P. Nicholas; and R. I. Ferguson. 1997. Measuring and defining biomodal sediments: Problems and implications. *Water Resources Research.* Vol. 33, no. 5, pp. 1178-1185.

Smith, V. H.; G. D. Tilman; and J. C. Nekola. 1999. Eutrophication: impacts of excess nutrient inputs on freshwater, marine, and terrestrial ecosystems. *Environmental Pollution.* Vol. 100, pp. 179-196. www.elsevier.com/locate/ envpol 53

In: Sewage Treatment: Uses, Processes and Impact
Editor: Anna Stephens and Mark Fuller
ISBN 978-1-60692-959-9

Chapter 12

DISPOSAL OF MUNICIPAL SEWAGE SLUDGE IN CHINA—A MINI REVIEW

***Hongbo Liu*[1,2*], *Liping Sun*[3], *and Rolf-Dieter Wilken*[1,4]**

[1]IWW-Rhine-Main, Justus-von-Liebig-Str.10
64584 Biebesheim am Rhein, Germany
[2]Shanghai Municipal Water Resource Development
and Utilization National Engineering Center, 200082 Shanghai, P. R. China
[3]Tianjin Institute of Urban Construction, 300384 Tianjin, P. R. China
[4]Institute of Geowissenschaft/Geologie,Johannes Gutenberg-Universität
Mainz 55099 Mainz, Germany

ABSTRACT

China produced about 5 million tons of dry municipal sewage sludge in 2007 and this number is increasing very fast; disposal and utilization of municipal sewage sludge is urgent and promising in China. This paper reviewed origins and sources of municipal sewage sludge, its production amount, disposal technologies, and its utilization fates/prospects. In particular sludge thickening, anaerobic/aerobic digestion, dewatering, drying and incineration technologies were illustrated with country specific principles and parameters; while energy replacement and land applications were elaborated as major utilization fates for municipal sewage sludge with practices done or is under operation in China. Disposal technologies and utilizations under research in China such as ultrasonic treatment, pyrolysis and gasification of municipal sewage sludge, hydrogen production through fermentation of municipal sewage sludge, producing activated carbon with municipal sewage sludge, producing construction and building materials out of municipal sewage sludge were also briefly introduced but not elaborated in this review.

Keywords: Municipal sewage sludge; Sludge disposal; Sludge utilization; Waste reuse.

---

* Corresponding author Email: lhb_7906@163.com; Tel: +49(0)6925490-8015 Fax: +49(0)6925490-8009

1. INTRODUCTION

Municipal sewage sludge refers to the biomass and/or chemical sludge discharged from municipal wastewater treatment processes. China was reported to have treated about 22 billion tons of municipal wastewater in 2007 and produced correspondingly about 5 million tons of dry municipal sewage sludge that year [1]; its municipal sewage sludge production is increasing very fast since more and more municipal wastewater is going to be treated according to the 11th Five-Year-Plan of the country [1]. As displaced valuable resource, municipal sewage sludge is a growing problem for Chinese municipal wastewater treatment plants (MWTP) partly because of its potential environmental hazards and partly because of lacking efficient disposal technologies and effective environmental management concepts and strategies [1-3]. Paradox that most Chinese MWTPs facing comes from the fact that they can not find a fate for such a big amount of valuable sludge "resource" either economically or technologically except theoretically [1-4]. Many efforts were made in terms of reliable technologies and feasible strategies by numerous researchers and engineers though, thanks to the country's preference to pollution control and waste reuse activities [5-16]. This review aims to summarize the main disposal technologies and strategies of municipal sewage sludge in China. In particularly, some key technical parameters and principals were discussed and recommended by comparing similar cases in China; also some reutilization strategies of municipal sewage sludge were discussed.

Table 1. Characteristics and discharging standards of municipal wastewater in China mg/L[1,4]

Item	Concentration	Discharging Standard[a]
COD	120~1000	60/50[b]
TN	30~80	20/15
TP	1.6~10	1/0.5
NH_4^+-N	20~60	8/5
SS[c]	80~400	20/10
NO_3^--N	<1	n/a[d]
NO_2^--N	<1	n/a

a. Discharging Standards for Chinese MWTPs (GB18918-2002)
b. First grade (A)/ First grade (B)
c. Suspended solids
d. Not available

2. ORIGIN AND PRODUCTION OF MUNICIPAL SEWAGE SLUDGE

Municipal wastewater is treated normally with biological and/or chemical processes in China [1, 4]. Sludge characters and correspondingly production amount varied with different wastewater treatment methods. Although municipal wastewater characteristics vary in different parts of China and depend on the ratio of industrial wastewater to domestic wastewater, characteristics of the municipal wastewater in China can be presented as in table 1.

2.1. Sludge Production of Biological Wastewater Treatment Process

There are two methods to estimate biological municipal sewage sludge production in China. The first one calculates sludge production on the basis of empirical statistics on sludge discharging equivalent per capita and thus is more experience based; the second one considers more on biochemical aspects such as sludge retention time, biomass yield coefficient but is also quite experienced based, especially when determining how much sludge would be produced by SS (see equation 1). Both methods are widely used in China though sometimes results of two methods do not match each other due to the fact that both methods are experienced based. A conservative method which chooses the bigger value from both calculation methods is recommended by many Chinese engineers [2, 3].

Empirical data for calculation of sludge production using the first method was listed in table 2; equation 1 demonstrated how to calculate sludge production using the second method.

Table 2. Parameters for estimating capital municipal sewage sludge production [2, 3]

Types of sedimentation tank		Retention time(h)	Volumetric load [$m^3/(m^2 \cdot h)$]	Capital sludge production (g/ca·d)	Water content (%)	Solid load [$kg/(m^2 \cdot d)$]
Primary		0.5~2.0	1.5~4.5	16~36	95~97	-
Secondary	Biofilm	1.5~4.0	1.0~2.0	10~26	96~98	≤150
	Activated Sludge	1.5~4.0	0.6~1.5	12~32	99.2~99.6	≤150

$$\Delta X = YQ\,(S_o - S_e) - K_d V X_V + fQ\,(SS_o - SS_e) \quad (1)$$

where : ΔX—Sludge production (kg/d)

V—Volume of bioreactors (m^3)

X—Concentration of mixed liquor suspended solid (gMLSS/L)

θ_C —Sludge retention time (d)

Y—Biomass yield coefficient ($kgVSS/kgBOD_5$, 0.4 ~ 0.8 in 20□)

Q— Average influent flow rate (m^3/d)

S_o—Influent biological oxygen demand, BOD_5 (kg/m^3)

S_e—Effluent biological oxygen demand, BOD_5(kg/m^3)

k_d—Decay coefficient (d^{-1})

X_V—Concentration of volatile mixed liquor suspended solid (gMLVSS/L)

f—Sludge production factor of SS, determined by experimental data; its fraud value is 0.5 ~ 0.7 if no experiment data is available (gMLSS/gSS)

SS_o—Influent SS (kg/m^3)

SS_e—Effluent SS (kg/m^3)

2.2. Sludge Production of Chemical Wastewater Treatment Process

Chemical municipal sewage sludge can be calculated by stoichiometric equations [6-8]. Iron salts and aluminum salts are the most used chemicals in Chinese MWTPs [7]; if necessary anionic/cationic or nonionic PAM (polyacrylamide) would also be used to support chemical treatment [6]. Sometimes calcium hydroxide was also used to treat municipal wastewater in China [7].

For most wastewater chemical treatment practices in China, phosphorus is the controlling parameter in municipal wastewater treatment processes i.e. when the effluent phosphorus can meet its discharging standard, so will COD in most cases [8]; but chemical nitrogen removal is extremely expensive and is not an option for developing countries such as China [6]. As more and more stringent standard has been put on nitrogen and phosphorus removal for wastewater discharging, a hybrid bio-chemical treatment for municipal wastewater treatment is sometimes necessary [8]. In this case, chemical treatment is targeted to remove phosphorus (and part of the COD as well, but not designed to), and biological treatment is targeted to remove nitrogen and COD [7, 8]. Sludge production for the hybrid process is understandably more than sole biological treatment [8]. Operational data in Shanghai Bailonggang MWTP (one million m^3/d, biological and chemical sludge is not separated) showed that when the chemical is iron or aluminum salt, sludge amount increased 40%~75% comparing to sole biological treatment; sludge amount increased 150% ~ 500% when the chemical is calcium hydroxide [2]. Operational data in another MWTP indicated that if chemical were dosed after biological secondary clarifier (biological and chemical sludge were thus separated), sludge amount increased 20% ~ 35% when the chemical is iron or aluminum salt comparing to sole biological treatment; sludge amount increased to 130% ~ 145% when the chemical is calcium hydroxide [8].

3. Principles of Municipal Sewage Sludge Disposal

China is a big territory; its regional economics and environmental conditions vary, so should sludge treatment and disposal technologies. But still some common fundamental principles for sludge treatment and disposal can be summarized as follows:

1) Municipal sewage sludge should be volume reduced, stabilized and much less hazardous after treatment in all regional economical and environmental conditions; strategic sludge reutilization is recommended in the long run [2, 9, 10, 12].
2) Municipal sewage sludge can be used as fertilizer, as building materials, as fuel, or it can be sent to landfill for disposal [3, 11, 13]. Specific disposal process should be selected based on technological and economical feasibility studies [14]. Sludge treatment technologies should be selected according to its final applications: if the municipal sewage sludge will be used as fertilizer, then it should be stabilized first and then make sure the product is harmless [15]; if the fertilizer is fed manually, sludge dewatering is recommended [16]; the sludge can be fed without dewatering if a mechanization feeding measure is adopted [12]; when the sludge fertilizer has to be

stored for a long time before application, a heat drying technology is recommended [16].

3) When the municipal sewage sludge is applied as fertilizer, its content of harmful substances should be consistent with existing national standards [5]. Municipal sewage sludge contains heavy metals and pathogenic bacteria, parasite eggs, and other harmful substances [5, 11]; when it is applied as farming fertilizer, strict measures should be made to ensure that all sources of hazardous substances such as industrial sewage will meet the latest existing standards before discharged into the municipal sewage pipelines [5]. At the same time, measures should also be taken to detect hazardous substances in municipal sewage sludge according to current existing standards [5].
4) Wastewater derived from the sludge treatment process should be returned to wastewater treatment units. Normally sludge contains more pollutants, and their concentration is higher than the original wastewater [16]. Therefore, wastewater derived from sludge treatment process should be returned to wastewater treatment units and be treated together with influent wastewater in order to avoid secondary pollution. If necessary, wastewater derived from sludge treatment process should be subject to chemical phosphorus removal treatment before it enters into biological wastewater treatment units, otherwise the latter might be overloaded in terms of phosphorus [10].
5) Odorous gas derived from sludge treatment process such as H_2S, organic ammonium should be collected and treated properly [2, 14].

4. Disposal Technologies

4.1. Sludge Thickening Technology

Both gravity thickening and mechanical thickening are widely used in China [17, 18]. Considering sludge thickening is a very mature technology. The following text focuses on specific technical aspects in China only.

4.1.1. Gravity Thickening

According to applications of gravity thickening in China (table 3), a satisfactory concentrated sludge can be achieved only when sludge retention time (SRT) for the thickening tank is greater than 12 hours (12h) and its solid load is in the range of 30~60 kg/(m^2·d) [2]. Under these operational conditions, water content of the municipal sewage sludge can be thickened from 99.2%~99.6% to 97%~98%.

Some technical measures are also necessary to make sure a sound operation of the thickening tank.

1) A scum remover is always recommended to install because SRT of the thickening tank is long enough to generate anaerobic environment for growth of anaerobes and can thus produce methane, H_2S and/or sometimes nitrogen, which floats sludge and

forms scum at surface of the thickening tank; scum will be washed into wastewater stream and harms wastewater treatment units if it is not removed in time.

Table 3. SRT and solid loads of gravity thickening in some Chinese MWTPs with good performance

Name of MWTP	SRT (h)	Solid load ($kg/m^2 \cdot d$)
Changzhou Chengbei MWTP [2]	14~18	40
Xuzhou MWTP [3]	26.6	38.9
Tangshan Nanbao MWTP [2]	12.7	26.5
Huzhou MWTP [2]	33.9	33.5
Xining MWTP [3]	24	46
Fuyang MWTP [2]	16~17	38

2) Wastewater derived from disposal of sludge out of biological phosphorus removal processes should be subject to phosphorus removal measures before returning to wastewater treatment units. Biological phosphorus removal is achieved by phosphorus removing bacteria (PAO), which can uptake much more phosphorus in aerobic conditions than releasing phosphorus in anaerobic conditions [3]. Waste sludge from biological phosphorus removal processes is thus very phosphorus enriched that when it is subject to thickening, much phosphorus would release into the sludge derived wastewater due to possible anaerobic environment during thickening process. Returning wastewater derived from disposal of sludge out of biological phosphorus removal processes to wastewater treatment units without any pre-treatment could trigger failure of the biological removal process by overloading.

4.1.1. Mechanical Thickening

Belt concentrator, spiral press concentrator and drum concentrator have been used in some Chinese MWTPs [2, 3, 17, 18]. As different manufactures have different specifications on their products, mechanical thickening should adhere to manufactures' instructions and/or test results strictly [18].

4.2. Sludge Digestion

Sludge digestion is an indispensable process for sludge stabilization. Both aerobic digestion and anaerobic digestion find practices in China [19-22]. Whether an anaerobic digestion process or an aerobic digestion process should be adopted depends on characteristics of the sludge, environmental demands, engineering conditions and final fate of the disposed sludge [2]. Generally, sludge digestion process should be both economical feasible and easy-to-be-maintained [3]. Anaerobic digestion has more applications in China than aerobic digestion due to the following reasons: (1) it is cheaper both in construction investment and in operation cost; (2) biogas is a by-product of sludge anaerobic digestion and can be utilized as a valuable fuel; and (3) it needs less construction area than aerobic sludge digestion technology. However, biogas triggered explosion of some of the anaerobic reactors were also reported in China [1], which makes process security an important aspect to consider during selection of sludge digestion technologies. On the other hand, aerobic sludge digestion technology is more expensive both in investment and operational cost; it needs bigger construction area, and its performance fluctuates with seasonal temperature variations and thus has fewer applications in China. But aerobic sludge digestion has also the following advantages: (1) aerobic digested sludge is normally more stabilized than anaerobic digestion; (2) less or no odorous gas produced during aerobic digestion and its operation is much safer than anaerobic digestion; (3) aerobic digestion can avoid phosphorus release to the sludge derived wastewater, and thus can return it to wastewater treatment units without necessity of chemical phosphorus removal. Taking process economics into consideration, aerobic sludge digestion is generally used in small MWTPs with wastewater scale less than 50,000 m^3/d in China [2, 3]. Finally, no digestion measurement is required for disposal of waste sludge derived from extended aeration wastewater treatment processes in China [3].

4.2.1. Anaerobic Sludge Digestion

The following is some technical aspects of anaerobic sludge digestion in China.

1) Anaerobic digestion can be carried out both with mesophilic (33~35°C) bacteria and themophilic (53~55°C) bacteria [3]. Considering the cost for heating and maintaining temperature of anaerobic reactors, most anaerobic sludge digestion applications in China are mesophilic, i.e. operation temperature is maintained in the range of 33~35°C [2].
2) Although anaerobic digestion reactors with two independent stages are believed to be more efficient and more economical [2, 3], many applications in China use one-stage anaerobic digestion [2, 3, 20]. Studies showed that two-stage anaerobic digestion could not improve efficiency as in theory if same SRTs were adopted as in one-stage anaerobic stage digestion [20]. Actually the economic balance between saving energy in the second stage digester and cost for more complicated maintenance determines the selection of one-stage digestion or two-stage digestion [2].
3) Since carbon nitrogen ratio of waste sludge from the secondary sedimentation tank is as low as 5 when primary sedimentation tank is installed and thus is not easy to realize efficient anaerobic digestion, normally it is digested together with the primary sludge [2, 3].
4) Volume of the anaerobic digester can be calculated with equation 2 or equation 3 with specific parameters provided by some Chinese experts [3].

$$V = Q_o \cdot t_d \tag{2}$$

$$V = \frac{W_s}{Lv} \tag{3}$$

where : t_d—SRT , 20 ~ 30d [3]

V—Volume of the digester, m^3

Q_o— Influent sludge volume (m^3/d)

L_V—volumetric load of volatile solid (kgVSS/(m^3·d)); 0.6 ~ 1.5kgVSS/(m^3·d) when the sludge is gravity thickened and less than 2.3 kgVSS/(m^3·d) when the sludge is mechanic thickened [3].

5) Sludge derived biogas should be utilized as fuel to fire boiler, electricity gas engines etc. When it is used to fire electricity gas engine, measures should be taken to remove hydrogen sulfide [2, 3].
6) Considering the risk of possible explosion triggered by biogas, safety measure should be taken in any pipelines, reactors, or equipments dealing with biogas. Technology measures include water sealing, back fire stopper, explosion proof electricity lines and equipments, and biogas leaking detector [2, 3].

4.2.2. Aerobic Sludge Digestion

Aerobic sludge digestion application is still few in China [19, 21, 22]. The following contents are based on both theoretical calculation and operation experiences.

1) Optimal SRT for aerobic digestion of secondary sludge is 10~15d in most practices [3]; while optimal SRT for mixture of both primary and secondary sludge is 15~20d in most cases [3], although some studies showed that it should be 15~25d [3]. Optimal volatile solid volumetric load for aerobic digesters depends also on sludge thickening technology since sludge concentration after gravity thickening is less than that after mechanic thickening [17, 18]. Optimal volatile solid volumetric loads for gravity thickened sludge (water content 96%~98%) and mechanic thickened sludge (water content 94%~96%) are 0.7 ~ 2.8 kgVSS / (m^3·d) and 1.4 ~ 4.2 kgVSS / (m3 • d) respectively [3]. Some aerobic sludge digestion scenarios have been proposed and their theoretical performances are listed in table 4.
2) Temperature in the aerobic digester will be higher than the raw sludge due to the fact that aerobic sludge digestion is a heat releasing biochemical reaction. When the raw sludge is 15°C in temperature, temperature in the aerobic digester is normally around 20°C [19]. Studies showed dramatic dropping of aerobic sludge digestion efficiency when temperature in the digester is lower than 20°C [3]. Insulating or heating measures therefore should be taken when atmosphere temperature is lower than 15°C; otherwise, a much longer SRT should be adopted.
3) Dissolved oxygen (DO) concentration of 2 mg/L is the minimum value to maintain sufficient aerobic respiratory demands of activated sludge [21]. In most Chinese MWTPs, Operational DO for aerobic sludge digestion is higher than 2 mg/L [2, 3].
4) Air demand for mixing sludge is always bigger than for biochemical respiration if air blowers are used in the aerobic sludge digestion [3]. According to application experiences and literatures, air demand for oxidizing secondary sludge is 0.015 ~ 0.02 m^3 air/(m^3 tank capacity·min), while demand for mixing the sludge is 0.02 ~ 0.04m3 air/(m^3 tank capacity·min); air demand for oxidizing primary sludge is 0.025 ~ 0.03m^3 air/(m^3 tank capacity ·min), while demand for mixing the sludge is 0.04 ~ 0.06m^3 (m^3 tank capacity ·min) [2, 3].
5) When mixing power and oxygen supply is provided by surface mechanical aerators, capacity of the aerator is 20 ~ 40W / (m^3 tank capacity) [3].

6) In south China, the aerobic sludge digester can be operated without any cover if no odor problems are found [3]; in north China, a cover is normally needed together with heating system [3].

Table 4. Some aerobic sludge digestion scenarios and their theoretical performances [2]

Scenario Number / Parameters	1	2	3	4	5	6	7	8	9	10
Dry solid mass of raw sludgekgSS/(d)	100	100	100	100	100	100	100	100	100	100
SRT (d)	20	20	20	20	20	10	10	10	10	10
Water content (%)	98	97	96	95	94	98	97	96	95	94
Volume of raw sludgem^3/(d)	5.0	3.3	2.5	2.0	1.7	5.0	3.3	2.5	2.0	1.7
Ratio of volatile solid(%)	70	70	70	70	70	70	70	70	70	70
Mass of volatile solidkgVSS/(d)	70	70	70	70	70	70	70	70	70	70
Volume of the digester(m$^{3)}$	100	67	50	40	33	50	33	25	20	17
Volatile solid volumetric load (kgVSS/(m^3·d))	0.7	1.05	1.40	1.75	2.10	1.4	2.10	2.80	3.50	4.20

4.3. Mechanical Dewatering of Municipal Sewage Sludge

Based on applications in China, filter press and centrifugation are the most popular dewatering technologies for municipal sewage sludge treatment [23-27]; some of the application cases were demonstrated in table 5 [2]

Table 5. Some of the application cases of MWTP sludge dewatering in China

MWTP	Type of influent sludge	Type of sludge dewatering device	Water content of influent sludge (%)
Sichuan MWTP	Activated sludge[a]	Vacuum press filter	95.8
Liaoyang MWTP	Activated sludge	Chamber filter	98.1
Beijing MWTP1	Activated sludge chemical sludge	Frame press filter	96 ~ 97
Beijing MWTP2	Activated sludge after floating	Belt press filter	93 ~ 95
Haerbin MWTP	Activated sludge after floating	Frame press filter	94 ~ 97
Beijing MWTP3	Primary sludge	Centrifugation	91.2 ~ 92.7
Shanghai MWTP1	Activated sludge	Vacuum press filter	95.8 ~ 98.7
Shanghai MWTP2	Activated sludge	Frame press filter	97.0
Wuxi MWTP	Activated sludge	Frame press filter	97.4

a. Both primary and secondary sludge.

Key parameters that matter selection of sludge dewatering devices include sludge resistance, viscosity, particle size of influent sludge, and water content request of outlet sludge cake [24]. Operational experiences in China show that when water content of the influent sludge is greater than 98%, there are operational problems for most dewatering devices, or the outlet sludge cake is not satisfactory in terms of water content and collection [25]. Chemical conditioning is used in most dewatering applications of municipal sewage sludge in China [24-26]; chemicals include polyacrylamide, aluminum salt and iron salt. Sometimes, digested sludge is washed by treated wastewater before chemical conditioning in order to save chemical dosage amount [25]; whether this practice is necessary or not depends on alkalinity of the digested sludge [2].

Transportation of dewatered sludge is normally realized by transmission belt conveyors, screw conveyor, and/or pipelines [3].

4.3.1. Press Filter

Types of press filter used in China include belt press filter, frame press filter, chamber press filter and micro porous press filter [23]. Water content of sludge cake after press filter is normally 75~80% [23]. Other parameters such as working and flushing pressure, sludge load, filtration period varies with different manufactures.

4.3.2. Centrifugation Devices

Vortex centrifugation devices are widely used in China [25]. Although theoretically the bigger the separation factor is, the better the sludge will be dewatered, it is not economical to increase the separation factor once it has reached a threshold value, after which dewatering efficiency of sludge increases only slightly with increasing separation factor while energy consumption and noise increase dramatically [25]. Also risks of sludge floc damage increase when the centrifugation speed is high enough to damage sludge flocs by shearing and thus decrease recovery rate of sludge. Table 6 shows currently used separation factors in some of the Chinese MWTPs using configuration sludge dewatering process [2, 3].

Table 6. Currently used separation factors in some of the Chinese MWTPs

MWTP	Type of influent sludge	Separation factor (G)
Shenzhen Binhe MWTP	Activated sludgea	2115
Luoyang Jianxi MWTP	Activated sludge	2115
Yizheng MWTP	Activated sludge	1700
Shanghai Caoyang	Activated sludge	1224
Yunnan Gejiu MWTP	Activated sludge	1450
Wuhan Tangxunhe	Activated sludge	2950
Liaoning Huludao	Activated sludge	2950
Shanghai Bailonggang	Chemical sludge	3200
Hongkong Angcuanzhou	Chemical sludge	3200

a. Both primary and secondary sludge.

Although separation factors vary with different sludge characters, almost no applications use a separation factor higher than 3000G for activated sludge in China [24]. A higher separation factor is adopted for chemical sludge or primary sludge due to the fact there are less organic matters in them, which changes the threshold separation factor [24].

Cases when the influent sludge has long fibers that can block discharging holes of the centrifugation devices, a sludge cutter should be installed before it entering into the dewatering unit. The sludge cutter is used to guarantee the treated sludge fiber smaller than 8 mm, according to experiences [2].

4.4. Drying of Municipal Sewage Sludge

Sludge disposal and utilization experiences in China show that sludge drying is necessary in many cases especially when sludge utilization is the target [28-31]. Natural sludge drying can save energy and reduce operation cost; however, due to its requirements on land, climate conditions and its potential pollution to the environment, not many application cases have been practiced in China yet [30, 32-34]; on the other hand, sludge is artificial dried in most of the applications using heating system [28, 29, 31]; some MWTPs such as Dalian MWTP, Qinhuangdao MWTP, Xuzhou MWTP use artificial heating system to dry sludge before it is used as an ingredient to compound fertilizer [2, 3].

Both direct heating system and indirect heating system have been used in Chinese MWTPs for drying municipal sewage sludge [28, 29, 31]. Table 7 demonstrated some technical specifications of both systems [2, 3].

Table 7. Application of typical heating systems in municipal sewage sludge drying

MWTP	Shanghai Shidongkou MWTP	Tianjin Xianyanglu MWTP
Location	Shanghai	Tianjin
Wastewater scale (10,000 m3/d)	40	45
Wastewater treatment process	Unitank	A/O
Year of first operation	2003	2004
Sludge scaleT/(d)	64	73
Drying device	Fluidized bed dryer (direct heating)	Disc dryer (indirect heating)
Water content of influent sludge(%)	70	75
Water content of outlet sludge(%)	≤10	<10
Fuel for drying device	Dried sludge	Biogas/natual gas

4.5. Incineration of Municipal Sewage Sludge

In China, primary sludge has 55% ~ 70% organic matter in mass; secondary sludge has 70%~80% [35]. Normally organic content in anaerobic digested sludge is only around 35%~50%, which explains why the digested sludge has lower calorific value [36, 37].Test results from Chinese MWTPs showed that calorific value of dry sludge is 30% ~ 60% of standard coal [2, 36], which means it has thermo contributions to incinerators but is technically difficult to be incinerated or burned alone. Calorific values of some Chinese municipal sewage sludge and typical fuels are listed in table 8 [3]. Business as usual practice of incineration of municipal sewage sludge in China is co-fired with domestic waste or coals [36-39]. In other words, it is mixed and burned with fossil fuels in waste incinerators, cement kilns, and/or boilers of power plants.

Table 8. Calorific values of some Chinese municipal sewage sludge and typical fuels

Sludge and/or fuels		Calorific value (kJ/kg)		
		Dewatered	Dried	Absolute dry
Fuels	Standard coal			29300
	Wood			19000
	Peat			18000
	Coal refuse			≤12550
Sludge	Primary sludge			10715~18920
	Secondary sludge			13295~15215
	Primary + Secondary sludge			12005~16957
Shanghai Shidongkou MWTP	Primary + Secondary sludge			11078~15818
Beijing Gaobeidian MWTP	Raw material			9830~14360
	Digested sludge			11120
	Digested + thickened sludge			10980~11910
Tianjin Jizhuangzi MWTP	Primary + Secondary sludge	559 (75% Water content)	12603 (6.8% Water content)	13823
	Fermented primary + Secondary sludge	1346 (75% Water content)	13873 (7.8%) Water content	15257
Tianjin Dongjiao MWTP	Primary + Secondary sludge	1672 (75% Water content)	12895 (7.7% Water content)	14187
	Fermented primary + Secondary sludge	1718 (75% Water content)	13134 (7.4% Water content)	14375

Since calorific value of sludge is quite close to coal refuse, in China sludge is incinerated in similar equipments as of coal refuse fired incinerators, with circulating fluidized bed incinerator as the most used technology [37].

Table 9 demonstrates some technical aspects of sludge incineration process in Shanghai Shidongkou MWTP [36]; as in other cases, circulating fluidized bed incinerator is used.

Table 9. Sludge incineration process in Shanghai Shidongkou MWTP

Name of MWTP	Shanghai Shidongkou MWTP
Location	Shanghai China
Wastewater scale(10,000 m3/d)	40
Wastewater treatment process	Unitank
Year of starting operation	2003
Sludge scalem(3/d)	213 (dewatered sludge)
Type of sludge incinerator	circulating fluidized bed incinerator
Water content of feed sludge(%)	≤10
Ash productiont/(d)	Around 42
Type of fuel	Dried sludge
Pre-heating temperature (°C)	136
Incinerator temperature (°C)	≥850
Retention time of flue gas	>2 s

Specific measurements should be taken to dispose sludge incineration fly ash due to its content of heavy metal and radioactive matters [36-38]; fuel gas generated during sludge incineration should also be properly treated before discharging [36]. In many cases, heat release from sludge incineration is used to generate electricity, which is financially supported by selling the electricity with reasonably higher prices to national or regional grids than coal fired electricity [1-3, 37]. When sludge treatment scale is not big enough to generate economic electricity for selling, waste heat from sludge incineration is encouraged to be used for other purposes such as drying sludge or firing boilers [37].

4.6. Land Application of Municipal Sewage Sludge

Besides using as fuel, another major might-to-be practice of municipal sewage sludge utilization in China is land application [39-42]. As there are plenty of nutrients such as organic matter, nitrogen, phosphorus, and other trace metals in the municipal sewage sludge, it can be used to improve soil quality. China is a large agricultural country; due to the extensive application of chemical fertilizers, soil quality especially organic content of soil declines [40]. Thus land application of municipal sewage sludge is promising. Due to potential risks of heavy metals and other toxic substances in sludge, sludge land application is tested by many MWTPs in China but normally only in trial period [2, 3, 39]. Since people concern that once soil is accidentally polluted by sludge, there will be serious consequences because land restoration from heavy metal pollution is very difficult and incomplete by state-of-art technologies, the country still has a very long way to go in terms of researches and studies on land applications of sludge.

Table 10 shows one land application case of sludge from a real MWTP. It is reported that heavy metal and other hazardous materials in sludge has been controlled in accordance with corresponding national standards in this case [5, 41].

Table 10. Land application of sludge from Fuyang MWTP [2]

Name of MWTP		Fuyang MWTP
Location		Fuyang, Hangzhou China
Wastewater treatment scale (10,000m^3/d)		2
Wastewater treatment process		Oxidation ditch
Year of starting operation		1999
Sludge scale (t/d)		3
Water content of sludge for land application (%)		80±2
Land application	Sludge application method	Mixing with original soil
	Plants	Trees
	Growth of plants	Comparable with chemical fertilizer application

Sludge land application experiences in China must abide by all its relevant national environmental standards, especially those related to heavy metal and other toxic substances both in wastewater, in sludge, and/or in soil [5, 41]. At the same time, measures must be taken to monitor and control heavy metals and other toxic substances in soil [5, 41-43].

4.7. Other Utilizations of Municipal Sewage Sludge

Other more interesting disposal technologies and utilization ways of municipal sewage sludge are also under research in China. Disposal and utilization scenarios includes ultrasonic treatment [44], pyrolysis and gasification of municipal sewage sludge [13, 35, 37], hydrogen production through fermentation of municipal sewage sludge [39], producing activated carbon with municipal sewage sludge [35, 37], producing construction and building materials out of municipal sewage sludge [39]. Unfortunately these technologies and concepts are not at industry level in China and thus will not be elaborated in this review.

CONCLUSIONS

Municipal sewage sludge amount is huge in China. Due to lacking of sufficient technical and economical reliably technologies, municipal sewage sludge is considered to be valuable only in concept. Aim of disposal of municipal sewage sludge focuses on volume reduction, organic stabilization, and harm reduction. Utilization of municipal sewage sludge is in the way but still at very preliminary stage.

Currently China has accumulated many country specific experiences on municipal sewage sludge disposal technologies such as thickening, dewatering, digestion, and drying. Fuel application and land application are two of the major directions the country is trying to reuse its waste municipal sludge with minor occasions also using as building materials.

ACKNOWLEDGMENTS

Financial support from DAAD (German Academic Exchange Service) is sincerely acknowledged.

REFERENCES

[1] Association of China Water Industry. Statistic report of Chinese water industry market 2007. Http://www.h2o-China.com.

[2] Shanghai Institute of Municipal Engineering Design and Research. *Code for design of outdoor wastewater engineering* (GB50014-2006) China Planning Publishing Press, 2006, (01): Beijing.

[3] Southwest China Institute of Municipal Engineering Design and Research Design *Manual for Water and Wastewater* (version 2; volume 1, 5, and 9). China Construction Industry Publishing Press, 2000, (10): Beijing.

[4] China Ministry of Environmental Protection. *Discharging Standards for Chinese MWTPs* (GB18918-2002). Http://mep.gov.cn.

[5] China Ministry of Environmental Protection. *Control Standards for Pollutants in Sludge for Agricultural Use* (GB4284-84). Http://mep.gov.cn.

[6] Tang Hongyan, Li Guozhi, Li Ming, Wei Xuefeng. Enhancement of Primary Treatment in Municipal Sewage by Chemical Coagulaiton. *Safety and Environmental Engineering*. 2008, (02): 64-68.

[7] Wang Wenchao, Zhang Hua, Zhang Xin. Chemical Phosphorus Removal in Municipal Wastewater. *Water Sciences and Engineering Technology*, 2008, (01): 14-16.

[8] Nian Dong, Wang Jiawei, Liu Lichao, Zhou Jun, Gan Yiping, Wang Hongchen. Research on Efficiency and Operating Cost of Chemical Phosphorus Removal. *Water and Wastewater Engineering,* 2008, (05): 7-10.

[9] Liang Peng, Huang Xia, Qian Yi . Research Progress on Sludge Reduction Technologies. *Technigues and Equipment For Enviro.poll.cont*, 2003, (01): 44-52.

[10] Wang Hongyuan. Technologic Analysis for Sludge Treatment and Disposal in Municipal Wastewater Treatment Plant. *Guangzhou Chemical Industry*, 2007, (04): 71-74.

[11] Li Jie, Xiong Biyong, Zhang Jie. Disposal of Urban Wastewater Sludge and Sustainable Development of Agriculture. *Water and Wastewater Engineering*, 2003, (09): 23-26,100.

[12] Pan Luting, Wang Jian, Shu Yubao, Wu Lei. A Literature Review of Research Progress in Sludge Reduction and Reutilization. *Environmental Pollution and Control*, 2008, (04):73-77.

[13] Ma Na, Chen Ling, Xiong Fei. Disposal and reuse of city sludge in China. *Ecological Environment,* 2003, (01): 92-95.

[14] Sun Changle. Disposal and Effective Utinization of Sludge in Urban Districts. *Journal of Fuyang Teachers College* (Natural Science), 2008,(01):52-54.

[15] Shi Xiong. Mehods to Treat Municipal Sludge. *Journal of Hebei Polytechnic Universityn*(Natural Science Edition), 2008,(01):128-132.

[16] Chu Jinyu, Yu Jiexiang, Zhang Bo, Hua Yinqun, Yao Jichen. Analysis on the Characteristics of Sludge in Zhenjiang City. *Journal of Anhui Agricultural Sciences*, 2008, (04):1503-1504.

[17] Zeng Ke, Li Pingxian, Li Bin, Pu Qiuping. Application of the New Sludge Thickening Technology in Municipal Sewage-disposal Plant. *Journal of Safety and Environment*, 2002, (01): 57-58.

[18] Hu Fengping, Huang Xiaodong, Wang Linyuan, Li Weimin, Wang Tao, Deng Rongsen. Study on Low Concentration Excess Activated Sludge Thickening by Cavitation Air Floatation (CAF). *Water and Wastewater Engineering*, 2006, (06): 31-34.

[19] Chen Yan, Han Hongjun. Experimental Research on Aerobic-Anoxic Sludge Digestion. *Journal of Harbin University of Commerce* (Natural Sciences Edition), 2007, (04): 405-407, 414.

[20] Yin Shouqian. Discussion on Project Technology of Sludge Anaerobic Fermentation. *Railway Occupational Safety Health and Environmental Protection*, 2008, (1): 26-28.

[21] Zhou Chunsheng, Yun Xiao, Buu Hai. Study of Biodegradability on Aerobic Digestion of Excess Activated Sludge.. *Chongqing Environmental Science*, 1995, (01), 27-30, 39.

[22] Zhu Yishu, Yu Guanghui, He Pinjing, Shao Liming. Improvement of Sewage Sludge Aerobic Digestibility by Ultrasonic Pretreatment. *Chinese Journal of Environmental Engineering*, 2008, (05):690-693.

[23] Yuan Quan, Sun Bin. Application of Belt Filter Press in Sludge Dewatering. *Journal of Filter and Separator*, 2005, (01): 31-33.

[24] Wang Tao. Research Progress of Sludge Thickening and Dewatering Technology. *China Environmental Protection Industry*, 2008, (02):32-35.

[25] Gao Qiong, Yang Shian; Wu Chunjiang. Experience Summary on Operation of Decanter Centrifuges. *China Water and Wastewater*, 2006, (22): 92-95.

[26] Gao Jianlei, Yan Yixin, Wu Jian-ping, Wan Yaoqiang. Sludge Dewatering Property in Municipal Wastewater Treatment Plant. *Environmental Science and Technology*, 2008, (02): 108-111, 131.

[27] Song Xianqiang, Lei Hengyi, Yu Guangwei, Bai Tao, Li Zhong, Deng Weicong. Study on Application of New Composite Coagulant to Improvement of Sludge Dewaterability. *China Water and Wastewater*, 2007, (13): 87-90.

[28] Yang Li, Wang Youhua. The Application of High Efficiency Agent of Sludge Dehydration. *Contemporary Chemical Industry*, 2007, (04): 420-423.

[29] Gu Ruihuan, Liu Yuzhong. Research on Concentrate and Dehydrate of City Sewage Treatment Plant Residual Sludge. *Henan Science*, 2008, (04): 475-477.

[30] Jiang Wenchao, Zhang Zhi, Qu Zhenxiao, Hu Jian, Chen Bin, Yu Yiping, Jiang Futong. Study on the Integration of Vertical Enhanced Filtration and Concentration and Natural Drying and Digestion on Sewage Sludge. *Chinese Journal of Environmental Engineering*. 2008, (08): 53-55.

[31] Weng Huanxin, Ma Xuewen, Su Minhua, Sun Feng, Liu Zan. Drying Sewage Sludge by Waste Heat from Flue Gas. *China Water and Wastewater*, 2008, (04): 58-61.

[32] Lv Yicheng; Li Tianqi, Jin Yaohua. Testing Study on the Odor Removal Engineering of the Tianshan Wastewater Treatment Plant. *Energy Environmental Protection*, 2007, (04): 45-46, 61.

[33] Li Juzhe. Analysis and Evaluation of Odor Pollution in Sewage Treatment Plant. *Pollution Control Technology*, 2008, (01):78-79.

[34] Wu Rongfang, Mi Zhongqiang. Study on the Application and Development of Ecological Deodorization Technology. *Journal of Anhui Agricultural Sciences*, 2008, (10):4231-4233, 4243.

[35] Guo Lixin, Liu Tianyuan, Luo Jinghui, Li Yanqiao. Research Development on Heat Treatment Technology of Sludge. *Journal of Changchun University of Science and Technology* (Natural Science Edition). 2008, (03): 53-56.

[36] Han Xiaoqiang, Chen Xiaoping. Commissioning of Shanghai Shi Dong Kou Dried Sewage Sludge Incinerator. *Boiler Technology*, 2006, (01): 77-80.

[37] Zhao Rui, Liu Dan, Li Qibin, Huang Mingxin, Hu Xiangfang. Prospect Discussion on Sludge Energy Utilization. *Journal of Anhui Agricultural Sciences*, 2008, (14):6004-6007, 6017.

[38] Xie Rongzhen. Study on Resource Reuse of Municipal Sewage Sludge. *Guangdong Chemical Industry*. 2008, (08): 52-54.

[39] Xie Xionghua, Yin Lingling, Ni Juan. Research of Sewage Sludge Comprehensive Utilization in Sewage factory. *Shanxi Architecture*, 2006, (04): 192-193.

[40] Song Jingyang. Land Application of Municipal Sewage Sludge. *Environmental Science and Management,* 1993, (03): 29-32.

[41] Sun Yuhuan, Luo Yongming, Yang Zhihai. Effect of Swage Sludge Application on Structural Diversity of Soli Microbial Communities. *Journal of Anhui Agricultural Sciences*, 2008, (12):5126-5127, 5131.

[42] Hu Jiukun, Xu Jinggang, Zhang Dan, Li Shuqin. Forecast for Disposals and Agricultural Reutilization of Sludge. *Journal of Northeast Agricultural University*, 2005, (06): 820-824.

[43] Zhang Longli, Liu Qing, Xu Zi, Liang Lina, Liji. Research on Effect of Compound Microbial Inoculant on Sewage Sludge Composting. *Chinese Journal of Environmental Engineering*, 2008, (02):266-269.

[44] Ma Shougui, Xu Honglin, Lü Xiaoping, Zhu Shuhui. Study on Pilot Experiment of Treating Excess Activity Sludge with Ultrasonic. *Chemical Engineering (China),* 2008, (02):46-49.

In: Sewage Treatment: Uses, Processes and Impact
Editor: Anna Stephens and Mark Fuller

ISBN 978-1-60692-959-9

Chapter 13

REMOVAL OF COLOUR AND TRACE ORGANIC MATTERS FROM RECYCLED WASTEWATER

***Linda Zou*[1]*and *Bo Zhu*[2]**
[1]Centre for Water Management and Reuse
University of South Australia, Mawson Lakes Campus
Adelaide SA 5095, Australia
[2]Institute for Sustainability and Innovation
Victoria University, Melbourne, Australia

ABSTRACT

Colour and non-biodegradable trace organics present in the treated wastewater from sewage treatment plants (STPs) can affect the options for disposal or recycling. Colour is sometimes a concern because its visual appearance affects public acceptance of using recycled water. In the present study, different Advanced Oxidation Processes (UV/O_3, UV/TiO_2 and UV/TiO_2/O_3) in laboratory scale systems were used to study the removal efficiencies of colour and trace organics from the treated effluent. The removal efficiencies of the processes were compared by the analysis of colour, A_{254} (the organics absorption at the wavelength of 254 nm), and total organic carbon (TOC). The effects of ozonation procedure and ozone dose were also investigated. The co-treatment of ozonation and photocatalytic oxidation increased TOC removal rate by nearly 50 % compared to the process in which ozonation was used as a pre-treatment for photocatalysis. This was due to the synergistic effects of photocatalysis and ozonation. The colour removal rate significantly increased when the O_3 dose was increased.

Apart from AOPs, adsorption methods using activated carbons were also investigated for colour removal from recycled water. TiO_2 nanoparticles were loaded on the surface of the granular activated carbons (GAC). This allows the GAC to be regenerated by UV irradiation without the thermal treatment. The adsorption results using a fixed bed column were reported, and the regeneration potential of the activated GAC was also discussed.

---

* Phone: +61 8 83025489; Fax: +61 8 83023386; E-mail address: linda.zou@unisa.edu.au

INTRODUCTION

The available water resources in many areas are decreasing whereas the wastewater volume is constantly increasing. The reuse of treated effluent can be an effective means of coping with the water shortage. After the bulk of organic material (measured as BOD or COD) and other contaminants are removed by biological treatment from a sewage treatment plant (STP), the treated wastewater is usually discharged to the ocean. There are significant opportunities to reuse a larger amount of this treated wastewater. Many uses of water do not require treatment to drinking water standard, and the use of recycled water for non-drinking purposes is a net benefit to the community and a minimization of detrimental discharges to the environment. New residential, commercial and industrial developments provide the greatest opportunities for the distribution of recycled water. The recycled water can be used for toilet flushing, garden watering and open space irrigation.

However, barriers to the reuse of the treated wastewater do exist, particularly if the treated wastewater contains contaminants, which may adversely affect human or environment health, or because it is aesthetically unappealing (i.e. odorous or visually unattractive). Colour and non-biodegradable trace organics present in the treated wastewater from Sewage Treatment Plants (STPs) have negative impact on the public acceptance of recycle water. Coloured recycled water will be perceived as dirty and not fit for its purpose even though it is. As more stringent water quality standards for colour and total organic matter are introduced, improved public acceptance of water recycling will be essential.

Since trace recalcitrant organic compounds in wastewaters, such as organo-halogens, organic pesticides, surfactants, and colouring matters, are resistant to the conventional chemical and biological treatments, Advanced Oxidation Processes (AOPs) are being studied as an alternative to the traditional methods and will probably constitute the best option in the near future. AOPs are based on the generation of very reactive species such as hydroxyl radicals ($OH^{\bullet}$) that oxidise a broad range of organic pollutants quickly and non-selectively [1]. In general, a combination of several methods enhances free radical generation, thus giving high treatment efficiency compared with individual treatment [2].

Among AOPs, the photocatalytic process using heterogeneous photocatalysts such as TiO_2 appears as an emerging destructive technology leading to the total mineralization of most of the organic pollutants to CO_2, water and mineral acids. When a semiconductor like TiO_2 is irradiated with light having energy greater than the band gap, an electron is excited from the valence band to the conduction band, creating a hole and an electron (Figure 1).

The hole is an oxidizing agent, and the electron is a reducing agent. In the generally accepted mechanisms for the photocatalytic process, the hole can react with water to produce the hydroxyl radical ($OH^{\bullet}$) and the electron can reduce molecular oxygen, hydrogen peroxide or some other oxidizing agent in the solution. The reactive hydroxyl radicals act as powerful oxidizing agents responsible for removal of the hazardous components from the water [3, 4].

The following equations describe the oxidation and reduction reactions [1, 5-8]:

$$TiO_2 + h\nu \rightarrow e^- + hole^+ \tag{1}$$

$$H_2O \rightarrow H^+ + OH^- \tag{2}$$

$$OH^- + hole^+ \rightarrow H^+ + OH^\bullet \quad (3)$$

Oxidation of the organic reactant:

$$R + OH^\bullet + O_2 \rightarrow \text{products } (CO_2, H_2O, \text{etc.}) \quad (4)$$

Although heterogeneous photocatalysis is gaining importance in the area of wastewater treatment since the process may use atmospheric oxygen as the oxidant, can be carried out under ambient conditions, it is not yet in full scale application due to its low oxidation rate.

Ozone is a strong oxidizer having high reactivity with organic compounds and has been used for various aspects of water treatments. It is known that ozone can degrade organic pollutants (R) by direct electrophilic attack in acidic or neutral conditions [9]:

$$O_3 + R \rightarrow ROX \quad (5)$$

However, direct attack by molecular ozone (commonly known as ozonolysis) is a selective reaction which results in the formation of carboxylic acids as end products. These oxidized organic products (ROX) cannot be further oxidized by molecular ozone. it has been reported that in conventional ozonation process, most of the organic compounds are not completely degraded by ozone alone and sometimes toxic intermediates are produced [10]. Therefore, a combination of photocatalyst with ozone is a reasonable choice for the treatment of hardly degradable organic compounds since the organic compounds are expected to decompose more quickly and achieve complete mineralization to carbon dioxide, water, or inorganic ions. The combination of TiO_2 photocatalyst with ozone has been successfully used for the degradation of organic substances in aqueous solutions such as formic acids, aniline, monochloroacetic acid and pyridine, humic acids, and dibutyl phthalate [2, 11-14]. A synergetic effect of these two oxidation methods was observed from the greatly enhanced mineralization rate of organic substances.

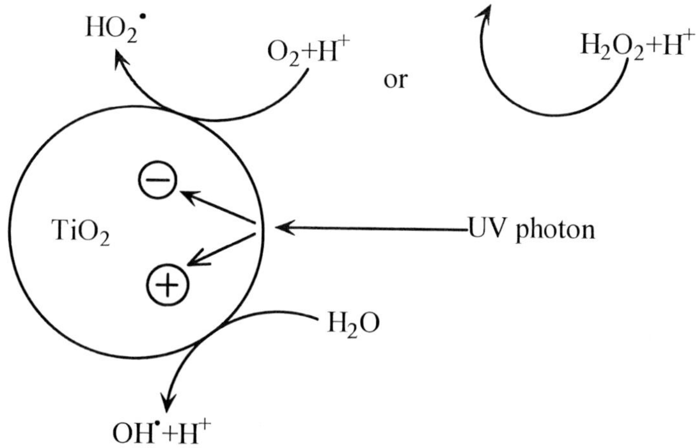

Figure 1. Illustration of the oxidation and reduction reactions in a semiconductor TiO2 exposed to UV light.

Apart from AOPs, activated carbon (AC) adsorption has been used successfully for the advanced (tertiary) treatment of municipal and industrial wastewater. Despite the efficacy of the adsorption process, the main disadvantage of such methodology is its non-destructive character; as a result, the saturated AC must be regenerated before reuse. Several methods, based on either desorption or decomposition, have been used to regenerate activated carbon, the thermal desorption being the most common process [15]. Photocatalytic oxidation can be used as an alternative to regenerate spent AC adsorbent and destroy organic adsorbates simultaneously [16]. Coating TiO_2 photocatalyst on AC can integrate the advantages of both techniques [17, 18]. On one hand, AC works as the supporting materials of nanosized TiO_2 photocatalyst and enriches the pollutants and intermediates around the TiO_2; on the other hand, the TiO_2 photocatalyst can decompose the pollutants and results in regenerating the AC in situ.

AOPs and AC adsorption appear to be the most promising technologies for developing real applications in water treatment. However, nearly all the studies have been conducted in artificial solution using only one organic compound as model molecule [19]. Investigations in actual wastewater containing both natural organic matter (NOM) and residuals from biological treatment process, therefore, are very important for developing processes for real applications.

In the present work, the actual wastewater from Sewage Treatment Plant (STP) as feed water was treated by a combined ozonation and photocatalysis process or AC adsorption to reduce colour and trace organics, and therefore improve the clarity of water for reuse. The objective of this study is to target the end of treatment colour at an acceptable level of 30 – 40 Pt-Co units. This makes the colour in the treated water almost invisible, so that it is fit-for-purpose as non-potable recycled water.

EXPERIMENTAL AND METHODS

Materials and Sample Preparation

The water sample (Figure 2a) used for this study was non-chlorinated effluent sampled from a Melbourne's Sewage Treatment Plant (STP) by the authorised operator. TiO_2 pellets (Degussa, Germany) were used as a photocatalyst. They contain pure TiO_2 (83.2 % anatase and 16.8 % rutile), with a S-content of < 20 ppm and a Cl-content of the order of 0.1 wt. %. No further elements are present in contents above 0.05 wt. %. The pellets are in a small pillar shape having the size ranging from 2.5 – 5.3 mm long and with a diameter of 3.7 mm.

The BET surface area of the pellets was measured to be 48 m^2/g, using a TriStar 3000 porosity analyzer (Micromeritics, USA). The TiO_2 nanoparticles of size 4 – 5 nm or smaller (TEM graph not shown here) were synthesized by hydrolysis of Titanium butoxide ($Ti(OC_4H_9)_4$, 97%, Aldrich). The TiO_2/GAC absorbent was prepared by coating TiO_2 nanoparticles onto the surface of GAC (Activated Carbon Technologies Pty Ltd, Australia) and was then heat treated at 200°C.

Figure 2. Colour of a typical STP non-chlorinated effluent sample (a) and tap water (b).

Apparatus and Procedures

The photocatalytic oxidation and ozonation experiments were carried out on a laboratory scale recirculating photocatalytic ozonation system (Figure 3). The photocatalytic reactor was made of stainless steel with baffles. Ozone was generated from pure oxygen by an ozone generator (SOZ-6G, A2Z Ozone Systems INC., USA). For each photocatalytic treatment (with or without ozonation), the effluent (1.5 L) was pumped from the tank into the reactor (with 240 g TiO_2 pellets) at a circulation flow rate of 0.6 L/min and was irradiated by UVA (NEC Blacklight UVA lamps) for 40 min. In the case of ozonation only, the effluent was treated by ozone under circulation and UVA irradiating conditions. For the experiments that ozonation was used as pre-treatment, the ozone was bubbled in the effluent tank at an ozone input flow rate of 0.4 L/min for 6 min without circulating and then, photocatalytic treatment was followed for 40 min. For the co-treatments of photocatalysis and ozonation, ozonation started at the beginning of the photocatalytic oxidation process under circulation and UVA irradiating conditions.

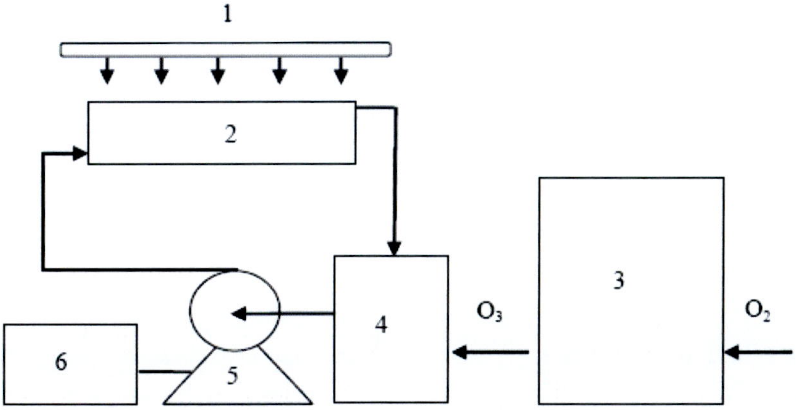

Figure 3. Schematic diagram of the installation setup for the photocatalytic ozonation system. 1. UV panel; 2. Photocatalytic labyrinth flow reactor; 3. Ozone generator; 4. Water tank; 5. Circulating pump; 6. Pump drive.

Adsorption experiment was conducted using a fixed bed column (Figure 4). 50 g adsorbent (GAC or TiO_2/GAC) was used to treat 1.0 L effluent sample for 60 min at a circulation flow rate of 0.6 L/min. The spent TiO_2/GAC was removed and photocatalytic regenerated by UV alone, or a combined UV with H_2O_2 or ultrasound, and was then tested for colour removal using a static adsorption method. All tested adsorbent samples were oven dried prior to the recovery capacity test by static adsorption experiments. For each recovery capacity test, 5 g regenerated adsorbent was used to treat 50 mL effluent sample for 60 min. The treated water samples were then analysed and compared with those treated by the virgin adsorbent.

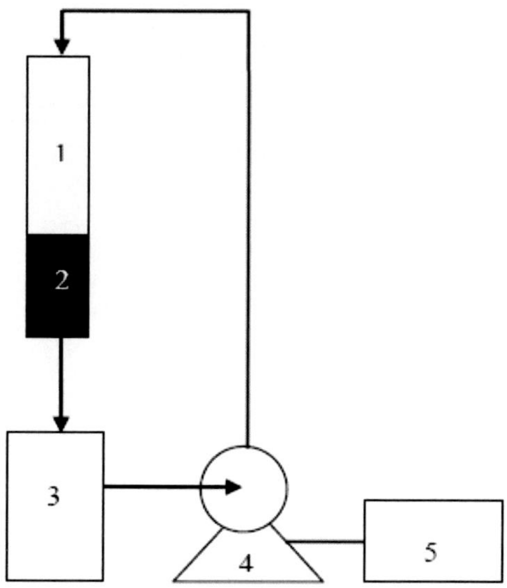

Figure 4. Schematic diagram of the fixed bed adsorption system. 1. Adsorption column; 2. Fixed bed of adsorbent; 3. Storage tank; 4. Circulating pump; 5. Pump drive.

Analyses

High performance size exclusion chromatography (HPSEC) was used to analyse the profile of colour-causing organic compounds in the effluent sample. HPSEC separates organic constitutes based on a differential permeation process, according to molecular weight (size) [20]. Organic compounds is adsorbed on a porous resin and then differentially eluted. The resultant chromatogram is based on the measurement of UV absorbance at 260 nm. In this study, a Waters 2690 separation module and Waters 996 photodiode array detector were used. The separation column used is Shodex KW 802.5 (Shoko, Japan) using a 0.1 M phosphate buffer solution. The flow rate was 1 ml/min and the injection volume was 100 μL.

Measurements of true colour (Method 8025, HACH) and A_{254} (Method 10054, HACH) were carried out with a spectrophotometer (HACH DR5000, USA). Colour is a useful parameter indicating dissolved humic substances in water. All samples were first filtered using 0.45 μm membrane filters (Toyo Roshi Kaisha, Ltd., Japan) and test results were measured at 455 nm and reported as true colour (Pt-Co units). The presence of organic constituents in water was indicated by measuring the absorption of the filtered water sample at a wavelength of 254 nm against organic-free water, and results were reported in absorbance per centimetre (cm^{-1}). The total organic carbon (TOC) was determined by an Aurora 1030W wet chemical oxidation TOC analyser (O.I. Analytical, USA) calibrated with potassium hydrogen phthalate.

RESULTS AND DISCUSSION

Characterization of Effluent

The untreated STP non-chlorinated effluent sample was characterized by general parameters such as true colour, A_{254}, TOC (Table 1), and by the analysis of colour-causing organic compounds using HPSEC (Figure 5). The HPSEC profile of colour showed that the molecular weights of the colour-causing organic compounds in the non-chlorinated effluent sample were in the range of 400 – 10,000. The origins of these organic compounds included natural organic matter and residuals from biological treatment process.

Table 1. Characterization of STP non-chlorinated effluent sample

Parameter	Average value
pH	6.7
TOC (total organic carbon), mg/L	22
True colour, Pt-Co Units	116
A_{254} (UV absorption), cm^{-1}	0.385

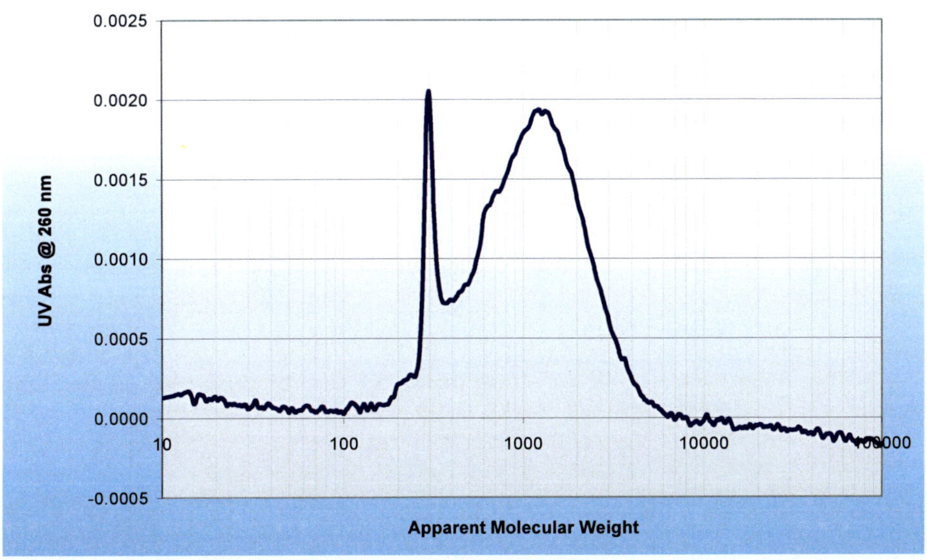

Figure 5. HPSEC profile of colour in the untreated effluent sample.

Comparison of Different Oxidation Processes for Colour, A_{254} and TOC Removal

The efficiencies of different processes (UVA/O_3, UVA/TiO_2 and $UVA/TiO_2/O_3$) for the reduction of colour, A_{254} and TOC are compared in Figure 6. The colour removal rates of processes with ozone (UVA/O_3 and $UVA/TiO_2/O_3$) were much higher than that of the process without ozone (UVA/TiO_2), indicating that the ozonation did have positive effect on the colour removal. For A_{254} and TOC reduction, however, the processes in the presence of TiO_2 catalyst achieved better performance compared to the process without TiO_2. This indicates that the removal efficiency of organic compounds can be significantly improved by the photocatalytic process. The combined treatment of photocatalysis and ozonation ($UVA/TiO_2/O_3$) showed the best performance in the reduction of colour, A_{254} and TOC. It is also found from Figure 6 that the TOC removal rate for the $UVA/TiO_2/O_3$ treatment was considerably higher than the sum of that using UVA/TiO_2 and UVA/O_3 as separate processes, which implies a synergistic effect of photocatalysis and ozonation [2, 11, 12].

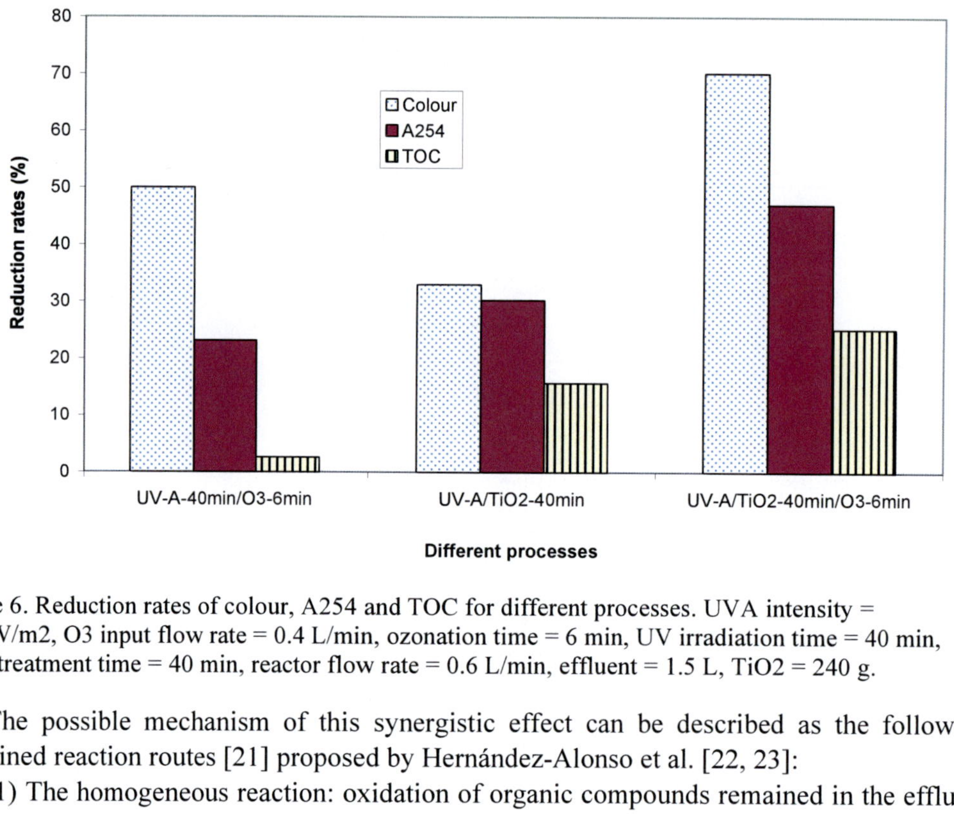

Figure 6. Reduction rates of colour, A254 and TOC for different processes. UVA intensity = 11.1 W/m2, O3 input flow rate = 0.4 L/min, ozonation time = 6 min, UV irradiation time = 40 min, TiO2 treatment time = 40 min, reactor flow rate = 0.6 L/min, effluent = 1.5 L, TiO2 = 240 g.

The possible mechanism of this synergistic effect can be described as the following combined reaction routes [21] proposed by Hernández-Alonso et al. [22, 23]:

(1) The homogeneous reaction: oxidation of organic compounds remained in the effluent sample (pH ~ 6.7) occurred through a homogeneous reaction of direct attack by dissolved ozone (Eq. 7).

(2) The heterogeneous catalytic reaction: in the presence of TiO_2, adsorption of ozone molecules on the TiO_2 surface occurred [24], and the adsorbed ozone reacted with the TiO_2 surface to generate ozonide radicals ($O_3^{\bullet-}$) [22], which rapidly reacted with H^+ in the solution to generate $OH^{\bullet}$:

$$O_3 + e^- \rightarrow O_3^{\bullet-} \tag{6}$$

$$O_3^{\bullet-} + H^+ \rightarrow HO_3^{\bullet} \tag{7}$$

$$HO_3^{\bullet} \rightarrow O_2 + OH^{\bullet} \tag{8}$$

(3) The heterogeneous photocatalytic reaction with oxygen: dissolved oxygen acted as electron trap to capture TiO_2 conduction band electron and generated $O_2^{\bullet-}$ radicals, which are very active and can produce OH radicals in the following possible ways [25]:

$$O_2 + 3e^- \rightarrow 3O_2^{\bullet-} \tag{9}$$

$$2O_2^{\bullet-} + 2H^+ \rightarrow 2HO_2^{\bullet} \tag{10}$$

$$2HO_2^{\bullet} \rightarrow H_2O_2 + O_2 \tag{11}$$

$$H_2O_2 + O_2^{\bullet-} \rightarrow OH^{\bullet} + HO^- + O_2 \tag{12}$$

(4) The heterogeneous photocatalytic reaction with ozone: in the presence of TiO_2 under UV illumination, adsorbed ozone acted as a very strong electrophilic agent to generate OH radicals through $O_3^{\bullet-}$ reaction pathways (Eq. (8) – (10)) by trapping photo produced electrons [11, 26].

It can be seen that OH radicals were generated in both photocatalytic reaction routes ((3) and (4)). A greater number of OH radicals were produced when TiO_2 is irradiated in the presence of ozone due to the higher ability of ozone to capture electrons photogenerated on the surface of TiO_2 [2, 22]. Therefore, the colour and TOC removal efficiencies were greatly enhanced by the synergistic effect between ozonation and photocatalysis.

The effect of ozonation procedure (as pre-treatment or co-treatment) on the reduction of colour, A_{254}, TOC was also investigated (Table 2). For the pre-treatment process, the effluent sample was continuously bubbled with ozone at an ozone input flow rate of 0.4 L/min for 6 min. This was followed by a 40 min of photocatalytic reaction under UVA irradiation with a mean intensity of around 11.1 W/m^2. In the co-treatment experiment, ozonation was conducted together with the photocatalysis for the first 6 min of the treatment. All the operating conditions for the co-treatment procedure were the same as those used for the above pre-treatment experiment.

Table 2. Effect of ozonation as co-treatment versus as pre-treatment

Treatment	Reduction rate (%)		
	Colour	A_{254}	TOC
Co-treatment of O_3 (6 min) and UVA/TiO_2 (40 min)	70.2	47.1	25.2
O_3 pre-treatment (6 min) + UVA/TiO_2 (40 min)	65.5	39.4	17.2

Table 2 showed significant differences in the efficiencies of colour, A_{254}, and TOC removal between two treatments. For the reduction of colour and A_{254}, the co-treatment method was slightly better than the pre-treatment procedure. For the reduction rate of TOC, however, the co-treatment method was nearly 50 % higher than that of pre-treatment procedure. This indicates that the co-treatment of photocatalysis and ozonation is more efficient for the mineralization of organic matter in the effluent sample compared to the pre-treatment procedure, which is again attributed to the synergistic effect between photocatalysis and ozonation, as discussed above.

Effect of Ozonation Time on Colour Removal

The effect of the ozone dose on the colour removal efficiency of UVA/TiO_2/O_3 was determined by varying the ozonation time from 0 to 6 min at an ozone input flow rate of 0.4 L/min. The other operating conditions were the same as above. It can be seen from Figure 7 that the ozone dose had significantly impact on the efficiency of the process for colour removal. The colour removal rate increased with an increase in the ozonation time, with a great increase by increasing the ozonation time from 2 to 4 min. Increasing of ozone dose can increase the chances of direct oxidation of organics and the number of OH radicals, thus accelerating the oxidation process [27].

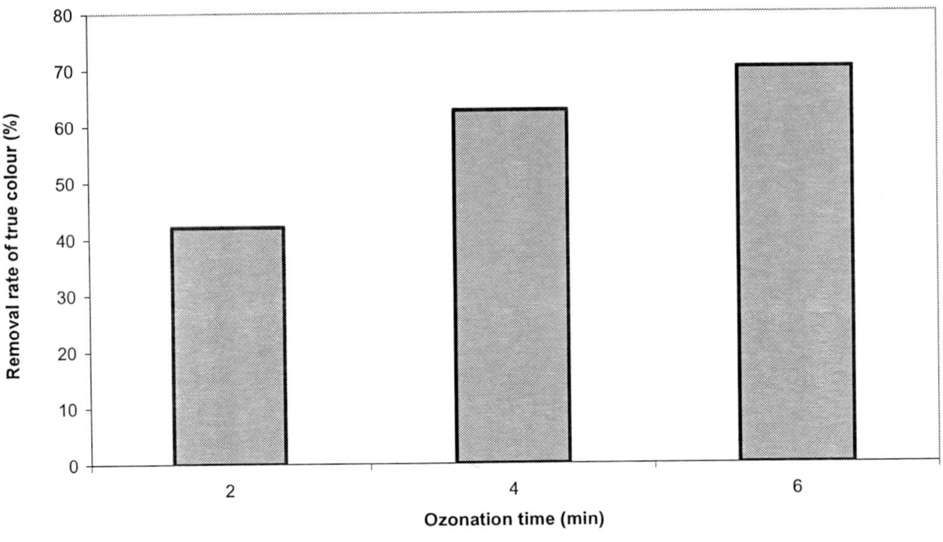

Figure 7. Variation of colour removal rate with ozonation time. UVA intensity = 11.1 W/m2, TiO2 treatment time under UV irradiation = 40 min, O3 input rate = 0.4 L/min, reactor flow rate = 0.6 L/min, effluent = 1.5 L, TiO2 = 240 g.

Colour and A_{254} Removal by TiO_2/GAC Sdsorption

The efficiency of the TiO_2/GAC adsorbent for colour and A_{254} removal was investigated using a fixed bed column (Figure 4) and the results are shown in Figure 8. More than 90 % of colour and A_{254} were removed after 60 min treatment using TiO_2/GAC adsorption. This demonstrated that the adsorption method using TiO_2 coated GAC was effective for the reduction of trace organics from recycled water.

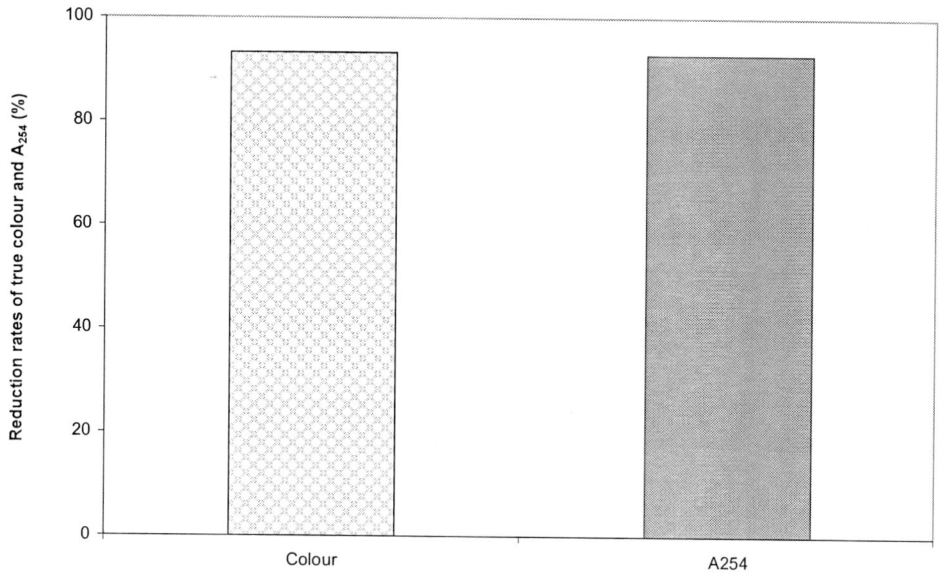

Figure 8. Colour and A254 removal by adsorption method, treatment time = 60 min, circulation flow rate = 0.6 L/min, effluent = 1.0 L, TiO2/GAC = 50 g.

Photocatalytic Regeneration of Used TiO_2/GAC

After the original TiO_2/GAC used the ten adsorption cycles (for each cycle, 1.0 L effluent was treated for 60 min), the used TiO_2/GAC was photocatalytic regenerated by different processes (UVC, UVC/H_2O_2 and UVC/ultrasound). The UVC irradiation was supplied by six 18 W NEC Germicide UVC lamps. In this study, the H_2O_2 solution (1 vol %) was prepared from 35 wt % H_2O_2 solution (Aldrich). Ultrasonic irradiation treatment was carried out in an ultrasonic bath (U-lab Instruments, Australia). Static adsorption experiments were used to evaluate the colour removal capacity of the regenerated TiO_2/GAC.

Table 3 compared the efficiencies of the original and regenerated TiO_2/GAC absorbents for colour removal.

Table 3 Comparison of original and regenerated TiO_2/GAC for colour removal

TiO_2/GAC sample	Regeneration time (h)	Colour removal (%)
Original	0	68.2
UVC regenerated	16	65.4
UVC/H_2O_2 regenerated	4	66.4
UVC/ultrasound regenerated	0.5	69.2

It can be seen that the adsorption capacity of the used TiO_2/GAC for colour removal was nearly recovered after 16 h UVC irradiation compared to that of the original one. This indicated that used TiO_2/GAC could be effectively regenerated by photocatalytic oxidation under UVC irradiation. The regeneration time could be significantly reduced with an assistance of H_2O_2 or ultrasound, especially when ultrasound was used. The colour removal capacity of used TiO_2/GAC reached the same level as that regenerated by UVC irradiation (16 h) after a shorter time of regeneration (4 h) in the presence of H_2O_2. The adsorption capacity of used TiO_2/GAC for colour removal was completely recovered after only 30 min of regeneration by a combined process of ultrasound and UV irradiation.

It is well known that the degradation of organic matters in photocatalytic oxidation process is mainly due to the formation of hydroxyl radicals ($HO^{\bullet}$). Under UV irradiation, the addition of hydrogen peroxide can significantly enhance the rate of generation of $HO^{\bullet}$ radicals [28, 29], and thus accelerate the photocatalytic oxidation reaction. Similar to UVC/H_2O_2, ultrasonic irradiation can produce active species such as $H^{\bullet}$ and $HO^{\bullet}$ radicals through the cleavage of dioxygen and water molecules [30]. $H^{\bullet}$ radicals are capable of attacking the organic compounds in water and can also interact with the oxygen present in the system to form peroxide radicals ($HO_2^{\bullet}$) and H_2O_2. The greatly enhanced efficiency of photocatalytic regeneration of used TiO_2/GAC by ultrasonic treatment observed in this study may be attributed to the synergistic effect between photocatalysis and ultrasound [31, 32]. This synergistic effect may be mainly due to the increase in the number of $HO^{\bullet}$ radicals [30] and the accumulation of hydrogen peroxide [33, 34] in the system.

CONCLUSION

This study presented the results of treating actual STP effluent using different AOPs (UVA/O_3, UVA/TiO_2 and UVA/TiO_2/O_3) and an adsorption method. The results indicated that with UV irradiation, a combined ozonation and TiO_2 photocatalytic process (UVA/TiO_2/O_3) has been efficient in the reduction of the most important parameters that account for the environmental impact such as colour, A_{254} absorbance and TOC. The overall process efficiency in TOC removal was significantly improved because of the synergistic effect of the two methods. The targeted 30-40 Pt-Co colour units can be achieved within 40 min. The adsorption method using TiO_2 coated activated carbon could remove more than 90 % of colour remained in the STP effluent within 1 h treatment time. The used TiO_2/GAC absorbent can be effectively regenerated by the photocatalytic process. The efficiency of the photocatalytic regeneration was significantly enhanced with the aid of H_2O_2 or ultrasonic treatment.

ACKNOWLEDGMENTS

This study is part of research project funded by Victoria Water Industry Smart Water Fund and Nanotechnology Victoria in Australia. The authors thank Dr Chris Chow (Australia Water Quality Centre, Adelaide) and Mr Noel Dow (Institute for Sustainability and Innovation, Victoria University) for their assistance in instrumental analysis.

REFERENCES

[1] Legrini, O., Oliveros, E., and Braun, A. M., Photochemical processes for water treatment.*Chem. Rev.* 1993;93:671-698.

[2] Wang, S., Shiraishi, F., and Nakano, K., A synergistic effect of photocatalysis and ozonation on decomposition of formic acid in an aqueous solution. *Chemical Engineering Journal* 2002;87:261-271.

[3] Goswami, D. Y., Engineering of solar photocatalytic detoxification and disinfection processes. *Advanced in Solar Energy* 1995;10:165-209.

[4] Blake, D., Link, H., and Eber, K., Solar photocatalytic detoxification of water. Advanced in Solar Energy 1995;7:167-210.

[5] Schiavello, M., ed. Photocatalysis and Environment: Treads and Applications (NATO *Science Series*: C). Dordrecht: Kluwer Academic Publishers;1988.

[6] Ollis, D. F. and Al-Ekabi, H., eds. *Photocatalytic Purification and Treatment of Water and Air.* Amsterdam: Elsevier;1993.

[7] Karkmaz, M., Puzenat, E., Guillard, C., and Herrmann, J. M., Photocatalytic degradation of the alimentary azo dye amaranth: Mineralization of the azo group to nitrogen. *Applied Catalysis B: Environmental* 2004;51:183-194.

[8] Herrmann, J. M., Guillard, C., and Pichat, P., Heterogeneous photocatalysis: an emerging technology for water treatment. *Catalysis Today* 1993;17:7-20.

[9] Agustina, T. E., Ang, H. M., and Vareek, V. K., A review of synergistic effect of photocatalysis and ozonation on wastewater treatment. Journal of Photochemistry and *Photobiology C: Photochemistry Reviews* 2005;6:264-273.

[10] Yang, Y., Ma, J., Qin, Q., and Zhai, X., Degradation of nitrobenzene by nano-TiO2 catalyzed ozonation. *Journal of Molecular Catalysis A: Chemical* 2007;267:41-48.

[11] Sánchez, L., Peral, J., and Domènech, X., Aniline degradation by combined photocatalysis and ozonation. *Applied Catalysis B:* Environmental 1998;19:59-65.

[12] Kopf, P., Gilbert, E., and Eberle, S. H., TiO_2 photocatalytic oxidation of monochloroacetic acid and pyridine: influence of ozone. *Journal of Photochemistry and Photobiology A: Chemistry* 2000;136:163-168.

[13] Kerc, A., Bekbolet, M., and Saatci, A. M., Sequential Oxidation of Humic Acids by Ozonation and Photocatalysis. *Ozone Sci. Eng.* 2003;25:497 - 504.

[14] Li, L., Zhu, W., Chen, L., Zhang, P., and Chen, Z., Photocatalytic ozonation of dibutyl phthalate over TiO2 film. *Journal of Photochemistry and Photobiology A: Chemistry* 2005;175:172-177.

[15] Wankat, P. C., *Rate-Controlled Separations.* New York Elsevier; 1990. p.

[16] Liu, J., Crittenden, J. C., Hand, D. W., and Perram, D. L., Regeneration of adsorbents using heterogeneous photocatalytic oxidation. *Journal of Environmental Engineering* 1996;122:707-713.

[17] Crittenden, J. C., Suri, R. P. S., Perram, D. L., and Hand, D. W., Decontamination of water using adsorption and photocatalysis. *Water Research* 1997;31:411-418.

[18] Khan, A. Y., Mazyck, D. W., and Wu, C. TiO_2 coated activated carbon: a regenerative technology for water recovery., in *32nd International Conference on Environmental Systems.* 2002. San Antonio, TX.

[19] Augugliaro, V., Litter, M., Palmisano, L., and Soria, J., The combination of heterogeneous photocatalysis with chemical and physical operations: A tool for improving the photoprocess performance. *Journal of Photochemistry and Photobiology* C: Photochemistry Reviews 2006;7:127-144.

[20] Chow, C., Fabris, R., Wilkinson, K., Fitzgerald, F., and Drikas, M., Characterising NOM to access treatability. *Water* 2006;33:74-85.

[21] Zou, L. and Zhu, B., The synergistic effect of ozonation and photocatalysis on color removal from reused water. *Journal of Photochemistry and Photobiology A: Chemistry* 2008;196:24-32.

[22] Hernandez-Alonso, M. D., Coronado, J. M., Javier Maira, A., Soria, J., Loddo, V., and Augugliaro, V., Ozone enhanced activity of aqueous titanium dioxide suspensions for photocatalytic oxidation of free cyanide ions. *Applied Catalysis B: Environmental* 2002;39:257-267.

[23] Hernández-Alonso, M. D., Coronado, J. M., Soria, J., Conesa, J. C., Loddo, V., Addamo, M., and Augugliaro, V., EPR and kinetic investigation of free cyanide oxidation by photocatalysis and ozonation. *Res. Chem. Interm.* 2007;33:205-224.

[24] Bulanin, K. M., Lavalley, J. C., and Tsyganenko, A. A., *Infrared Study of Ozone Adsorption on TiO2* (Anatase). 1995. p. 10294-10298.

[25] Okamoto, K., Yamamoto, Y., Tanaka, H., Tanaka, M., and Itaya, A., Heterogeneous photocatalytic decomposition of phenol over TiO_2 powder. *Bull. Chem. Soc. Jpn.* 1985;58:2015-2022.

[26] Tanaka, K., Abe, K., and Hisanaga, T., Photocatalytic water treatment on immobilized TiO2 combined with ozonation. *Journal of Photochemistry and Photobiology A: Chemistry* 1996;101:85-87.

[27] Zou, L. and Zhu, B. Colour in the recycled water: how to remove it?, in *Enviro 08, Australasia's Environmental and Sustainability Conference and Exhibtion.* 2008. Melbourne.

[28] Gogate, P. R. and Pandit, A. B., A review of imperative technologies for wastewater treatment II: hybrid methods. *Advances in Environmental Research* 2004;8:553-597.

[29] Pera-Titus, M., Garcia-Molina, V., Banos, M. A., Gimenez, J., and Esplugas, S., Degradation of chlorophenols by means of advanced oxidation processes: a general review. *Applied Catalysis B: Environmental* 2004;47:219-256.

[30] Kotronarou, A., Mills, G., and Hoffmann, M. R., Ultrasonic irradiation of p-nitrophenol in aqueous solution. *Journal of Physical Chemistry* 1991;95:3630-3638.

[31] Davydov, L. and Smirniotis, P. G., Quantification of the primary processes in aqueous heterogeneous photocatalysis using single-stage oxidation reactions. *Journal of Catalysis* 2000;191:105-115.

[32] Davydov, L., Reddy, E. P., France, P., and Smirniotis, P. G., Sonophotocatalytic destruction of organic contaminants in aqueous systems on TiO_2 powders. *Applied Catalysis B: Environmental* 2001;32:95-105.

[33] Théron, P., Pichat, P., Guillard, C., Pétrier, C., and Chopin, T., Degradation of phenyltrifluoromethylketone in water by separate or simultaneous use of TiO_2 photocatalysis and 30 or 515 kHz ultrasound. *Physical Chemistry Chemical Physics* 1999;1:4663-4668.

[34] Mrowetz, M., Pirola, C., and Selli, E., Degradation of organic water pollutants through sonophotocatalysis in the presence of TiO_2. *Ultrasonics Sonochemistry* 2003;10:247-254.

In: Sewage Treatment: Uses, Processes and Impact
Editor: Anna Stephens and Mark Fuller
ISBN 978-1-60692-959-9

Chapter 14

PHOTODEGRADATION OF OCTYLPHENOL USING SIMULATED AND NATURAL SUNLIGHT RADIATION

Mariana Neamțu[a*]***, Dana-Melania Popa***[b] ***and Fritz H. Fimmel***[c]
[a]Technical University of Iaşi, Faculty of Industrial Chemistry, Mangeron, Romania
[b]University of Agricultural Sciences and Veterinary Medicine,
Doctoral Studies Department, Biotechnologies Field, Cluj-Napoca, Romania
[c] Water Chemistry, Engler-Bunte-Institute,
University of Karlsruhe, Karlsruhe, Germany

ABSTRACT

The photolysis of octylphenol (NP) in pure water and surface water was investigated using a solar simulator and natural sunlight in the absence/presence of dissolved organic matter (DOM), HCO_3^-, NO_3^- and Fe(III) ions. The effects of different parameters such as irradiation source, initial pH, initial concentration of substrate, temperature, and the effect of hydrogen peroxide concentration on photodegradation of octylphenol in aqueous solution have been assessed.

The results indicate that the oxidation rate increases in the presence of H_2O_2, Fe(III) and DNOM. The real solar irradiation proves to be more efficient than solar UV simulated irradiation, but the best OP removal, both as percentage and time, is realised by the medium-pressure mercury lamp (UV-MP). In addition, the disappearance of the estrogenic activity of octylphenol during irradiation using YES test was investigated. Based upon the YES test results, there was a strong decrease of estrogenic activity of octylphenol after 8 h sunlight irradiation in the presence of hydrogen peroxide.

INTRODUCTION

Alkylphenol ethoxylates (APEs) belong to the group of non-ionic surfactants and are widely used in cleaning products and as industrial process aids. The spectrum of applications

* E-mail: mariana.neamtu@web.de

ranges from dispersing agents in paper and pulp production to emulsifying agents in latex paints and pesticides formulations, flotation agents, industrial cleaners (metal surfaces, textile processing, and food industry), cold cleaners for cars, and household cleaners [1].

The annual worldwide production of APEs is about 390 000 tons [2]. The typical structure of the APEs – an alkylphenol hydrophobe part and the substituted long chain of repeating ethylene oxide units as the hydrophilic moiety – is responsible for the surfactant activity. The ethoxylate chain can have 1 to 100 repeating units. The longer the chain, the more water-soluble (hydrophilic) is the compound. The alkyl group is a branched nonyl, octyl, or dodecyl chain. Alkylphenols were first found to be estrogenic (estrogen-mimicking) in the 1930s, but more recent research has highlighted the implications of these effects.

Metabolites of alkylphenol ethoxylates are generally more toxic than the original compounds, exhibiting estrogenic activity and they accumulate in organisms, with bioconcentration factors varying from ten to several thousand, depending on species, metabolite and organ [3]. Octylphenol ethoxylates are biodegraded by removal of ethoxy groups, producing less biodegradable products: octylphenol mono- and di- ethoxylates, octylphenoxy acetic and octylphenoxypolyethoxy acetic acids, and octylphenol. These metabolites frequently persist through sewage treatment and in rivers.

Octylphenol (OP) represents an important metabolite in the biodegradation of octylphenol ethoxylates (OPEs), which are released into the aquatic environment mainly via sewage treatment plant effluents. They are known to be taken up by organisms in the inland water system and to be stored in their fatty tissues. Due to its chemical structure does not easily biodegrade. It was suggested that their residues released in rivers can have toxic effects on aquatic life. Chronic exposure to those compounds can result in detrimental estrogenic effects on aquatic biota.

The most potent alkylphenol compound was found to be also OP [3], which was active in many assays at 0.1 μM, and was able to stimulate responses similar to those produced by estradiol at concentrations only 1000 times higher than those of estradiol. Concerning fish reproduction, OP was investigated in a full life cycle test with zebra fish (*D. rerio*). OP caused significant reduction in juvenile growth and prolongation of the time until first reproduction at the highest test concentration of 38 μg/l [2]. At this concentration, egg-laying capacity of the females and fertility of the males were significantly affected.

The growth of cultured human breast cancer cells is affected by octylphenol at concentrations as low as 0,1 μM (20μgL^{-1}) [4]. Estrogenic effects have also been shown on rainbow trout hepatocytes, chicken embryo fibroblasts and a mouse estrogen receptor. Studies on UK rivers receiving sewage effluent have shown that an estrogenic pollutant is present, as male rainbow trout placed in these rivers produce a female egg yolk protein. Similar observations have been reported in male roaches from selected rivers in Japan [5].

For that reason, OP was included in the 11 substances list, proposed as "priority substances under review", in the Decision of the European Parliament and of the Council (from 16.1.2001, Brussels), establishing the list of priority substances in the field of water policy [6]. Octylphenol is a solid that is soluble in water with 19 mg/L at 22°C, and its $\log K_{OW}$=4.12. This $\log K_{OW}$ value indicates a moderate potential for bioaccumulation in aquatic flora and fauna [7]. Available data indicate that octylphenol has a low solubility in water but will be adsorbed by organic matter in soil, sediments and mud. Both nonyl- and octylphenol are primarily found in sediments. Organisms that live in the sediments and organisms that live of these are vulnerable groups considering exposure to octylphenol [8].

The environmental levels of alkylphenolic compounds are probably high enough to be affecting the hormonal control systems of some organisms. It is also possible that human health could be being affected. Human exposure to APs can result by different routes, like inhalation, ingestion of contaminated foods and dermal absorption. The oral intake may be significant, e.g. via seafood, water supply, food contaminated through leaching from packaging or pesticide formulation applications [9].

Based on the environmental relevance of APEOs and their poor biodegradability, the main purpose of this work was to study the photochemical behaviour of OP in aqueous solution exposed to simulated UV solar and real solar radiation. Fundamental parameters related to OP transformation by UV have been determined. Additionally, the prediction of degradation rates in purified water and in natural waters using raw water characteristics was studied.

The photochemical degradation of OP was studied in lab-scale experiments, using real and simulated sunlight for irradiation. The effects of different reaction parameters such as initial pH, initial concentration of substrate, reaction temperature, and effect of hydrogen peroxide concentration on photodegradation of octylphenol in aqueous solution were assessed. The octylphenol photolysis was also investigated in the presence / absence of dissolved organic matter, HCO_3^-, NO_3^- and Fe(III) ions. The disappearance of the estrogenic activities of OP during irradiation using YES test were also performed.

2. Experimental Section

2.1. Reagents

Octylphenol was obtained from Aldrich, as a commercially available product (CAS no. 1806-26-4) and was used without further purification. The main characteristics of OP are presented in Table 1.

Hydrogen peroxide solution (30%) and all organic components (p.a.) were of analytical grade and obtained from Merck, Germany. Phenylglyoxylic acid was purchased from Fluka (Deisenhofen) in p.a. quality. All chemicals used for solutions (buffer, eluents, etc.) were reagent grade and were used without further purification.

NOM was taken from Lake Hohloh, a brown water lake in the south west of Germany (Hohlohsee, Black Forest). The high concentration of DOC was mainly of humic substances type. The sampling day was 15 July 2004 for Lake Hohloh water. River Rhine water was sampled on February, 20th, in 2005. Subsequently, the samples were filtered (0.45 mm, cellulose nitrate) and stored in the dark at 4 °C. The measured basic characteristic data of the non-diluted brown Lake Hohloh water and River Rhine water are given in Table 2.

Hydrogen peroxide solution (30%) and all organic components (p.a.) were of analytical grade and obtained from Merck, Germany. Phenylglyoxylic acid was purchased from Fluka (Deisenhofen) in p.a. quality. All chemicals used for solutions (buffer, eluents, etc.) were reagent grade and were used without further purification.

Table 1. Properties of octylphenol

Property/specification	Octylphenol
CAS registry number	1806-26-4
Synonyms	4-(n-octyl) phenol; p-octylphenol; 1-(p-hydroxyphenyl) octane
Molecular formula	$C_{14}H_{22}O$
Appearance	white crystalline powder
Molecular weight (g/mol)	206.33
Melting point (°C)	41
Boiling point (°C)	150 at 4 mm Hg
Specific gravity (g cm^{-1})	0,961
Solubility (mg L^{-1})	19 at 22°C
log K_{OW}	4.12
log P_{OW}	5.50
Flash point	> 110 °C
Organic carbon – water partition coefficient (K_{OC}), L/kg	2 740
Solid – water partition coefficient in suspended matter (Kp_{susp}) L/kg	274
Suspended matter – water partition coefficient ($K_{susp\text{-}water}$) m^3/m^3	69.4
Fish bioconcentration factor (BCF_{fish}) L/kg	634
Biomagnification factor in fish (BMF_1)	1

Table 2. Characteristic data of non-diluted brown water (Lake Hohloh) and Rhine surface water, filtered through 0.45 mm pore size cellulose nitrate membrane

	Lake Hohloh	River Rhine
pH	3.54	7.36
Electrical conductivity (μS/cm)	58	362
DOC (mg/L)	30.23	2.57
AOX (μg/L)	80.42	13.62
Chloride (mg/L)	0.9	30.4
Nitrate (mg/L)	0.2	7.8
Sulfate (mg/L)	1.6	21.7
Spectral absorption λ_{254nm} (m^{-1}) λ_{436nm} (m^{-1})	 123.7 m^{-1} 7.3	 67.2 m^{-1} 3.7 m^{-1}
ρ (Fe) (μg/L)	378.05	<2
ρ (Ca) (mg/L)	0.532	41.15
ρ (Mg) (mg/L)	0.166	6.906

NOM was taken from Lake Hohloh, a brown water lake in the south west of Germany (Hohlohsee, Black Forest). The high concentration of DOC was mainly of humic substances type. The sampling day was 15 July 2004 for Lake Hohloh water. River Rhine water was sampled on February, 20th, in 2005. Subsequently, the samples were filtered (0.45 mm, cellulose nitrate) and stored in the dark at 4 °C. The measured basic characteristic data of the non-diluted brown Lake Hohloh water and River Rhine water are given in Table 2.

2.2. Sample Preparation

Solutions containing a given octylphenol concentration ($2.55*10^{-5}$ mol/L) were prepared by adding an appropriate volume of stock solution in methanol ($12.75*10^{-3}$ mol/L of OP) to ultra pure water (original conductivity of 0.056 μS/cm) to get the final concentration. All was stirred for 15 minutes to ensure complete dissolution. After stirring, the sample was filtered through a 0.45 μm Cellulose Nitrate Filter. Canonica et al [10] found that the methanol content of aqueous solutions (0.5–2%) had no significant influence on the lifetimes of the triplet state of aromatic ketones. For experiments with model solutions containing NOM, filtered Hohloh Lake water and the stock solutions were mixed to get the same final concentration.

2.3. Solid-Phase Extraction

Solid-phase extraction (SPE) using Oasis HLB extraction cartridges, ($V = 6\ cm^3$, sorbent mass m = 200 mg, Waters, U.S.A.), was performed prior to HPLC and YES analyses. Conditioning of the cartridge was done with 5 mL methanol followed by equilibration with 4 mL water. Sample volumes of 200 mL were sucked through the cartridges through Teflon tubes at a flow rate of 3 to 4 mL/min.

Loaded cartridges were washed with 2 mL of a methanol : H_2O mixture (5 : 95 v/v) and dried by application of vacuum. Elution was performed with 2 mL dichloromethane (CH_2Cl_2). The eluent was dried in a stream of N_2 at 40°C. After drying, the residue was dissolved in 1 mL ethanol to give a concentration factor of 200. The procedure is described in detail elsewhere [11].

2.4. Irradiation Experiments

The samples were irradiated using a solar UV simulator (Oriel Corp., Stratfort, CT) with additional filters of the type WG 295, 6 mm (Schott Glaswerke, Mainz, Germany) and an atmospheric attenuation filter (Oriel Corp., Stratfort, CT) installed in the radiation beam to match the spectrum of the Xenon lamp to the typical solar UV-spectrum. The scheme of the solar simulator used for irradiation experiments is presented in Figure 1.

The radiation source was a 1000-W Xe short-arc lamp. The spectral irradiance of the simulated solar radiation with the WG 295 filter and the atmospheric attenuation filter in the radiation beam was determined by polychromatic actinometry. The spectrum of the solar UV simulator contained relatively small portions of visible light compared to real sunlight, while the cut-of wavelength in the UV-range was situated nearly at the same wavelength as in real sunlight (Figure 2).

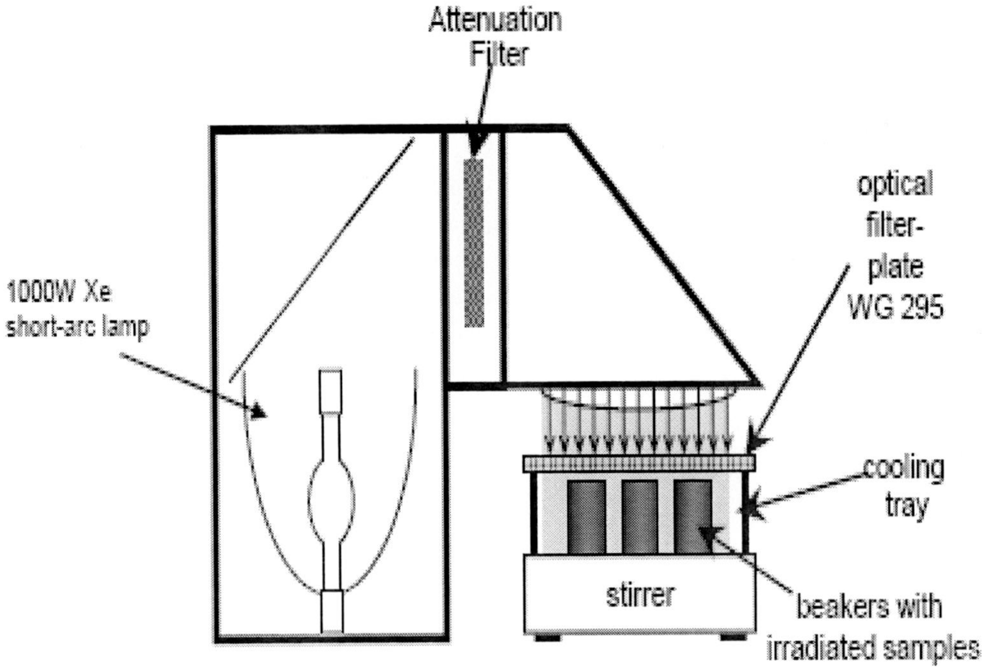

Figure 1. The scheme of the solar UV simulator used for irradiation experiments.

The intensities in the UV range of the Solar UV Simulator are 1.3 times higher than those of the real sunlight. The incident photon flux was determined by chemical polychromatic actinometry using phenylglyoxylic acid in AcN : MilliQ water = 3:1 (v/v). The solution functions for the incident radiation between 290 and 400 nm as an actinometer by the photochemical decarboxylation of phenylglyoxylic acid to benzaldehyde and carbon dioxide [12]. The photon flux in the UV range (290≤k≤400 nm) of the solar UV simulator measured by polychromatic actinometry with phenylglyoxylic acid was $3,97*10^{-5}$ Einstein/m^2s in the wavelength range 300 – 400 nm. The wavelength dependence of the quantum yield / k for the chemical actinometer (phenylglyoxylic acid) in AcN:MilliQ water = 3:1 (v/v) is given in Defoin et al. [12].

The samples were also irradiated on the building's roof, for catching real solar radiation. The experiments were carried out on typical July summer days (t = 31°C), under clear sunny conditions.

The total reaction volume was 500 mL in all experiments performed using solar UV simulator, as well as in the ones using real solar radiation.

Other irradiation sources used were a low-pressure mercury lamp (UV-LP), of type TNN 15/32, and a medium-pressure mercury lamp (UV-MP), of type TQ 150, respectively. The lamps were provided by *Heraeus Noblelight*, Kleinostheim, Germany. In both cases the reaction volume was 700 mL.

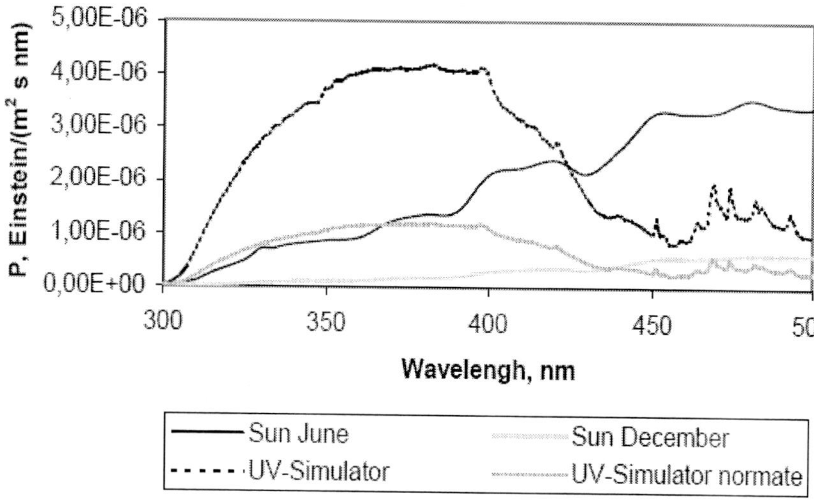

Figure 2. The spectral irradiance of the sun and UV-solar simulator [14].

2.5. Sample Analysis

The analysis of octylphenol and degradation products was performed by reversed-phase high-performance liquid chromatography (RP-HPLC) using HPLC system HP 1100 with Chemstation software coupled to a HP 1100 fluorescence detector (Agilent Technologies, Waldbronn, Germany). Separations were carried out with a Zorbax RX-C18 column (dimensions 125 mm × 3 mm, 5 μm; Rockland Technologies Inc., USA). The injection volume was 20 μL. The mobile phase was a mixture of MilliQ-water (A) and HPLC-grade AcN (B). At a flow rate of 1 mL/min, a gradient elution was performed starting with 70% B which went up linearly to 90% B in 15 min and to 100% B in another 5 min (post time: 5 min). Detection wavelengths were 227 nm (excitation) and 305 nm (emission) at a photomultiplier (pmt) gain of 13. For quantification, 5 standards for external calibration were used. The 4-OP standards concentrations were: 12.5μM; 25μM; 50μM; 75μM; 100μM. The stock standard solutions were made in 100% AcN and the dilutions there from, to get the standards, were made in AcN:H_2O mixture (50:50 v/v). In order to assess the precision of the HPLC method, the lowest and the highest mix standards were analysed thrice each ($n = 3$) during each sample analysis cycle, and the average values for the two standards were used for the calibration curve. More details are presented elsewhere [11].

UV/Visible absorption spectra of brown water from Lake Hohloh and River Rhine were recorded using a Varian Cary 50 Conc spectrophotometer, equipped with a quartz cell having a path length of 1 cm. The spectral absorbance was measured including baseline correction with a scan rate of 600 nm/min and a data point interval of 1 nm.

2.6. Yeast Estrogen Screen (YES) Bioassay

The YES test of Routledge and Sumpter [13] was used to measure estrogenic activity of irradiated water samples. Standard stock solutions were made in ethanol, and all serial dilutions were carried out in ethanol. The procedure for measurement of the total estrogenic activity was carried out by serially diluting each (100μL) sample concentrate in ethanol across 12 wells in a 96-well plate (NUNCLON^{TM}). Each dilution series was initiated by placing 200μL of the sample concentrate in the first well of a single row. 100μL was transferred to the second column and mixed with 100μL of ethanol. The process was repeated across the row. Aliquots of 5μL of each concentration were transferred to a 96-well plate. Yeast cells were grown in the Routledge/Sumpter medium to OD_{620}=1.0 cm^{-1}. Aliquots (195μL) of the seeded assay medium (medium containing recombinant yeast) and the chromogenic substrate CPRG were then dispensed to each sample. Each plate contained at least one row of blanks (assay medium only), as well as a standard curve for 17β-estradiol. The plates were sealed with autoclaved tape and incubated at 32°C for 3 days, after which absorbance readings were taken at 540 nm (using a Victor^3 Multilabel reader). Relative potencies of test samples were determined only when dose-response curves were parallel to that of 17β-estradiol. For YES analysis the irradiated samples were concentrated to give a concentration factor of 200.

3. Results and Discussions

3.1. Photodegradation of Octylphenol Using a Solar UV Simulator

3.1.1. The Influence of pH on Photodegradation of Octylphenol

In analogy with some other phenols it can be presumed that deprotonated OP reacts about an order of magnitude faster than the undissociated compound. The fraction of the dissociated species depends on the dissociation constant and on pH of the aqueous solution [14]. Therefore the influence of pH on dissolved octylphenol removal using a UV solar simulator was studied at pH 6.58 and 8, determined initially and without any modifications or control of the pH during the process. The initial pH of a 2.55 10^{-5} mol/L octylphenol solution was pH 6.58. The rate of OP decay increases as the initial pH of reaction solution rises from 6.58 to 8, as shown in Figure 3.

The conversions of about 16.74% and 14.26% were achieved after 8h from the start of the reaction when the solutions pH were 8 and 6.58, respectively. Octylphenol removal at the highest rate constant occurs in alkaline medium. This is probably due to the fact that in alkaline medium the molar absorption coefficients of the molecule are slightly higher than at pH 6.58. Similar results were reported for degradation of 2-chlorophenol and nonylphenol [14, 15].

It is known that the rate of hydroxyl radicals' formation is pH dependent and increases when more alkaline conditions are used [15]. In strongly alkaline solution OH• is rapidly converted to its conjugate base $\bullet O^-$:

$\bullet OH + OH^- \rightarrow \bullet O^- + H_2O$ $\quad$ $k = 1.2\ 10^{10}$ L/mol s and pK_a (•OH) = 11.9.

$\bullet O^- + H_2O \rightarrow OH^- + \bullet OH$ $\quad$ $k = 1.8\ 10^{6}$ L/mol s [16]

In its reaction with organic molecules OH• behaves as an electrophile whereas $•O^-$ is a nucleophile. Thus •OH readily adds to unsaturated bonds but $•O^-$ does not. However, both forms of the radical abstract H from C-H bonds.

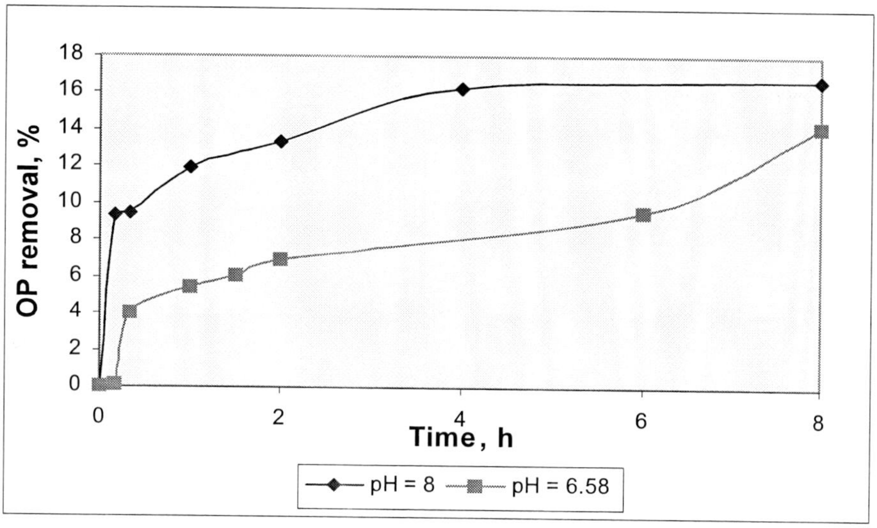

Figure 3. Effect of pH upon photodegradation of octylphenol. Initial conditions: 2.55*10-5 M of octylphenol, 15°C, under solar UV simulated irradiation.

3.1.2. The Influence of the Reaction Temperature on Photodegradation of Octylphenol

The influence of reaction temperature on the OP removal under simulated solar irradiation was studied at 15 and 25°C (Figure 4).

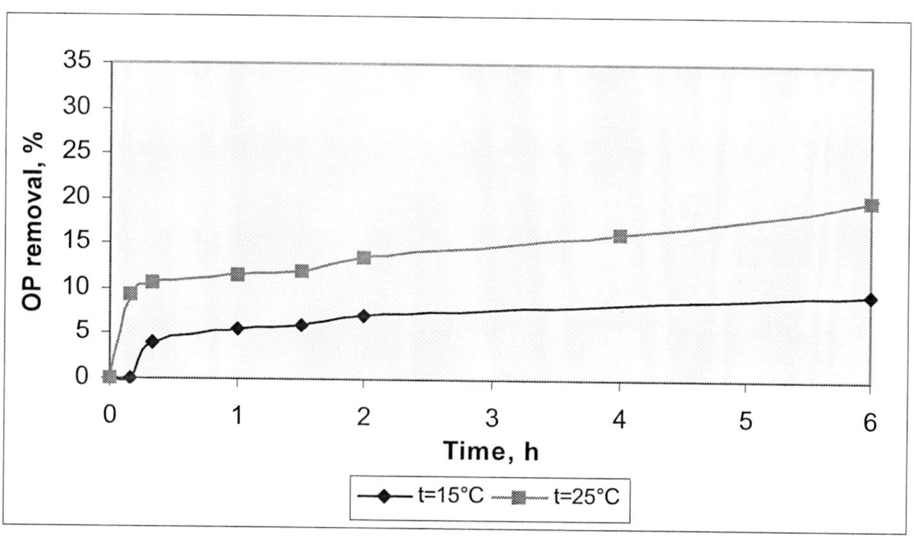

Figure 4. Effect of temperature upon photodegradation of octylphenol. Initial conditions: 2.55*10-5 M of OP, pH 6.58, under solar UV simulated irradiation.

After 6h of reaction at 15°C the removal was 9.5%. As expected, by increasing the temperature to 25°C, a large increase in OP conversion was observed. After the same reaction time at 25°C, 20% of octylphenol removal was obtained. To simulate better the natural conditions, for further investigation of the process a temperature of 15°C was selected.

3.1.3. The Influence of the Initial Concentration on Photodegradation of Octylphenol

The effect of initial concentration was studied at 25.5μM and 15μM of octylphenol (Figure 5).

After 6h reaction time, the removal was 9.5% for 25.5μM and 12.67% for 15μM of OP. The corresponding first-order rate constants were 0.0174 h^{-1} and 0.020 h^{-1}, respectively.

It was expected that the degradation rate constants from the first-order kinetics were higher for lower initial concentrations of OP than for higher ones. A possible explanation could be the complete absorption of the incident photon flux by higher initial OP concentrations over shorter path length and the lower transmission of the samples at higher concentration of octylphenol (inner filter effect). So the absorbed photons flux of the lower and higher initial compound concentrations were probably not proportional to the initial concentration of OP. Similar results were reported for photochemical degradation of other pollutants [16].

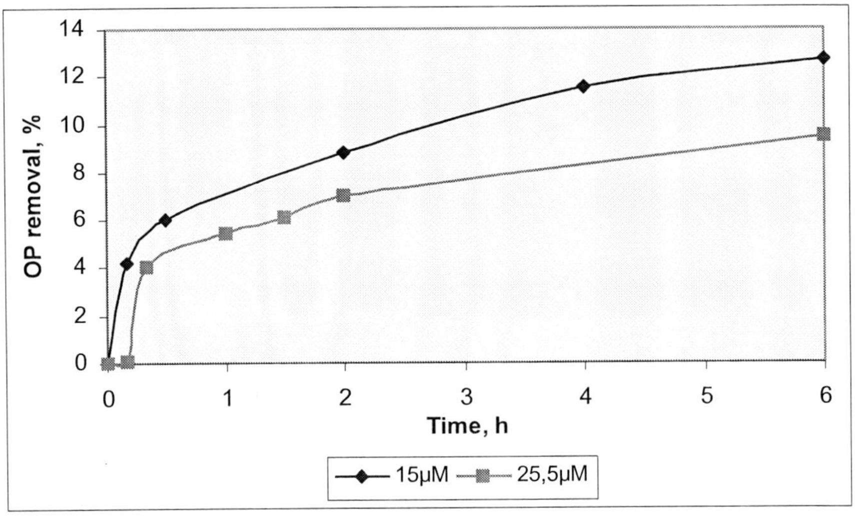

Figure 5. Effect of initial concentration upon photodegradation of octylphenol. Initial conditions: pH 6.58, 15°C, under solar UV simulated irradiation.

3.1.4. The Influence of Hydrogen Peroxide Concentration on Oxidation of Octylphenol

The kinetic studies of catalytic oxidation with different hydrogen peroxide concentrations were carried out at pH=6.58, with an initial concentration of 25.5μM of octylphenol and at t=15°C. The results are presented in Figure 6.

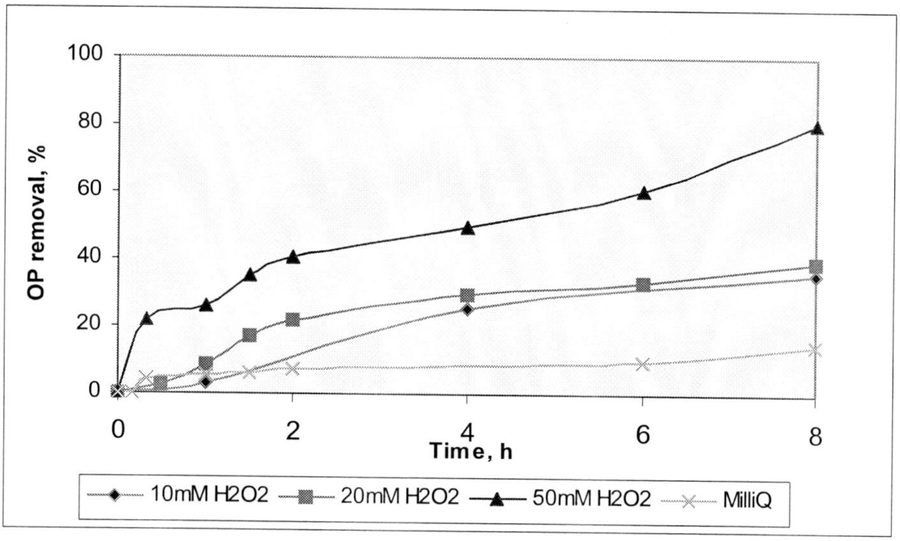

Figure 6. Effect of hydrogen peroxide concentration upon photodegradation of octylphenol. Initial conditions: 2.55*10-5 M of OP, pH 6.58, 15°C, under solar UV simulated irradiation.

After 4h of reaction time at a dose of 10 mmol/L H_2O_2, 25.16% of OP removal was achieved. Addition of 20 mmol/L H_2O_2 did not much improve the degradation, the removal of OP was 29.53%. With a concentration of 50 mmol/L H_2O_2, almost 50% of octylphenol was removed. As expected, the increase in hydrogen peroxide concentration accelerated the octylphenol degradation at the beginning of the reaction. This can be explained by the effect of the OH• radicals produced (photolysis of H_2O_2) [17], leading to rapid disappearance of OP. The major kinetic law of OP degradation can be expressed by equation (6):

$$-\frac{d[OP]}{dt} = (k_d + k_{OP,OH}[HO\bullet]ss)[OP] \tag{6}$$

where, [OP] represents the OP concentration and [HO•]ss denotes the hydroxyl radical concentration (mol L^{-1}); k_d and $k_{OP,OH}$ represent the direct photolysis rate constant and the rate constant for the reaction of OP with the •OH radicals, respectively. At the very low concentrations of OP as used in this study, a pseudo-first order kinetics was observed (7):

$$-\frac{d[OP]}{dt} = k_1[OP] \tag{7}$$

The observed pseudo-first order rate constants (k_1, s^{-1}) were calculated from the experimental data by using linear regression $\ln(C_0/C)$ versus time plots with all regression coefficients higher than 0.9. The calculated results are presented in Tables 3 and 4.

Table 3. Kinetic parameters for photolytical degradation of octylphenol using a UV solar simulator. Initial conditions: $2.55*10^{-5}$ M OP, pH 6.58, 15°C

Experiment	UV-Solar Simulator	
	k, h^{-1}	$t_{1/2}$, h^{-1}
in absence of common water constituents	0.017	39.83
$1.55*10^{-5}$ M OP	0.020	34.65
pH 8	0.024	28.52
25°C	0.036	19.36
10 mM H_2O_2	0.043	16.15
20 mM H_2O_2	0.053	13.00
50 mM H_2O_2	0.128	5.41

Table 4. Kinetic parameters for photolytical degradation of octylphenol using a UV solar simulator and real solar irradiation. Initial conditions: $2.55*10^{-5}$ M OP, pH 6.58

Experiment	UV-Solar Simulator		Solar radiation	
	k, h^{-1}	$t_{1/2}$, h^{-1}	k, h^{-1}	$t_{1/2}$, h^{-1}
50 mM H_2O_2	0.128	5.41	0.272	2.55
NO_3^-	0.075	9.30	0.130	5.31
HCO_3^-	0.021	33.81	0.108	6.42
Fe(III)	0.018	38.94	0.124	5.57
Rhine River	0.018	39.83	0.104	6.64
in absence of common water constituents	0.017	39.84	0.134	5.16
Hohloh Lake	-	-	0.109	6.35

3.1.5. Effect of Common Water Constituents

Figure 7 shows the effect of HCO_3^-, NO_3^- and Fe(III) ions on photodegradation of octylphenol, under solar UV simulated irradiation.

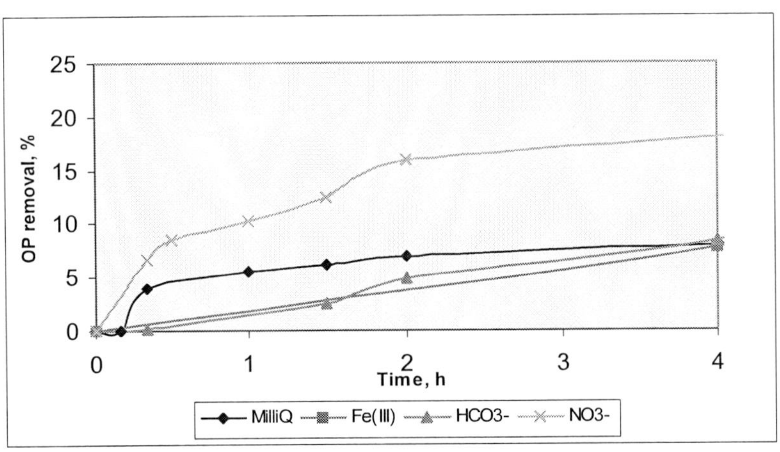

Figure 7. Effect of HCO_3^-, NO_3^- and Fe(III) ions upon photodegradation of octylphenol. Initial conditions: 25.5μM of OP, 15°C, 725 mg HCO_3^-/L, 61 mg NO_3^-/L, 100μg Fe(III)/L; under solar UV simulated irradiation.

HCO_3^-. The removal of OP was slower in the presence of hydrocarbonate ions. After 4h of reaction time the removal was only 8.4%. It is known that carbonate as well as bicarbonate ions act as •OH scavenging species (Eqs (10) and (11)).

$$HO\bullet + HCO_3^- \rightarrow HO^- + HCO_3^{\bullet -} k=8.5*10^6 \text{ L mol}^{-1} \text{ s}^{-1} \tag{10}$$

$$HO\bullet + CO_3^{2-} \rightarrow HO^- + CO_3^{\bullet -} k=3.9*10^8 \text{ L mol}^{-1} \text{ s}^{-1} \tag{11}$$

The oxidation potential of the generated carbonate radical anion has been shown to be less positive than that of the •OH radical, introducing, hence, selectivity as far as the compounds to be degraded are concerned [18, 19]. Consequently, the omnipresent bicarbonate or carbonate ions can compete successfully for •OH radicals, especially at low loads of organic matter.

NO_3^-. The best degradation rate of octylphenol was obtained in the presence of 61 mg NO_3^-/L. After 4h of reaction time, about 18% of OP removal was achieved. Nitrate absorbs light in the UV range and acts as an inner filter for the UV light [20, 21]. Nitrate is also able to form •OH radicals, therefore it is expected that the degradation rate of photolysis will increase. Figure 8 shows the results for the addition of 61 mg NO_3^-/L on the degradation rate of OP ($k = 0.0745 \text{ h}^{-1}$). Similar results were also reported by other authors [20, 21, 22]. The increase in photolytic degradation rate can be explained via reaction of the organic micropollutants with •OH radicals generated during the photolysis of nitrate, as shown bellow [23] (Eqs (12) to (16)):

$$NO_3^- \xrightarrow{h\nu} NO_2^- + O \tag{12}$$

$$NO_3^- \xrightarrow{h\nu} O\bullet^- + NO_2\bullet \tag{13}$$

$$2NO_2\bullet + H_2O \rightarrow NO_2^- + NO_3^- + 2H^+ \tag{14}$$

$$O + H_2O \rightarrow 2HO\bullet \tag{15}$$

$$O\bullet^- + H_2O \rightarrow HO\bullet + HO^- \tag{16}$$

Fe(III). The removal of OP obtained in the presence of 100 μg Fe(III)/L was 7.68%, after 4h of reaction time. The hydrolysis of Fe(III) salts in aqueous solutions is a complicated and pH dependent process. Fe(III)-aquo complexes are known to undergo photolysis through an internal electron transfer upon irradiation with wavelengths λ>300 nm, yielding Fe(II) and hydroxyl radicals (eq. 9).

$$Fe(OH)^{2+} \xrightarrow{h\nu} Fe^{2+} + HO\bullet \tag{17}$$

At pH ≤ 5 at least four different Fe(III) ions, which well characterized, coexist in aqueous solution: Fe^{3+}, $Fe(OH)^{2+}$, $Fe(OH)_2^+$ and the dimer $Fe_2(OH)_2^{4+}$. $Fe(OH)^{2+}$ is the dominant species [14]. The rate of degradation depends on the age of the ferric solution and the concentration of $Fe(OH)^{2+}$, the most photoreactive species in terms of •OH radical formation.

3.1.6. Effect of Dissolved Natural Mater (DNOM)

Typically aromatic and coloured DNOMs are fairly photoactive [24, 25], and as expected, highly coloured DOM possesses the highest reactivity toward OP. Non-diluted water from Lake Hohloh was applied, derived from a "black water" environment that has high aromatic contents and high molar extinction coefficients at 254 nm. The other source of DOM was the River Rhine water that has lower aromatic contents and lower molar extinction coefficients at 254 nm.

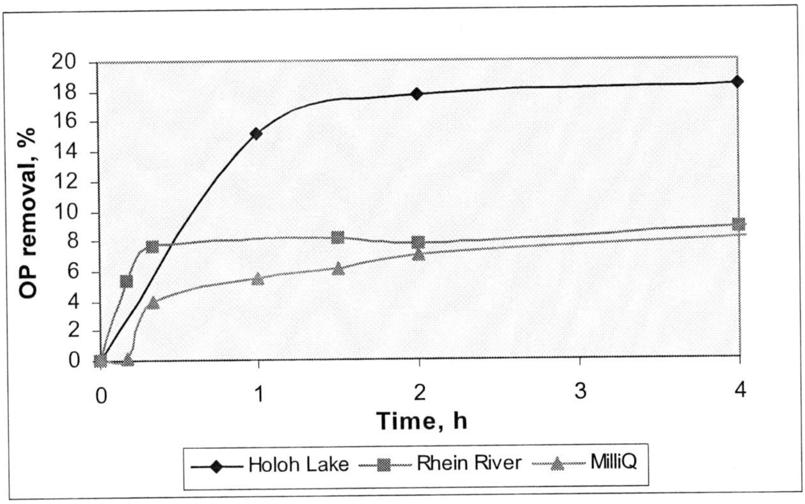

Figure 8. Effect of DNOM upon photodegradation of octylphenol. Initial conditions: $2.55*10^{-5}$ M of OP, 15°C, under solar UV simulated irradiation.

The degradation rate of OP increases by irradiation of the Lake Hohloh sample, of high DOC content (Figure 8). When the experiments were conducted in the river Rhine water, the degradation of OP was faster than in the experiments with DOM-free water (MilliQ water).

The degradation of OP in the two natural waters examined is expected to be primarily due to oxidation reactions triggered by the reactive species such as singlet oxygen and hydroxyl radicals generated through photo-processes involving excited states of DOM. The light absorption by DNOM leads to the formation of photooxidants such us singlet oxygen ($^1O_2$) [26], superoxide anion ($\bullet O_2^-$) hydroxyl radicals ($\bullet OH$), solvated electrons, excited triplet states, peroxyl radicals (ROO•) [24, 27] which are able to react with OP. It is also possible to assume a direct reaction of excited DNOM with octylphenol by electron transfer or energy transfer.

2.2. Photodegradation of Octylphenol Under Solar Irradiation

2.2.1. Effect of Common Water Constituents

One of the objectives of this study was to simulate the degradation process of OP in real surface waters and under natural conditions. For that reason, aqueous solutions of OP were exposed to real sunlight, on open air, on top of the building. Natural conditions included continuous irradiation at noon, summer sun, with an average air temperature of 31°C.

The effects of common water constituents (HCO_3^-, NO_3^-, Fe(III) ions and dissolved natural mater (DNOM)) on OP degradation are shown in figures 9, and 10.

As presented in Figure 9, OP removal has approximately the same tendency in all cases. The slowest degradation occurred in the presence of HCO_3^- (78% OP removal), while in the other reactions ca. 80% of OP was removed.

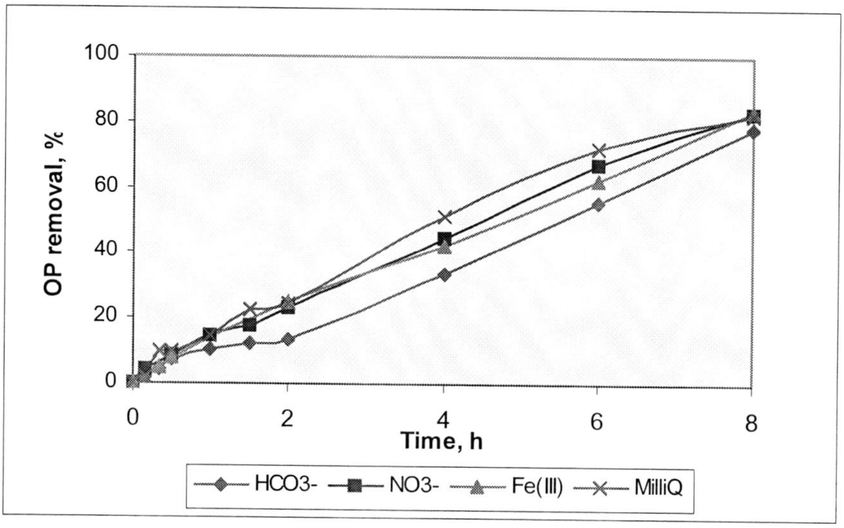

Figure 9. Effect of HCO3$^-$, NO3$^-$ and Fe(III) ions upon photodegradation of octylphenol. Initial conditions: 25.5μM of OP, natural sunlight, 725 mg HCO3$^-$/L, 61 mg NO3$^-$/L, 100μg Fe(III)/L.

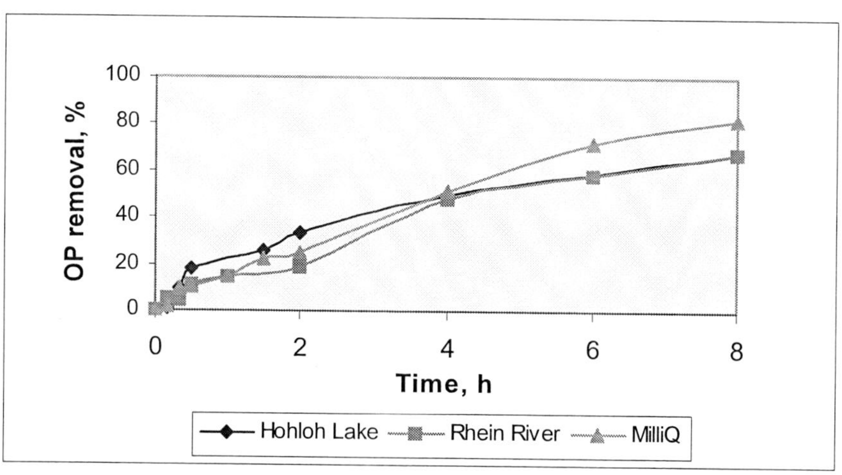

Figure 10. Effect of DNOM upon photodegradation of octylphenol. Initial conditions: 2.55*10-5 M of OP, solar light.

The octylphenol was degraded faster in MilliQ water than in solutions containing DNOM (Figure 10), which confirms that DNOM is the primary light-absorbing species in aquatic systems, the optical properties of natural waters being affected by the breakdown of this matter. It is also known that highly coloured waters actually tend to produce less singlet oxygen ($^1O_2$), a photooxidant, for a given amount of absorbed light.

Comparing the reaction rates by real and simulated solar irradiation, the results show that real solar radiation is more efficient in removing OP than the solar UV simulator (Figures 11 and 12).

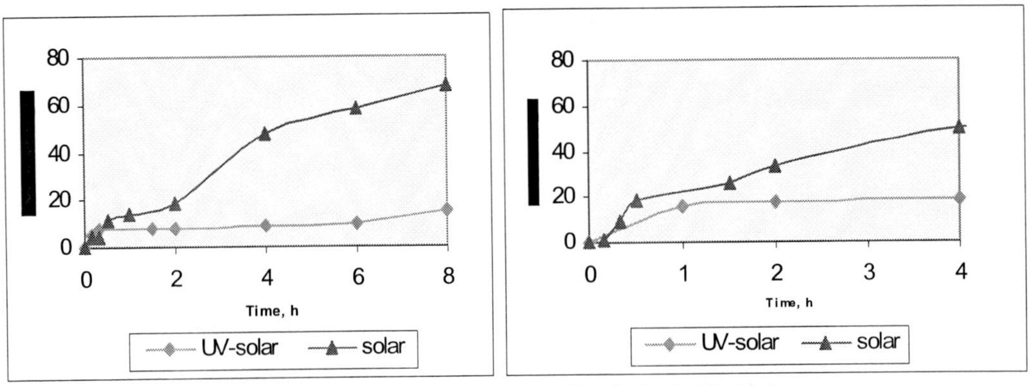

(a) – in River Rhine water; (b)– in Lake Hohloh water.

Figure 11. Effect of irradiation source upon photodegradation of octylphenol. Initial conditions: 2.55*10-5 M of OP.

As shown in figures 11 and 12, in all cases, the reaction rates increased by irradiation with real sunlight, compared to the solar simulator. Figure 12(e) presents similar results obtained by adding 50mM H_2O_2 to the initial solution. The comparison with the results of the other experiments (Figure 12(a) to 12(d)) allows the conclusion that the oxidation effect of hydrogen peroxide (in an appropriate amount) is stronger than the photolysis effect of the irradiation source.

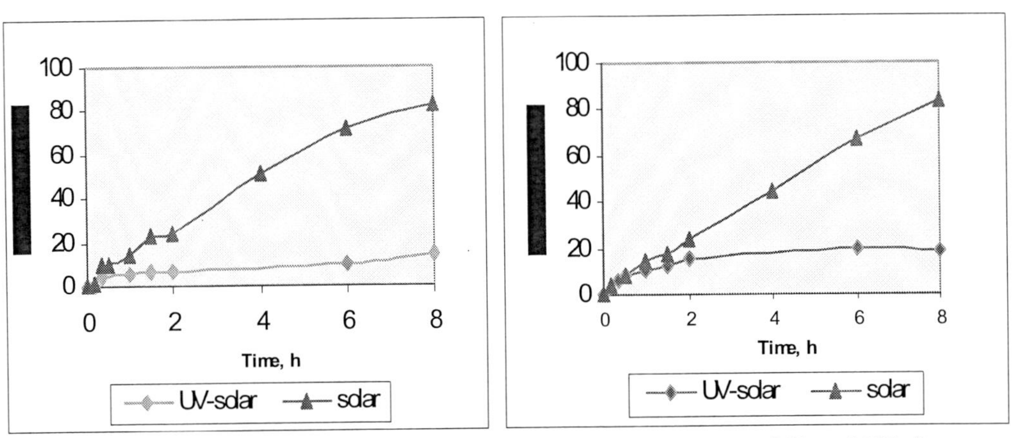

(a) in absence of common water constituents; (b) in the presence of 61 mg NO3-/L;

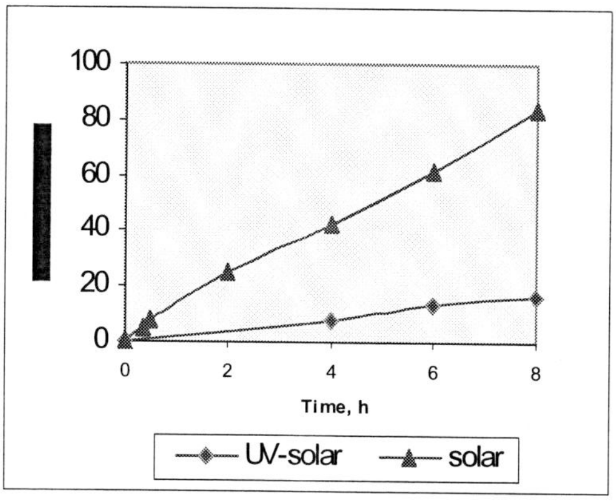

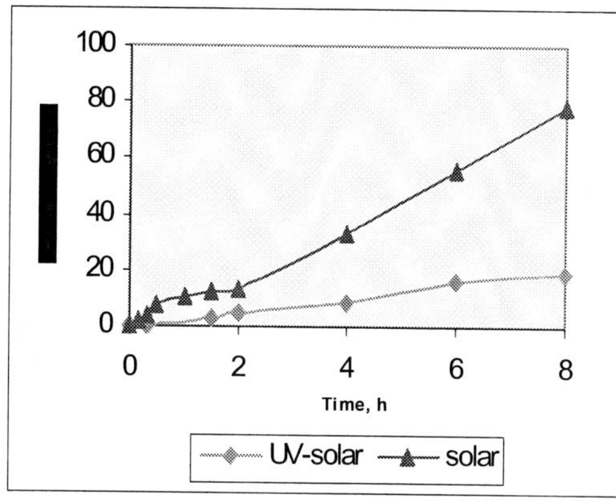

(c) – in the presence of 100 μg Fe(III)/L; (d) – in the presence of 750 mg HCO_3^-/L;

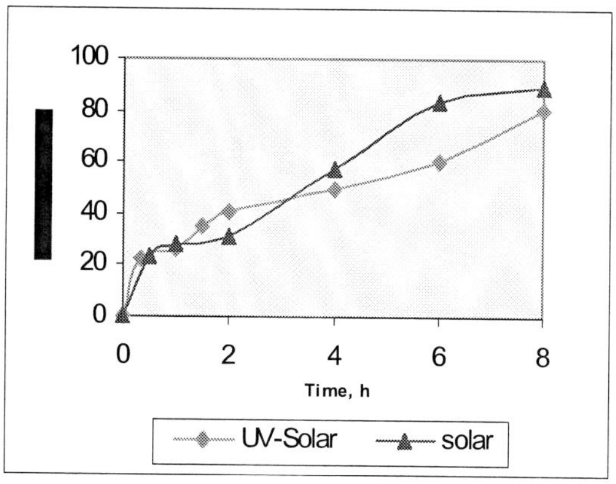

(e) – in the presence of 50 mM H_2O_2 .

Figure 12. Effect of irradiation source upon photodegradation of octylphenol. Initial conditions: 2.55*10-5 M of OP, pH=6.58.

3.3. Effect of the Irradiation Source on Photodegradation of Octylphenol

As presented in figure 13, the characteristics of the irradiation source plays a significant role in OP photodegradation, meaning that, after 2h of irradiation, the slowest response is given by UV-solar simulator (6.98% OP removal). Meanwhile 24.43% of OP degrades by natural solar irradiation. The low-pressure mercury lamp (UV-LP) is much more efficient, removing 80.79% of OP in 2h. The fastest reaction, of 94.59% OP removal after only 1h, was registered by medium-pressure mercury lamp (UV-MP).

UV-LP provides almost monochromatic UV radiation at λ of 253.7 nm, while UV-MP supplies polychromatic emission ranging from UV over VIS to IR region of the electromagnetic spectrum. Due to the difference in the wavelength range of the two lamps, their effect upon OP was to be expected.

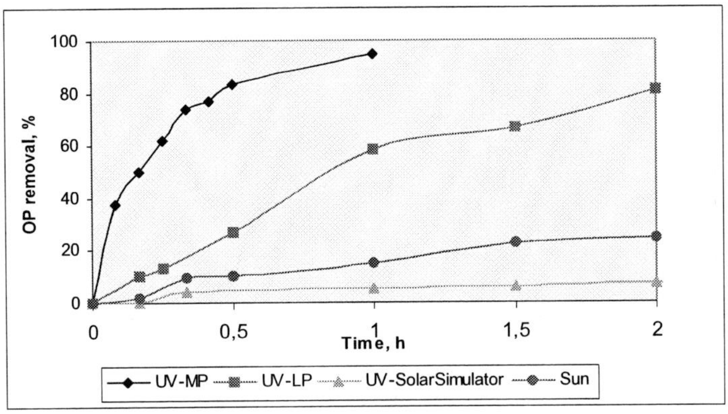

Figure 13. Effect of irradiation source upon photodegradation of octylphenol. Initial conditions: 2.55*10-5 M of OP.

3.4. Products of Octylphenol Photodegradation

The intermediates generated during the degradation process were analysed by HPLC, according to the retention times and the UV spectra based on comparison with standards. HPLC analysis showed the disappearance of OP along with formation of photoproducts (Figure 14).

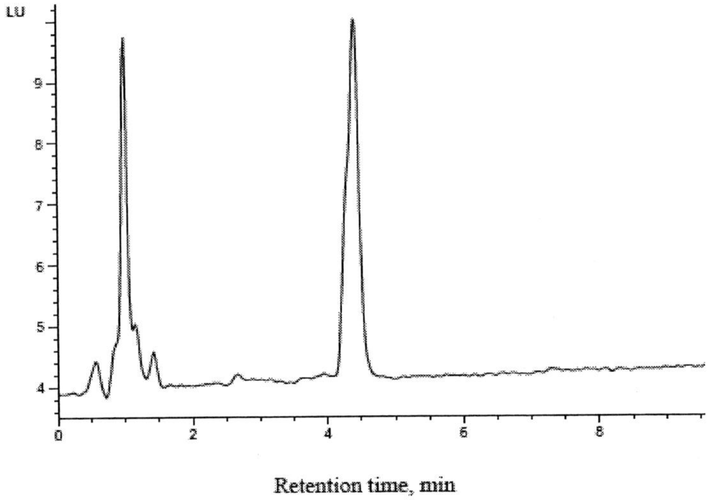

Figure 14. HPLC chromatogram of octylphenol and its intermediates. Sample after 4h irradiation using LP Lamp.

Literature reports that the first step of degradation is the decomposition of OP to phenol and C1-C3 acid fragments, formed by oxidation with hydroxyl radicals generated in the H_2O_2 photolysis reaction. Although available analytical information does not allow the prediction of the •OH radical attack site, considerations concerning the previous, indicate that the primary step involves the hydrogen abstraction from the aliphatic chain. The shortening of the aliphatic chain during the degradation process leads to various phenolic products.

3.5. Disappearance of Estrogenic Activities

Degradation of OP and disappearance of its estrogenic activities are very important in water treatment. The YES test is an assay system in which the human estrogen receptor (hER) is expressed in yeast in a form capable of activating transcription of a promoter carrying ERE, in an estrogen-dependent manner [28].

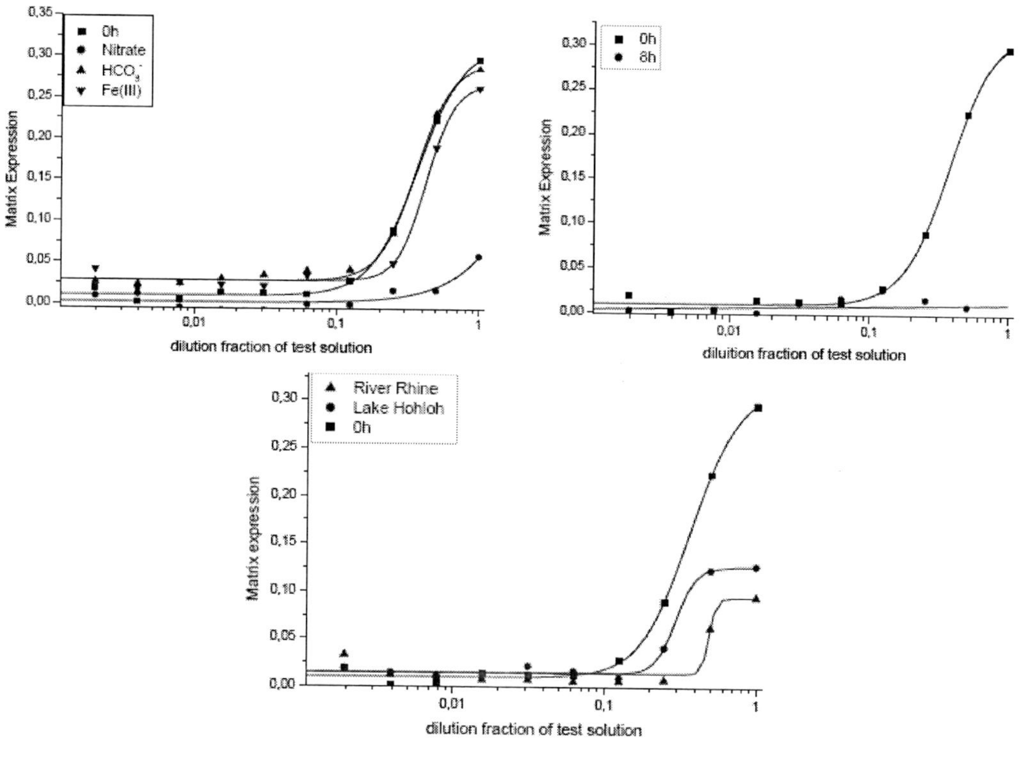

Figure 15. Yeast screening of estrogen activity for experiments: (a) – under solar irradiation in the initial presence of 61 mg NO3-/L, 100μg Fe (III)/L, 750 mg HCO3-/L (b) – under solar simulated irradiation in the presence of 50 mM H2O2, (c) – under solar radiation, in the presence of Rhine River and Hohloh Lake water. Irradiation time 8h.

Activation of the receptor, by binding of ligand, causes expression of the reporter gene *lacZ* resulting in the production of enzyme β-galactosidase. The enzyme is secreted into the medium, where it is detected indirectly after metabolization of the substrate chlorophenol red β-D-galactopyranoside (CPRG). The yellow CPRG is converted to a red product which is detected by measuring absorbance at 540 nm. Cell vitality and growth are monitored by measuring absorbance at 640 nm [28, 29]. The results show that after 8h of irradiation the estrogenic activity of OP remained approximately constant for experiments conducted using the UV solar simulator in the presence of carbonate ions, whereas the estrogenic activity decreased for River Rhine and Hohloh Lake water samples after sunlight irradiation and it practically disappeared when in the initial solution 50 mM H_2O_2 were added (figure 15).

Using E-screen, OP exhibits the highest activity with a relative potency compared to 17-β-estradiol of 10^{-4} followed by *p*-nonylphenol, butyl- and pentylphenols with relative potencies of 10^{-5} and 10^{-4}, respectively [2].

The presented results confirm that the yeast screen is highly sensitive. There are no problems from variable background. The reproducible results are rapidly obtained and facilitate the application as a routine assay for screening the estrogenic activity of chemicals.

Conclusion

This chapter has shown that the degradation of octylphenol in aqueous solution by simulated and solar irradiation is fairly slow.

It was clearly shown that pH, initial concentration of OP, reaction temperature, concentration of hydrogen peroxide, common water constituents (HCO_3^-, NO_3^- , Fe(III) ions and dissolved natural matter), as well as the type of irradiation source affect the phototransformation processes. This underlines the importance of common water constituents. The effect of changes in pH on photooxidation also should be investigated more intensively since this is of environmental importance due to the acidification of lakes and rivers.

The results indicate that the oxidation rate increases in the presence of hydrogen peroxide and nitrate ions. Also the surface waters (River Rhine and Lake Hohloh) improve OP removal, compared to ultra pure water. The real solar irradiation proves to be more efficient than solar UV simulated irradiation, but the best OP removal, both as percentage and time, is realised by the medium-pressure mercury lamp (UV-MP).

The data presented in this study confirm again that OP is a persistent intermediate in NPEs degradation. OP does not degrade adequately in sewage treatment plants, and does persist to a far extend in the environment. OP breakdowns form toxic intermediates that may have estrogenic effects at current environmental concentrations. That is enough reason for drinking water to be analysed for all alkyphenolic compounds, particularly in case raw water is extracted from a river. Advanced treatment should be introduced if the total alkylphenolic compound content is above the safety level.

References

[1] Thiele, B.; Günther, K.; Schwuger, M. J. (1997) Alkylphenol Ethoxylates: Trace Analysis and Environmental Behavior, *Chem. Rev.*, 97, 3247-3272.

[2] Lintelmann, J.; Katayama, A.; Kurihara, N.; Shore, L.; Wenzel, A. (2003) Endocrine Disruptors in the environment (IUPAC Technical Report), *Pure Appl. Chem.*, Vol. 75, No. 5, 631-681.

[3] Warhurst, A. M. – *An Environmental Assessment of Alkylphenol Ethoxylates and Alkylphenols*, January 1995.

[4] OSPAR Commission 2003 (2004 Update) – *Hazardous Substances Series*, Octylphenol.

[5] Furuichi, T.; Kannan, K.; Giesy, J. P.; Masunaga, S. (2004) Contribution to known endocrine disrupting substances to the estrogenic activity in Tama River water samples from Japan using instrumental analysis and in vitro reporter gene assay, *Water Research* 38 4491-4501.

[6] Amended proposal for a Decision *Of The European Parliament And Of The Council establishing the list of priority substances in the field of water policy* (presented by the Commission pursuant to Article 250 (2) of the EC Treaty); Brussels, 16.1.2001, COM (2001) 17 final; 2000/0035 (COD).

[7] Aaboe, R.; Haoya, A. O.; Edeskar, T. – Leaching of phenols from tyre shreds in a noise barrier, Norvegian Roads Recycled Materials R&D Program.

[8] Servos, M. R. (1999) Review of the Aquatic Toxicity, Estrogenic Responses and Bioaccumulation of Alkylphenols and Alkylphenols Polyethoxylates, Water Qual. Res. J. Canada, Vol. 34, No. 1, 123-177.

[9] Ferrara, F.; Fabietti, F.; Delise, M; Funari, E. (2005) Alkylphenols and alkylphenols ethoxylates contamination of crustaceans and fishes from the Adriatic Sea (Italy), *Chemosphere*, vol. 59, Issue 8, 1145-1150.

[10] Canonica, S.; Hellrung, B.; Wirz, J. (2000) Oxidation of phenols by triplet aromatic ketones in aqueous solution. *J. Phys. Chem. A*. 104, 1226-1232.

[11] Goel, A.; Mueller, M.; Sharma, M.; Frimmel, F. H. (2003) Biodegradation of nonylphenol ethoxylate surfactants in Biofilm Reactors. *Acta hydrochim. Hydrobiol.* 31 (2), 108-119.

[12] Defoin, A.; Defoin-Straatmenn, R.; Hildenbrand, K.; Bittersmann, E.; Kreft, D.; Kühn, H. J. (1986) A new liquid phase actinometer: Quantum yield and photo-CIDNP study of phenylglyoxylic acid in aqueous solution. *J. of Photochemistry* 33, 237-255.

[13] Routledge, E. J.; Sumpter, J. P. (1996) Estrogenic activity of surfactants and some of their degradation products assessed using a recombinant yeast screen. *Environ. Toxicol. Chem.* 15 (3) 241-248.

[14] Neamtu, M.; Frimmel, F. H. (2006) Photodegradation of endocrine disrupting chemical nonylphenol by simulated solar UV-irradiation, *Science of the Total Environment*, 369 (1-3), 295-306.

[15] Miller, J. S. Rose (2005) Bengal-sensitized photooxidation of 2-chlorphenol in water using solar simulated light. *Wat. Res.* 39, 412-422.

[16] Santos, A.; Yustos, P.; Quintanilla, A.; Rodriguez, S.; Garcia-Ochoa, F. (2002) Route of the catalytic oxidation of phenol in aqueous phase. *Appl. Catl. B: Environ.* 39, 97-113.

[17] Mazellier, P.; Leverd, J. (2003) Transformation of 4-tert-octylphenol by UV irradiation and by an H_2O_2/UV process in aqueous solution, *Photochem. Photobiol. Sci.*, 2, 946-953.

[18] Oppenländer, T. Photochemical purification of water and air. Advanced Oxidation Processes (AOPs): Principles, Reaction Mechanisms, *Reactor Concepts*. Wiley-VCH, Weinheim, Germany. 2003.

[19] Legrini, O.; Oliveros, E.; Braun, A. M. (1993) Photochemical processes for water treatment. *Chem. Rev.* 93, 671-698.

[20] Hessler, D. P.; Gorenflo, V.; Frimmel, F. H. (1993) UV degradation of atrazine and metazachlor in the absence and presence of H_2O_2, bicarbonate and humic substances. *J. Wat. SRT – Aqua* 42, 8-12.

[21] Schindelin, A. J.; Frimmel, F. H. (2000) Nitrate and Natural Organic Matter in aqueous solutions irradiated by simulated sunlight. *ESPR- Environ. Sci. and Pollut. Res.* 7(4), 205-210.

[22] Sharpless, C. M.; Linden, K. G. (2001) UV Photolysis of nitrate: effects of natural organic matter and dissolved inorganic carbon and implications for UV water disinfection. *Environ. Sci. Technol.* 35, 2949-2955.

[23] Sörensen, M.; Frimmel, F. H. (1997) Photochemical degradation of hydrophilic xenobiotics in the UV/H_2O_2 process: influence of nitrate on degradation rate of EDTA, 2-amino-1-naphtalenesulfonate, diphenyl-4-sulfonate and 4,4′-diaminostilbene-2,2′-disulfonate. *Wat. Res.* 31 (11), 2885-2891.

[24] Chin, Y.-P.; Miller, P. L.; Zeng, L.; Cawley, K.; Weavers, L. K. (2004) Photosensitized degradation of Bisphenol A by dissolved organic matter. *Environ. Sci. Technol.* 38, 5888-5894.

[25] Canonica, S.; Freiburghaus, M. (2001) Electron-rich phenols for probing the photochemical reactivity of freshwaters. *Environ. Sci. Technol.* 35, 690-695.

[26] Haag, W. R.; Hoigne, J. (1986) Singlet oxygen in surface waters. 3. Photochemical formation and steady-state concentrations in various types of waters. *Environ. Sci. Technol.* 20, 341-348.

[27] Canonica, S.; Jans, U.; Stemmler, K.; Hoigne, J. (1995) Transformation kinetics of phenols in water: photosensitization by dissolved natural organic material and aromatic ketones. *Environ. Sci. Technol.* 29, 1822-1831.

[28] Routledge, E.; Sheahan, D.; Desbrow, C.; Brighty, G. C.; Waldock, M.; Sumpter, J. P. (1998) Identification of estrogenic chemicals in STP effluent. 2. In vivo responses in trout and roach. *Environ. Sci. Technol.* 32, 1559-1565.

[29] Beresford, N.; Routledge, E. J.; Sumpter, J. P. (2000) Issues Arising When Interpreting Results from an in Vitro Assay for Estrogenic Activity. Toxicol. *Appl. Pharmacol.* 1623 22-33.

In: Sewage Treatment: Uses, Processes and Impact
Editor: Anna Stephens and Mark Fuller
ISBN 978-1-60692-959-9

Short Communication

WASHING OF INCINERATION RESIDUE OF SEWAGE SLUDGE WITH ACID AND COAGULATION-SEDIMENTATION TREATMENT USING THE WASHINGS

Seiichi Ishikawa, Satoe Ito, Yoshio Eguchi and Naoko Ueda
Faculty of Environmental Engineering, The University of Kitakyushu,
Hibikino 1-1, Wakamatsu-ku, Kitakyushu, 808-0135, Japan

ABSTRACT

To use sewage sludge, the acid-washing of its incineration residue and the coagulation-sedimentation treatment using the washings were performed in comparison with water supply sludge and red clay incineration residues. The concentration level of total phosphorous (T-P) (221-4,790 mg/l) in the solutions of (1+1) H_2SO_4 extraction from both sludges was higher than those of COD (10.7-127 mg/l) and total nitrogen (T-N) (0.2 mg/l>). As for metals, the concentration level was as follows: Al (874-2,900 mg/l) > Fe (777-1,130 mg/l) > Ca (71.5-614 mg/l) > Mg (21.2-271 mg/l) > Mn (37.7-148 mg/l) > Zn (5.72-139 mg/l) > Cu (3.37-42.7 mg/l) > Ni (2.32-6.64 mg/l) > total chromium (T-Cr) (0.27-2.68 mg/l) > Sb (0.01 mg/l>).

The concentration levels of Cd, Pb and Se were those of effluent standards. In the coagulation-sedimentation treatment at pH 6.0 using the acid-washings, max. 10.0-23.5% of COD, max. 22.9-27.2% of T-N and max. 96.5% of T-P were removed. The remaining rate of T-P using the washings of sewage sludge incineration residue was over 100% because the washings contained much T-P. In the same treatment conditions, the remaining rates of Al, Ca, Cu, Mg, Mn, Ni and Zn were over 100%. However, the concentration levels of these metals in treated water were in conformity with the effluent standards.

Keywords: Sewage sludge, Incineration residue, Acid-extraction, Acid-washing, Coagulation-sedimentation.

1. INTRODUCTION

As for the organic resources in sewage sludge and water supply sludge, methane fermentation (Ochi et al., 2005; Yasui et al., 2005), composting and the utilization as soil-improving material (Imagawa, 1992; Mori, 1992; Onodera et al., 1994; Sato, 1992) are performed. As for the inorganic resources, these sludges are utilized as brick and road material (Hiraga, 1994; Iwashita et al., 1992; Momoi, 1992). Takahashi et al. had reported the recoveries of phosphorous (P), a drain resource (Sato et al., 1994), and aluminum (Al) from sewage sludge incineration residue with alkali (Takahashi, 2004; Takahashi et al., 2001). These sludges are incinerated or melted when they are utilized as inorganic resources. In this time, hazardous substances such as heavy metals are also concentrated together and it becomes a problem in sludge utilization.

The acid-washing is usually performed when heavy metals are removed from incineration residue (Kida et al., 2005). We reported the effect on the coagulation-sedimentation using the recovered coagulant from water supply sludge with acid (Ishikawa et al., 2007; Ishikawa et al., 2006). As a result, COD, total nitrogen (T-N), total phosphorous (T-P) and pesticides were more effectively removed than with a commercial $Al_2(SO_4)_3$ or polychlorinated aluminum (PACl). Sewage sludge and water supply sludge naturally contain Al and iron (Fe) at higher concentration level because they contain the coagulant used and soil. Then, we performed the acid-washing of these incineration residues to remove heavy metals and to recover valuable materials such as P. Farther, we examined to reuse the acid-washing solution as a coagulant.

2. MATERIAL AND METHODS

2.1. Samples

Dried sewage sludge (Photo. 1) was purchased from a municipal sewage treatment plant. Water supply sludge (75% water content) (Photo. 2) was purchased from a municipal water purification plant (PACl and $Al_2(SO_4)_3$ are used). Red clay (Photo. 3) was purchased from Mt. Takato in Kitakyushu city.

The water content of sludge or red clay was measured by drying for 2 h at 105□ (Japan, 1982). Further, their ignition losses were measured by heating for 1 h at 600±25□ (Japan, 1982). These incineration residues were used as test samples.

The sample water for coagulation-sedimentation treatment was prepared by the filtration of the mixture of sewage sludge and water with 1 μm Whatman GF/B.

2.2. Reagents and Instruments

Sulfuric acid was used Wako Pure Chemical Ind. Co. for toxic metal analysis. The standard solutions of metals etc. were prepared by the dilution of 100 or 1,000 mg/l Saishin Syoji Co. Certi Prep ICP XSTC-22 with 0.1N HNO_3. Other reagents were special grade reagents. The water was purified using a Millipore Milli-Q Ultra-pure Water System.

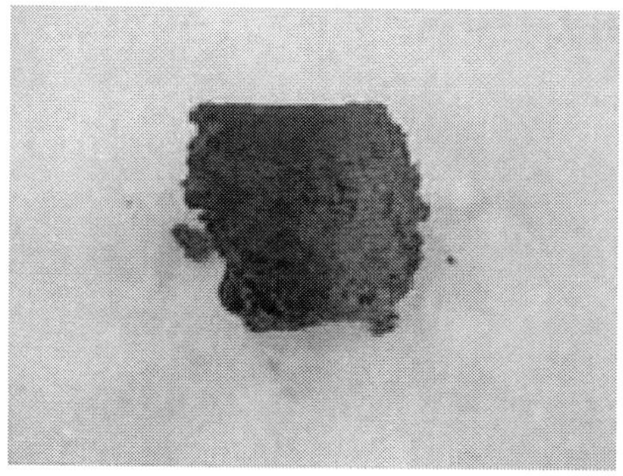

Photo 1. Sewage sludge (dried).

Photo 2. Water supply sludge (dewatered)

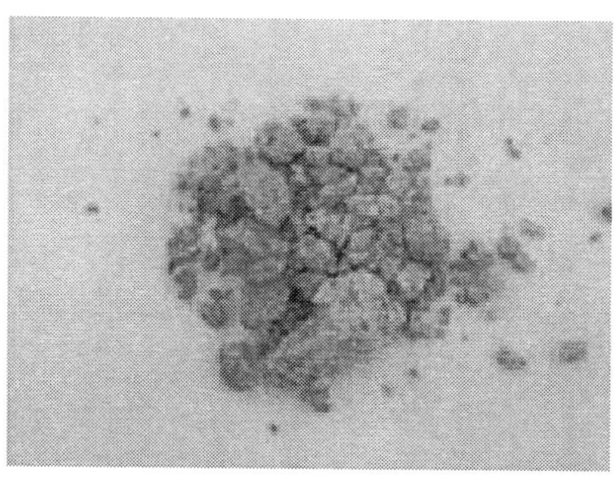

Photo 3. Red clay.

Water cohesion reaction tester was a Miyamoto Riken Ind. Co. JT-6D. Total organic carbon (TOC) analysis was performed using a Shimazu Co. TOC-VCSH and AUTO SAMPLER ASI-V. Total nitrogen (T-N) analysis was performed using a Tokyo Kasei Ind. Co. TCI-NOX 1000, GASTORR GT-102, VISIBLE DETECTOR S-3250 and AUTO SAMPLER SS-3600. Spectrophotometer was a Hitachi Co. U-2000A and atomic absorption spectrophotometer was a Valian Technologies Japan Co. Spectr AA 220.

2.3. Analytical Methods

pH, COD, TOC, T-N and T-P were measured by the methods of JIS K 0102 (Japan Industrial Standards Committee, 1998). Metals etc. were measured by atomic absorption spectrophotometry (Japan Industrial Standards Committee, 1998) after the dilution with 0.1N HNO_3.

2.4. Acid-Extraction

Each 5 g of incineration residue was extracted by dipping for 30 min in 25 ml of (1+1) H_2SO_4. The residue was washed with purified water after the filtration with 1 μm Whatman GF/B and the filtrate was made up to 100 ml.

2.5. Acid-Washing

Each 3 g of incineration residue was dipped for 30 min in 10 ml of 1N H_2SO_4. The mixture was filled in a 1 cm i.d. glass column having a tap and the H_2SO_4 solution was flowed out. Farther, the residue was washed with 40 ml of purified water and the filtrate was made up to 50 ml. This procedure was performed 3 times and 3 acid-washing solutions were prepared.

2.6. Coagulation-Sedimentation

Each 1 l of sample water was placed in each 1 l beaker. Each 1 or 10 ml of the acid-extraction solution was added and pH was adjusted at 6. The coagulation-sedimentation was performed using a water cohesion reaction tester in the condition: stirring for 1 min at 150 rpm, stirring for 10 min at 50 rpm and standing for 15 min. The treated sample water was filtrated with 1 μm Whatman GF/B and the filtrate was measured for each item.

3. Results and Discussion

3.1. Acid-Extraction

To examine the heavy metal content in sewage sludge and water supply sludge residues and the utilization of the acid-extraction solution as a coagulant, extracted we from each residue with (1+1) H_2SO_4. Each concentration was shown in Figure 1. The weights of the extracts from sewage sludge, water supply sludge and red clay residues were 20,100, 18,600 and 5,180 mg/l, respectively.

COD and T-N concentrations were 10.7-127 and 0.2 mg/l >, respectively. On the other hand, T-P concentration levels for sewage sludge (4,790 mg/l) and water supply sludge residues (221 mg/l) were much higher than that for red clay residue (6.4 mg/l).

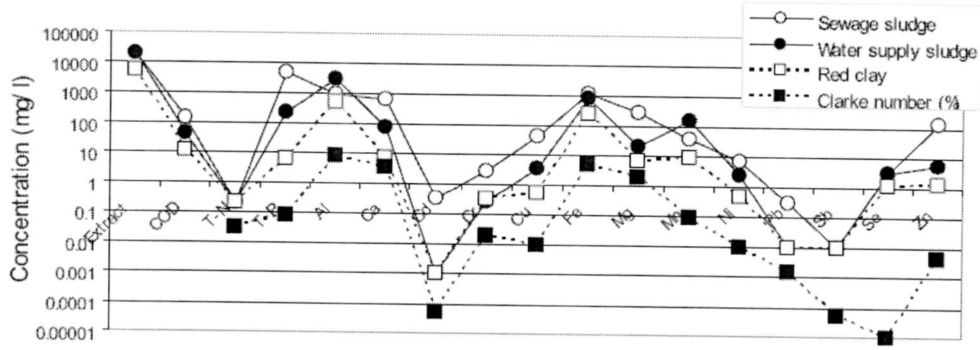

Figure 1. Concentration of each item in the acid-extraction solution from each incineration residue.

As for heavy metals, the concentration levels of Al and Fe were 18-2,900 and 244-1,130 mg/l, respectively. And Ca (7.17-614 mg/l), Mg (6.93-271 mg/l), Mn (8.54-148 mg/l), Zn (1.41-139 mg/l), Cu (0.50-42.7 mg/l), Ni (0.45-6.64 mg/l), T-Cr (0.27-2.68 mg/l) and Sb (0.01 mg/l>) followed in turn. The concentrations of Cd (0.001-0.32 mg/l), Pb (0.01-0.29 mg/l) and Se (3 mg/l>) were same level as those of the effluent standard values. The proportions of Ca, Mg, Pb and Zn to the other metals in water supply sludge and red clay residues were smaller than those of the Clark number (). The proportion of Cu to the other metals in sewage sludge and water supply sludge residues and the proportion of Mn to the other metals in water supply sludge and red clay residues were larger than those of the Clark number, respectively.

3.2. Acid-Washing

To remove the heavy metals in sewage sludge and water supply sludge residues, which were hazardous for the residue recycling, were these residues washed 3 times with 1N H_2SO_4. Figure 2 showed the concentrations of TOC, T-P and heavy metals in each acid-washing solution.

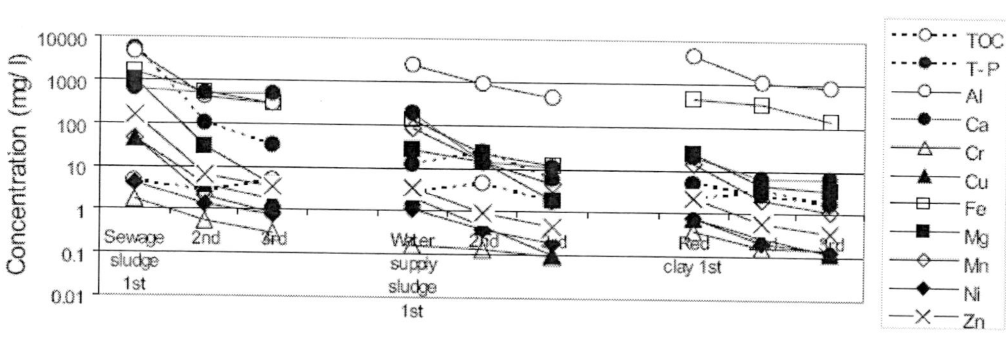

Figure 2. Concentration of each item in the acid-washing solution with 1N H_2SO_4.

The concentration levels of TOC in the 3 times acid-washing solutions for the 3 residues were almost same (1.63-4.99 mg/l). T-P in each acid-washing solution for water supply sludge and red clay residues, which contained a little of T-P, changed at low concentration levels of 12.2→5.68 mg/l and 4.92→1.68 mg/l, respectively. On the other hand, T-P in each acid-washing solution for sewage sludge residue, which contained much T-P, changed at high concentration level of 5,560→105→31.3 mg/l. Takahashi et al. extracted P with alkali (Takahashi, 2004; Takahashi et al., 2001). The acid-extraction of P was also possible.

Al was much contained in these samples. The concentrations of Al for sewage sludge and water supply sludge residues, which contained Al as coagulant, changed into 4,450→287 mg/l and 2,390→464 mg/l, respectively. The concentration of Al for red clay residue changed into 4,600→847 mg/l and Al contained in red clay residue was more hardly extracted than that in the other residues. Fe (127→11.7 mg/l) and Ca (193→11.2 mg/l) in water supply sludge residue were more easily extracted than those in the other residues (sewage sludge: Fe 1,620→302 mg/l, Ca 666→482 mg/l; red clay: Fe 444→137 mg/l, Ca 20.6→6.46 mg/l). Mg in sewage sludge residue (1,040→3.76 mg/l) was more easily extracted than that in water supply sludge (26.3→1.89 mg/l) and red clay residues (23.5→3.35 mg/l). The concentrations of T-Cr, Cu, Ni and Zn were below 0.758 mg/l after 3 washings except for those of Cu (46.9→1.72 mg/L) and Zn (151→3.34 mg/L) for sewage sludge residue.

3.3. Coagulation-Sedimentation Treatment

The removal effects of organic pollutants, nutrients and metals on coagulation-sedimentation were investigated using the acid-extraction solution shown in *3.1.* Figure 3 and 4 show the appearance removal effects, where the treatments were performed in the prepared sample water at pH 6 using 1 or 10 ml of each acid-extraction solution. Table 1 shows the reference data using a commercial coagulant or the recovered coagulant (Ishikawa et al., 2007; Uchimura et al., 1988).

The maximum removal ratios of COD using the acid-extraction solutions from sewage sludge, water supply sludge and red clay residues were 10.0, 23.5 and 7.9%, respectively. These values were smaller than those using $Al_2(SO_4)_3$, PACl and the recovered coagulant. However, the maximum removal ratios of T-N using the acid-extraction solutions were 27.2, 22.9 and 29.8%, respectively and many ratios were larger than those using the other coagulants etc.. The maximum removal ratios of T-P using the acid-extraction solutions from water supply sludge and red clay residues were 96.5 and 58.7%, respectively and the value using the acid-extraction solution from water supply sludge residue was larger than those using the other coagulants etc. except for using the recovered coagulant. The removal ratio of T-P was largely different by the amount of the solution. As the acid-extraction solution from sewage sludge residue contained large amount of T-P, the appearance remaining ratio exceeded 100%.

As for metals, the concentration levels of Al, Ca, Mg, Mn, Ni and Zn in the treated water became higher than those in the original prepared water in the case of aiming at the removal of COD, T-N and T-P at pH 6. Especially, the increase was remarkable in Al, Cu, Mn and Zn for sewage sludge residue and in Al and Mn for water supply sludge and red clay residues. However, the concentration levels of all measured metals in the treated water were lower than their effluent standard values. Each concentration of Cd, T-Cr, Pb and Sb in both the original prepared water and the treated water was below each detection limit.

Table 1. Comparison of maximum removal ratio with other coagulants etc.

Coagulant etc.	Maximum removal ratio (%) COD	T-N	T-P	pH
Recovered coagulant from water supply sludgea	66.8	29.9	98.9	6.0
Al2(SO4)3 a	48.7	20.3	47.3	6.0
PACla	52.7	12.4	52.1	6.0
PAClb	-	31.2	77.8	5.3
PACl+ activated carbonb	-	27.1	74.6	5.3
AlK(SO4)2 b	-	18.4	74.1	4.6
FeCl3 b	-	16.7	81.0	5.8
FeCl3+ activated carbonb	-	43.2	89.7	5.6
FeCl3+ high test hypochloriteb	-	10.3	75.9	7.9
FeCl3+ CaOb	-	34.9	75.9	9.6
Organic polymer coagulantb	-	3.8	3.5	8.7
Acid-extraction solution from sewage sludge residuec	10.0	27.2	+6.0	6.0
Acid-extraction solution from water supply sludge residuec	23.5	22.9	96.5	6.0
Acid-extraction solution from red clay residuec	7.9	29.8	58.7	6.0

[a] Sample water: Inflow of sewage treatment plant; Amount of coagulant: Recovered coagulant 5 ml/l (29.0 mgAl/l), $Al_2(SO_4)_3$ 30 mg/l, PACl 50 mg/l (Ishikawa et al., 2007).

[b] Sample water: The solution extracted from sewage sludge; Amount of coagulant: 100 mg/l (Uchimura et al., 1988).

[c] Sample water: The solution extracted from sewage sludge; Amount of coagulant: Acid-extraction solution from sewage sludge residue 10 ml/l (8.7 mgAl/l, 11.3 mgFe/l), acid-extraction solution from water supply sludge residue 10 ml/l (29.0 mgAl/l, 7.8 mgFe/l), acid-extraction solution from red clay residue 10 ml/l (5.2 mgAl/l, 2.4 mgFe/l).

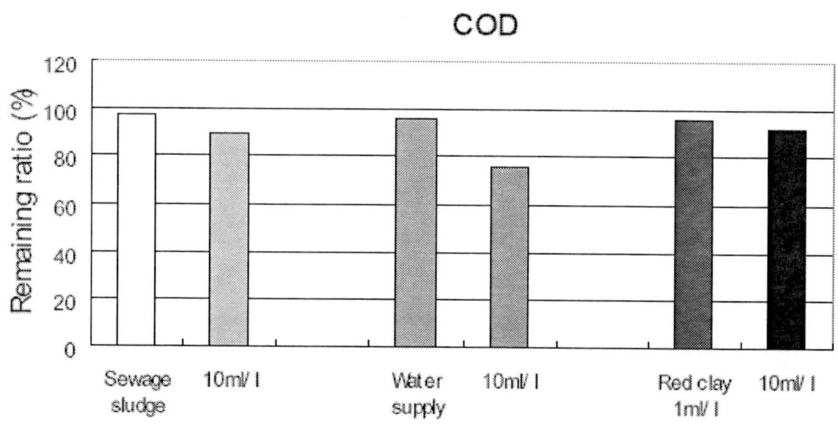

Figure 3. (Continued).

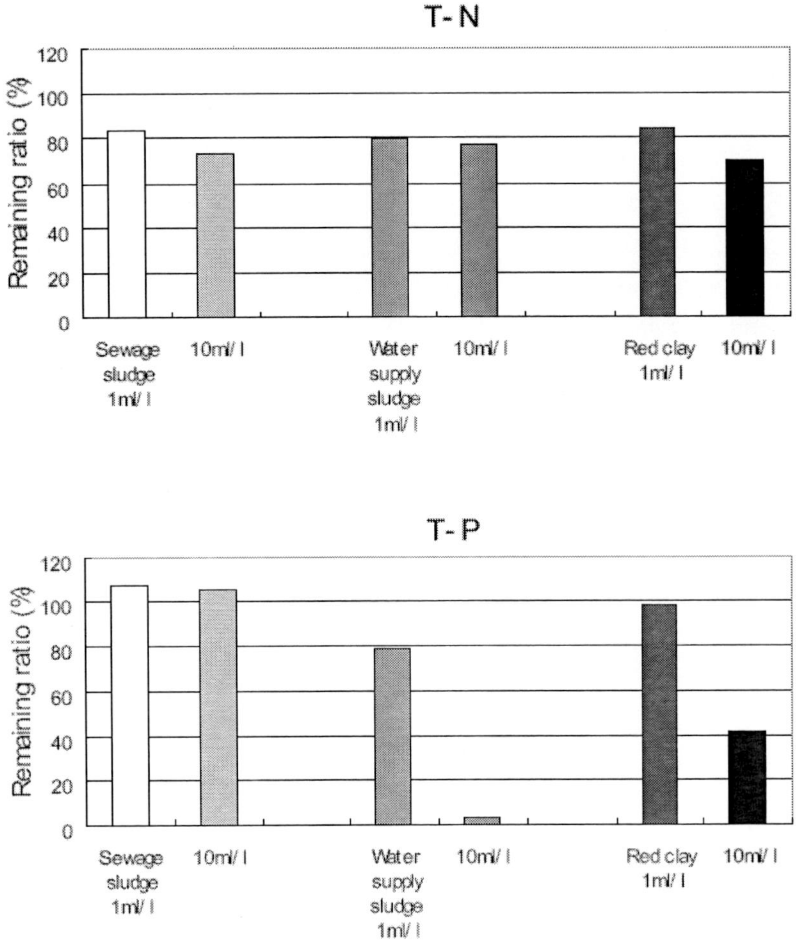

Figure 3. Effect of coagulation-sedimentation with each acid-extraction solution (COD, T-N and T-P).

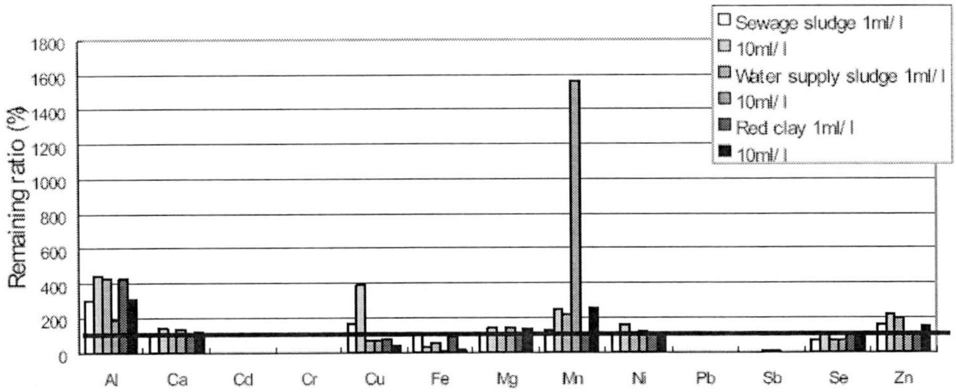

Figure 4. Effect of coagulation-sedimentation with each acid-extraction solution (metals).

References

Hiraga, N., 1994. The effective utilization of sewage sludge incineration ash in Osaka City- The production of the water permeability brick utilizing sewer resources-. *Regeneration and Utilization* 17(65), 12-17.

Imagawa, M., 1992. Influence on the growth of sweet corn and cabbage in the mineral soil field and the soil by using the compost from the lime containing sewage sludge. *Regeneration and Utilization* 15(55), 42-50.

Ishikawa, S., Ueda, N., Okumura, Y., Iida, Y., Baba, K., 2007. Recovery of coagulant from water supply plant sludge and its effect on clarification. *Journal of Material Cycles and Waste Management* 9, 167-172.

Ishikawa, S., Ueda, N., Okumura, U., Iida, Y., Higuchi, M., Naetoko, E., Tokunaga, Y., Baba, K., 2006. Removal efficiency for pesticides on coagulation and sedimentation using coagulant recovered from water supply sludge. *J. Japan Soc. Water Environ.* 29(10), 653-658.

Iwashita, H., Kamata, O., 1992. The present condition and subjects on the utilization of sewage sludge as construction materials. *Regeneration and Utilization* 15(55), 34-41.

Japan Industrial Standards Committee, 1998. JIS K 0102. Japanese Standards Association, Tokyo.

Japan Society on Water Pollution Research, 1982. Guideline for Lake Environmental Survey. Tokyo, pp. 151.

Kida, M., Uchida, N., Sakai, S., 2005. The amounts of the metals extracted from molted slag and soil with HCl. Proceedings of the 16th Annual Conference of The Japan Society of Waste Management Experts, pp. 654-656.

Momoi, K., 1992. The subjects on the melting furnace technology of sewage sludge and the utilization as construction materials. *Regeneration and Utilization* 15(57), 17-24.

Mori, T., 1992. For the new development on the utilization of sewage sludge for land application. *Regeneration and Utilization* 15(57), 10-16.

Ochi, S., Ozaki, M., 2005. Influence of kitchen waste inflow on the medium and high temperature anaerobic digestive treatments. *Journal of Japan Sewage Works Association* 42(515), 89-98.

Onodera, M., Hirai, Y., 1994. Effect of the utilization of grain compost (lime containing sewage sludge) on the growth of garden crop. *Regeneration and Utilization* 17(62), 40-47.

Sato, H., 1992. Effect of the utilization of dewatered sludge on the soil in Toyohashi City. *Regeneration and Utilization* 15(55), 51-56.

Sato, K., Ito, H., Sekine, T., Hirano, Y., 1994. The fundamental investigation on the recovery technology of valuables in sewerage. Annual Report of Japan Institute of Wastewater Engineering Technology, pp. 86.

Takahashi, Y., 2004. Recovery of phosphorus and aluminum from incinerator ash of sewage sludge. *J. Japan Soc. Water Environ.* 27(3), 181-187.

Takahashi, Y., Yoshida, K., Sone, Y., 2001. Study on the effective recovery of the valuables in sludge incineration ash. *Journal of Japan Sewage Works Association* 38(468), 181-192.

The Editorial Committee for Handbook of Inorganic Chemistry, 1968. *New Handbook of Inorganic Chemistry*. Gihodo, Tokyo.

The Special Committee for the examination of Construction Materials, 1992. The manual of sewage sludge utilization as construction materials. *Regeneration and Utilization* 15(56), 63-69.

Uchimura, Y., Ishikawa, S., Yasuda, K., 1988. Model test on the leachate at a sea reclamation waste disposal site (□). *Liq. Waste Treat.* 29(6), 365-370.

Yasui, H., Komatsu, K., Matsuhashi, R., Ohashi, A., Harada, H., 2005. LCA on the anaerobic digestive process and the utilization of digestion gas considering area characteristic. *Environmental Engineering Research* 42, 171-182.

INDEX

A

B

C

D

F

I

J

K

L

M

N

O

P

Q

R

T

U

V